IMAGE ANALYSIS
AND PROCESSING II

IMAGE ANALYSIS AND PROCESSING II

Edited by

V. Cantoni

Pavia University
Pavia, Italy

V. Di Gesù

Palermo University
Palermo, Italy

and

S. Levialdi

Rome University
Rome, Italy

PLENUM PRESS • NEW YORK AND LONDON

Library of Congress Cataloging in Publication Data

International Conference on Image Analysis and Processing (4th: 1987: Palermo, Sicily)
  Image analysis and processing II / edited by V. Cantoni, V. Di Gesù, and S. Levialdi.
      p.   cm.
  "Proceedings of the Fourth International Conference on Image Analysis and Processing, held September 23-25, 1987, in Palermo, Italy" — T.p. verso.
  Includes bibliographical references and index.
  ISBN-13: 978-1-4612-8289-1   e-ISBN-13: 978-1-4613-1007-5
  DOI: 10.1007/978-1-4613-1007-5
  1. Image processing — Congresses. I. Cantoni, V. II. Di Gesù, V. III. Levialdi, S. IV. Title. V. Title: Image analysis and processing two. VI. Title: Image analysis and processing 2. TA1632.I552   1987   88-9416
006.3'7 — dc19   CIP

Proceedings of the Fourth International Conference on Image Analysis and Processing, held September 23-25, 1987, in Palermo, Italy

© 1988 Plenum Press, New York
Softcover reprint of the hardcover 1st edition 1988
A Division of Plenum Publishing Corporation
233 Spring Street, New York, N.Y. 10013

PREFACE

This book contains the proceedings of the 4th International
Conference on Data Analysis and Processing held in Cefalu'
(Palermo, ITALY) on September 23-25 1987.

The aim of this Conference, now at its fourth edition, was to
give a general view of the actual research in the area of
methods and systems for achieving artificial vision as well
as to have an up-dated information of the current activity
in Europe.

A number of invited speakers presented overviews of statistical
classification problems and methods, non conventional architectures, mathematical morphology, robotic vision, analysis
of range images in vision systems, pattern matching algorithms
and astronomical data processing. Finally a survey of the
discussion on the contribution of AI to Image Analysis is
given.

The papers presented at the Conference have been subdivided
in four sections: knowledge based approaches, basic pattern
recognition tools, multifeatures system based solutions, image
analysis-applications.

We must thank the IBM-Italia and the Digital Equipment Corporation for sponsoring this Conference.

We feel that the days spent at Cefalu' were an important step
toward the mutual exchange of scientific information within
the image processing community.

V. Cantoni
Pavia University

V. Di Gesu'
Palermo University

S. Levialdi
Rome University

CONTENTS

MULTIFEATURES AND SYSTEM BASED SOLUTIONS

INVITED LECTURES

MORPHOLOGICAL OPTICS

J. Serra

Ecole Nationale Supérieure des Mines de Paris
35, rue St. Honoré, 77305 Fontainebleau, France

Key words : Mathematical Morphology, optics, picture processing, irreversibility, machine vision, artificial intelligence, lattice theory.

Summary :

Image analysis methods are first classified into four groups of theories related to optics, after which the associate hypotheses, the structures, the laboratory equipment and the mathematical framework for one of them, better known by the term of morphological optics are studied in detail. Conclusions are drawn as to the role played by hypotheses in image analysis.

1 - INTRODUCTION : OPTICS AND IMAGE ANALYSIS

When one considers the volume and diversity of literature devoted to "picture processing", it would appear that this term refers to the nebulae of theories and processes used to describe spatial phenomena (two- or three-dimensional). However, a closer examination shows that some of the methods are closely related, although this closeness is not recognized. A distinction is then made between 4 groups of analyses, distributed among 4 main ideas (Fig. 1). In the conclusions to this paper, we shall return to the classification problem to better situate Mathematical Morphology within the total context.

In more general terms, the four methods are related to current developments in optics. Classically, the term "optics" designates that branch of physics the first purpose of which is the study of human vision, but which is finally used to develop theories on the nature and the behavior of light. Optics is part of physics, since in parallel to each of its mathematical developments, optics is used to invent equipments adapted for the visual field (microscopes, photo-electric cells, lasers, etc...). Moreover, thanks to

Lineur	Data treatment
Signal processing Convolution, Fourier analysis Tomography	For stereology For multivariate analysis For taxonomy
Morphological	Syntactical
Morph. filtering Integral geometry Random sets Skeleton	Extraction of primitives Grammars of contours Hough transform

Fig. 1. The four major approaches in Image Analysis.

its mathematical theories, the science of optics has overflowed the strict domain of human vision (for example, by extending the visible spectrum to all electromagnetic radiation) and has expanded the formalism itself. The history of optics in the XIXth century illustrates this point : FRESNEL discovered a series of wave-related phenomena, but he interpreted them in terms of geometry, which is somewhat inadequate when it comes to distinguishing between longitudinal or transverse waves. Some time later, GREEN invented an appropriate vectorial formalism, which justified FRESNEL's results, and laid the groundwork for MAXWELL and his famous synthesis between electrical and light waves, etc...

It is worth noting that picture processing methods tend to be organized similarly to optical theory, i.e., with an initial emphasis on vision, the same dialectic distinctions between theory and instrumentation, and the same gradual breaking-up of the approach, moving from the "seen" to the "unseen", via generalization of the mathematical framework.

In this paper, I wish to describe in detail the filiation insofar as it relates to Mathematical Morphology.

2 - MORPHOLOGICAL OPTICS DERIVED FROM VISION

In vision, a distinction should be made between geometric and morphological optics. It is well-known that the theories of geometric optics depend initially on the postulate that a system is identified when it is possible to predict the image of a luminous point. The complete field of vision is then deduced by superposition of elementary transforms. Often, there is an even stronger hypothesis, namely : "the image of the point is itself a point". Obviously, in such a process the convolution and point transforms of Euclidean space (homothetics, rotations, affinities, etc...) play a major rôle. The usefulness of the linearity property need not be demonstrated : when an image is taken, one sums the images to attenuate the background noise ; short-sighted correction lenses deconvolute the retinal image, etc...

Linearity also occurs in acoustics. Indeed, the intensity of sounds, when one leaves aside considerations of phase, combine arithmetically. When several sources emit sound at the same time, the hearing process accommodates all the vibrations, and, to a certain extent, isolates and

compares them. If this were not the case, there would be no
orchestras. Since preserving the ratios among the sound
sources is necessary for an intelligent understanding of the
sound scene, all amplifiers (or transmitters) are required to
comply with the relative proportions of the source origins,
i.e., in mathematical terms, they must be _linear_.

However, visual signals combine differently.
Objects in space generally have three dimensions, which are
reduced to two dimensions in the photograph or on the retina.
In this projection, the luminances of the points located
along a line oriented directly away from the viewer are not
summed, because most physical objects are not translucent to
light rays, in the same way that they would be to X-rays, but
are opaque. Consequently, any object which is seen hides
those that are placed beyond it with respect to the viewer :
this self-evident property is a basic one.

In fact, it serves as a starting point for
Mathematical Morphology (SERRA √1986≠), for, whenever we wish
to describe quantitatively phenomena in this domain, a set
approach must be used. Stating that A is in front of B is
equivalent to asserting that we see the visible contour of B
minus, in the set sense, that of A. Stating that A hides B is
equivalent to saying that the contour of B is included in the
contour of A, etc...etc... A morphological description, i.e.
a description of their shapes, must primarily use _portions_ of
space. When transformations are involved, they must apply
globally, i.e., they cannot be reduced to simple
juxtapositions of point transformations (just like _gestalt_-
psychology when it deals with human vision).

Now, the set $P(R^n)$ of subsets of Euclidean space is
equipped with an order relation, called inclusion, such that
any family X_i of elements of $P(R^n)$ admits a smallest upper
bound, called the union, and denoted $\cup X_i$, and a greatest
lower bound, called the intersection (the dual of the union),
and denoted $\cap X_i$. In the same way that the theory of geometric optics puts its emphasis on the transformations which
commute with addition, morphology naturally stresses the
transformations, or mappings $\psi : P(R^n) \rightarrow P(R^n)$ that are
related to the basic structures of $P(R^n)$. Thus they will be
developed :
  - either from those which preserve inclusion, i.e.

$$X \supset Y \Rightarrow \psi(X) \supset \psi(Y) \qquad X,Y \in P(R^n)$$

these transformations are designated as _increasing_ transformations ;

  - or from those which commute with the union, i.e.,

$$\psi(\underset{i}{\cup} X_i) = \underset{i}{\cup} \psi(X_i) \qquad X_i \ . \ P(R^n)$$

which are called _dilations_. (The dual operation which
commutes with the intersection is called an _erosion_). It will
be noted, moreover, that the three classes are not
independent : each of the latter two can generate the first
class.

3 - IRREVERSIBILITY AND MORPHOLOGICAL DISCOURSE

The parallel that could be drawn with linear methodologies ceases at this point, insofar as there is a major problem. When we say that in geometric optics we improve a fuzzy picture, making it sharp, we are expressing the point of view of the spectator. The physicist would tend to feel that nothing had been gained since it is always possible to revert from the sharp to the fuzzy picture : both of them contain exactly the same amount of information. The implied linear process is <u>reversible.</u> We are well aware, since the works of SHANNON [1949] and of WIENER [1949], of the emphasis that signal processing lays on the notion of information content of a message. This high level of interest is all the more justified when one considers questions related to transmission (amplifiers, broadcasting, etc...).

In computer vision, what are we seeking - to transmit information or rather to assimilate it ? Reversibility is acceptable when we improve the images that provide the input to a system, as with the case of spectacles for shortsighted people. It is also acceptable to encode images for transfer to processing devices, as is the case when the retinal image is transferred to visual cortex of the occipital lobe. But beyond this point ? The brain does not add a third "eye", which would then look at the visual zones and be observed itself. The chain stops there, and with it the notion of reversibility. Recognition of an object simply means that all the rest has been eliminated from the scene. This is a definitive irreversible operation.

The tool created within the framework of mathematical morphology satisfies this property. The simplest dilation, the union of a set X and its translate X_h can only lose information : from $X \cup X_h$, one cannot backtrack to identify X. The question then arises as to how we can spread out successive losses among a series made up of dozens of transformations, so that the result converges to a single aim ? This is the central question of morphology ; in various forms and in particular expressions this question occurs frequently. Thus :

i) since we no longer have the structure of a group, which proved to be useful in the case of geometric optics (the usual similarities and convolutions), what are the conditions required in order that the composition product of two morphological operators remains in the same class as (at least) one of them ? If the answer is no such conditions exist, does this mean that the composition takes on a new meaning ? If such conditions exist, how can we interpret the composition ? For example, the product of two dilations is yet another dilation, but a dilation followed by an erosion leads to a product which has the characteristics of neither. Thus, as the various possible combinations between dilations and their complementary operations take place, there arises an infinite world of <u>new</u> mappings which can then be concatenated with each other : openings, closings, size distributions, morphological filters, ultimate erosion, skeletons, conditional bisector, and many others.

ii) Among all the possible "new species", how do we go about selecting them ? The answer is to have a set of reference properties on which to base the assessment. Thus, since there are sequences of operations, it is advisable to apply the first set of criteria to the choice of comparison between inputs and outputs, and to direct one's interest to the mappings, which may be :

- extensive, i.e., such that $\psi(X) \supset X$ (or anti-extensive if $\psi(X) \supset X$),
- over- or underpotent, depending on whether $\psi\psi \supset \psi$ or $\psi\psi \supset \psi$,
- and especially idempotent, when $\psi\psi = \psi$.

The last-mentioned concept is a particularly powerful one. Its meaning is that a chain of processes, symbolized by ψ has acted as far as was possible, and that there is no point in pursuing the iteration. Idempotence acts as a buffer to the propensity for ψ to lose information. Filters, skeletons and ultimate erosions are operators that fall into this class, and which have a very high degree of internal consistency.

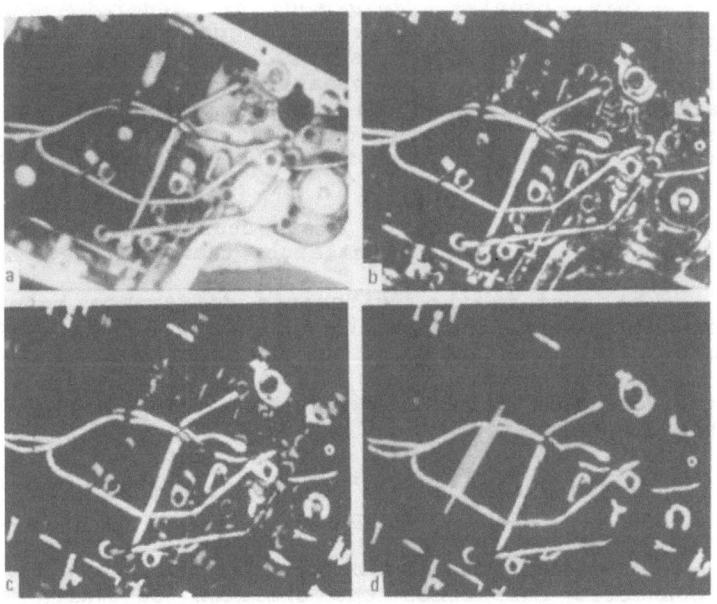

Fig. 2. Use of mathematical morphology in pattern recognition. A morphological algorithm checks automatically whether the two small brackets (see arrows in Fig. 2a) are in place. With the current dedicated systems, the full treatment lasts approximately 1 s.

The first set of criteria, although useful, are not the only ones. It may be required of ψ that it be isotropic, or that it satisfy a given anisotropy, or be self-dual, or preserve connectivity or homotopy, or that it act, or not act, selectively on a given class of sets known in advance, etc...

4 - FROM SENTENCES TO WORDS

The fact that we know all the "good" mappings does not ensure that we have worked our way through the problem, and does not give any indication as to how we should manage a sequence of morphological operations with a given aim in view. In a sense, it is a lexicon in a language the words of which are the key basic transformations and in which sentences are the sequences we are trying to construct. We now raise the question as to how the sentences are constructed. Let us approach this question via two specific examples.

4.1. <u>First example : computer-aided vision in the automobile industry</u>

In this first example, a camera is observing the under-side of an automatic transmission box (Figure 2a) and has to check out automatically if the two safety brackets are correctly positioned.

This problem, proposed by General Motors to STERNBERG [1982] led the latter to suggest the following approach. By observing that these brackets are less visible as brackets and more visible as breaks in pipe-structures, it is the latter phenomenon that will be extracted first. The pipes that are of interest to the case are long, narrow, objects with contrast of clear against a dark background. Hence the following strategy :

a) open image 1a by a structuring element (in this case a cuboctahedron) not much larger than the pipes ;

b) perform the arithmetic difference between the initial picture and its opening (Figure 2b). All bands that are larger (e.g., the contours of the motor) are eliminated, as well as the narrow bands which are darker than their surroundings. This first filtering was based both on contrast and on width parameters ;

c) place a threshold on the transformation 2b. The result appears in light grey and in dark grey in picture 2c ;

d) in this binary image, identify and retain only those connected components the skeleton of which have a length greater than or equal to a given $\geq l_0$. The result is marked in a light shade in 2c and the subset identified in the filtering is in light shade in 2d ;

e) isolate the target zone (the grey triangle in 2d) by a triangular opening restricted to the centre of the picture ;

f) extract the cords parallel to the small side (direction α) and with lengths lying between given values d_1 and d_2 ;

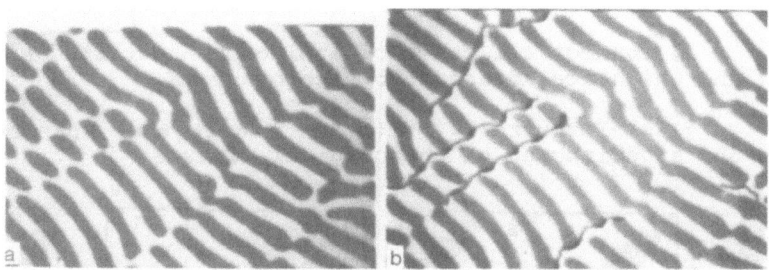

Fig. 3. (a) Defect lines in an oriented eutectic material (G×200). A typical situation where no geometric primitive exists. (b) Result of the treatment shown in Fig. 4.

g) dilate this band in the direction α (the dark grey zone in 2d) somewhat like a pastry rolling pin ;

h) open the pin with a segment running perpendicular to α with a length somewhere between that of the pin and that of the two ends, in which the residuals designate the positioning of the brackets. If either of the brackets is missing, this means that the corresponding end of the pin will be missing too.

4.2. Second example : quantitative microscopy

We now move over from macro- to micro-scopy, from grey-tone image analysis to binary structures, and from robotics to a study of composite materials.

What we are observing in Fig. 3 is a cross-section perpendicular to its axis of a lamellar eutectic with an oriented solidification. It was produced by C.E.N.G. in Grenoble ; they wanted to estimate the number and the size distribution of fault lines per unit area of the surface. In order to solve this problem, LANTUEJOUL [1978] noted that the eye "sees" these faults because it is able to detect alignments in the extremeties of certain lamellae, and the brain makes a mental interpolation. Hence, the strategy, the steps of which are set out in Figure 4 :

a) theshold the image corresponding to Figure 3a (set X) ;

b) skeletonize the eutectic lamellae so as to be able to identify their extremities (X_1) ; actually this "skeleton" is a homotopic thinning (ROSENFELD and KAK [1976]) (SERRA [1982]) ;

c) clip the skeleton without reducing its length (X_3) ;

d). extract the extreme points (Y) of the skeleton (X_3), from the rest of the set X_4 ;

e) determine by homotopic thickening the zones of influence of the connected components of the set $Y \cup X_4$ (i.e:,X_5) ;

f) connect all zones of influence associated with a point, and which are adjacent to each other (X_6) ;

g) skeletonize X_6, with the result that appears in Figure 3b, superimposed on the original picture.

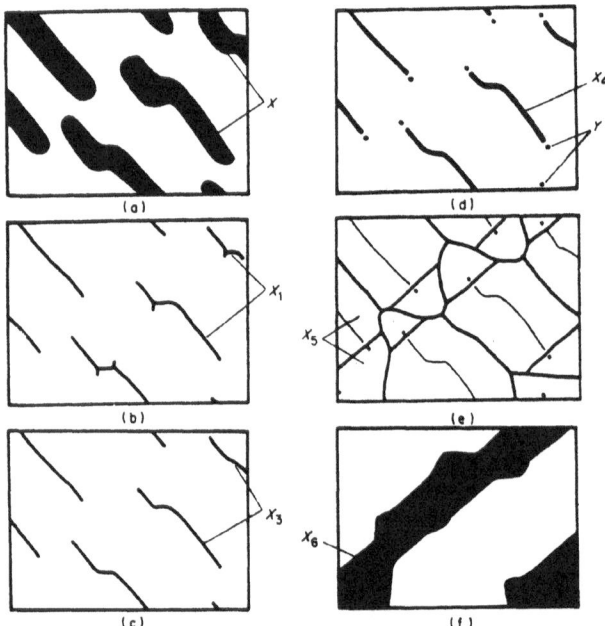

Fig. 4. The successive steps between the initial image (a) and the final results (b).

5 - HANDLING LOSS OF INFORMATION

Since we have noted that all morphological operations are irreversible, the sequence will depend on the items we decide to eliminate at each stage. The question arises as to how we should go about this.

5.1. First rule

One can use three possible ways to reduce a structure :

a - by spatial reduction : by limiting the zones of interest to subsets of the initial field (e.g., by eliminating those particles that cut across the field edge) ;

b - by reducing the definition. Images produced by the sensor are numerical : from what stage is it not better to binarize them ? and possibly to reduce their resolution ?

c - by overall-local reduction : what is presented initially for analysis is a field, i.e., an image defined in a rectangular frame that one modifies by bounded operators and identical for all points in the field. Such processes can lead to isolating a portion of the picture, that is then considered as being a whole, surrounded by a vacuum extending to infinity, and which thus can be dealt with globally.

It is advisable to bear these three ways in mind when proceeding with any analysis, thus ensuring that each will be tested in each case, and to the best effect. A significant example of their use is given in PRETEUX [1984], in the domain of numerical radiology.

5.2. Second rule

The second rule relates to the order of operations. Sometimes certain elementary operations can be exchanged (e.g., to fill in holes, or to eliminate particles that hit the field edge). Two operations can be switched when they do not undergo interplay. The ideal situation would be that every stage in a complex operation verifies this principle. The gestalt approach would most certainly break the processes into a series of criteria, the intersection of which would be the definition of the object under analysis. In such cases, there is an optimum order for the transformations, i.e., the order which minimizes the total processing time, starting with those steps that make the largest reduction in time, in the connotation of (a) and (b) above. F. MEYER [1979], in his study of Feulgen coloured cancer cells uses two simple criteria : the optical density and the maximum inscribed circles of the nuclei, which are mutually independent. They are also independent of the rest of the process, which is much more sophisticated, than these first two steps.

In this case, it proves advantageous to begin with the two rapid criteria. In fact, three stains visible in Fig. 7a are either wide or dense. two of them correspond to suspected cells, while the third (top right) is an artefact. The remainder of the process, which will eliminate the artefact, will only be brought to bear on these three stains.

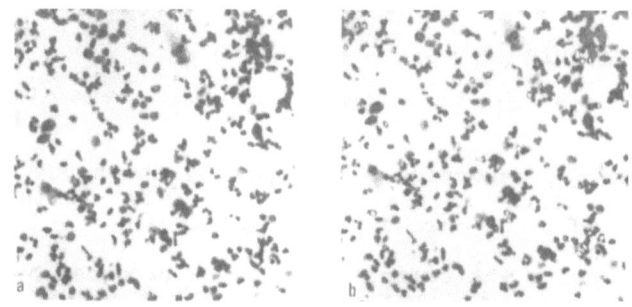

Fig. 5. (a) Feulgen stained cervical smear (J.S. PLOEM, Un. Leiden). (b) A first pre-treatment allows to concentrate very soon on the only possible cancerous cells of the smear.

5-3 Third and fourth rules

This priority approach rule is still applicable when the stages are not independent. We can identify two cases here :

- either the dependence is an aim built into the stages of analysis. For example, in the case of the lamellar eutectic, the extraction of the end-points presupposes the skeleton step, followed by the removal of the spurs. Thus, there are three distinct stages but they come in a preset order, which cannot be changed. A good knowledge of the main groups of this sort is highly necessary for persons wishing to use morphological mathematics, especially if they wish to build expert systems using this subject-matter ;

- or the dependence is forced. In this case, the first transformation must be that which least deteriorates the information needed for the second transformation. In the first example given earlier, we could have tackled the problem with a length criterion for the pipes ; on a grey-tone image, this would have led to an extraction of crest lines and we would have lost thereby the proximity information needed to detect widths. It was thus more appropriate to begin with the openig, as was done.

5.4. Fifth rule

It sometimes occurs that this rule of seeking minimum destruction of the information is impoossible, and we now set out a well-known example provided by F. MEYER [1980]. The purpose of the analysis was to separately study chromatin structure in the five distinct lymphocytes, shown in Fig. 8a. It is difficult, and at the same time very time-consuming, to directly separate them out on a grey-tone picture. However, this image accepts a threshold process fairly well, and the resulting concavities obtained in the binary image (6b) then enable good segmentation by watershed techniques (6c). Unfortunately, the initial information embedded in the grey shades has been lost by this process. In order to recover it,

we must use the connected components of Fig. 6c as templates
which are then applied successively to Fig. 6a . Thus, we have
both a global local process and a logical loop in the total
processing sequence. This back-tracking approach is quite
common in morphology.

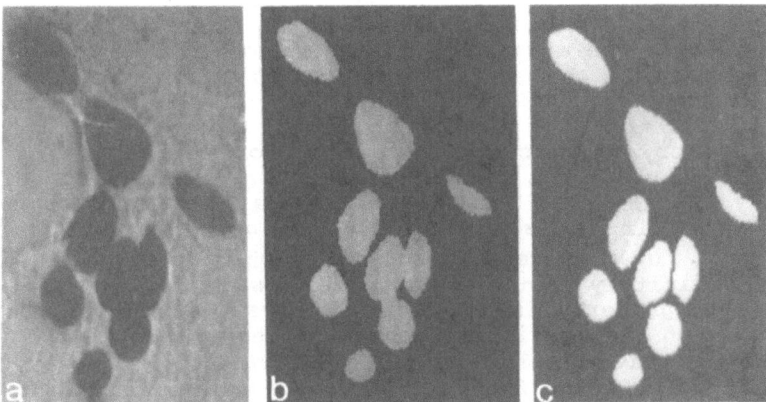

Fig.6 . An example of a logical loop in a morphological treatment: (a) lymphocytes; (b) their binary masks; (c)
segmentation of the binary mask; each of these five particles is the used as a local window acting on Fig. 8a.

5.5. Individualization of stages and idempotence

We can note that since we have presented the two
examples of section 4, we have been using the term
"processing stage", as if the choice were a self-evident one.
Let us now look more closely at what is meant by processing
stage. Does it refer to the moment when the picture has gone
through a dedicated processor 50 times ? Or the moment it has
gone through it once ? Is the switch-over from "decimal
image" to "binary image" (or the reverse) a stage ?

Let us refer back to our two examples. According to
the first example, it would appear indeed that each stage
corresponds to the introduction of a new parameter : e.g.,
the radius of a ball for the initial opening, followed by the
threshold level, then the minimum length for the skeleton,
the dimensions of the triangle, and finally the size of the
central strip. Although this is perfectly true, it only
comprises a partial answer to the question raised, since the
second example, strictly speaking, does not depend on any
parameter at all : the only parameter would be a threshold
level, but in the present case the initial image is
practically binary. However, the recognition process is
subdivided into two major steps : a search for the end points
and a reconstruction of their alignment. Between these two
stages, the seven operations we listed are carried through
individually (stages (a) to (g), Fig. 4).

13

If we interpret the thresholding as the transformation of a function $f_1(X)$ into another function it restores a function $f_2(X)$ which assumes the value 0 to 100% at point x according to whether $f(x)$ is greater than, lesser than or equal to a given level t_0. If one iterates the operation over f_2, there are no further changes. Moreover, f_2 is an overall indicator that enables us to pursue the process in terms of 2D-sets. This is what happens when one makes a skeleton as in stage b. Once again, if we iterate the algorithm, the skeleton of the skeleton is itself, without change. In stage c, likewise, there are no further changes, since the removal of spurs + rebuilding of the main branch, has no effect with further iterations. In stage d also, this is true, since if one seeks the extreme points of the extreme points, it still gives the same points. Finally, stages e, f and g still retain the same property. Each stage, if it is repeated before going on to the next stage will not produce anything new : when the step is done, it is definitely finished. However, within a stage, this particular property is not true : if we start up a thinning process when the previous thinning process is only half done, the image that results will be modified. Similarly, the property is not true for a group of stages taken together. If we take the final skeleton as the input for all the process, the new extreme points will bear no relation to the previous ones and indeed the process will diverge.

We leave it to the reader to check that the same logic applies to the case of the motors and calls for no special comments, provided the stages g and h (that we distinguished for pedagogical reasons) are grouped together. Only stage b is somewhat suspicious, because if we iterate it, it will give the initial image minus the residuals, i.e., it will remain open. However, if we iterate once more, we shall then have the residuals.

In fact, we could have taken any morphological analysis and the conclusion would have been the same. Thus we see that there exist steps in any process chain, such as the sequence of transformations carried out when one goes from one step to the next - which we call ψ, will satisfy the relationship :

(1) $\psi \circ \psi = \psi$

We have seen before (sect. 2) that the logical property expressed by the relationship (1) is idempotence. The physical role it plays is considerable. When we wish to dissect a perceptive process, with a view to programming it on a computer, we are led to thinking in terms of sequences of idempotent transformations.

Hence the notion that the various stages of a morphological process we have been discussing are autonomous, even when they are not independent.

6 - WHAT IS A MORPHOLOGICAL PROCESSOR ?

Since the onset of the 80's, we have witnessed a spectacular development of morphological processors. Doubtless for commercial reasons, certain manufacturers announce a strong filiation with the mother theory, while others prefer to ignore the connexion. Anyway, the theory does exist and implies a strong common denominator for all these systems.

In fact by using the rules described in section 5, and a few others out of this text (which relate to (anti-) extensive, but non-increasing, transformations in particular), we can decompose any morphological process into a series of dilations, and complementations. Indeed, these very dilations can be reduced to a smaller number in different ways, that one need only take as basic material. We can therefore define a morphological processor as being any computer-based system which provides :

i) a set of dilation primitives, which act as the letters, on the basis of which

ii) other modules can elaborate "words", or even complete composite processes (i.e., "sentences").

Such a system can be designed in terms of software (Noesis'Morpholog package) or, on the other hand can be entirely hard-wired, i.e., integrated into circuits (specific to a given application, for instance) ; all intermediate mixtures of software and hardware can also be envisioned.

The choice of a set of basic dilations, from which all other dilations will be developed, followed by complex morphological operations, is obviously highly important. Up to the present time the manufacturers have always limited the primitives they present to translation invariant dilations, characterized by structuring elements. For Leitz texture analysers, the size of the basic elements varies from 1 to 100 pixels ; in the Noesis Morpholog package or in the Numelec system, they are obtained as sub-sets of the elementary 7 pixel hexagon ; in the Machine Vision Inc. systems, they are reduced to a pair of points that may be considerably separated from each other. It is fairly easy to imagine other structuring alphabets (e.g., square ones, 3-D ones, ...) which would be optimal for other technological domains. Note that some of these alphabets are based on cellular logics (i.e., comparisons between one pixel and its immediate neighbours, PRESTON and DUFF [1985]µ but not all of them. In association, several techniques are able to accelerate the implementation of the dilation primitives : "operator pipelines", simultaneous processing of various sectors of the image, and for the convex structuring elements, recursive algorithms.

The quality factors of a morphological processor can be assessed also in terms of the <u>kinds of dilations</u> that the processor can easily perform. Now, in the simplest case, i.e., the 2D-binary case, there are three of which are used very frequently. There are the translation invariant operators, or structuring elements, the operators which are translation invariant in a given zone (the eroded part of the image field) and identically null beyond them. Also, there are the disks for geodesic operations (LANTUEJOUL, BEUCHER [1981]). Moreover, for any given space E (digital grids, Euclidean space, etc) there is the associate lattice of the partitions of this space which becomes important whenever we wish to simplify an over-segmented scene, and finally, the lattice of the sub-graphs of the digital functions defined over E. Computer systems should be evaluated not only in terms of the beauty of their architecture, but also in terms of the capability of the system to tackle all these lattices and dilations. If one makes the assessments from the latter point of view, it will be seen that not all morphological processors are equivalent.

One must then add on, the transformation sequences. From this point of view, the real degree of the power of a morphological processor is not to be attached so much to what we termed the words, such as skeleton, ultimate erosions, etc... but rather in the panoply of application software packages (sentences). This is only natural and logical. Geometric optical devices are mostly used to transmit images, i.e., images which are to be analysed, and reduced by morphological optical devices. As for all robots, their value is linked to hereditary or acquired memory, with respect to the cases they can effectively resolve, or the scenes they can "understand".

9 - CONCLUSION

To conclude, we shall examine again the classification of image analysis techniques into four categories ; this classification served as the starting point for the study (Fig. 1).

First there is the class of linear methods. As the name suggests, these methods develop image transformations which preserve addition, and beyond addition, the notion of group structure, thereby making reversibility an important feature. Their scope of application is enormous. We touched on them earlier when we discussed geometric optics, but they will also be found in other signal processing domains, in radiological tomography, in cartography (by kriging, for example). We have already seen how morphology, insofar as it puts the emphasis on irreversibility, takes the opposite course from that of the linear approach.

The second group emphasizes the measurement approach. Here, we consider that a picture is well described when there is a large number descriptors (volumes, histograms, diameters,...). These numbers are then reduced to a set of essential values, which, for example, have a stereological meaning, or which identify factor axes, or which may be grouped together by some taxonometric method. In contrast to this, morphology is more directed to mappings of a set E onto itself than onto R or onto R^+ ; it is, moreover, much more sequential : mapping ψ_3 follows ψ_2 which follows ψ_1, etc... whereas, the measurements usually derive in parallel from the initial image.

Lastly, we will place under the heading "syntactic" those methods which decompose the picture to be analysed into a family of simple shapes, such as disks, convex polyhedra, contour arcs, and then go on to describe each object as a particular organization of these primitive templates, using more or less deterministic rules of proximity or succession. These approaches are much more global than those undertaken in mathematical morphology, which, on the contrary, seek sequences of local operators. They are also more discrete, and finally, they use the calculation of optima more than morphology does.

The fact that these 4 methods differ vastly from each other and are sometimes contradictory in no way prevents them from co-existing. Experimenters have known this for a long time. It is never illegal to perform an image subtraction when applying morphological techniques, nor, once a specific object has been extracted by an ultimate erosion, to subject it to measurements for statistical or stereological computation. It would be more precise to state that the first stages of an image processing operation tend to be linear (summing, auto-focus, image enhancement by deconvolution, ...) ; then followed by morphological processing, and finally by statistical computations or syntactic recognition processes. This particular approach may be reduced to one or two stages. Human vision follows this procedure exactly : the lens, the retinal nerve endings, and the various brain stages are assigned to distinct operations, which follow one another (this does not exclude flash backs and logical loops).

However, there is a more profound link between the methods. All of them restore explicitly and sometimes visually, a sort of "object" which is more or less hidden and more or less spatial. The scanner 'slice' in tomography, shows something that was invisible ; a stereological calculation estimates the specific three-dimensional surface of a porous medium using just a few plane slices (in this case the restoring function relates to a number). A certain sequence of morphological operators will extract the cancerous cells specifically from an otherwise dense smear background. The Boolean random model which is invoked to deduce and explain a given crystalline growth interprets structures seen...

Finally, in these situations, more than in the result, it is the procedure followed that was made intelligible, thus enabling it to be "taught" to an image analysis device. We have already seen that, traditionally, optics as part of the study of vision, leads to hypotheses relative to the nature of light. Here, although we are still

dealing with optics, the hypothesis is not in the same
domain, and the postulates deal with an abstraction of the
processes of vision, i.e., on what we call a <u>mental image</u>. A
certain method will reduce the concept of the cancer cell to
a sequence of morphological operators, another method will
reduce it to a list of measurements followed by a
discriminant analysis, or yet another to a table of standard
shapes, etc, etc... ; i.e., the hypotheses vary, but in every
case, "it is not objects themselves which are stored but
mental images of the actual things, and in a form such that
the "mind" can recall them immediately".

       Could my famous compatriot, Saint Augustine, ever
have imagined that, if we were one day to reformulate his
question ("but how are mental images formed ?") in terms of
operational hypotheses, we would come to a world with
vision-endowed automats ? Quis dicit ?

REFERENCES

CHERMANT J-L., COSTER M. (1985) : Précis d'analyse d'image C.N.R.S.,
Paris, 521 p.
LANTUEJOUL Ch. (1978) : Détection automatique des lignes de défaut dans
les eutectiques lamellaires, rapport interne, 22 p.
LANTUEJOUL Ch., BEUCHER S. (1981) : On the use of the Geodesic metrics in
Image Analysis, J. of Micr. Vol. 121, pt. 1, pp. 39-49.
LAY B., LANTUEJOUL Ch. (1986) : Morphology User's Guide, Ecole des Mines
de Paris, and Noesis, 84 p.
MATHERON G. (1967) : Eléments pour une théorie des Milieux Poreux,
Masson, Paris, 166 p.
MATHERON G. (1975) : Random sets and Integral Geometry, Wiley, 275 p.
MEYER F. (1979) : Iterative Image Transformations for an Automatic
Screening of Cervical Smears (Journal of Histochemistry and
Cytochemistry, Vol. 27, n° 1, pp. 128-135).
MEYER F. (1980) : Empiricism or Idealism...("Pattern Recognition in
Practice", E.S. Geselma and L.N. Kanal, Editors, North-Holland Publishing
Co ; pp. 21-33).
PRESTON K., DUFF M. (1985) : Modern Cellular Automata Plenum Press, N.Y.,
300 p.
PRETEUX F. (1984) : Détection automatisée de différents organes en coupes
scanner (Colloque Image, GRETSI CESTA), 6 p.
ROSENFELD A., KAK A.C. (1976) : Digital Picture Processing, Ac. Press,
N.Y., London.
SERRA J. (1982) : Image Analys and Mathematical Morphology, Ac. Press,
600 p.
SERRA J. (1986) : Eléments de théorie pour l'optique morphologique,
Doctor Thesis, Univ. P. and M. Curie, Paris, 217 p.
SERRA J. (1987) : Advances in Mathematical Morphology, by Ac. Press.
STERNBERG S.R. (1982) : Cellular Computers and Biochemical Image
Processing, Proc. us-France Seminar Biochemical Image Processing,
Springer Verlag, Naw-York.

IMAGE PROCESSING ARCHITECTURES

M.J.B.Duff

Department of Physics and Astronomy
University College London
Gower Street, London WC1E 6BT

INTRODUCTION

It has seldom been disputed that the conventional von Neumann computer architecture is inappropriate for image analysis. As increasingly
elaborate image processing tasks were attempted and as the size and resolution of the images to be processed increased, so also did it become more
and more apparent that adequate processing speeds could never be achieved
by systems employing only a single processor. This paper reviews past and
current solutions to the problem of combining many processors into a uni-
fied system and also discusses the general considerations affecting the
choice of multiprocessor architectures for image analysis computing.

THE FIRST TWENTY YEARS

The history of the development of image analysis computers cannot be
traced much further back than the late 1950s. At this time, papers were
appearing describing the structure of the retina of the frog[1] and the cat[2]
and a variety of psychological studies began to suggest how the human
brain might be handling visual data. Conventional computing had made its
way into research laboratories where data analysis, particularly image
data analysis, was becoming a very laborious task, especially in the newly
developing field of high energy particle physics. Finally, the not always
reliable electronic valve was rapidly being displaced by the smaller and,
eventually, cheaper and more reliable transistor, leading eventually to
integrated circuit technology and its potential for incorporation into
laboratory built, special purpose computers.

Two of the pioneers in image computer design who responded to these
influences were M Golay[3] and S Unger[4,5] and their design proposals contained many of the more significant features to be found in subsequent
systems, even though many of these systems emerged independently in
research laboratories distributed around the world, their inventors not
having been aware of the earlier pioneering efforts. Even today, image
processing literature is scattered widely through an enormous range of
technical journals, often relating more to an image processing application
field than to the engineering principles of the system itself; it is
therefore perhaps not so surprising that the subject has developed in such
a piecemeal way.

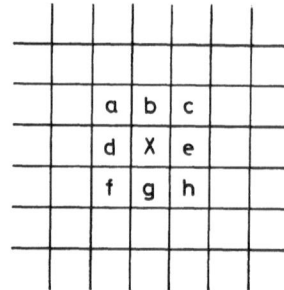

Fig. 1. Local neighbourhood operators.

The array architecture suggested by Unger and the pipeline proposed
by Golay both owed their success to their suitability for implementing
local neighbourhood operations. If **a** to **h** and **X** represent the pixel
values (grey levels) in a region of the image (see Fig. 1), then if an
operation is performed such that the pixel value **X** is changed to **X**** where

$$X^* = F(a, b, c, d, e, f, g, h, X)$$

and **F** is any logical or arithmetic function, then **F** is a local neighbourhood operator. The field of influence of **F** may extend beyond the 3 by 3
neighbourhood surrounding the pixel **X**, but this does not substantially
alter the definition.

In a parallel local neighbourhood operation, the same function **F** is
applied simultaneously at every pixel, the neighbourhood values being the
original set of grey levels in every case (i.e. values before **F** has been
applied anywhere in the image). This prevents indeterminate results
occurring as a consequence of small variations in the speed of the circuits in the processors (in a multiprocessor system) or non-uniform processing as a consequence of the scanning direction as the processor visits
different parts of the image.

An image pipeline processor facilitates a local neighbourhood operation by using a system of shift registers to present a single processor
with data from the defined neighbourhood (Fig. 2a). The full 3 by 3
neighbourhood of the first pixel **Xf** in the N by M image to have a complete
neighbourhood will be in the nine storage elements of the processor
exactly (2N + 3) clock cycles from the time the first pixel **a** is scanned.
The new value for the last 'meaningful' pixel **Xl** is computed **NM** cycles
after the start (Fig. 2b).

Thus with a single processor, images are processed at a rate 1/(**NMc**)
per second, where **c** is the length of the clock cycle. Note the assumption
that the processor can access the nine elements of pixel data and carry
out the required computation in a time not exceeding 1 clock cycle.

However, the stream of processed pixels begins to emerge (2N + 3)**c**
seconds after the first pixel has been scanned, so it would be possible to
cascade many such processors in order to carry out a sequence of operations[6], the delay introduced by each stage then being only (2N + 3)
seconds (rather than **NMc** seconds, the time taken by any one stage to process the complete image). Inevitably, as each stage of processing is carried out, the size of the array of correctly computed data shrinks by 2
pixels in each array dimension, unless steps are taken to restore the

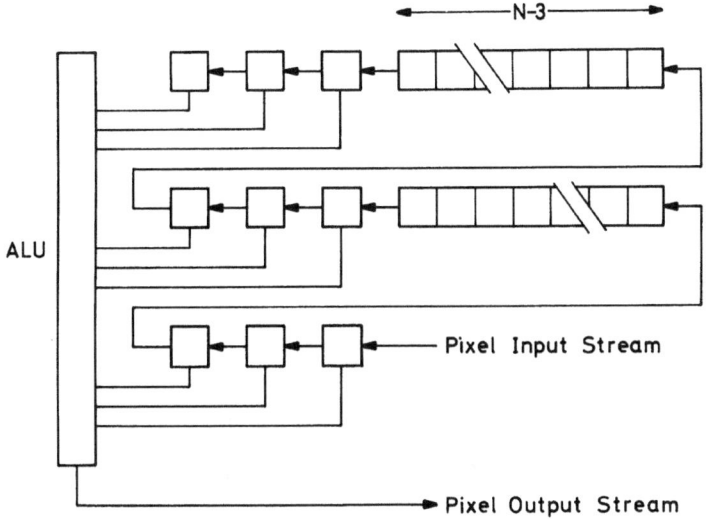

Fig. 2a.   The image pipeline.

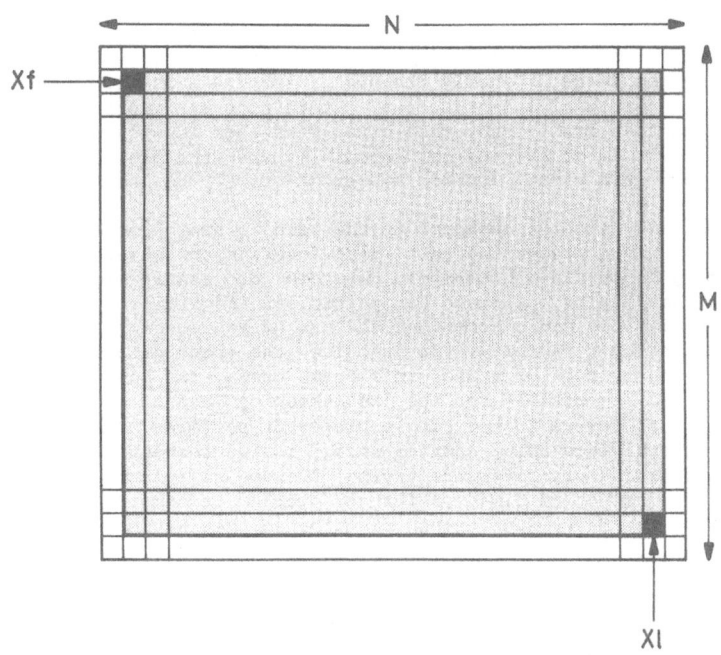

Fig. 2b.   Area (shaded) of computed image in which all pixels
            have a complete neighbourhood.

boundary by, for example, repeating the outer edge of the array in the form of a border duplicating the edge of the true image. If this is done and if the boundary conditions can be established without involving extra computing time, the first completely processed image emerges from an S stage processor after a time delay L (the 'latency') given by

$$L = \{(S - 1)(2N + 3) + NM\}c$$

Various practical difficulties are encountered with cascaded pipelines, such as the need to make available the results of earlier computations at processors further along the cascade. Nevertheless, the general principle is sound and cascaded pipelines have proved to be very effective systems for batch-processing streams of images.

The alternative, cellular logic array, approach to implementing local neighbourhood operations has been to assign a simple, one-bit processing element (PE) to every pixel in the image and to allow every PE to pass information between itself and its eight (sometimes only four) neighbours. Data is stored locally with each PE; instructions are broadcast to every PE so that each performs the same instruction simultaneously. Several systems have been constructed in this way[7-10] and much research has gone into the design of powerful algorithms which take advantage of the processor's array structure.

Although these two basic designs have appeared in a variety of different forms, the underlying principles have remained unchanged for many years. Cellular logic arrays, despite their simpler processing elements, are inherently faster than even the larger cascaded pipelines, except when the processing required is so simple that process times are dominated by data I/O times. The fundamental difference is that the cascaded pipeline has, at the most, one processor for each sub-process in the computational algorithm whereas the cellular logic array can have one processor for every pixel in the image. Usually the number of pixels is very much larger than the number of sub-processes. It is also an important feature of the pipeline that its latency will often mean that it will not be suitable for real-time image processing, especially when the processing is required as part of a visual control system.

A conceptually simpler approach to the need to employ more than one processor on an image processing task is to interconnect several processors and memory units via a high-speed bus. The processors might be identical or some may be optimised for particular image processing functions[11]. Ideally, one might hope that the use of P processors would speed up the processing by a factor P. In practice, the speed gain is often nearer to the square root of P and, in extreme cases, can be inversely related to P. It is possible for the 'housekeeping' and communication overheads to be so severe that a single processor performs more effectively than several together. A major hazard is bus contention, where several processors/memory units are trying to exchange information via the bus at the same instant of time. A further problem is memory contention, arising because the data, even though distributed over several memory units, may be required simultaneously by more than one processor.

A fourth major class of specialized processors comprises systolic arrays[12]. These have been described as 'pipelining arrays' since they are usually in the form of a two-dimensional mesh of identical, simple PEs, which differ from the more familiar cellular logic arrays in that vectors of data are repetitively clocked through the array, often from more than one direction at a time. Thus one vector stream might be passing through the array from North to South whilst, at the same time and in step, another stream passes from East to West. Calculations are performed by each PE in the array where the streams intersect. Usually, systolic arrays perform fixed calculations, but more recent designs[13] have incorporated programmable elements.

In summary, four major classes of processor can be identified as characterizing the first two decades of image processor development. Particular systems might be easily recognisable as falling within one of the four classes but others owe allegiance to more than one class (e.g. GOP[14] has both a pipeline section and a bus-structured section). However, for many researchers involved in image processing, good service has been obtained from relatively simple systems involving little more than a single fast microcomputer, a framestore and some look-up tables. The framestore holds one or more input images which are then sampled pixel by pixel and statistical analysis performed on the pixel values. Images can be combined by computing on pairs of corresponding pixels in two images (so-called 'point' operations) and the look-up tables provide the possibility of changing pixel values so as to give more easily interpreted displays, often in colour. A natural development of such systems is to add an image pipeline operating on a 3 by 3 pixel window, thereby upgrading the 'smart framestore' into a pipeline processor and, in so doing, blurring the boundary between the processor classes. In a similar way, additional circuits for address computation can be added to facilitate image transformations (such as rotation). The ever-falling cost of integrated circuit design and manufacture is encouraging this type of system growth.

THE NEXT STEP

With circuit technology now becoming cheaper at a virtually logarithmic rate, researchers have been encouraged to experiment with novel assemblies of processors (particularly microprocessors) and memory. Increasing standardization has also played its part; it has become much easier to interface between all the elements of a new computer architecture. Those of us who are old enough to have cut our computing teeth on electronic valve circuits which could be relied on neither to flip nor to flop on demand, look with envy at today's designers who can, to all intents and purposes, work at a keyboard and never suffer the agonies of picking up a soldering iron by the hot end.

Certain general principles and design aims are beginning to emerge. Assuming that circuit technology (NMOS, CMOS, gallium arsenide, etc.) is not under discussion, since designers will usually use the fastest technology available to them within the constraints of cost and proven reliability, then attention can be focussed on the architectural issues. The situation can be summarised thus: we need to decide what is the optimum type and number of processing elements (PEs) to be employed, how the memory units (MUs) should be distributed and how the MUs and PEs should be connected amongst themselves and to external systems.

Processing Element Complexity

Usually, in a parallel processing multiprocessor system, the cost of including large numbers of PEs will put a limit on the cost of each one individually and hence an upper limit on its complexity. The majority of systems still operate in a single instruction stream, single data stream (SIMD) mode in which one controller serves all the PEs, causing each to execute the same instruction at the same time. An exception is the cascaded pipeline where the stages operate independently (apart from I/O synchronization) and therefore need a separate controller with every PE.

The fundamental difficulty is to decide whether to have a few PEs of high complexity and power or more PEs with lower individual complexity. However, in general, it would seem that since it is difficult to achieve a linear speed-up with increasing numbers of PEs, it is more effective to increase the performance of an individual PE by a factor N than to attempt to employ in a parallel processing assembly N PEs of the original low performance.

When the required computation does not demand high precision (e.g. in many instances of binary image processing), much of the complexity of a processor can be wasted. Thus a 32-bit ALU-based microprocessor might perform no faster on binary images than a simple PE with merely a single-bit ALU. Conversely, in the reverse situation, 32-bit calculations which would involve at least 32 single-bit sub-operations can take an unacceptably long time in a single-bit PE, whilst it is unlikely that the 32-bit PE would be as much as 32 times more expensive than the single-bit PE. Clearly, optimisation of the PE complexity has to take into account the expected mix of single- and multiple-bit operations the system will be required to perform and this is, in practice, an unknown quantity.

Unfortunately, the choice of PE complexity is becoming more important as a result of the current trend, which is to employ very large numbers of PEs in each system and to attempt to pack as many as possible into each large scale integrated circuit. So-called 'silicon real-estate' is at a premium; higher complexity PEs imply fewer PEs per chip and hence higher unit cost. At some point, a threshold of acceptability is reached and the design becomes unacceptable in terms of its cost. In the real world, the cost-performance ratio may be attractive but the actual cost may be too high.

Connection Schemes

The effectiveness of a cellular logic array can be, and usually is, drastically reduced when the number or distribution of PEs does not match the image data array. Not only does each PE have to serve more pixels, but the neighbour relationships in the data are not all maintained amongst the PEs. Thus if a relatively small PE array is scanned sector by sector through a large image, problems arise at the edge of each sector. Also, potentially powerful operations involving propagation through adjacent PEs lose much of their power when sector-limited.

Similarly, cascaded pipelines become less satisfactory when the number of processor stages is much smaller than the number of algorithmic steps to be performed; the throughput of sequences of images is impeded whilst data is recycled through the reprogrammed processor loop.

But even when economics (or the size of the problem) permit a one-to-one relationship between PEs and pixels or between processor stages and algorithmic steps, there will more often than not be a requirement for moving data between distant processors in order to effect parts of the computation and this will usually involve stepping through chains of adjacent processors.

A superficially attractive but unrealistic approach is to try to provide connections between all pairs of processors so that there are direct paths for data no matter where it must go. Not only would the number of connections be impossibly large but so also would be the problem of routing the data unless each processor were to be individually controlled. Two compromises have been proposed and tested. In the first, the interconnection structure of the conventional two-dimensional mesh is extended into D dimensions, where the number of PEs is N and $N = 2^D$. Each PE is

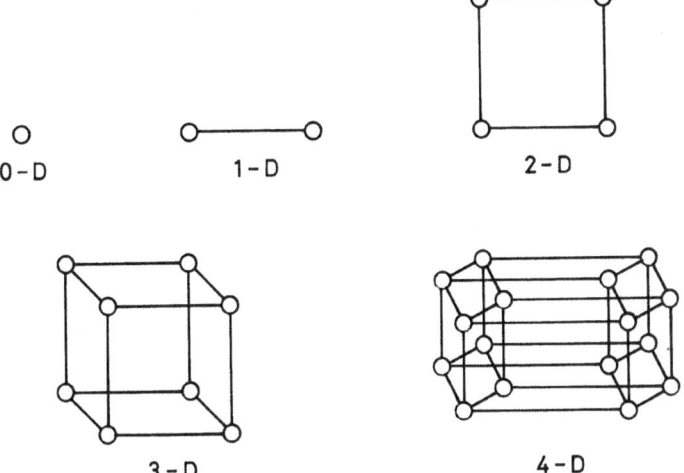

0-D   1-D   2-D

3-D   4-D

Fig. 3.   Hypercube connections.

connected to **D** PEs. The connections are between nearest neighbours in the
multidimensional space (see Fig. 3). It is easy to visualize the connection pattern by regarding each progression into a higher dimension as
being a replication of the set of PEs in the lower dimension. Thus the
0-dimensional array is a single PE; this is repeated in the 1-dimensional
array and the first PE makes a connection to its replica; in the 2-
dimensional array, the 1-dimensional pair is replicated and corresponding
PEs are connected together; in the 3-dimensional case (and similarly for
higher dimensions), if the address of a PE is (x,y,z), it will connect
with PEs whose addresses differ by 1 in only one dimension. For example,
the PE $(1,0,1)$ will connect with the three whose addresses are $(0,0,1)$,
$(1,1,1)$ and $(1,0,0)$ respectively. In effect, the path between any pair of
PEs will involve at most **D** steps, where **D** is the logarithm (to base 2) of
the total number of PEs in the array. This compares with a maximum number
of **N** steps in an **N** by **N**, two-dimensional mesh-connected array.

A fundamental defect in this so-called 'hypercube' connectivity is
that, although all pairs of PEs are only **D** or less steps apart, the paths
intersect and the effective bandwidth for data travelling simultaneously
between different pairs of PEs is greatly reduced. Practical systems have
to provide queuing facilities at every node where paths can cross.

An alternative approach is to structure the connection scheme so as
to facilitate the execution of commonly required algorithms. If an often
used algorithm involves data being moved between a particular pair of PEs,
then a direct path is provided. The most familiar example of this
approach is the two-dimensional mesh (relating to local neighbourhood
operations). The more recent pyramid architecture extends the mesh into a
stack of layers, each being half the linear size of the one below it.
Every PE connects with its four 'siblings' (the PEs to the North, South,
East and West) in its own layer, to its four 'children' in the layer below
and to its 'parent' in the layer above. This structure provides a natural
connectivity for implementing quadtree algorithms for multiresolution processing and also interconnects pairs of PEs in the lowest (largest) layer
by not more than 2 log **N** steps when that layer has **N** by **N** PEs. However,
as with the hypercube, bandwidth limitations are restrictive when

simultaneous connections between many pairs of PEs are required. Pyramid
architectures and their limitations are discussed thoroughly in [15,16].

All-to-all connectivity between processors becomes more of a possibility when the relatively small multiple instruction stream, multiple
data stream (MIMD) systems are considered. These systems, with typically
16 to 64 processors on a common bus or set of busses (e.g. the POLYP system[17]) are more usually concerned with intermediate- or high-level processing in which there is no commonly occurring algorithmic structure
which can be emulated in the computer architecture. Processing power is
obtained by evenly distributing the task between a large number of processors. The processors communicate amongst themselves and with distributed
memory by means of one or many buses and much of the skill in using the
system must be applied to avoiding bus contention as well as to task partitioning. But, in any case, the number of processors is small compared
with the number found in the SIMD arrays which are used more for low-level
processing, so multiple communication channels can reach nearer to the
all-to-all ideal without implying the need for excessively costly networks. Connection schemes in general and their relative merits are discussed in detail in [18].

Number of Processing Elements

It is perhaps worthwhile at this point to pause and reflect as to
what is the purpose of the quest for new architectures. There are two
major problems in image analysis and the hope is that the new and specialised computer architectures will help to solve them both. The problems
are:

1. Despite their apparent simplicity, many tasks in image
    analysis seem to remain beyond understanding; help is
    needed in designing algorithms that will do what is wanted.

2. Even when it is known how to perform a particular task, the
    process often takes far too long when existing computers
    are used; more powerful and more efficient processors are
    needed.

The second problem is an obvious target for computer architects, but so
also should be the first. A craftsman will refer to 'the right tool for
the job'; the 'right tool' does not only make the job easier but also will
sometimes give a clue as to how the job can be done. In the same way, it
has been found that computer architectures which relate closely to data
structures and/or algorithm structures will also guide the algorithm
designer into productive patterns of thought. The challenge, therefore,
is to develop an architecture which will not only make available computing
power in abundance but will also structure its components in a way which
will stimulate its effective use.

Given this dual motivation, it is clear that it would not be wise to
attempt to construct a system merely by stringing together a vast array of
processors on a common bus. In the first place, such a system would not
relate to the structure of an algorithm and would not give a lead as to
how it could be programmed in an efficient way; secondly, it would present
serious bus-contention problems. It is extremely unlikely that a system
of this sort would demonstrate the required power since most problems
would not distribute evenly between the available PEs. On the other hand,
if the bus were to be partitioned at each PE so that it became a form of
shift register permitting data to be stepped simultaneously between all
adjacent pairs of processors, then the system would have merit. Furthermore, the linear array could then be seen as matching a row of data in a

two-dimensional image and task-partitioning would relate to the data structure rather than to the algorithmic structure. CLIP7A is a system which has just been constructed to this pattern[19,20] although the main purpose in this case is to provide a powerful machine which will be used to emulate other architectures, rather than to represent a finished design in its own right. The philosophy behind the CLIP7A research programme has been to treat the linear array as an SIMD system but to allow the PEs a small degree of local autonomy. In particular, each PE incorporates an activity bit (allowing it to not respond to a broadcast instruction), a local address calculator (so that data can be fetched from different locations in each PE), and a register which determines the connections between each PE and its own local neighbourhood. Thus data calculated in an earlier part of the program modifies how each PE responds to later instructions. In this way it is hoped that the now familiar techniques for programming SIMD arrays will be gradually extended to deal with MIMD arrays, arguing that a more drastic jump would lose the skills already acquired rather than building upon them. It remains to be seen whether these expectations are to be fulfilled.

In summary, whilst it can be agreed that large numbers of highly interconnected processors will be needed if substantial speed gains are to be achieved, noting that it is seldom possible to increase speed in proportion to the number of processors employed, it is also sadly becoming obvious that programming difficulties are increasing along with the complexity of the system. These difficulties can be so severe that the increase in system complexity can even result in a drop in performance. It is therefore not unreasonable to ask whether any further improvements will be possible or whether some other line of attack should be tried.

OTHER APPROACHES

The early studies of neural networks did not in themselves produce much in the way of useful principles for the design of image analysing computers, even though the fact that the studies were in progress did seem to stimulate work on cellular logic arrays. The present day advances in semiconductor technology are beginning to encourage a new look at some of this earlier work since a large factor acting against emulation of neural networks had always been the virtual impossibility of approaching their complexity and researchers had always had a suspicion that the complexity might be an essential component of their design. An argument putting forward this view appears in [21].

Similar ideas are incorporated in newly appearing systems such as the Connection Machine[22] and the Boltzmann Machine[23] and it is interesting and encouraging to observe that I. Aleksander's WISARD system[24] is proving to be remarkably effective as a trainable pattern recogniser (and hence image recogniser) now the low cost of memory is permitting large assemblies to be constructed.

The various components of the argument are as follows: systems in general are minimised, for reasons of cost and efficiency, so that each system is cut down to the bare necessities for the job it is required to do; redundancy, if included at all, is added in controlled amounts in order to give more reliable performance. In contrast, neural networks appear to have an enormous degree of 'overkill' in that the number of neurons in a mammalian brain, for example, and the number of connections between the neurons are both orders of magnitude larger than might be thought necessary. At the same time, whenever a complex system is put together, its behaviour almost invariably surprises its designer in at least some respects; the system exhibits certain abilities which were not

contemplated when it was being built. Does this suggest that the
stringent minimization might be counterproductive and that it would be
advantageous to deliberately overcomplicate the system beyond the requirements of its performance specification?

A suggestion of this type must seem like a scientific heresy but,
nevertheless, is worthy of consideration. There would clearly be no point
in randomly throwing together circuits in the hope that something useful
might emerge (rather in the tradition of the monkey playing with a type-
writer and just happening to produce a passage from Shakespeare). The
essence of the proposal is that a large, multiply interconnected network
should be assembled and that the assembly should have at least some attributes which are in excess of the minimum that is both required and
thoroughly understood. If the philosophy is correct, then in exercising
the understood parts of the system, the user would begin to learn about
the remaining parts. Naive though this approach might seem at first
sight, it has already proved valuable in the understanding of cellular
logic arrays.

An obvious difficulty inherent in over-complex systems is that they
tend to be unprogrammable. Boltzmann machines, WISARD, the Perceptron and
other complex processors all incorporate some degree of training or learning rather than explicit programming. In the main, the functions of the
elements of the system stay constant but the parameters of the functions
(threshold level, for example), are adjusted as the processing proceeds.
It may be that the future lies in a further stage of sophistication in
which the functions of the elements also change as the process advances.
This is certainly the case when processors are all placed on a bus or in a
recirculating pipeline, but is not normal practice when the processors are
PEs in a densely interconnected network. Of course, to some extent the
distinction between a variable function and a function whose parameters
are variable, is unreal. This is demonstrated in local neighbourhood
operations such as the ranking functions where the operation can select
the maximum, median or minimum grey-level in the neighbourhood, deciding
which by adjusting the ranking parameter. The resulting image operations
(grey expand, 'median filtering', grey shrink) are extremely different.
It is interesting to speculate how one would interact with an array in
which the ranking function parameters adjusted locally on the basis of the
values of the local image data being processed. Similar reasoning lies
behind the operation of the Boltzmann machines in which the elements are
threshold gates and the thresholds locally adjust.

Whatever is the way forward, the ever-present danger is that the systems being built will become uncontrollable, not in the science fiction
manner in which the machine turns on its inventor, but by presenting such
an obscure interface to the programmer that programming becomes an intellectual impossibility. Another way in which progress can be impeded is if
a self-training system succeeds in training itself to perform a particular
task but does not reveal how it did it in an intelligible way. An attempt
to combat this weakness is made in the more useful 'expert systems' which
have a built-in explanation procedure which can be interrogated at will.
Even so, it has been bemoaned that the explanations produced are themselves often quite incomprehensible.

CONCLUSION

Scientific research never advances knowledge at a steady rate and the
study of image analysis computer architectures is no exception to the
rule. It would seem that we have reached a turning point where the conso-
lidating developments in the design of processors based on two-dimensional

28

mesh-connected arrays and pipelines are probably near to an end and that new thinking will now be necessary. One possible way forward is to explore highly complex assemblies of densely interconnected processors; another is to look more closely at the potential of self-programming and variable function arrays. It will not be altogether surprising if the way ahead involves some combination of both these approaches, together with other ideas still waiting to be discovered.

REFERENCES

1. J. Y. Lettvin, H. R. Maturana, W. S. McCulloch, and W. H. Pitts, What the frog's eye tells the frog's brain, **Proc. IRE** 47:1940-1951 (1959).
2. D. H. Hubel and T. N. Wiesel, Receptive fields, binocular interaction and functional architecture in the cat's visual cortex, **J. Physiol.** 160:106-154 (1962).
3. M. J. E. Golay, Apparatus for Counting Bi-nucleate Lymphocytes in Blood, U.S. Patent 3,214,574 (1965).
4. S. H. Unger, A computer oriented toward spatial problems, **Proc. IRE** 46:1744-1750 (1958).
5. S. H. Unger, Pattern detection and recognition, **Proc. IRE** 47:1737-1752 (1959).
6. S. R. Sternberg, Cytocomputer real-time pattern recognition, **in:** "Proc. 8th Auto. Imagery Pattern Recog. Symp.", pp. 205-214 (1978).
7. M. J. B. Duff, D. M. Watson, T. J. Fountain, and G. K. Shaw, A cellular logic array for image processing, **Patt. Recogn.** 5:229-247 (1973).
8. M. J. B. Duff, Review of the CLIP image processing system, **in:** "Proc. Nat. Comp. Conf.", pp. 1055-1060 (1978).
9. S. F. Reddaway, The DAP approach, Infotech State of the Art Report on Supercomputers, Infotech Ltd., Maidenhead (1979).
10. K. E. Batcher, Design of a massively parallel processor, **IEEE Trans.** C-29:836-840 (1980).
11. M. Kidode, H. Asada, H. Shinoda, and S. Watanabe, Image processing unit hardware implementation, **in:** "Real-Time/Parallel Computing", M. Onoe, K. Preston Jr. and A. Rosenfeld, ed., Plenum Press, New York, pp. 279-296 (1981).
12. H. T. Kung and C. E. Leiserson, Systolic arrays (for VLSI), **in:** "Sparse Matrix Proceedings 1978", I. S. Duff and G. W. Stewart, ed., Society for Industrial and Applied Mathematics, pp. 256-282 (1979).
13. H. T. Kung, Systolic algorithms for the CMU Warp processor, **in:** "Proc 7th Int. Conf. on Pattern Recognition, Montreal", pp. 570-577 (1984).
14. G. H. Granlund, The GOP parallel image processor, **in:** "Digital Image Processing Systems", L. Bolc and Z. Kulpa, ed., Springer-Verlag, Berlin, pp. 200-227 (1981).
15. V. Cantoni and S. Levialdi, eds., "Pyramidal Systems for Computer Vision", Springer-Verlag, Berlin (1986).
16. L. Uhr, "Multi-Computer Architectures for Artificial Intelligence: Toward Fast, Robust, Parallel Systems", John Wiley & Sons, New York (1987).
17. W. G. Griswold, P. H. Bartels, R. L. Shoemaker, H. G. Bartels, R. Maenner, and D. Hillman, Multiprocessor computer system for medical image processing, **in:** "Intermediate-Level Image Processing", M. J. B. Duff, ed., Academic Press, London, pp. 267-286 (1986).
18. H. J. Siegel, "Interconnection Networks for Large-Scale Parallel Processing: Theory and Case Studies", Lexington Books, Lexington, Mass. (1985).
19. T. J. Fountain, "Processor Arrays: Architectures and Applications", Academic Press, London (1987).

20. T. J. Fountain, K. N. Matthews, and M. J. B. Duff, The CLIP7A image processor, **IEEE Trans. PAMI** (Submitted for publication).
21. M. J. B. Duff, Complexity, **in**: "Intermediate-Level Image Processing", M. J. B. Duff, ed., Academic Press, London, pp. 307-314 (1986).
22. W. D. Hillis, "The Connection Machine", MIT Press, Cambridge, Mass. (1985).
23. D. H. Ackley, G. E. Hinton, and T. J. Sejnowski, A learning algorithm for Boltzmann machines, **Cognitive Sci.** 9:147-169 (1985).
24. I. Aleksander, W. V. Thomas, and P. A. Bowden, WISARD - a radical step forward in image recognition, **in**: "Sensor Review", pp. 120-124 (July 1984).

ROBOTIC VISION

Alexander I. Boldyrev, and Vitaly I. Rybak

V. M. Glushkov Institute of Cybernetics
Ukrainian Academy of Sciences
Kiev, USSR

ABSTRACT

Some results of research and development concerning 2-D and 3-D robot vision systems carried out in V.M.Glushkov Institute of Cybernetics are presented. The problems of 3-D object model construction and algorithms of its structure elements extraction from greyscale and range-finder data are discussed. An example of 2-D industrial vision system with its soft- and hardware is offered to illustrate practical implementation of the results achieved. Research and development are provided by means of simulation and a subsequent test of methods and algorithms on real data. This approach is implemented in specialized hard- and software complex.

1. INTRODUCTION

Machine vision is considered to be the most long-term and informative mean of robot adaptation to operations in small batch production and hostile environments. At the moment commercial vision systems are aimed at solving very simple tasks in a specific conditions. Most of them have SRI vision module[1] as a prototype where class and position of plane individual object are calculated according a binary image. The module uses such invariant integral features as square of object, perimeter, number of holes, etc. Naturally, the result vitally depends on image quality, binarization conditions of input data and relative position of objects in a field of view. Greyscale 2-D vision systems become more commercial lately[2]. They provide weakening requirements to the quality of input data.

However perception of 3-D objects and scenes is the main field of application. 3-D vision systems are now still in research stage. Both specialized systems for welding, assembly, sorting parts out and all-purpose systems are developed. Difficulties in practical implementations are caused by huge quantity of input data, cost limitations and absence of theoretical results in 3-D object perception.

Recognition 2-D and 3-D objects is distinguished qualitatively due to much diversity of images. For example incompliteness of object data on the images occurs not only because of noise but also because of mutual overlapping of objects. Good progress in 3-D vision systems development depends mainly on the solution of interrelated problems of 3-D object mo-

del construction and methods of its structure elements extraction from
the image.

2. 3-D OBJECT MODEL

Our investigations of visual data processing are carried out on the
basis of complimentary approach to the 3-D scene perception[3]. It consists
in simulteneous consideration and addition of data processing which represents combination of greyscale and range-finder raster images. Thus the
level of brightness and 3-D coordinates of the visible object points correspond to each image pixel. This approach supposes a more reliable solution of a recognition task.

It is then expedient to consider a fragment of the object surface
(or patch) which has all points with the same set of features and relations
as a structure element of the object model. The aggregate of patches is
the model of the whole object.

The overwhelming majority of modern industry production presents
parts with surfaces having not more than second order. Therefore we are
focusing that class of objects at the first stage of our investigations
concerning 3-D vision systems[4]. We are using a set-theoretical mode of
patches and objects description. Surface equations are interpreted as predicates with two meanings and take part in set-theoretical operations for
patch description.

Displacement, rotation and scaling are typical transformations in
industrial environment. Therefore the invariant features and relations
for our object model are as follows:
  - patch surface type and types of patch restrictions;
  - set-theoretical patch descriptions;
  - neighborhood relations;
  - mutual orientation of coordinate systems connected with patches.
Besides that coefficients of surface canonical equations and mutual coordinate systems positions are invariant to the transformations of displacement and rotation.

Outlines of patches are objectively present in greyscale and range-
finder images as changes of gradient. Therefore the conventional sequence
of operations is applicable: edge detection; image segmentation; features
measurement; segments subdivision into subsets that correspond to the 3-D
objects; recognition and calculation of objects spatial location.

3. EDGE DETECTION AND IMAGE SEGMENTATION OF GREYSCALE
   AND RANGE-FINDER DATA

There are a great deal of methods and algorithms of edge detection.
Most of them are aimed at processing greyscale data only. A universal algorithm for preprocessing of greyscale and range-finder data has been proposed in[5]. It admits parallel implementation the most time-consuming steps
of processing. The algorithm scans image by filter of $l*l$ ($l=3,..,8$) elements and subdivides a local section of image into subsections with
a straight line. All elements of each subsection are approximated by plane
and the section is classified as "roof" or "step". Image scanning is carried out with overlapping of sections.

Segmentation process consists of joining up compatible neighbor subsections in homogeneous regions that correspond to patch projections[6].
Two subsections are considered to be compatible if they have a mutual

fragment of subdividing lines, and one of the following three conditions
is satisfied:

1) $(A\tilde{A} + B\tilde{B} + 1)/(A^2 + B^2 + 1)(\tilde{A}^2 + \tilde{B}^2 + 1) < T_1;$
   $(A - \tilde{A})(i + \tilde{i}) + (B - \tilde{B})(j + \tilde{j}) + 2(\tilde{D} - D) < T_2,$

2) $(A\tilde{A} + B\tilde{B} + 1)/(A^2 + B^2 + 1)(\tilde{A}^2 + \tilde{B}^2 + 1) > T_3;$
   $Ai + Bj - D > \tilde{A}i + \tilde{B}j - \tilde{D};$
   $A\tilde{i} + B\tilde{j} - D < \tilde{A}\tilde{i} + \tilde{B}\tilde{j} - \tilde{D},$

3) $(A\tilde{A} + B\tilde{B} + 1)/(A^2 + B^2 + 1)(\tilde{A}^2 + \tilde{B}^2 + 1) > T_3;$
   $Ai + Bj - D < \tilde{A}i + \tilde{B}j - \tilde{D};$
   $A\tilde{i} + B\tilde{j} - D > \tilde{A}\tilde{i} + \tilde{B}\tilde{j} - \tilde{D},$

where (i,j), $(\tilde{i},\tilde{j})$ – raster coordinates of the section centers; A, B, D
and $\tilde{A}$, $\tilde{B}$, $\tilde{D}$ – coefficients of approximating planes; T_1, T_2, T_3 – assigned
thresholds.

4. APPROXIMATION OF RANGE-FINDER IMAGE PIXELS

During a recognition process it is necessary to determine what patch
of the object model is best-fit to the segmented set of range-finder
pixels. Straightforward solution consists in a selection of shift and rotation meanings for coordinate system connected with patches. It is un-
realizable in practice because of huge quantity of calculations. Therefore
it appears more expedient to search a quadric surface that is the best-fit
to the set of 3-D points and then to compare this result with a set of
patches in all object models. It is shown in[7] that it is rather simple to
do this in case of not high level of noise in 3-D coordinates measurement.
However, used in[7] a least squares method gives physically impossible results in practice. The statement of the task and its solution providing
3-D perception physics are proposed[8]. The solution is aimed at application
of parallel computing. It is shown that approximation task can be formulated as four subtasks with a subsequent choice of the best results:

1) minimization of functional $R = \sum\limits_{i=1}^{i=m}(b_{11}x_i^2 + b_{22}y_i^2 + b_{33}z_i^2 + 2b_{12}x_iy_i +$

   $+ 2b_{13}x_iz_i + 2b_{23}y_iz_i + 2b_{14}x_i + 2b_{24}y_i + 2b_{34}z_i + b_{44})^2$ (1)

   with a constraint system $b_{13}x_i + b_{23}y_i + b_{33}z_i + b_{34} \leq 0$ and $b_{33} = 1;$

2) minimization of functional (1) with a constrain system

   $b_{13}x_i + b_{23}y_i + b_{33}z_i + b_{34} \geq 0$ and $b_{33} = 1;$

3) minimization of functional (1) when $b_{33} = 0;$

4) minimization of functional $R = \sum\limits_{i=1}^{i=m}(b_1x_i + b_2y_i + b_3z_i + b_4)^2.$ (2)

Coefficients b_{pq} and b_p in expressions (1) and (2) correspond to the equations of quadric surfaces and planes that are defined in coordinate sys-
tem attached to raster image; x_i, y_i, z_i, $i = \overline{1,m}$ – 3-D coordinates of pixels
that belong to the homogeneous region.

The first and second tasks are the tasks of nonlinear programming
with linear constraints. For its solution effective algorithm is proposed[8]. The algorithm is aimed at application of parallel computing structures. The third and fourth tasks are reduced to the searching of matrix
eigen-values.

5. EXPERIMENTS IN 3-D VISION

Our first results in 3-D vision systems have been achieved still in 1977 when the first soviet model of autonomous "hand-eye" type robot was developed[9,10]. A vision system of that model recognized real scenes of polyhedrons. Those were disposed on supporting plane. Thus only one image was sufficient for calculation of 3-D coordinates of objects.

Research in recognition of objects with quadric and plane surfaces was carried out on the simulated greyscale and range-finder data[6]. Testing of algorithms was conducted using the example of the scene shown in Fig.1. "Hummer" type object was recognized. The model of the object was represented as a set of surface coefficients and geometry relations. Synthesized ideal greyscale (64 levels) and range-finder images 240*240 pixels were distorted by noise with Gaussian and even distribution. It was considered that objects were located 80-120 cm from image plane.

Recognition of object was done through comparison of surface parameters received in the approximation process with the meanings of model surface parameters. The result of recognition process was considered positive if there was coincidence of three neighboring patches with model parameters. Experiments shown that for practical implementation it is necessary to increase reliability of algorithms and do additional research both for synthesized and real data.

Fig.1

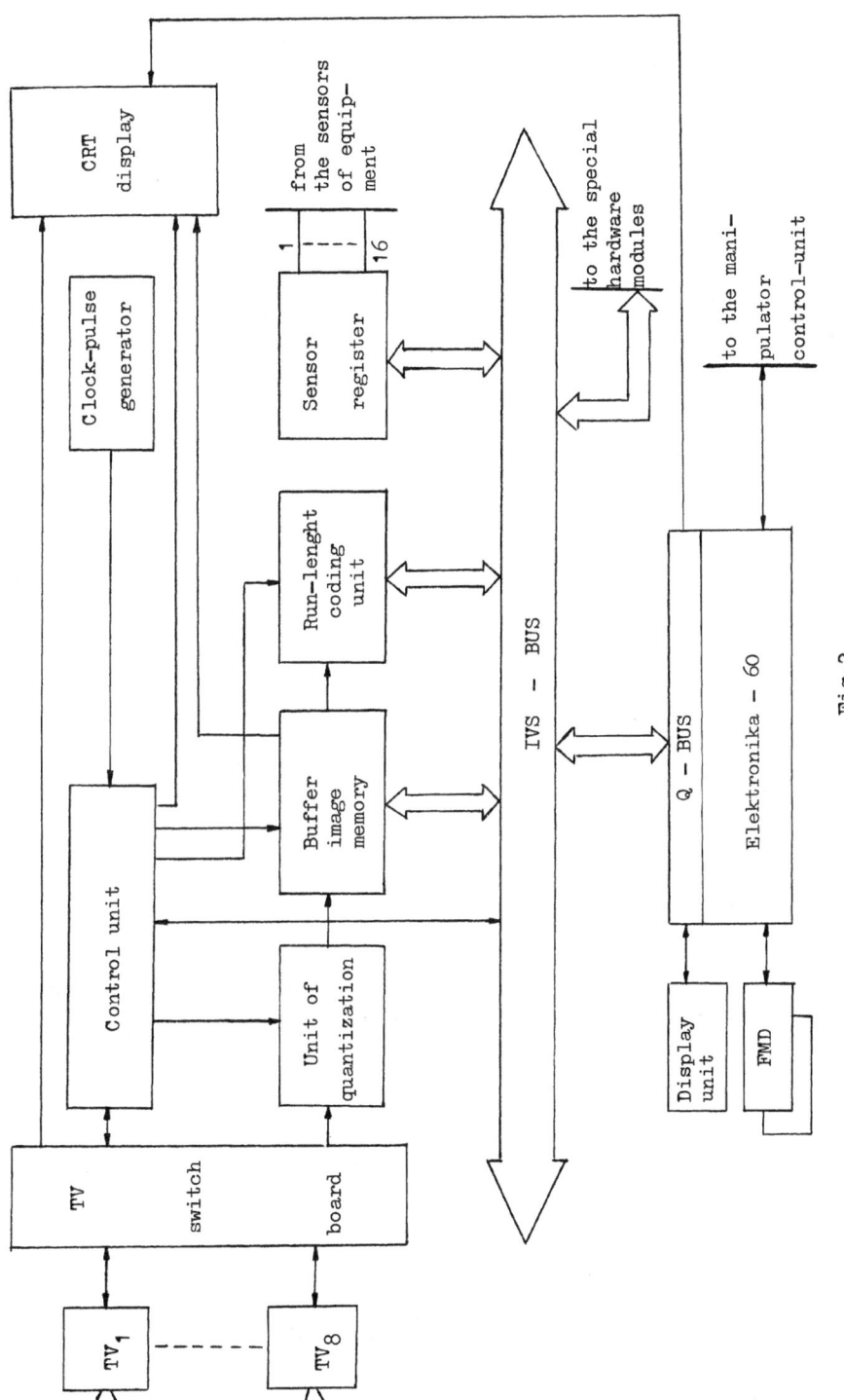

Fig.2

35

6. INDUSTRIAL ROBOTIC VISION

In our institute we have developed the industrial vision system (IVS) (Fig.2) for application some our results in industry[11]. IVS uses the set of invariant image features like SRI vision module[1] but it has more powerful hard- and software and what is the most important thay can be improved.

The first variant of IVS includes microcomputer Elektronika-60 and image input-output module with the switch-board of 8 TV cameras. The module can operate with greyscale images 512*512 pixels. A number of grey levels is determined by a number of bit-slice plates in image buffer memory. It can be increased veryeasily by plugging in new plates and using the corresponding analog-digital converter. The first variant of software is intended for processing binary images of 256*256 pixels and consists of TV camera calibration, IVS testing, teaching and recognition sections each of which can be chosen and managed by the user in dialog mode.

During the calibration process IVS calculates coefficients of transformation matrix between coordinates on image and object planes. It greatly simplifies a problem of invariant features extraction because it permits to make all calculations in integer numbers.

Fig.3

Constraction of 2-D object model is made by the well-known method teaching-by-show. The model structure corresponds to the recognition method that will be used. As object models are represented in object space casual or deliberate changes of TV camera location don't require new iteration of teaching. Each object of given class is presented to the system some times for calculation of features middle meanings and standard deviations. It is possible to change the set of models. The system can't store more then 10 models because of memory limitation (64K byte).

Testing IVS includes standard software microcomputer Elektronika-60 and special means for testing applied software and image input-output module.

The object recognition section consists of different variants of the nearest neighbor, decision tree, maximal click methods with the godograph to be used as well. These methods allow recognition not only individual but partially overlapped plane objects if they have special regions like holes etc.

Reliability tests of recognition methods were conducted on the set of industrial objects, figures from 0 to 9 and some letters (Fig.3).

The probability of recognition error was estimated[12] when all the experiments gave correct results. For objects and letters in Fig.3 this estimation was $4 \cdot 10^{-2}$ and for figures $2 \cdot 10^{-6}$.

The second variant of IVS provides overcoming constraints of the first one and has a hardware implementation of the most time-consuming algorithms. Those are adaptive thresholding, segmentation of images into homogeneous regions and calculations of feature meanings[13]. In contrast to the first variant of IVS where run-lenght image coding is used the second one includes algorithms and hardware that processe greyscale raster images. Algorithms are implemented in serial production microprocessor sets, VLSI and developed by us VLSI on the basis of array gates devices.

Adaptive thresholding and histogram devices are made as autonomous modules with their own vector on system bus. The histogram meanings and new thresholds are calculated during $\frac{1}{2}$ TV frame time period or 1/50 s.

Image segmentation features extraction are made by another device during 1/25 s. The device is intended for processing images of 256*256 pixels with 256 homogeneous regions. This limitation is caused by a quantity of VLSI (64) for associative memory that is necessary for speeding up image processing. However the structure of device allows its easy modification for processing image of 512*512 and 1024*1024 pixels.

7. METHODS AND MEANS OF ROBOTIC VISION RESEARCH

Modelling techniques have become conventional in the field of vision system research as in other fields where mathematization of task represents severe complications. In our institute we also develop a hardware and software complex for investigation of different approaches, algorithms and devices by simulation and real modelling[14].

Complex hardware consists of minicomputer SM-4, microcomputer Elektronika-60 and image input-output module (Fig.4). SM-4 and Elektronika-60 have the same set of instructions and compatible operation system RAFOS and FODOS.

Image input-output module includes buffer image memory of 512*512*8 bit, controller, thresholding device, standard TV camera and monitor. Arbitrary window of TV image is stored in buffer image memory and is read out to the CRT monitor in real time rate. An analog comparator is used for quantization of image in two levels if the module works with binary images. Greyscale data are the output of integral 6-bit analog-digital converter. The image input-output module recieves from the computer the following: meanings of thresholds, window size and position, steps along lines and columns of the image. Buffer memory can be randomly accessed and used in this case as an additional computer memory. The module is connected with SM-4 or Elektronika-60 through standard interface.

Software of research complex represents a set of modules opened for including and changing new modules. It allows to build up a different configurations for solution of specific tasks using simulated or real data. Modules are written in PASCAL and ASSEMBLER languages.

8. CONCLUSION

Research and development results in the fields of 2-D and 3-D robotic vision systems are presented. The problems of 3-D object model construction and algorithms of the model features extraction are discussed. An industrial vision system and its two variants are described.

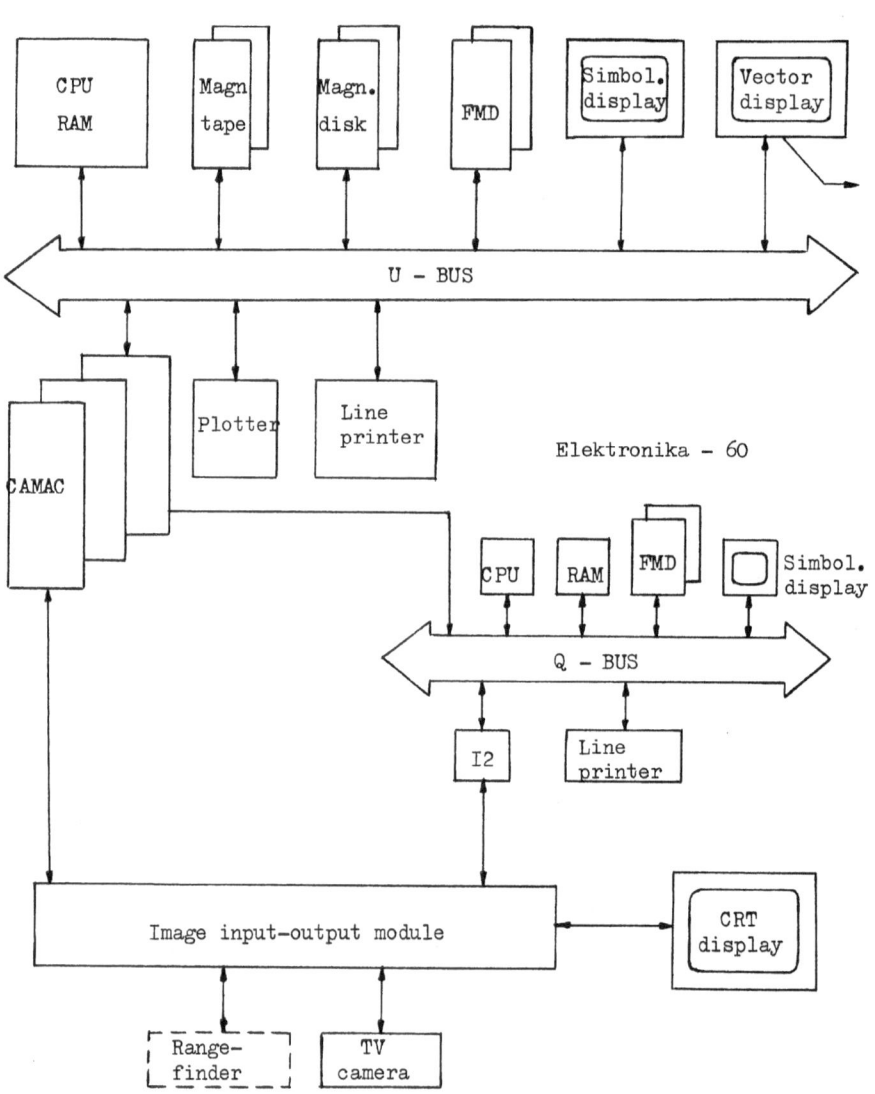

Fig. 4

REFERENCES

1. G. Gleason, G. Agin, A modular vision system for sensor-controlled manipulation and inspection, Proc. 9th Int. Symp. on Industr. Robot
ISIR-79, Tokyo, 1979, 57-70.
2. R. N. Nagel, Robotics: A state-of-the-art, Proc. 4th Jerusalem Conf.
on Inform. Technology, Jerusalem, Israel, May 21-25, 1984, 566-577.
3. G. L. Gimel'farb, V. I. Rybak, A complimantary processing of 3-D scene
greyscale and range-finder data, in: The problems of robot and artificial intelligence theory, Kiev, Inst. Cybern. Ukr. Ac. Sci., 1976,
66-75 (in Russian).
4. A. I. Boldyrev, A construction of 3-D object model boundered by plane
and quadric surfaces for description of 3-D scenes, in: The automatization of mathematic texts processing and robot construction problems, Kiev, Inst. Cybern. Ukr. Ac. Sci., 1979, 63-76 (in Russian).
5. A. V. Khomenok, An universal filter for greyscale and range-finder image
processing, 2nd All-Union Conf. on Robotic Systems. Abstracts, Part 1,
Minsk, 1981, 117-119 (in Russian).
6. V. I. Rybak, A. I. Boldyrev, A. V. Khomenok, Algorithms of visual data
processing in robot-manipulator vision perception systems, in: Robo-
tization of assembly processes, (Editor D. E. Ohocimskij), Moskow,
Nauka, 235-249 (in Russian).
7. G. L. Gimel'farb, V. M. Krot, Using of linearization space for matching
of curve line or surface that approximate given set of points, in:
The problems of robot and artificial intelligence theory, Kiev,
Inst. Cybern. Ukr. Ac. Sci., 1976, 52-65 (in Russian).
8. A. I. Boldyrev, An approximation of 3-D points of range-finder data by
plane and quadric surfaces in modelling system of "hand-eye" type
robots. Methods and algorithms, in: The problem of robot and artificial intelligence theory, Kiev, Inst. Cybern. Ukr. Ac. Sci., 1980,
13-27 (in Russian).
9. V. M. Glushkov, V. I. Rybak, The problems of active system construction
for robot environment perception, 4th All-Union Symp. on Theory and
Princ. of Robot-manip. Arrang., Abstract, Sec. 1, Tol'yatti, 1976,
29-32 (in Russian).
10. G. L. Gimel'farb, V. I. Rybak, V. I. Shul'ga, A description of 3-D
scene composed of simple polyhedrons by its image, Kibernetika,
1979, No.3, 73-76 (in Russian).
11. V. I. Rybak, A. I. Boldyrev, et al., Hard- and software of the first
variant of robot industrial vision system, in: Intel. robots and
pattern recogn., Kiev, Inst. Cybern. Ukr. Ac. Sci., 1983, 3-24
(in Russian).
12. V. I. Rybak, L. F. Buyalo, A. I. Velikoivanenko, An investigation of
classification validity and accuracy measurement of object location
by industrial vision system, in: Intel. robots and pattern recogn.,
Kiev, Inst. Cybern. Ukr. Ac. Sci., 1985, 3-10 (in Russian).
13. V. I. Rybak, A. I. Boldyrev, et al., A hardware implementation of image
preprocessing algorithms, Preprint 85-89, Kiev, Inst. Cybern. Ukr.
Ac. Sci., 1985, 20p. (in Russian).
14. V. I. Rybak, Methodological questions and technical and mathematical
facilities of design automation of system for external world perception of robot-manipulator, Kibernetika, 1979, No.2, 66-72 (in Russian).

UNDERSTANDING UNCONVENTIONAL IMAGES

T. Kasvand

Department of Computer Science
University of Ottawa
Ottawa, Ont., K1N-6N5, Canada

ABSTRACT

A computational strategy is suggested for images obtained from a laser range finding scanner. The scanned scenes may contain many objects of arbitrary sizes and in arbitrary positions and orientations. No a priori information is available on scene contents but the scenes are assumed to be "man-made". The computations are mathematically very simple but many steps have to be carried out to discover scene contents. The procedures are practical only if appropriate hardware is designed. The paper also discusses philosophical aspects of image understanding as applied to range images.

INTRODUCTION

It has been stated that human and other biological vision systems are approximately 1 000 000 000 years "old". The number of "evolutionary experiments" (species and individuals) during these billion years must have been formidable. Of course, biological evolution is not directly comparable to scientific research and development, but a billion years is a very long "lead time". Even though there are many different "designs" for biological vision systems, basically they see gray level and colour images using two eyes. Consequently, the images that our eyes can "absorb" we may call "conventional". As a consequence of the billion years of evolution, our vision is so advanced that we are not even aware of the extreme complexities of the problem. We simply take seeing and understanding of what we see for granted. Only with the advent of "computer vision" have we suddenly found that the vision problem is complex indeed.

Instead of covering the "old ground" again by trying to find different methods of processing gray level and colour images, it may be interesting to select a different image processing and understanding environment. Many such "new" images are available, for example, images from invisible spectral bands from radar to X-rays and gamma rays, electron microscope and laser range finder or depth images, and so on. Some of the new features are of the "conventional" type, such as more colour bands, some are "different" since the features give material properties, others may be called "unconventional", consisting of pixel spatial positions. An interesting question is, what else would one like to measure directly which offers truly novel features useful for image understanding?

The new images, first of all, allow us to see "strange new worlds in old light" (photographs and displays), but the new sensors also give us directly information for which we have evolved elaborate "computing machinery" in order to extract this information from the "conventional" images. The "depth" or "range" image from a (laser) range finder is a very good example. Range scanners can be designed to give the conventional gray and colour images as well as a distance image which is in registration with the "conventional" images (Rioux). Our vision uses about a dozen methods of detecting depth, ranging from "preprocessing" to elaborate "knowledge based" methods (Gibson). It may be conjectured that in order to simplify the image understanding problems, we would like to have ever more "comprehensive" scanners that measure even more point (pixel) features. However, valuable these point features may be in specific applications, such a wish reduces to an absurdity by asking for too much. For example, a natural scene scanner that can distinguish between pixels belonging to trees, cars, cats, gas stations, and everything else one encounters in an environment, is hardly feasible. Thus, somewhere there is a compromise between what can be usefully sensed and what has to be extracted by computations in order to provide "material" upon which image understanding may be based.

We use words like "knowledge", "understanding", "recognition", "memory", and so on, without defining their actual meanings. These are terms with which we describe our understanding of some of our innate information processing capabilities or other behaviour. Philosophers have argued about such problems for ages. For the present it may suffice to "define" the image understanding problem as follows: "For a mechanism to function adequately in its environment, it must be able to distinguish between old and new situations (recognize), to acquire new information (learn), and refer the old cases to its remembered (stored) experiences (knowledge) as a function of what it senses (sees) in its surrounding". Whether the image understanding system is a piece of specialized hardware or a computer program does not alter the fundamental problems, only the physical realizations differ. In a brief article it is impossible to review all the endeavors that involve various aspects of "image understanding". However, a perspective may be obtained by mentioning the very basic ideas only.

1) In the "self-organizing" approach a solution to the image understanding problem becomes automatic. We do not need to know "how the machine does it"! Basically, it is a question of designing a machine that somehow modifies (reorganizes) itself as a function of the impressed signals and rewards and punishments metered out for correct or incorrect responses. (Rosenblatt, Kohonen, Hecht-Nielsen, Tech. com.). The approach has been inspired by studies of biological systems which seem to accomplish this feat. There is an important aspect which should be noted. Whatever these systems use as "knowledge", this knowledge is acquired via the same sensors and processes that are used in recognition. An "external agent" may have designed the machine and influenced its "learning" methods and controls the environment, but he cannot define the "knowledge" beyond these limits.

2) In the "expert" approach "all the fine details" are processed and "every piece of knowledge" is identified beforehand and integrated into some form of "network of categorical and conceptual dependencies", graphs, rules, grammars, etc. "On the bottom" of even very abstract methods one always finds models of some forms of the objects to be recognized since the information must be stored somehow. However, it is well known that the meaning of a part of a scene usually depends on the context or on how the other parts in the scene have been interpreted. To attempt to specify all the fine details and interrelations between the details for all conceivable scenes is hardly possible. A detailed description and criticism of classical pattern recognition techniques has been given by Watanabe. In the "expert" approach the "knowledge" is given to the system by an "external

agent". Consequently, there is no guarantee whatsoever that this "external knowledge" (be it rules, graphs, grammars, or whatever) can be "matched to" what the sensors and subsequent processing actually can produce. This mismatch between the externally given knowledge, which usually is on too high a level, and what the image processing actually can "deliver" is one of the usual causes of failure, besides a likely combinatorial explosion.

3) Between these two extremes above there is a large variety of methods intended for "small closed worlds" or for very limited problems requiring few "models" or "little knowledge". This, of course, is a perfectly valid and reasonable approach for most practical problems in order to produce an economical solution. The methods are very specialized and extremely varied (Fu, Perkins, Rosenfeld, Besl and Jain, Jain and Hoffman). In addition, many other "boundary conditions" are used (isolated objects, clear background, known object size, etc.) to simplify the problem. This a very realistic view but, unfortunately, these solutions can neither be generalized nor combined to solve increasingly complex image understanding problems. An argument of the form "if we can solve problems A, B, C, ..., Z, then all we need is a supervisory program that selects the right problem solver" simply shifts the image understanding problem to the supervisory program. The fundamental questions in image understanding are still unanswered:

1) What form should the "knowledge" take?
2) Which experiences should be acquired and stored (learning)?
3) How should the machine acquire its knowledge, and via what "channels"?
4) How much a priori (externally given) information should be given?
5) How should the knowledge be used?
6) At what levels in the processing should knowledge be applied?
7) Are there "physical laws", for example, resembling those of classical or statistical mechanics, which could be applied? Etc.

The present study suggests some answers to these questions. By definition, the scene could contain a very large number of objects of any size, shape, orientation, etc., where any combinatorial approach will "explode". Initially, the machine has no a priori knowledge, i.e., the machine must learn and then use its learnt knowledge. The range image is used as an example (Kanade).

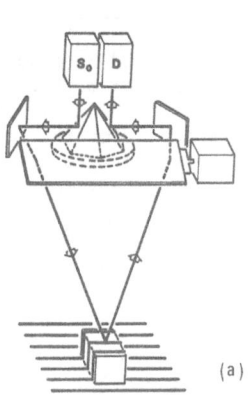
(a)
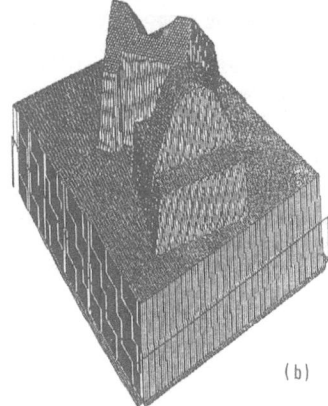
(b)

Figure 1: A range finding scanner (a) as designed by Rioux and a scanned scene (b) of some "industrial parts". In (a) So is a laser and D is a linear array of detectors. In (b) the Z-axis is upwards and the X-axis is to the left and downwards. A mosaic of processed results is given in Figure 3 for the top edge of the cylindrical region in (b).

THE 3D RANGE IMAGE

The so-called range or 3D image obtained from a laser range finder scanner (Fig.1a) is an image of distances from the scanner to the first surface encountered by the laser beam. The design of the scanner guarantees a rather large depth of focus. After correcting the signal for the geometry of the scanner and after isometric conversion, the range image is given as $Z(X,Y)$ where Z represents the distance from a reference plane and X,Y are the pixel coordinates (Fig.1b). The scales for X, Y, and Z may be assumed to be the same. A surface element in $Z(X,Y)$ is sometimes called a Monge patch. There are certain problems with range images. The scanner illuminates the scene from one point in space while it "sees" the scene from another point in order to have a parallax for range computation. This generates certain "invisible regions" where the scene is either not illuminated or not seen by the sensor. The image coordinates are initially deflection angles of mirrors. During the conversion to isometric form ($Z(X,Y)$) the various invisible regions are approximated by planes. These "nonexistent planes" confuse the image and complicate the analysis. In the future the pixels in the invisible regions will be coded as "invisible". However, whether the range image is given in terms of deflection angles and invisible pixels or as an isometric image does not alter the basic problem, namely, given one $Z(X,Y)$ image, on the premise that we know nothing about the scene content, what constitutes a reasonable computational procedure?

The n-dofels

Since we do not know what objects to expect in the image $Z(X,Y)$, any approach based on matching of complex models cannot be the starting point. The models either have to be very primitive, such as first or second degree surfaces, local pixel configurations corresponding to edges and corners, or we have no models at all, and we fall back onto the basic description of a surface element ("surfel") as defined in differential geometry (Kreyszig). Maximally, a surfel can have 8 independent parameters or degrees of freedom, namely, the position (X,Y,Z), the direction angle of the normal (N) and the maximum and minimum surface curvature (K1,K2) directions U1 and U2, and the surface curvatures K1 and K2. The question now is how to compute the number (n) of the degrees of freedom (dof) of each picture element (el). Let us call this n-tuple of geometrical pixel parameters the "n-dofel". A purely "bottom-up" approach is taken since no models can be assumed. The following computations have given reasonable results:

1) Compute the surface gradient images $Zx(X,Y) = dZ/dX$ and $Zy(X,Y) = dZ/dY$, and smooth the results if necessary. Compute the magnitude of the gradient $Zm = Sqrt(Zx*Zx+Zy*Zy)$ and the length Nm of a vector in the direction of the normal N, $Nm = Sqrt(Zx*Zx+Zy*Zy+1)$, where Sqrt indicates the square root and * indicates multiplication. The local surface normals are given in terms of direction cosines as $N(Nx,Ny,Nz)$ or in terms of angles as $N(A,B)$, where A is the orientation of the surface pixel on the XY plane with respect to the X-axis, and B is its tilt with respect to the Z-axis. $Nx = -Zx/Nm$, $Ny = -Zy/Nm$, $Nz = 1/Nm$, $A = Arctan(-Zy,-Zx)$, and $B = Arctan(1,Zm)$, Fig.2a.

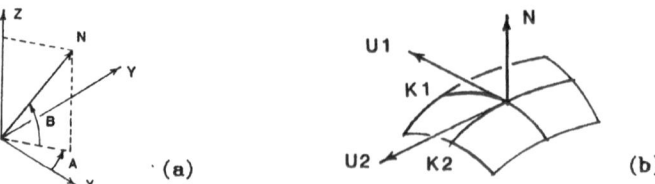

(a) (b)

Figure 2: (a) The surfel normal N, its orientation A and tilt B. (b) Surfel normal N and the surface curvatures K1 and K2 in directions U1 and U2.

Thus, of the 8 degrees of freedom three (X,Y,Z) are given and two angles (A,B) are obtained from first-order differencing, i.e., we now have a 5-dofel (X,Y,Z,A,B). If we know beforehand that the scene consists of planar surfaces only, then the above 5-dofel is sufficient for further processing. This 5-dofel is also sufficient if we want to approximate all the curved surfaces in the scene by planar patches.

2) Compute the local maximal surface curvature K1 and the minimal surface curvature K2, including their directions C1 and C2 in the XY plane or vectors for N, U1, and U2 (Fig.2b). These parameters are much more difficult to obtain numerically and they always involve second-order differences. An error analysis, which is standard practice for numerical procedures, implies that the formulas given in texts on differential geometry are very sensitive to noise. Instead, the derivation procedure for the formulas for these parameters was simulated in the computer, in order to have some control over the process and to obtain error measures (Kasvand). K1 and K2 are scalar quantities for which the signs (+ or -) can be assigned arbitrarily. Here convex surfaces, as seen from the scanner, have been chosen to have positive curvatures. Hence, concave surfaces will have negative curvatures. The Gaussian curvature Kg = K1*K2 and the mean curvature Km = (K1+K2)/2. A good measure for flatness is $|K1|+|K2|$. Now we have a "9-dofel" of (X,Y,Z,-A,B,C1,C2,K1,K2) but there are only 8 true degrees of freedom since only the three vectors (N,U1,U2) are required to define the orientation of a surfel. However, for the moment we may consider all these nine parameters as "independent" since only selected subsets are used.

The computations outlined in (1) and (2) have given numerical values for the "geometry" of each pixel. The position of each pixel was known at the start. Thus, we now know, X, Y, Z, A, B, K1, K2, C1, and C1, where C1 is the direction of the maximal curvature K1, and C2 that of the minimal curvature K2 on the XY plane.

In the computation of any "local feature" we encounter a basic paradox. The larger the local area, the better we are able to compensate for noise in Z(X,Y). The noise usually is of three types, namely, real noise from the scanning process and subsequent data conversions, that caused by spatial and amplitude quantization, and "computational noise" from local differencing. However, the larger the local area over which we carry out the computations for a local feature, the greater are the chances of including areas with different characteristics. This situation is very noticeable at discontinuities in Z(X,Y). A "discontinuity detector", however, is only a partial answer suitable only for detecting significant "jump edges" (Kasvand). "Ordinary" edges cannot be distinguished until later. The immediate conclusion is that the computation of any local feature will produce erratic results in unknown locations in the image. The only information available to us is the size (diameter Ds) of the local operator and the number (Nd) of sequential local operations that have been carried out. We also know that the cross section (thickness) of the "erratic region" could be as wide as the local operator (Ds). The erratic region is spread by each subsequent local operation. If all the local operations are of the same size (Ds) and there are Nd sequential operations, then the maximal width De of the erratic region is De = k*(Ds-1)*Nd where k=1. Usually the spreading effect De "tapers off" allowing us to use k<1. This argues strongly against the use of large-area "local" operators. In fact, local operator methods "self-destruct" due to De, implying that small area (for example, 3 by 3) operators followed by nonlinear smoothing ("relaxation") are preferable.

Local coherence

The computations so far were based on the concepts developed in differential geometry. Except for the difficulties of obtaining reliable numerical values for the parameters, the methods used are quite formal. The geometrical parameters of each surface picture element (surfel), or the "9-dofels", consist of (X,Y,Z,A,B,K1,K2,C1,C2), which is "overdetermined". The information available now consists of the original range image and the set of registered images produced by preliminary processing, i.e.,

    Z(X,Y) = original range image (given)
    A(X,Y) = orientation angle image of normal vector
    B(X,Y) = tilt angle image of normal vector
    K1(X,Y) = maximum surface curvature image
    K2(X,Y) = minimum surface curvature image
    C1(X,Y) = direction angle image of K1
    C2(X,Y) = direction angle image of K2

The geometrical information on each surfel is now fairly complete, but it will contain a certain degree of errors which increase as a function of the amount of sequential processing. Hence, it is unlikely that numerical results for third and higher order derivatives of Z(X,Y) can be relied upon. In addition, the scanner can produce a registered gray level image G(X,Y) and also registered images for red, green, and blue. The information is still pixel based.

Given that the 9-dofels described above are the only information available, what would be a logical next step? Since the contents of the scene are unknown, all we really can do next is to search for some form of regional coherence among the n-dofels. However, what form this coherence takes is unknown. The approach follows the classical methods where objects are assumed to consist of regions or boundaries between the regions. It is unlikely that scene understanding can be achieved without extracting some regional characteristics and the interrelationships between these regions. Normally a distinction is made between "edge" detection and "region" segmentation. The detection of pixels belonging to edge regions (edge pixels) is based on some form of difference among spatially adjacent pixels while the pixels belonging to regions (region pixels) are detected by their uniformity. Since edges are boundaries between regions, these two approaches thus try to solve the same problem from opposite directions. The question now is, of course, which is the better procedure?

Edges, by their very nature, are rather thin and elongated regions between any two areas in the image that they are supposed to separate. In an edge image the edges are represented as "strings of edge pixels". The edge pixels have been detected by some form of local differencing, possibly followed by thresholding, if a binary edge image is wanted. Since the decisions are local and do not involve context, the continuity of these "strings of edge pixels" cannot be guaranteed. In practice, edge detection produces only a scattering of individual edge pixels which must be "joined" to form continuous edges. Since edge shape is unknown, classical model based methods, such as the Hough transform (Ballard), cannot be used. Various "relaxation procedures" and other methods may be tried but they do not guarantee continuity of edges. Thus, edge detection cannot guarantee continuity of edges, assuming that edge continuity is needed.

If regions (local homogeneous areas) in the image have been detected reliably, then the boundaries between the spatially adjacent regions are the edges. The edges are now always continuous and also closed curves, except at image boundaries. However, if the edges are represented as pixels in an edge image, small or thin areas can be "lost" since their pixels are converted to edge pixels. Continuity of edge pixels remains, but the edges may not always form closed curves. Thus, edge detection via region segmentation always produces continuous edges, but there is no guarantee that the edges are in any way "perfect" or even located in correct places in the image. Of course, region segmentation suffers from its own difficulties, i.e., sensitivity to various thresholds resulting in too few regions, too many regions, small erratic regions.

In the more generalized approach to coherence detection in range images the region and edge detection methods are combined. The procedure is an elaboration on methods used in remote sensing, consisting of clustering of pixel features, cluster labelling, projection of the cluster labels back into the image space, and processing of the back projected label images. It should be noted that the mathematical rigour of the methods is becoming increasingly "amorphous". However, the nature of the "primitives" that emerge from this processing are determined by the process itself. The "external agent" can only define the processing strategy but not the primitives themselves. This approach guarantees "internal consistency" but will fail if the processes are too specific and aimed for a particular image only. Hence, image content is assumed unknown. The methods are explained by some examples.

The origami world. The word "origami" may be translated from Japanese as "the art of folding paper" in order to make variously shaped objects (polyhedra, birds, animals, etc.). Thus, the origami world (Kanade) only contains scenes of objects approximated by planar surfaces (and sharply defined edges). Whatever the objects, their number, size, position, orientation, and overlap, all the surfaces in the scene are flat planes representable by first-order equations. It will be assumed that the planes are relatively smooth and that the spatial resolution of the $Z(X,Y)$ image is adequate for processing. The first problem is to segment the $Z(X,Y)$ scene into planar facets and to detect the edges and the corners. The scene content is unknown. The processing steps are as follows, see Fig.3:

1) Given the original range image $Z(X,Y)$, compute the orientation angle image $A(X,Y)$ and the tilt angle image $B(X,Y)$ of the surface normal. A 3 by 3 operator should suffice for computing the dZ/dX and the dZ/dY images, possibly preceded and followed by some smoothing. Determine the maximal spread De of the erratic regions at the (unknown) edges.

2) Carry out a clustering procedure using all the pixels in $A(X,Y)$ and $B(X,Y)$. The number of clusters, which corresponds to the number of differently orientated planar facets in the scene is, of course, unknown. If the image is large, the number of pixels that have to be clustered may "overwhelm" a standard clustering package. The following procedure is suggested which gives better results the more pixels there are:

2a) Construct a two-dimensional histogram (decision space, clustering space, contingency table, Gauss sphere) $H(A,B)$. The A-axis represents the orientation of the surfel. The A-axis range is from 0 to 360 degrees and it is cyclic (angle 0 is the same as angle 360). The B-axis represents the tilt of the surfel. Its range is from 0 to 90 degrees and is not cyclic. Choose the same quantization for the A and B axes, say, 2 to 5 degree intervals, depending on the noise level in the data and desired "resolution capability". The histogram $H(A,B)$ may be treated as a gray level image.

2b) Detect and extract the "peaks" in $H(A,B)$ since they represent the variously orientated planar surfaces in $Z(X,Y)$. Note that histogramming "scrambles" the spatial information and one peak in $H(A,B)$ represents all the facets that have the same spatial orientation. Many different techniques may be used to extract the peaks. The method used consisted of some smoothing of $H(A,B)$ followed by dynamic thresholding (subtract smoothed $H(A,B)$ from $H(A,B)$ and threshold). The result is a binary version $Hb(A,B)$ of peaks. $Hb(A,B)$ may be processed as any binary image.

2c) Impress constraints on the Hb(A,B) image in order to separate certain situations, if desired. For example, near vertical surfaces (jump edges) may be forced into one class by setting all the pixels in Hb(A,B) to 1 for the corresponding quantization value of B.

2d) Label each isolated component in Hb(A,B) and "spread" the label numbers, since dynamic thresholding removes the "skirts" from the peaks in H(A,B). Call the resultant image Hl(A,B).

2e) Project the labels in Hl(A,B) back into image space, i.e., create an image Lh(X,Y) of Hl(A,B) cluster labels by determining into which cluster each pixel belongs. Pixels which do not belong to a cluster are zeroed.

2f) "Clean up" the Lh(X,Y) image using techniques similar to those in remote sensing after classification. However, one should be careful since the H(A,B) histogram has not included the Z values. Thus, two adjacent horizontal facets Fl and F2 will have the same label in Lh(X,Y). The boundary between Fl and F2 is usually narrower that the "erratic region" indicated by De previously. The width of the boundary region depends on the amount of "label spreading" allowed in step (2d). The simplest and crudest procedure is to zero each label in Lh(X,Y) which has a differently labelled neighbor in a 3 by 3 eight-connected region. Call this image Lcore(X,Y) which shows only the reliably classified pixels in Z(X,Y). Only the pixels belonging to these "core" regions should be used in analytic approximations of the planar facets. The Ll approximation is quite suitable (Abdelmalek).

3) The back projected cleaned cluster label image Lcore(X,Y) is next processed for detecting individual facets and the edges between the facets. The processing, basically, consists of the following steps.

3a) Set all the label values in Lcore(X,Y) to 1, resulting in Lb(X,Y), to which binary techniques may be applied. However, care has to be exercised since the boundary regions in Lb(X,Y) may not be very wide.

3b) The 0-valued regions in Lb(X,Y) represent edges. The width of these regions vary depending on the noise level in the dofel values and how successful the clustering and subsequent "label cleaning" operations were. Consequently, we should be prepared to handle "rather thick" edge regions in Lb(X,Y). These thick regions will be called "undefined". In brief, the edge and corner detection consists of:

(i)   Negate Lb(X,Y) giving Lb'(X,Y).
(ii) Erode Lb'(X,Y) to remove regions of width De, giving Lerode(X,Y).
(iii) Exclusive "or" Lb'(X,Y) with Lerode(X,Y) giving Le(X,Y), which is the "thick" edge image.
(iv) "Perfectly" thin (Tamura) Le(X,Y), giving an image called Ledge(X,Y).
(v)   Detect junctions (corners) and segments (boundaries between adjacent facets) using Ledge(X,Y) and label both. Junction and segment regions are 8-connected. This gives a labelled junction and segment image Ljs(X,Y). The junctions may be labelled negatively and the segments positively in order to make the representation more compact.
(vi) Label the 0-valued (4-connected) regions in Ledge(X,Y). This gives the facet label image Lfacet(X,Y).
(vii) Number the pixels on each segment sequentially to give an ordering to the pixels. Call the resultant image Lseq(X,Y).

In practice some smoothing may be needed and the thinning process should be made "more intelligent". Edges detected by other methods are "or-ed" into Lb'(X,Y) before step (ii). A better method of integrating edge detection and region segmentation is given in the next section.

48

In summary, the information obtained in the coherence detection process is contained in Lcore(X,Y), Lfacet(X,Y), Ljs(X,Y), and Lseq(X,Y) where,

Lcore(X,Y) = Cluster label image identifying the reliably detected pixels in Z(X,Y) and the corresponding cluster label numbers.
Lfacet(X,Y) = Facet labelled image where each detected facet in Z(X,Y) has been given a unique label number.
Ljs(X,Y) = Junction and segment labelled image where each junction (corner) and each segment (edge between any two adjacent facets) have been given unique label numbers.
Lseq(X,Y) = Pixel sequence number image establishing ordering of pixels on each segment.

The success of this sequence of operations depends mainly on the accuracy of the orientation A(X,Y) and tilt B(X,Y) values and the resolution chosen for the histogram H(A,B). The use of this information is described further on.

The man-made world. Many manufactured objects, but specially the parts out of which these objects are constructed, frequently have relatively simple surfaces in order to simplify manufacturing. The facets tend to be planar, cylindrical, conical, and spherical, i.e., second-order surfaces. Such facets are easily detected if the local surface curvature images K1(X,Y) and K2(X,Y) have been computed. K1 is the maximum value of local pixel surface (surfel) curvature and K2 is the minimum surfel curvature. The K1 and K2 values are scalars with an arbitrarily assigned plus or minus sign. As seen from the scanner, a convex surface is considered to have positive curvature and a concave one negative curvature. The directions in which K1 and K2 are measured are given by the orthogonal unit vectors U1 and U2, or by C1 and C2. C1 and C2 are the direction angles of K1 and K2 projected onto on the XY plane. C1 and C2 are orthogonal only when the surfel tilt is zero.

In terms of surface curvatures K1 and K2, a planar surface has zero curvature (K1=0 and K2=0), a sphere has constant and equal curvatures (K1 = K2 = constant) which are positive for "bulges" and negative for "bowls", a cylindrical surface has K1 = constant and K2=0, a cone has a linearly varying K1 while K2=0, and so on. The planar and spherical surfaces are "degenerate" since their directions cannot be determined. If one is searching for a prespecified surface in the image, the necessary conditions for K1, K2, and C1, C2, or U1 and U2 are predictable. However, in the present case scene contents are assumed unknown, except that the facets are assumed to be at most second-order surfaces.

In the Z(X,Y) image the location of the edges are, of course, unknown. However, an edge in Z(X,Y) has a very large value for ¦K1¦ and a much smaller value for ¦K2¦, where ¦.¦ indicates magnitude. For a straight edge K2=0. Thus, the surface curvature K1 is an excellent "edge detector" for range images. A sharp point or a deep and narrow hole will have large values for both ¦K1¦ and ¦K2¦. By utilizing this property, it is possible to treat region segmentation and edge detection as one procedure. As in the case of the origami world, coherence detection may be treated as a clustering problem, but since more information is available (the 9-dofels), the number of alternatives is increased.

In the absence of prior information on scene contents, the suggested strategy consists of segmenting the Z(X,Y) image into (i) planar facets, (ii) cylindrical and conical facets, (iii) spherical facets, and (iv) edges. This creates a set of parallel cases but now we know the types of regions being analyzed. The K1 and K2 histogram or decision space allows the segmentation of the Z(X,Y) image into the four basic categories. Actually there are seven categories since the surfaces can have either positive or negative curvatures.

1) Construct the K1,K2 histogram H(K1,K2). The curvatures K1 and K2 can be measured in terms of angles where the range is from −180 to + 180 degrees. Hence the H(K1,K2) space has finite ranges. In practice the minimum and maximum values for K1 and K2 for histogramming purposes should be made variable. There usually is no need to use different limits for K1 and K2. Hence, let the "extreme" curvature values for K1 and K2 be −Kex and +Kex.

2) Process the H(K1,K2) histogram as for the "origami" world (smoothing, dynamic thresholding) in order to create a binary version Hb(K1,K2). Each isolated region in Hb(K1,K2) indicates the type of surface found in the Z(X,Y) image. Thus, once the clusters in Hb(K1,K2) have been labelled and back projected, the regions formed by the back projected labels in the image have become automatically recognized. However, in order to extract the four or seven different regions mentioned above, it is better to impress some constraints onto Hb(K1,K2).

3) Apply constraints to Hb(K1,K2) such that the seven different surface types become unique regions or clusters. The constraints consist of:

(i) Create a connected isolated region in Hb(K1,K2) at the origin of the Hb(K1,K2) space (:K1: and :K2: less than some constant Tp). This region represents what is to be considered planar or flat in Z(X,Y).

(ii) Create a connected isolated region along the positive K1 axis in Hb(K1,K2) where +Kex > K1 > Tp. This corresponds to the positively curved cylindrical and conical surfaces in Z(X,Y).

(iii) Create a connected isolated region along the negative K1 axis in Hb(K1,K2) where −Kex < K1 < −Tp. This corresponds to the negatively curved cylindrical and conical surfaces in Z(X,Y).

(iv) Create a connected isolated region along the entire K2 axis for the extreme positive K1 values (K1=+Kex) in Hb(K1,K2). This corresponds to the convex edges in Z(X,Y).

(v) Create a connected isolated region along the entire K2 axis for the extreme negative K1 values (K1=−Kex) in Hb(K1,K2). This corresponds to the concave edges in Z(X,Y).

If desired, one may also create separate connected regions for the positive and negative "spherical" regions, but since such facets form relatively tight clusters in the H(K1,K2) space, this step is usually superfluous. The constrained version of Hb(K1,K2) is indicated by Hc(K1,K2).

3) Label the isolated regions in Hc(K1,K2) uniquely and identify the label numbers which correspond to the cases (i) to (v) in (2). Let Lf indicate the label number for flat regions (i), let Lcp and Lcn indicate the labels for the positive and negative cylindrical and conical regions (ii) and (iii), and let Lep and Len be the labels for the positive and negative extremes or edges (iv) and (v). The remaining label numbers will belong to various "spherical" regions. Call the labelled histogram Hl(K1,K2).

4) Back project specific cluster labels or label groups. In brief:

i) Create an image of flat surfaces by back projecting Lf only. The planar facets belong to the origami world and will be processed as already described. The origami world is a subset of man-made world.

ii) The regions in Z(X,Y) which consist of cylindrical and conical facets are obtained by back projecting Lcp and Lcn. These form another sub-category and have to be processed further to separate cylindrical and conical surfaces.

iii) Edge regions will be obtained by back projecting Lep and Len. The

edges are processed essentially as described in the origami world.
iv) The spherical facets are an end product. These may be back projected separately or all together, as desired.

Images of the "man-made" world thus segment into several categories according to surface curvature. The separate images can be processed in parallel. The categories are processed as binary images and are used as overlays to extract the 9-dofels belonging to each category. Surfaces of order higher than two are segmented into at most second-order patches, but the segmentation is influenced by the "label spreading" process. Of course, in practice there are quite a few other details that must be taken care of since the back projected results are not perfect.

The real world. In the natural environment untouched by man smooth planar and second-order surfaces exist but they are relatively rare compared to the essentially random or possibly "fractal" nature (Mandelbrot) of the natural scenes. On the micro-level one can find smooth surfaces (the surface of an apple, for example) and smooth surfaces occur on the macro-level (say, the shape of the ground one is walking on). Thus, if the size of the surface element (surfel) is chosen correctly, the processing methods des-cribed for the origami and man-made worlds may have some relevance.

A natural scene consisting of a range of mountains seen from above (digital terrain model) was processed as if it were an "origami world". The orientation and tilt histogram $H(A,B)$ segmented the mountain ranges into variously sloping mountain sides. When the "man-made world" model was applied using $H(K1,K2)$ and the extreme curvatures (edges) were extracted, the result showed the ridges, peaks, and valley bottoms. The origami model was applied to the flat regions with predictable outcome, i.e., the variously sloping mountain sides were obtained. In between the micro and macro levels it is likely that region segmentation becomes a texture detection problem. Experimental evidence is insufficient to draw many conclusions.

Whether the scenes are natural or man-made, there are other aspects to range or distance images besides trying to deduce shapes of surfaces. When one is navigating in a 3D world, the very close surfaces (or the ones moving rapidly towards the observer) are of primary concern since a collision is likely. Distance is given by the range image and hence the regions in the environment which are close are immediately known. Only these regions need further analysis. The regions in the range image that are "far away" need not be studied, while those in the intermediate range may be assigned "second priority". Motion detection in range images becomes similar to motion detection in binary images if sufficiently fast processing is avai-lable such that the remaining registration between the results in sequential images can be used. Thus, the range image can be used in many more ways than just for shape analysis.

Facet Characteristics, Adjacencies, and Edges

The initial processing of the $Z(X,Y)$ image was based on the assumptions that the surfaces in the image are relatively smooth but the contents of the image are unknown. The main results are given by the registered n-dofel images $Z(X,Y)$, $A(X,Y)$, $B(X,Y)$, $K1(X,Y)$, $K2(X,Y)$, $C1(X,Y)$, and $C2(X,Y)$. The next steps in the processing consisted of "coherency" or region and edge detection based on n-dofel classification. The interesting aspect is that we can impress "our particular opinion of the world" onto the segmentation process. If the "origami world model" is used, then the segmentation will give planar or first-degree facets and edges between these facets. If the "man-made world" model is used then the facets are at most of second degree. The results of segmentation are given by four images in registration with all the previously obtained images, i.e., Lcore(X,Y), Lfacet(X,Y), Ljs(X,Y), and Lseq(X,Y). Several points should be noted, since we are now slowly beginning to find answers to the questions: What should the knowledge in the machine consist of, and how should the machine acquire this knowledge?

1) The cluster labels in Lcore(X,Y) originate from various known decision spaces. Hence, each cluster can be considered as "recognized" in the classical pattern recognition tradition. Alternatively, every cluster in every decision space can be considered as an automatically formed "concept" or "an invariant piece of knowledge" about the world. The pixels in Z(X,Y) which have labels in Lcore(X,Y) can be considered "reliably" identified.

2) The Z(X,Y) image has been segmented into uniquely labelled facets as indicated by the image Lfacet(X,Y). The reliably identified pixels in each facet are those for which there are labels in Lcore(X,Y). The automatically formed "concepts" thus carry directly over to each facet. Facets which have few or no labels in Lcore(X,Y) compared to their size have to be considered unknown, requiring reprocessing at a more suitable spatial resolution or from a different view. It may be interesting to note that the "unknown" has become identified as unknown, and the machine can "direct its attention" to such regions on "its own accord". We may call such behaviour "curiosity"!

3) In the labelled junction (corner) and segment (edge) image Ljs(X,Y) the segments are exactly one pixel thick as a result of "perfect" thinning. Consequently, the detection of which facets are adjacent to which other facets becomes very simple, consisting of using a 3 by 3 local neighborhood centered on the segment labelled pixel in Ljs(X,Y) while the search is carried out in Lfacet(X,Y). Each segment has exactly two sides, i.e., a segment label can at most be used to relate its two adjacent facets. Each segment also has exactly two ends which may terminate on junctions (corners) or which may be free. The junctions are the corners between three or more facets. (Segment which form closed loops may occur, i.e., such closed segments have no ends at all.) The construction of annotated graphs or lists of facet, edge, and corner interrelationships is now a very straight forward procedure. The computations of average feature (n-dofel) values and statistics on these values for each facet is only a matter of "looking up" the appropriate n-dofel values corresponding to the label values in Lfacet(X,Y) since all the images are in registration. Use of this data for model matching was made by Oka et al. and Dhome and Kasvand.

4) The locations of the edges (segment pixels in Ljs(X,Y)) are relatively accurate, depending on how carefully the thinning process was controlled. However, the corresponding Z value cannot be obtained reliably from Z(X,Y) since edges can have two Z values depending on the facet to which the edge is considered to "belong". For example, at a discontinuity (jump edge) in Z(X,Y) the Z values on either side of the edge are very different. If the edges are used as space curves (say, piecewice linear approximations), the Z values have to be obtained from analytic approximations of the facets for the given X and Y values of the edges in the XY plane. The edges are one pixel thick and ordered sequence of (X,Y,Z) coordinates is obtained by using the sequence numbers in Lseq(X,Y) and the segment labels in Ljs(X,Y). The boundary shape for each facet can now be obtained since we know which segment belongs to which facet, and the facet may be rotated to produce a view along the normal of the facet. Two- and three-dimensional edge model fitting experiments were carried out by Yamada et al., and Hospital et al.

Fig.3 (continued). Column numbers top vertical, row numbers left horizontal. The codes vary according to image (amplitude for Z, K1, K2; angles for A, B, C1; labels for Lcore, Lfacet, Ljs; pixel sequences for Lseq). Dot (.) represents unclassified pixels or zeros, comma (,) is jump edge, star (*) is thin edge. Segmentation by a mixture of jump edge and origami world.

```
00000000000000000000000011111111        00000000000000000000000011111111
77777888888888899999999999900000000      77777888888888899999999999900000000
5678901234567890123456789012345 67        5678901234567890123456789012345 67
```

```
53 POOOOOOOOO,YYYYYYYYYYYYYXXXXW,PPPP      LLMMXXXYXX,PPPPPPPPPPPPPPPPO,HHUV
54 TROOOOOOOO,YYYYYYYYYYYYYYYYYV,PPPP      LLLLOWWWXX,PPQQQPPPPPPPPPPPO,HJUU
55 TSOOOOOOOO,ZZZZZYYYYYYYYYYYYV,PPPP      KKKKMWWWWX,PQQQQQQQQPPPPPPPH,IXVU
56 SSROOOOOO,ZZZZZZZZZZZYYYYYY,PPPPP       IJJJKSYXX,ZQQQQQQQQQQPPPPPP,XXXWU
57 SSSOOOOOO,ZZZZZZZZZZZZZZZZY,PPPPP       FHIIIPWXX,XQQRQQQQQQQPQPPPP,XXXXV
58 SSRPOOOOO,ZZZZZZZZZZZZZZZW,PPPPP        DFGGHLWXX,RRSRRQQQQQQQQPPPP,XXYYX
59 SRQOOOOOO,ZZZZZZZZZZZZZZZZW,PPPPP       DEEFFFWWW,TTSSRRRRRQQQQPPPP,XYYYX
60 RQOOOOOO,ZZZZZZZZZZZZZZZZZ,PPPPPP       DDDDD5YW,UTTTTSRRRRRQQQPPP,YYYYYX

              Z(X,Y)                                    A(X,Y)
```

```
53 45BOYYYXXX,JKLKKJJIHGGFEEDCB,ISYY       HFA13FGOJI,LZLKJJJJKJJJJPUMI,I70G
54 446EUYYXXX,KLNMMLKKJJIHGGFDC,OYYY       HIK70AGOGH,IZJKKJJJKKKKKTSKI,F42G
55 7559NYYYXX,LNOOONMMLLKKJIGE2,XYYY       MKICO4BODE,IZJJJKJJJJKKXPJH,EO6H
56 C657GYYYY,1NOQPPOONNMMMLKIG,YYYYY       TOKN5024A,ECZKJJJKJJJJJNYNJ,GDOBH
57 G866EYYYY,2PRSRQQQPOOONNMKI,YYYYY       USPN700CF,DCZKKKKKKKKKKTWKH,GBOEH
58 D867EYYYY,UTUTTSSSRQQQPPPMK,YYYYY       RTMN600FH,HMYKKKKKKKKKJXSIH,G8OIH
59 9668IZYYY,WWVVVUUTTSSRQQQOM,XXYYY       LNNN303GG,ISVKKKKKKKJJZOIG,F42IH
60 667CPZYY,XXXXWWWVVUUTSSSRP,XXXYYY       IJJ900BI,GJYOJJJJJJJJMZLH,HF16IH

              B(X,Y)                                    K1(X,Y)
```

```
53 GHJKHHHIFH,IOIHHHHHHHHHHHJIHH,GIHG      9E12248H88,73B5555555555EDA8,8FFE
54 HGFKIHHHHH,HJHHHHHHHHHHHJIHH,HIHG       ABC1226HF8,6365555555555EB98,EFFD
55 IHGLJIHAGH,FIHHHHHGHHHHILIHH,HHHH       9BA001CE88,2166555556658DA98,FFED
56 JIGFMJG8G,HIIGHHHHHHHHHKKIH,HHGHH       79AAHODB7,8806666665666DD98,BFFE8
57 KJHGQJHIG,HJHGHHHHHHHHHKJHH,HHHHH       6889GHFF7,77H6666666666EC98,CFFD5
58 LJGFRJEIH,GGIHHHGGHHHHHJHHH,HIHGH       5676GFFF7,10F6666666666ECA9,CFF24
59 JHGEOJDGH,HMJHHHGGHHHHIKHHH,HHHGH       4556FEC67,6FD666666555AEB97,EFFH3
60 HGFLJJJG,HGLHHHHGGHHHHJKIH,HHHHHH       444EEEF7,7HFB666666556DEB8,HEFFG4

              K2(X,Y)                                    C1(X,Y)
```

```
53 ...MMMMMM.FFFFFFFFFFFFFFFFFF.MM M       R*MMMMMMM*PPPPPPPPPPPPPPPPPP*SSSS
54 ....MMMMM.FFFFFFFFFFFFFFFFF..MM M        R*MMMMMMM*PPPPPPPPPPPPPPPPPPP*SSSS
55 ....MMMMM.F.....FFFFFFFFFFF.MMMM M       RR*MMMMMM*PP***PPPPPPPPPPPP*SSSSS
56 E...MMMMM...MMM...FFFFFFFF.MMMM          RR*MMMMMMM**VVV**PPPPPPPPPP*SSSSS
57 EE..MMMMM.MMMMMM.....FFFF.MMMM M         RRR*MMMMM*VVVVVVV***PPPPP*SSSSS
58 EE...MMM.MMMMMMMMMM....F.MMMM M          RRRR*MMMM*VVVVVVVVVVV***PP*SSSSS
59 EEEE.MMM.MMMMMMMMMMMM...MMMM M           RRRR*MMM*VVVVVVVVVVVVVV*P*SSSSS
60 EEEE.MMM.MMMMMMMMMMMMMMM.MMMMM M         RRRR*MMM*VVVVVVVVVVVVVVVV*SSSSSS

              Lcore(X,Y)                                Lfacet(X,Y)
```

```
53 .S.........E...............L....         .D........P................K....
54 .S........E................L....         .E........Q.................L....
55 ..S........E..TTT..........L....         ..F........R..DEF...........M....
56 ..S........5T...TT.........L....         ..G........5C...GH.........N....
57 ...S......U.....TTTT.....L....            ...H......C.......IJKL....O.....
58 ....S....U.........TTT..L.....            ....I....D.........MNO..P.....
59 ....S...U...............T.L.....          ....J....E................P.Q....
60 ....S...U...............6......            ....K...F.................6......

              Ljs(X,Y)                                  Lseq(X,Y)
```

Figure 3: Mosaic of one character per pixel alpha-numerically coded results for a 33 by 8 pixel region in Fig.1b, at the cylindrical region. (Continued)

LEARNING AND UNDERSTANDING

The rather lengthy bottom-up and open loop processing has automatically given the facets, edges between the facets, corners, facet and edge features, and all adjacencies between the facets, corners, and edges. The facets are at most second-order surfaces which can be approximated analytically. The edges can be represented as piece-wise linear curves in 3D space. The processing does not involve combinatorial searches since no models are required. Simulation experiments have verified that these results are obtainable on realistic scenes of man-made objects but the results are not withour errors (false facets and split edges). The size of the errors is within the "erratic region" De. Up to this stage practically all the image processing could be done by suitably designed serial and parallel hardware. Except for the initial design, the machine does not require an "external agent" for acquiring its "basic concepts" (recognized facets, corners, edges, etc.). The external agent can also be avoided in the integration of the basic concepts into composite descriptions of objects. An object is either shown in isolation or the machine is allowed to manipulate the object on its own accord to "see" the object at suitable spatial resolution and from desired view point. Poorly resolved and unknown regions were identified during the processing. An integrated external view description from all suitable points of view ("exo-sketch") can be created automatically. The exo-sketch is on a much "lower level" than abstract mathematical descriptions, computer aided design models, or even the "primal sketch".

Actually, a single fixed resolution image is a very meager starting point for comprehensive analysis. Even we cannot give a detailed description of an unknown object by just one "glimpse" of it. Proper analysis requires variable spatial resolution and multiple views, and the whole process has to be closed by a feedback loop from the intermediate results as well as from the requirements of exo-sketch construction. Whether such a behaviour is to be called "learning" and whether the subsequent use of the stored exo-sketches is called "understanding" is a problem identical to the one that originated the "Turing test" for machine intellignece.

CONCLUSIONS

A 3D laser range finder image was chosen as the example of an "unconventional" image since it gives geometrical information not normally available. Based on the assumption that the image contents are unknown, a lengthy sequence of processing steps was described and partly illustrated. The suggested computational steps have been verified and the results found reasonable for single fixed spatial resolution images. The method differs from existing techniques (Besl and Jain, Jain and Hoffman, Kanade) in that no a priori object models are required for image segmentation while a "world model" gives an initial meaning to the segments. The results can be, and have been, used successfully in model matching using several techniques. Attempts at integration of the results from varying resolutions and viewpoints (exo-sketch construction) has overwhelmed the to the author available hardware, programming, and computational resources.

The processes described are general for all range images and can be formulated in hardware. For practical use and rapid progress in image understanding research such hardware is necessary. Furthermore, it would be very interesting to construct a closed-loop system where a robot could manipulate the objects as a function of its visual requirements.

ACKNOWLEDGEMENTS

The author wishes to express his sincere thanks to all colleagues and to NSERC, DEE NRC, and CCRS in Canada, and ETL in Japan.

REFERENCES

Abdelmalek, N.N., L1 solution of overdetermined systems of linear equations, ACM Trans. Math. Software 6, 1980, pp. 220-227.

Ballard, D.H., Generalizing the Hough transform to detect arbitrary shapes, Pattern Recognition, 13(20), 1981, pp. 111-122.

Besl, P.J. and Jain, R.C., Three-dimensional object recognition, Computing Surveys, Vol. 17, No. 1, March 1985, pp. 75-145.

Dhome, M., and Kasvand, T., Polyhedra recognition by Hypothesis Accumulation, IEEE Trans. PAMI, Vol. 9, No. 3, May 1987, pp. 429-438.

Fu, K.S., Applications of Pattern Recognition, CRC Press Inc., Boca Raton, Florida, 1982.

Gibson, J., The Perception of the Visual World, Houghton Mifflin Co., Boston, Mass., 1950.

Hecht-Nielsen, R., Artificial neural systems technology, TRW Rancho Carmel AI Center, One Rancho Carmel, San Diego, CA 92128, Arpanet:Hnielsen@isi, June 9, 1986.

Hospital, M., Yamada, H., Kasvand, T., and Umeyama, S., 3D curve based matching method using dynamic programming, submitted for publication.

Jain, A.K. and Hoffman, R., Evidence-based recognition of 3D objects, Tech. Report: MSU-ENGR-86-013, Dept. of Computer Science, Michigan State University, East Lansing, MI 48824, 1986.

Kanade, T., Three-dimensional Machine Vision, Kluwer Acad. Publishers, 1987.

Kanade, T., A theory of origami world, Tech. Rep. CMU-CS-78-144, Computer Science Dept., Carnegie-Mellon University, 1978.

Kasvand, T., Surface curvatures in 3D range images, Proc. 8'th ICPR, Paris, 1986, pp. 842-845.

Kasvand, T., Step or jump edge detection by extrapolation, Proc. 8'th ICPR, Paris, 1986, pp. 374-377.

Kohonen, T., Self-Organization and Associative Memory, Springer-Verlag, 1984.

Kreyszig, E., Introduction to Differential Geometry and Riemannian Geometry, University of Toronto Press, 1968.

Mandelbrot, B., The Fractal Geometry of Nature, Freeman Press, San Francisco, 1982.

Oka, R., Kasvand, T., and Rioux, M., Cross-angle transform for viewer-independent recognition of 3D objects, IEEE Proc. CVPR, San Francisco, June 19-23, 1985, pp. 470-475/

Perkins, W.A., A model-based vision system for industrial parts, IEEE Trans. Vol. C-27, No. 2, pp. 126-143, 1978.

Rioux, M., Laser range finder based on synchronized scanners, Applied Optics, Vol. 23, No. 21, pp. 3837-3844, 1984.

Rosenblatt, F., Principles of Neurodynamics, Spartan Books, Washington, D.C. 1962.

Rosenfeld, A., Digital Picture Analysis, Springer-Verlag, 1976.

Tamura, H., A comparison of line thinning algorithms from digital geometry viewpoint, Proc. 4'th ICPR, Kyoto, 1978, pp. 715-719.

Tech. Comments: Computing with Neural Networks, Science, Vol. 235, pp. 1226-1229.

Watanabe, S., Pattern Recognition: Human and Mechanical, John Wiley & Sons, 1985.

Yamada, H., Hospital, M., and Kasvand, T., Rotation-invariant contour DP matching method for 3D object recognition, Proc. IEEE Internat. Conf. on Systems, Man, and Cybernetics, Atlanta, Georgia, Oct. 14-17, 1986. pp. 997-1001.

STATISTICAL PATTERN RECOGNITION:

THE STATE OF THE ART

Josef Kittler

Department of Electronic and Electrical Engineering
University of Surrey
Guildford GU2 5XH
United Kingdom

INTRODUCTION

Objects or events in the universe are perceived by biological systems as patterns. Pattern recognition is a process which assigns these sensory stimuli into perceptually meaningful categories. For the last four decades a considerable effort has been made to simulate the human pattern recognition capabilities by a machine. This quest for automation of pattern recognition processes is primarily driven by applications in computer vision for flexible manufacturing, speech recognition, text recognition, remote sensing, medicine and others. In computer vision for robots, for instance, the pattern recognition task may involve identification of object shape. In speech recognition the object categories may be words, phonemes or diphones and the sensory data on which classification is based could be vector quantised speech signal. In text recognition the objects of interest are characters and their groups forming words. Object categories in remote sensing relate to land cover and the sensory data are reflected energies in several spectral channels of the electromagnetic spectrum.

The field of endeavour concerned with automatic pattern recognition is relatively broad, covering all aspects of sensory data processing, culminating in pattern labelling. Thus it includes data acquisition and filtering, reconstruction, segmentation, measurement extraction as well as pattern classification. However in the context of this paper, pattern recognition will be considered in the narrow sense of pattern labelling. It will be assumed that patterns to be recognised have already been subjected to any necessary preprocessing and we shall not be concerned with relative merits of various techniques that could be used for that purpose.

The review of recent developments in pattern recognition here focuses on techniques relevant to the ultimate process of decision making which assigns patterns into their respective categories. The discussion will further be restricted to statistical approaches where pattern classes can adequately be modelled by means of probability distributions. Pattern recognition problems where structural models are more appropriate can be handled using syntactic methods. For a recent review of this topic see Thomason (1987). Efforts to combine statistical and structural models leading to hybrid systems were first reported by Fu (1986) and the recent interests are reviewed by Bunke (1987). The role of knowledge based approaches in classification is discussed by Chandrasekaran (1986).

The pattern recognition system is conventionally considered to comprise three major subsystems: sensor, feature selector, and classifier. The task of pattern recognition system design is primarily associated with the problem of feature selection and classification. In addition any design has to be assessed from the point of view of recognition performance. This involves estimation of classification error probability. The purpose of this paper is review recent advances in these three areas, namely in feature selection, classification and error estimation respectively. The ensuing sections are devoted to these respetive topics.

FEATURE SELECTION

In a statistical pattern recognition system each input pattern is represented by a vector of measurements which is considered to be a realization of some random pattern generating process. The mixture probability distribution characterising this random process is composed of class conditional components $p(\mathbf{x}|\omega_i), \forall i$ weighted by the a priori class probabilities $P(\omega_i)$ respectively. These measurements will in general contain more information than what is absolutely necessary to discriminate between patterns of different classes.

The purpose of feature selection is to remove any redundant and irrelevant information from these representation patterns and in the process reduce the dimensionality of the pattern recognition problem. The lowest possible dimensionality is required from the point of view of the classifier design.

The formulation of the feature selection problem requires the specification of a form of the mapping between the measurement and feature spaces, a feature selection criterion function, the desired dimensionality of the feature space, and an optimization algorithm. In addition to the above topics the research issues in feature selection include small sample effects.

The choice of mapping determines whether all available measurements or only their subset are used to derive pattern features. The majority of feature selection techniques which do involve some mapping of the complete measurement space into lower dimensional feature space (Biswas, 1981) are linear. The recent developments in the area of linear feature extraction have been concerned with techniques maximizing the average pairwise intersample distance based on the Euclidean metric which are related to the discriminant analysis.

A serious drawback of the classical discriminant analysis is that discriminatory information in class conditional covariance matrices is not considered. The implication of this defficiency is that in an m-class pattern recognition problem not more than m-1 features can be extracted. Extensions to the classical approach have been suggested by Fukunaga and Mantoch (1983), Okada and Tomita (1985) and Malina (1981) which permit more useful information to be extracted from the data. Similar aims are pursued by Duchene (1986) and Longstaff (1987).

Feature selection criterion function and their properties were active research topics in the late nineteen sixties and seventies. The aim was to find measures which are easier to compute than error probability, the natural criterion for feature selection, but are very closely related to it. Interests in this problem diminished when Boekee and Van der Lubbe (1979) showed that the various feature selection criterion functions in the literature are just special cases of the generalised divergence and therefore they do not have inherently different capacity for assessing the discriminatory potential of features. Recently research has focused on methods of estimating error probability itself. The progress in this area will be reviewed in a separate section dedicated entirely to this important subject.

Feature selection criterion functions are superior to error probability from the point of view of computational complexity when class conditional porbability distributions are parametric and in particular gaussian. Under the assumption of normality, the integral form of many class separability measures simplifies to simple parametric formulae defined in terms of the class mean vectors and covariance matrices. However, when the sample set size is small relative to the dimensionality of the feature space, estimates of covariance matrices can be unreliable and this can seriously affect the feature evaluation process. Small sample effects on covariance matrix estimates have recently been investigated by Peck and Van Ness (1982), Keating and Mason (1985) and Morgera (1986) who suggest alternative remedies to minimize this problem.

As pointed out earlier, the upper limit on the dimensionality of the feature space is normally imposed by the engineering considerations concerning the classification stage of the pattern recognition system. An important issue is however, whether the actual number of features used should equal this upper limit or whether it should be lower. In the first instance the number of required features is related to the intrinsic dimensionality of the data (Wyse et al, 1980). Intuitively, as more features are included, more discriminatory information is available and the performance of the pattern recognition system should monotonically improve. In practice however it has been observed that the probability of correct classification increases with the number of features only up to a certain point. Subsequent addition of further features results in system performance degradation. This behaviour is known in the literature as the peaking phenomenon. It is due to the training set being of finite size (Jain and Chandrasekaran, 1982, Foley, 1982). The practical guideline emerging from a number of studies of sample size to dimensionality ratio is that the number of training patterns should be about five times the number of features (Trunk, 1979, Kalayeh and Landgrebe, 1983).

CLASSIFICATION

The theoretical result underpinning the design of any pattern classification system is the Bayes decision rule (e.g. Devijver and Kittler, 1982) In its general form it specifies how best decisions about class membership of patterns can be made taking into account their probability distribution and any given loss function. The latter defines how misclassifications should be weighted depending on particular class assignment. In the majority of cases the zero one loss function is used whereby all misclassifications are weighted equally. In this particular case the Bayes decision rule simply states that an unknown pattern should be assigned to that class which is most probable.

Thus if we knew the a priori probabilities of classes and their probability distributions, making optimal decisions would be relatively simple. In practice the exact probability structure of patterns is not known. It can only be estimated from training patterns which may or may not be labelled. Depending on whether or not class labels are available, the classification system design involves supervised or nonsupervised learning respectively. The latter task is often cast as a clustering problem.

One of the first issues the designer has to consider whether the classifier should be implemented as a single stage system or in terms of a decision tree. In principle a single stage solution is optimal. However, if the estimated p.d.f. structure does not approximate the true density correctly and that situation can arise frequently due to paucity of the training data a multistage design can outperform its single stage counterpart. This can be explained by the large number of features needed by a single stage classifier to achieve class separation in multiclass problems which in turn causes the small sample problem discussed in the previous section to be more severe. Moreover, a tree classifier can simulate the human pattern recognition process to a greater degree and for this reason it is often favoured in certain application areas such as medicine. Also processing speeds are often higher as decisions are made in lower dimensional spaces using fewer features.

The difficulty with the decision tree approach is that it is largely heuristic. Mui and Fu (1980) recommended an interactive technique for the tree design. At each step a feature is chosen and the feature space partitioned by a hyperplane perpendicular to the feature axis and intersecting the axis at a suitable threshold. Sethi and Sarvarayudu (1982) determined the tree structure by maximizing the gain in average mutual information at each node. Li and Dubes (1986) developed a statistical test to determine when splitting a node would be beneficial. Dattatreya and Kanal (1985) advocated a tree structure where natural distinctions are carried out at the top of the tree and more subtle decisions are made at later stages when the number of candidate classes is much smaller.

Parametric Decision Rules

Once the general structure of the classification stage is decided on, each node of the decision tree implements a practical decision rule solving a particular decision task. Both, the decision rule inference and implementation are relatively simple, if the class conditional probability distributions of patterns are parametric.

The most important parametric form is the Gaussian. However the problem of verifying the validity of the assumption of data Gaussianness is far from trivial. Recent work on this topic is reported by Smith and Jain (1985) and Fukunaga and Flick (1986).

When the normality can safely be assumed, the decision rule design involves simply estimation of the distribution parameters. However, a comparative study of several parametric and nonparametric decision rules on high dimensional Gaussian data reported by Van Ness

(1980) suggests that due to small sample effects nonparametric methods may outperform parametric ones. The statistical inference of parameters should therefore rely on robust estimators discussed by Van Ness (1982), Keating and Mason (1985) and Morgera (1986) if the sample set to dimensionality ratio is small.

Nonparametric Decision Rules

Broadly speaking, nonparametric decision rules are derived from the Bayes rule by replacing the class conditional probability density functions (p.d.f.s) with their nonparametric estimates. Two important classes of p.d.f. estimators have gained popularity in pattern recognition: the kernel estimators and the k-nearest neighbour (k-NN) estimators. The former approach originally suggested by Parzen, involves placing a smoothing kernel at each observation in the training set. An estimate of the p.d.f. at a given point is then obtained by summing up the contributions of all these kernels at that point.

Recent efforts in kernel p.d.f. estimation addressed the problem of choosing the form (Devroye and Machell, 1985) and the region of influence of each kernel (window width). Two factors have a direct bearing on suitable window size: the number of features, d, and size, n, of the training set. For the uniform kernel Postaire and Vasseur (1982) recommend that the window width should be inversely proportional to $n^{\frac{c}{d}}$ where c is a constant from the open interval $(0, 1)$. The window size choice can be cross-validated by procedures developed by Van Ness (1980) and Hall (1983).

Postaire and Vasseur (1982) were also concerned with the computational complexity of kernel estimation methods. They consider the uniform kernel and propose a coarse quantization of the observation space which results in two benefits. The quantized patterns and therefore kernels can lie only in predetermined positions (cells) in a regular rectangular lattice and the number of kernels needed will be less than the number of patterns in the training sets. Their p.d.f. estimation method effectively precomputes the p.d.f. estimate everywhere in the observation space and consequently it is much faster than other implementations.

The k-NN approach on the other hand involves the computation of the volume occupied by the k-nearest neighbours of a pattern of interest drawn from among the elements of the training set. The method requires a careful decision about the value of k and the metric to be employed in order to obtain reliable estimates. The k-NN decision rule which assigns an unknown pattern to that class which receives a maximum vote from its k-nearest neighbours is a straightforward extension of the k-NN p.d.f. estimation method.

The performance of the k-NN classifiers can be considerably improved if non Euclidean metric is used in determining nearest neighbours. Ideally nearest neighbours should be selected from among samples having identical aposteriori class probability. Such samples are likely to be found in the subspace orthogonal to the direction of maximum change of the aposteriori probability function, that is to the direction of the gradient of the probability function (Short and Fukunaga, 1980, Fukunaga and Flick, 1984b). The distance of a pattern from point x consistent with this constraint on the corresponding aposteriori class probability value can be determined by projecting the pattern to the direction of the gradient and then measuring its distance from x in this one dimensional space. Fukunaga and Flick (1985) give the condition under which a quadratic distance becomes appropriate.

While for a given size of the training set the well known guidelines indicate a suitable value of k for the k-NN decision rule, a practical rule may for computational reasons employ much smaller values of k, even k=1. Any reduction in computational complexity however is achieved only at the expense of degraded performance. Devijver and Kittler (1980) considered how this performance loss can be restored so that even the 1-NN rule assymptotically approaches the Bayes error rate. They showed that this can be achieved by editing the training data set.

The computational complexity of the k-NN decision rule can be reduced by the application of efficient techniques of searching for nearest neighbours. Fast k-NN search algorithms have been developed by a number of authors. Kamgar-Parsi and Kanal (1985) proposed an efficient search technique based on the branch and bound algorithm. A parallel search method has been developed by Kumar and Kanal (1984). Miclet and Dabouz (1983) have a very fast method for finding approximate nearest neighbours. Kim and Park (1986) speed up the k-NN search process by the application of ordered partition of the search space.

Context

Perhaps the most dramatic developments in pattern classification have been witnessed on the front of contextual methods. The techniques discussed thus far base the decision about class membership of a pattern simply on the information contained in the features. Now patterns represent objects or events which seldom exist in isolation. For instance in text recognition individual characters are an integral part of larger objects such as words or sentences formed by character groups. Word dictionary and rules of grammar dictate which combinations of characters and implicitly which individual characters are possible. The a priori wordl knowledge or context can be used to help to disambiguate decisions based simply on noisy features of individual patterns.

The early developments in contextual classification were largely heuristic. In the last decade or so, serious attempts have been made to put these methods on a proper theoretical footing. Classification schemes that exploit context can be broadly divided into two categories: those based on the statistical compound decision theoretic approach, and methods maximizing the aposteriori probability (MAP) of joint labelling.

The compound decision rule approach has been proved very effective and feasible for classification of sequential patterns such as speech (Jelinek et al, 1982) where the Markovian properties can be assumed for the underlying pattern label process. With the exception of the Pickard random field model (Haslett, 1985), this approach does not extend naturally to image data. However, Devijver (1986) succeeded in developing a comparable 2-dimensional recursive compound decision scheme for images which can be modelled by a Markov mesh process.

Other contextual classification methods for 2-dimensional data concentrate primarily on local context (Swain et al, 1981, Kittler and Foglein, 1984, Kittler and Hancock, 1987, Hancock and Kittler, 1987). In these schemes global contextual information is incorporated by iteratively updating class probabilities using local context. The relationaship of this contextual classification approach and probabilistic relaxation is discussed in Kittler (1986) and Kittler and Foglein (1986).

The MAP approach, at any particular instance of time, assigns hard labels to all the objects in a one or two dimensional network. These labels are updated iteratively by considering local context only. In order to guarantee the global optimality of the label assignment, the stochastic optimization technique known as simulated annealing must be used for label updating. The updating schemes of Kittler and Pairman (1985) and Besag (1986) only guarantee that a local maximum of the joint aposteriori probability will be found.

Clustering

In many situations it may not be feasible to provide class labels for the training patterns. For instance, in remote sensing applications, ground truth data could be acquired only by visiting the imaged sites carrying out detailed land cover surveys at prohibitive costs. In the absence of labelled data, nonsupervised learning techniques have to be adopted for the classifier design instead of methods discussed thus far. A popular substitute is to analyse the structure of the mixture p.d.f. for modes using clustering. There is a vast body of literature on cluster analysis (Dubes and Jain, 1980) but the difficulty is that different clustering methods tend to identify different data structures. Dubes and Jain (1979) argue that this problem can be overcome first by establishing whether a data set exhibits any clustering tendency. All clusters determined by clustering should be validated. In this context, Smith and Jain (1984, 1985) suggest that clustering should be applied only if data can be shown to deviate from a uniform distribution and they developed a method of testing for uniformity in multidimensional data.

An exciting new development in data structure analysis is the conceptual clustering approach of Michalski and Step (1983) used successfully for knowledge organisation by Cheng and Fu (1984).

ERROR ESTIMATION

The performance of a pattern recognition system is characterised by error probability. There are basically two kinds of error of interest. The Bayes error probability quantifies the amount of overlap of class conditional probability distributions and it therefore expresses the theoretical limit on pattern recognition system performance. Its estimate is indicative of both the minimum achievable recognition error which then can be compared with the performance of a practical decision rule, and the discriminatory potential of features on which decisions about the class memebership of patterns is based. In contrast the system error probability specifies the future performance of a given pattern recognition system.

The basic approaches to error estimation are discussed in Toussaint (1974) and standard texts such as Duda and Hart (1973) and Devijver and Kittler (1982). The plug-in estimator, applicable to parametric densities, involves numerical evaluation of the error probability integral with the class distribution parameters replaced by their sample based estimates. The empirical error count estimate returns the number of test samples misclassified by the decision rule. More recent are the average conditional error estimators which employ the k-nearest neighbour or other nonparametric method to estimate the conditional error probability for pattern $\mathbf{x}$ which is then averaged over the mixture distribution $p(\mathbf{x})$. The advantage of this approach is a lower variance of the estimate and the fact that the conditional error averaging can be carried out using unclassified samples (see e.g. Hand, 1986a).

One of the key issues in classification error estimation is how to use observations most efficiently. From the point of view of pattern recognition system design, all the available data should be utilised for the decision rule development. However the use of the same pattern samples for the subsequent error estimation is known to lead to an optimistic bias. A number of schemes have been suggested to minimise this problem. Invariably, a small set of patterns is excluded from the design set and it is then used for testing. The leave one out method is a typical example of these schemes. From a set of L training patterns, one is selected for testing the decision rule based on the remaining L-1 patterns. The process is repeated for other patterns in the training set and errors averaged. The method is a special case of the so called rotation procedure where more than one patterns are retained at a time and used for testing.

Because of its central role in pattern recognition system design, the problem of error estimation has received considerable attention during the recent years. Attempts have been made to develop error estimators with improved asymptotic properties. Kittler and Devijver (1981) combined the empirical error count and averaged conditional error estimators to yield a method with even a lower asymptotic variance. Hand (1986b) considers the effect of the test sample distribution on the averaged conditional error estimator properties and derives an optimal distribution which leads to a smaller asymptotic variance. Multiclass error rates are studied by Devijver (1985) and Fukunaga and Flick (1984a).

The bias of average conditional error estimators in the multivariate normal case is discussed by Ganesalingam and McLachlan (1980). Snapinn and Knoke investigate the effect of smoothing on the trade off between estimator bias and variance (1985).

Efron (1982, 1983) explores the possibility of linearly combining two estimators , one known to be biased pessimistically and one optimistically. A judicial choice of the weighting coefficient can remove or at least reduce the bias.

Although average conditional error estimators have asymptotically lower variance than the empirical error count, their small sample performance does not always live up to the expectation. One cause in the degradation in performance is attributed to the failure to satisfy the assumption pertaining to the k-NN estimation scheme, namely that all the observations be drawn from a binomial distribution with parameter $p(e|\mathbf{x})$, the conditional error at point $\mathbf{x}$. In the small sample case the k-nearest neighbours are drawn from a finite volume of the pattern space where $p(e|\mathbf{x})$ varies and the resulting distribution is mixed binomial (Kittler and Devijver, 1982). Short and Fukunaga (1980) suggest a remedy whereby all the k-NN points $\mathbf{x}_i, i = 1, .., k$ are drawn so that $p(e|\mathbf{x}_i) = p(e|\mathbf{x})$. This is achieved by the application of a non-euclidean metric. Since the conditional probability $p(e|\mathbf{x})$ is least likely to vary along the hyperplane intersecting point $\mathbf{x}$ and perpendicular to the gradient of $p(e|\mathbf{x})$ at that point, all neighbours are first projected on to the direction of gradient and the nearest neighbours selected according to their projected distance from **x**. This new metric has been demonstrated to improve the small sample properties of the average conditional error estimator in experimental studies reported by Fitzmaurice and Hand (1987) who also advocate the use of the mean square error criterion of Snapinn and Knoke (1984) as a better overall measure of error estimator properties.

Perhaps the most remarkable developments in error estimation derive from the bootstrap idea of Efron (1982, 1983). Here the shortage of design data and the associated small sample problems are mitigated by using the design data primarily as a bootstrap sample. Additional data is generated from the bootstrap sample by resampling with replacement. The class labels of patterns generated by resampling may be randomised. Bootstrapping techniques have been applied successfully to problems of bias estimation and reduction by Efron (1982, 1983), Chernik et al (1985, 1986) and McLachlan (1980). A comparative study of bootstrap error estimators was carried out by Gong (1982) who found the 632 estimator of Efron to perform best of all.

REFERENCES

Besag, J., 1986, On the statistical analysis of dirty pictures, Journal of the Royal Statistical Society Series B, 48:259.

Biswas, G., Jain, A.K. and Dubes, R., 1981, Evaluation of projection algorithms, IEEE Trans Pattern Analysis and Machine Intelligence, 3:701.

Boekee, D.E. and Van der Lubbe, 1979, Some aspects of error bounds in feature selection, Pattern Recognition, 11:353.

Bunke,H., 1987, Hybrid methods in pattern recognition, in "Pattern Recognition Theory and Applications", P. Devijver and J. Kittler eds., Springer-Verlag.

Chandrasekaran, B., 1986, From numbers to symbols to knowledge structures: Pattern recognition and artificial intelligence perspectives on the classification task, in "Pattern Recognition in Practice", Vol.2, E.S. Gelsema and L.N.Kanal, eds., North Holland.

Cheng, Y. and Fu, K.S., 1984, Conceptual clustering in knowledge organization, Proc. First IEEE Conf. on Artificial Intelligence Applications, Denver, 274.

Chernick, M.R., Murthy,V.K. and Nealy, C.D., 1985, Application of bootstrap and other resampling techniques: Evaluation of classifier performance, Pattern Recognition Letters, 3:167.

Chernick, M.R., Murthy,V.K. and Nealy, C.D., 1986, Application of bootstrap and other resampling techniques: Evaluation of classifier performance, Pattern Recognition Letters, 4:133.

Dattatreya, G.R. and Kanal, L.N., 1985, Decision trees in pattern recognition, Technical Report TR-1429, Machine Intelligence and Pattern Analysis Laboratory, University of Maryland.

Devijver, P.A., 1986, Probabilistic labelling in a second order Markov mesh, in "Pattern Recognition in Practice", Vol.2, E.S. Gelsema and L.N.Kanal, eds., North Holland.

Devijver, P.A., 1985, A multiclass, k-NN approach to Bayes risk estimation, Pattern Recognition Letters, 3:1.

Devijver, P.A. and Kittler, J., 1980, On the edited nearest neighbour rule, Proc 5th International Conference on Pattern Recognition, Miami Beach, Florida, 72.

Devijver, P.A. and Kittler, J., 1982, "Pattern Recognition: A Statistical Approach", Prentice Hall, Englewood Cliffs, NJ.

Devroye,L. and Machell, F., 1985, Data structures in kernel density estimation, IEEE Trans. Patt. Anal. Mach. Intell., 7:360.

Dubes, R. and Jain, A.K., 1980, Clustering methodology in exploratory data analysis, in "Advances in Computers", Vol.19, M Yovits, ed., Academic Press.

Dubes, R. and Jain, A.K., 1979, Validity studies in clustering methodology, Pattern Recognition, 11:235.

Duchene, J., 1986, A significant plane for two-class discrimination problems, IEEE Trans. Pattern Analysis and Machine Intelligence, 8:557.

Duda, R.O. and Hart, P.E., 1973, "Pattern Classification and Scene Analysis", Wiley, New York.

Efron, B., 1982, The jackknife, the bootstrap, and other resampling plans, Society for Industrial and Applied Mathematics, Philadelphia, PA.

Efron, B., 1983, Estimating the error rate of a prediction rule: Improvement on cross-validation JASA, 78:316.

Fitzmaurice, G.M. and Hand, D.J., 1987, A comparison of two average conditional error rate estimators, Pattern Recognition Letters, (to appear).

Foley, D.H., 1982, Consideration of sample and feature size, IEEE Trans. Inf. Theory, 18:618.

Fu, K.S., 1986, A step towards unification of syntactic and statistical pattern recognition, IEEE Trans. Patt. Anal. Mach. Intell., 8:398.

Fukunaga, K. and Flick, T.E., 1984a, Classification error for a very large number of classes, IEEE Trans. Patt. Anal. Mach. Intell., 6:779.

Fukunaga, K. and Flick, T.E., 1984b, An optimal global nearest neighbour metric, IEEE Trans. Patt. Anal. Mach. Intell., 6:314.

Fukunaga, K. and Flick, T.E., 1985, The 2-NN rule for more accurate risk estimation, IEEE Trans. Patt. Anal. Mach. Intell., 7:107.

Fukunaga, K. and Flick, T.E., 1986, A test of Gaussian-ness of a data set using clustering, IEEE Trans. Patt. Anal. Mach. Intell., 8:240.

Fukunaga, K. and Mantock, J.M., 1983, Nonparametric discriminant analysis, IEEE Trans. Patt. Anal. Mach. Intell., 5:671.

Ganesalingam, S. and McLachlan, G.J., 1980, Error rate estimation on the basis of posterior probabilities, Pattern Recognition, 12:405.

Geman, S. and Geman, D., 1984, Stochastic relaxation, Gibbs distributions, and the Bayesian restoration of images, IEEE Trans. Patt. Anal. Mach. Intell., 6:721.

Gong, G., 1982, Cross-validation, the jackknife and the bootstrap: Excess error estimation in forward logistic regression, PhD Thesis, Stanford University Technical Report No. 80, Department of Statistics.

Hall, P., 1983, Large sample optimality of least squares cross-validation in density estimation, Annals of Statistics, 11:1156.

Hancock, E.R. and Kittler, J., 1987, A list driven contextual decision rule, Proc. 5th Scandinavian Conf. Image Analysis, Stockholm, 555.

Hand, D.J., 1986a, Recent advances in error rate estimation, Pattern Recognition Letters, 4:335.

Hand, D.J., 1986b, An optimal error rate estimator based on average conditional error rate: Asymptotic results, Pattern Recognition Letters, 4 :347.

Haslett, J., 1985, Maximum likelihood discriminant analysis on the plane using a Markovian model of spatial context, Pattern Recognition, 18:287.

Jain, A.K. and Chandrasekaran, B., 1982, Dimensionality and sample size consideration in pattern recognition practice, in "Handbook of Statistics", Vol.2, P.R.Krishnaiah and L.N.Kanal eds., North Holland.

Jelinek, F., Mercer, R.L. and Bahl, L.R., 1982, Continuous speech recognition: statistical methods, in "Handbook of Statistics", Vol.2, P.R.Krishnaiah and L.N.Kanal eds., North Holland.

Kalayeh, M.M. and Landgrebe, D.A., 1983, Predicting the required number of training patterns, IEEE Trans. Patt. Anal. Mach. Intell., 5:664.

Kamgar-Parsi, B. and Kanal., L.N., 1985, An improved branch and bound algorithm for computing k-nearest neighbors, Pattern Recogntion Letters, 3: 7.

Keating, J.P. and Mason, R.L., 1985, Some practical aspects of covariance estimation, Pattern Recognition Letters, 3:295.

Kim, B.S. and Park, S.B., 1986, A fast k nearest neighbor finding algorithm based on the ordered partition, IEEE Trans. Patt. Anal. Mach. Intell., 8:761.

Kittler, J., 1986, Compatibility and support functions in probabilistic relaxation, Proc 8th Intern. Conf. Pattern Recognition, Paris, 186.

Kittler, J. and Devijver, P.A., 1981, An efficient estimator of pattern recognition system error probability, Pattern Recognition, 13:245.

Kittler, J. and Devijver, P.A., 1982, Statistical properties of error estimators in performance assessment of recognition systems, IEEE Trans. Patt. Anal. Mach. Intell., 4:215.

Kittler, J. and Foglein, J., 1984, Contextual classification of multispectral pixel data, Image and Vision Computing, 2:13.

Kittler, J. and Foglein, J., 1986, On compatibility and support function in probabilistic relaxation, Comput. Vision, Graphics and Image Processing, 34:257

Kittler, J. and Hancock, E.R., 1987, Contextual decision rule for region analysis, Image and Vision Computing, 5:145.

Kittler, J. and Pairman, D., 1985, Contextual pattern recognition applied to cloud detection and identification, IEEE Trans. Geosci. Remote Sens., 23:825.

Kumar, V. and Kanal, L.N., 1984, Parallel branch and bound formulation for AND/OR tree search, IEEE Trans. Pattern Analysis and Machine Intelligence, 6:768.

Li, X. and Dubes, R., 1986, Tree classifier design with a permutation statistic, Pattern Recognition, 19:229.

Longstaff, I.D., 1987, On extensions to Fisher's linear discriminant function, IEEE Trans. Pattern Analysis and Machine Intelligence, 9:321.

Malina, W., 1981, On an extended Fisher criterion for feature selection, IEEE Trans. Pattern Analysis and Machine Intelligence, 3:611.

McLachlan, G.J., 1980, The efficiency of Efron's bootstrap approach applied to error rate estimation in discriminant analysis, J. Statist. Comp. Simul. 11:273.

Miclet, L. and Dabouz, M., 1983, Approximative fast nearest-neighbour recognition, Pattern Recognition Letters, 1:277.

Michalski, R.S. and Step R.E., 1983, Automated construction of classification: Conceptual clustering versus numerical taxonomy, IEEE Trans. Pattern Analysis and Machine Intelligence, 5:396.

Morgera, S.D., 1986, Linear, structured covariance estimation: An application to pattern classification for remote sensing, Pattern Recognition Letters, 4:1.

Mui, J.K. and Fu, K.S., 1980, Automated classification of nucleated blood cells using binary tree classifier, IEEE Trans. Pattern Analysis and Machine Intelligence, 2:429.

Okada, T. and Tomita, S., 1985, An optimal orthonormal system for discriminant analysis, Pattern Recognition, 18:139.

Peck, R. and Van Ness, J., 1982, The use of shrinkage estimators in linear discriminant analysis, IEEE Trans. Pattern Analysis and Machine Intelligence, 4:530.

Postaire, J.G. and Vasseur, C., 1982, A fast algorithm for nonparametric probability density estimation, IEEE Trans. Pattern Analysis and Machine Intelligence, 4:663.

Sethi, I.K. and Sarvarayudu, G.P.R., 1982, Hierarchical classifier design using mutual information, IEEE Trans. Pattern Analysis and Machine Intelligence, 4:441.

Short, R.D. and Fukunaga, K., 1980, A new nearest neighbour distance measure, Proc 5th International Conference on Pattern Recognition, Miami Beach, Florida, 81.

Smith, S.P. and Jain, A.K., 1984, Testing for uniformity in multidimensional data, IEEE Trans. Pattern Analysis and Machine Intelligence, 6:73.

Smith, S.P. and Jain, A.K., 1985, An experiment on using the Friedman-Rafsky test to determine the multivariate normality of a data set, Proc. IEEE CVPR Conf, San Francisco, 423.

Snapinn, S.M. and Knoke, J.D., 1984, Classification error rate estimators evaluated by unconditional mean square error, Technometrics, 26:371.

Snapinn, S.M. and Knoke, J.D., 1985, An evaluation of smoothed classification error rate estimators, Technometrics, 27:199.

Swain, P.H., Vardeman, S.B. and Tilton, J.C., 1981, Contextual classification of multispectral image data, Pattern Recognition, 13:429.

Thomason, M.G., 1987, Advances in structural methods, in "Pattern Recognition Theory and Applications", P. Devijver and J. Kittler eds., Springer-Verlag.

Toussaint, G.T., 1974, Bibliography on estimation of misclassification, IEEE Trans. Inf. Theory, 20:472.

Trunk, G.V., 1979, A problem of dimensionality: A simple example, IEEE Trans. Pattern Analysis and Machine Intelligence, 1:306.

Van Ness, J., 1980, On the dominance of non-parametric Bayes rule discriminant algorithms in high dimensions, Pattern Recognition, 12:355.

Wyse, N., Jain, A.K., and Dubes, R., 1980, A critical review of intrinsic dimensionality algorithms, in "Pattern Recognition in Practice", E.S. Gelsema and L.N.Kanal eds., North Holland, 415.

PATTERN MATCHING IN STRINGS*

Maxime Crochemore

Université Paris Nord
Av. J.B. Clément
F - 93430 VILLETANEUSE

Dominique Perrin

Université Paris 7
L.I.T.P.
2, Place Jussieu, F - 75221 PARIS

INTRODUCTION

The problem of pattern recognition in strings of symbols has received considerable attention. In fact most formal systems handling strings can be considered as defining patterns in strings. It is the case for formal grammars and especially for regular expressions which provide a technique to specify simple patterns. Other kind of patterns on words may also be defined (see for instance [23], [4]) but lead to less efficient algorithms.

The recognition of patterns in strings is related to the corresponding problem on images. Indeed, some algorithms on words can be generalized to two-dimensional arrays (see [17] for instance). Also extracting the contour of a two-dimensional object leads to a one dimensional object which can again be considered as a string of symbols. The patterns recognized in this string give in return valuable information on the two dimensional object. More geneally string processing algorithms can be used on curves, that is to say one dimensional objects in the plane or the space

In this paper, we present a new method for pattern matching in strings. As mentioned above, pattern matching in strings can be handled by standard and well-known techniques from automata theory. This formalism uses the so-called regular expressions. It allows to express logical disjunction, concatenation and iteration. As advocated very early in [27], such expressions can be compiled or interpreted producing a recognition algorithm that simulates the behaviour of a finite automaton. Many implementations of this method are now available especially in connexion with the UNIX operating system.

As a particular case of pattern matching in strings, the problem of string matching simply consists in locating a given word, the *pattern* within another one, the *text*. It has been extensively considered and is still an active field of research. This study is both concerned with optimization for practical purposes and also with theoretical considerations. For practical purposes, it is interesting to develop algorithms which are fast, require few memory, limited buffering and operate in real time. This may be especially meaningful in the case of words coding images or contours since in this case, the pattern can be very long. From the theoretical point of view, it is interesting to know the lower bounds achievable in such a problem. Those lower bounds correspond to various quantities such as time and space and it is not always possible to optimize them all at the same time. Another theoretical incentive comes from the theory of programing. Indeed, string-matching algorithms have lead to computer programs which are now considered as paradigms in the theory of program development. Finally extensions of the string matching problem have been studied. In particular several approximate string matching algorithms have been proposed ([20], [28]).

---

* This work has been supported by PRC Math-Info.

The classical algorithms can be divided in two families. Roughly speaking, in the first one the pattern x is considered as fixed and the text t as variable. The converse point of view is adopted in the second family. The first family of algorithms contains the well-known algorithms of Knuth, Morris & Pratt (KMP) on the one hand and of Boyer & Moore (BM) on the other (see [18], [7]). Those algorithms were studied and improved by several authors (see [1], [13], [15], [16], [24], [29]. The second family is based upon the notion of a suffix tree due to Weiner [30]. An efficient algorithm to compute suffix trees was devised by McCreight [21]. Later on this construction was superseded by the suffix automaton construction (see [5], [9], [10]).

The new algorithm presented here belongs to the first family. From the practical viewpoint, its merits consist in requiring only constant additional memory space. It can therefore be compared with the algorithm of [15] but it is faster and simpler. From the theoretical viewpoint, its main feature is that it makes use of a deep theorem on words known as the critical factorization theorem due to Cesari, Duval, Vincent (see [8], [11], [19]). It is also amusing that the new algorithm can be considered as a compromise between KMP's and BM's algorithms.

The paper is divided in three sections. In the first one, we first recall some definitions. We state the critical factorization theorem. We give a new proof of this result which leads to an efficient algorithm to find critical factorizations. This result algorithm requires the computation of the suffix of a word that is maximal for alphabetic ordering. In the second section, we present an efficient algorithms to solve this problem. This algorithm is not really new (see [6], [12], [25]) but is included here for the sake of completeness. Finally, in the third section, we present our new string matching algorithm. We do not prove here formally the correctness of the algorithm. This will be done in a forthcoming publication.

1. THE CRITICAL FACTORIZATION THEOREM

Let A be a finite alphabet and let A* be the set of words on the alphabet A. We shall denote by ε the empty word. We denote by $|w|$ the length of a word w. Thus $|\varepsilon| = 0$. We write $w[i]$ the i-th letter of the word w.

Let x be a word on A. We say that a pair (u,v) of words on A is a *factorization* of x if x = uv. The word u is called a *prefix* of x and v is called a *suffix*. A prefix of v is called a *factor* of x. A word u which is both a proper prefix and suffix of x is called a *border* of x. An *unbordered* word is a nonempty word x which admits no nonempty border.

Each order on the alphabet A extends to an *alphabetical ordering* on the set A*. It is defined as usual by $x \le y$ if either x is a prefix of y or if

$$x = l\,a\,r \,, \quad y = l\,b\,s$$

with a,b two letters such that a < b.

We say that an integer $p \ge 1$ is a *period* of a word x if

$$x[i] = x[i{+}p]$$

whenever both sides are defined. In other terms, p is a period of x if two letters of x at distance p always coincide. We will often designate by "the" period of x the smallest period of x. It is denoted by p(x). One may verify that p(x) is the difference of $|x|$ and the length of longest border of x.

Given a factorization (u,v) of x, the *repetition* at (u,v) is the minimal length of a non empty word w such that

(i) w is a suffix of u or conversely u is a suffix of w
(ii) w is a prefix of v or conversely v is a prefix of w

It is denoted r(u,v). One has always the inequalities $1 \leq r(u,v) \leq |w|$. More accurately, one may verify that

$$r(u,v) \leq p(w)$$

The following result is due to Cesari, Vincent and Duval (see [19] for precise references).

THEOREM (Critical Factorization Theorem). *For each word x there exists at least one factorization (u,v) of x such that*

$$r(u,v) = p(x)$$

A factorization (u,v) such that r(u,v) = p(x) is called a *critical factorization* of x. For instance, the word

$$x = abaabaa$$

has period 3. It has three critical factorizations, namely

(ab, aabaa)   (abaa, baa)   (abaab, aa)

There exist several available proofs of this result. All of them lead to a more precise result asserting the existence of a critical factorization with a cutpoint in each factor of length equal to the period of x. A weak version of the theorem occurs if one makes the addition assumption that the inequality

$$p(x) \leq 3|x|$$

holds. Indeed, in this case, one may write $x = l\,w\,w\,r$ where $|w| = p(x)$ and w is chosen minimal among its cyclic shifts. This means, by definition, that w is a Lyndon word (see [19] or also section 2). One can prove that a Lyndon word is unbordered. Consequently the factorization (lw, wr) is critical. This version is the argument used in [15] to build a string matching algorithm using restricted memory space.

In the sequel we shall be interested in computing a critical factorization of a given word. Among the existing proofs, one relies on the property that if x = ayb with a,b two letters, then a critical factorization of x either comes from a critical one for ay of from yb [19]. This leads to a quadratic algorithm. Another proof given in [11] relies on the notion of a Lyndon factorization. It leads, via the use of a linear string matching algorithm to a linear algorithm to compute a critical factorization.

We present here a new proof of the critical factorization theorem that leads to a relatively simple linear algorithm which also uses only constant additional memory space.

THEOREM. *Let $\leq$ be an alphabetical ordering and let $\sqsubseteq$ be the alphabetical ordering obtained by reverting the order $\leq$ on A. Let x be a word on A. Let v (resp. v') be the maximal suffix of x according to the ordering $\leq$ (resp. $\sqsubseteq$). Let x = uv = u'v'. If $|v| \leq |v'|$ then (u,v) is a critical factorization of x. Otherwise (u', v') is a critical factorization of x.*

As an example, the theorem gives for the word x = abaabaa the critical factorization (ab, aabaa).

The proof relies of the following lemma.

LEMMA 1. *Let* v *be the maximal suffix of* x *and let* $x = uv$. *Then no nonempty word is both a suffix of* u *and a prefix of* v .

<u>Proof</u> Let w be a suffix of u and a prefix of v. Let $v = wt$. By the definition of v we have $wv \leq v$ and $t \leq v$. The first inequality can be written $wwt \leq wt$ and this implies $wt \leq t$. The second one can be written $t \leq wt$. We obtain $t = wt$ whence $w = \varepsilon$. $\square$

Proof of the theorem. First observe for later use that the intersection of the orderings $\leq$ and $\subseteq$ is the prefix ordering. Equivalently if

$$w \leq w' \qquad \text{and} \qquad w \subseteq w'$$

then w is a prefix of w'.

We first rule out the case where the word x has period 1, that is to say when x is a power of a single letter. In this case any factorization of x is critical. We now suppose that $|v| \leq |v'|$. The other case is symmetrical. Let us prove first that $u \neq \varepsilon$. Let indeed $x = ay$ with a in A. If $u = \varepsilon$ then $y \leq x$ and $y \subseteq x$. Thus y is a prefix of x whence $p(x) = 1$ contrary to the hypothesis.

Let r be the repetition at (u,v). By the lemma, we cannot have $r \leq |u|$ and $r \leq |v|$. Since v is maximal, it cannot be a factor of u. Hence $r > |u|$ since $r \leq |u|$ would imply $r > |v|$. Let z be the shortest word such that v is a prefix of zu or conversely zu is prefix of v. Then $r = |zu|$. We distinguish two cases according to $r > |v|$ or $r \leq |v|$.

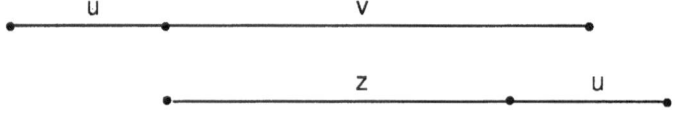

Figure 1.1 $r > |v|$

Case 1. u is not a factor of v. The integer $|uz|$ is a period of uv since uv is a prefix of $uzuz$ (see Figure 1.1). Since $|zu|$ is the repetition at (u,v), the period of uv cannot be shorter than $|zu|$. This proves that the factorization (u,v) is critical.

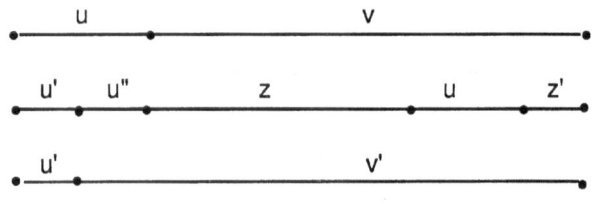

Figure 1.2. $r \leq |v|$

Case 2. u is a factor of v. Since $|zu|$ is the repetition at (u,v) we only need to prove that $|zu|$ is a period of x. Let $u = u'u''$ and $v = zuz'$. We have by the definition of v' the inequality

$$u''z' \subseteq v' = u''v$$

70

hence $z' \subseteq v$. By the definition of v, we also have $z' \leq v$. By the observation made at the beginning of the proof these two inequalities imply that z' is a prefix of $v = zuz'$. Hence z' is a prefix of a long enough repetition of zu. Since $x = uzuz'$, this shows that $|uz|$ is a period of x. $\square$

The computation of a critical factorization relies, as a consequence of the previous result, on the localization of a maximal suffix. This is described in the next section.

2. MAXIMAL SUFFIXES

In this section we shall describe an algorithm to compute the maximal suffix of a word. It is a consequence of several known algorithms that this computation can be done in linear time. One may use the suffix tree construction [21] or also the Lyndon factorization [12]. We describe here an algorithm which essentially is the same as the one in [12] but slightly more simple.

We use a fixed ordering on the alphabet A. We want to know what happens to the maximal suffix of a word x when x is replaced by xc with c in A. This will lead to a left to right algorithm to compute the maximal suffix.

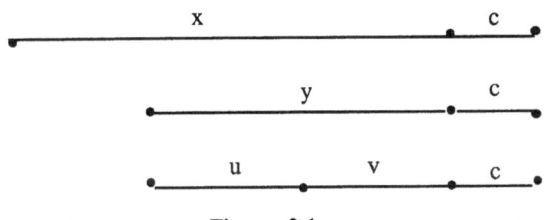

Figure 2.1.

Let y be the maximal suffix of x. Clearly the maximal suffix of xc cannot be longer that yc. Let v be a suffix of y such that $vc \geq yc$. Since on the other hand $v \leq y$ the only possibility is that v is a prefix of y. Hence v is a border of y. We shall therefore keep track of the longest border of y. Since its length is equal to $|y| - p(y)$, we shall keep in a variable p the shortest period of the word y. We shall then be able to compare step by step the word vc with the word y, according to the letter a such that va is a prefix of y.

We now present the algorithm more formally. One may use Figure 2.2 to follow the notation. Let $x = x[\,1\,]x[\,2\,] \ldots x[\,n\,]$ be a word of length n. We assign initially

$$i \leftarrow 0 \,;\, j \leftarrow 1 \,;\, k \leftarrow 1 \,;\, p \leftarrow 1$$

We denote
$$a = x[\,i+k\,] \,;\, b = x[\,j+1\,] \,;\, c = x[\,j+k\,]$$
and we iteratively perform the following action while $j+k \leq n$:

Case 1. If $(a < c)$ then $\{i \leftarrow j, j \leftarrow i+1 \,;\, k \leftarrow 1 \,;\, p \leftarrow 1\}$
Case 2. If $(a = c)$ then $\{$if $(k = p)$ then $\{j \leftarrow j+k \,;\, k \leftarrow 1\}$ else $\{k \leftarrow k+1\}\}$
Case 3. If $(c < a)$ then $\{j \leftarrow j+k \,;\, k \leftarrow 1, p \leftarrow j-i\}$

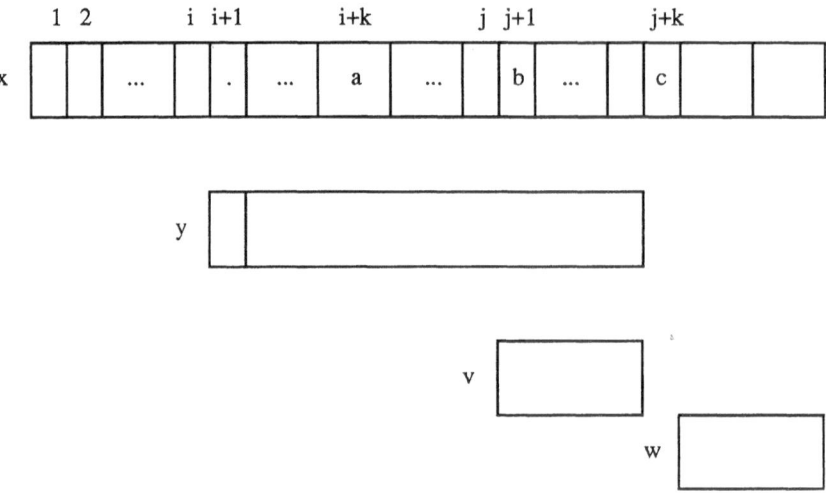

Figure 2.2

We shall prove that the final value of x [i+1] ... x [n] is the maximal suffix of the word x. We shall also prove that the number of steps before the algorithm stops is at most 2 n. To this end let

$$y = x\,[\,i{+}1\,]\,x\,[\,i{+}2\,] \dots x\,[\,j{+}k{-}1\,]$$
$$v = x\,[\,j{+}1\,]\,x\,[\,j{+}2\,] \dots x\,[\,j{+}k{-}1\,]$$
$$w = x\,[j{+}k{+}1] \dots x\,[\,n\,]$$

We shall consider the alphabetic ordering $\subseteq$ induced by the reversed order on the alphabet. We shall also use Lyndon words corresponding to this order. By definition, a word l belongs to the set L of Lyndon words if it is minimal for the order $\subseteq$ among the set of its suffixes. This is of course equivalent to say that l is unbordered and maximal among its suffixes for the order $\leq$. Thus the list of Lyndon words on two letters a,b with $a < b$ begins with a,b,ba,baa,bba,baaa,bbaa,bbba,baaaa,babaa,bbaaa,bbaba,.....

In the proof of the algorithm, we shall use several facts concerning Lyndon words among which the fact that a word y is maximal among its suffixes for the order $\leq$ iff it is of the form

$$y = l^e\, v$$

where l is a Lyndon word for $\subseteq$, $e \geq 1$ and v is a prefix of y. We shall also use the well-known fact that if l,m are Lyndon words with $l \subset m$ then lm is also a Lyndon word (see [11] or [19]).

We prove the correctness of the algorithm by showing that the following two conditions remain true after each execution of the body in the main loop.

(i)   x [r] x [r+1] ... x [n] ≤ x [i+1] ... x [n]   (1 ≤ r ≤ i)

(ii)   y = l ^{e} v where l is a Lyndon word of length p for ⊆, e ≥ 1 and v is a prefix of l.

We have initially i = 0, y = a, v = ε so that both conditions trivially hold true. When j+k > n, the conditions imply that y is the maximal suffix of x.

Let us consider the execution of one step of the main loop.

Case 1. The assignments are equivalent to

$$y \leftarrow b , v \leftarrow \varepsilon$$

so that condition (ii) is trivially satisfied. Let us prove that condition (i) is also satisfied. The hypothesis a < c implies that y < vc. A suffix s of x beginning before the (old) value of i satisfies s < ycw. Since y < vc we also have ycw < vcw. Hence s < vcw as required. A suffix beginning between the old value of i and the new one is of the form tcw with t a suffix of y. By the remark made earlier, we have t < y. If t is not a prefix of y, then tcw < ycw. Otherwise, we have t = l ^{f}v with 0 ≤ f ≤ e-1. Since l ^{f}v < vc we have tcw < vcw.

Case 2. Condition (i) is trivially satisfied since the value of i is not modified. If k = p then we have l = vc and we replace (y,v) by (yc, ε). Otherwise we change v to vc because vc is a proper prefix of l. Hence condition (ii) is also satisfied.

Case 3. Condition (i) is trivially satisfied. The assignments are equivalent to

$$y \leftarrow yc , v \leftarrow \varepsilon$$

Indeed, when vat is a Lyndon word with v,t ∈ A*, a ∈ A and a > c, then for any e ≥ 0 the word (vat)^{e}vc is again a Lyndon word. Indeed, since l is a Lyndon word we have b ≥ a. Therefore, we have v ⊂ c and thus vc is a Lyndon word. Since l ⊂ vc, we obtain l ^{e} vc ∈ L. Thus condition (ii) is also satisfied.

This proves the correctness of the algorithm follows.

We now evaluate its time complexity. It is clearly proportional to the number of times that the body of the main loop is executed. Let us verify that this number is bounded by 2 n. For this, we remark that the sum i+j+k strictly increases at each step. This is clear in cases 2 and 3 where it increases by one unit. In case 1, i+j+k is replaced by 2j+2. But since the inequality i+k ≤ j constantly holds (see Figure 2.2) we constantly have i+j+k ≤ 2j. Hence i+j+k increases at least by two units in case 1. Since

$$2 \leq i+j+k \leq 2 n$$

the above argument shows the claim. Consequently the algorithm is linear in time. Moreover it requires only constant extra memory space.

3. STRING MATCHING

Among all string matching algorithms developped up to now, two of them are particularly famous and efficient. They are known as Knuth, Morris & Pratt's algorithm and Boyer & Moore's algorithm. Let us briefly explain how they work.

Let x be the word which is searched for inside an a priori longer word t. We consider string matching algorithms computing all occurrences of x in t. All algorithms first check whether x appears at the left end of t and repeat this process at increasing positions. The word x is thus shifted to the right until it reaches the right end of t. Shifts must be as long as possible in order to save time.

In Knuth, Morris & Pratt's algorithm the letters of x are checked against the letters of t from left to right until the right end of x is reached if x occurs at that position or until a mismatch is met. If y is the longest prefix of x recognized at that position then the shift is made according to both a period of y and the letter of t that causes the mismatch. Hence, a shift function, whose domain is the set of prefixes of x, is precomputed.

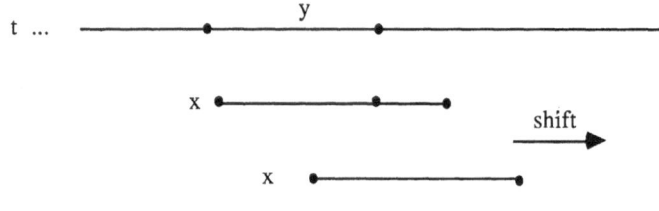

Figure 3.1

In Boyer & Moore's algorithm the letters of x are scanned from right to left and x is shifted according to both the periods of its suffixes and the letter of t that causes a possible mismatch. Proceeding in that way increases the length of shifts in the average. For instance, a letter of t which does not occur in x leads to a shift of x after the position of this letter.

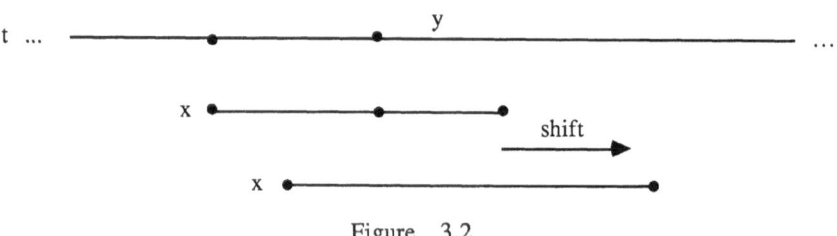

Figure 3.2

During the search for x in Boyer & Moore's algorithm the result of some comparisons is forgiven and it is not obvious at all that the whole algorithm is linear in time.

In fact, tricky versions of this algorithm succeed in doing the whole process using a number of comparisons bounded by twice the length of t. (See [13]).

The same bound holds for Knuth, Morris & Pratt's algorithm. The two algorithms greatly differ on the minimum number of letter comparisons. It is $|t|$ for Knuth, Morris and Pratt's algorithm while it becomes $|t|/|x|$ for Boyer & Moore's algorithm. Both algorithms use linear additional memory space for shift functions on the word x.

We describe an algorithm which matches the letters of x and t in both directions. The starting point is given by a critical factorization (x_l, x_r) of x. The algorithm uses an integer $q(x_l, x_r)$ defined as follows :

$$q(x_l, x_r) = \min \{ |w| / x \text{ proper suffix of } x_r w\}$$

The algorithm does not use any shift function precomputed on x but only the integers $p(x)$ and $q(x)$. Thus, the total extra memory space used by this algorithm is constant which makes possible its general implementation even in a low level language.

To discover the occurrences of x in t, the algorithm is conceptually divided in two successive phases. The first phase consists in searching for x_r only. The letters of x_r are scanned from left to right. When a mismatch is found during this phase, the fact that the factorization (x_l, x_r) is critical allows us to shift x at least after the mismatch.

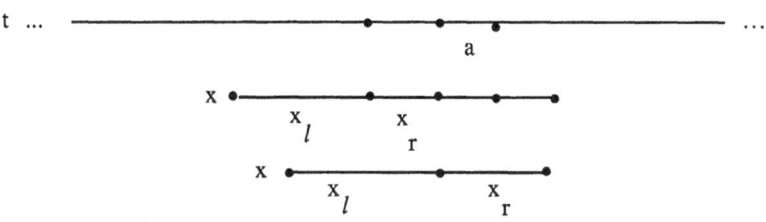

Figure 3.3

When an occurrence of x_r is found in t, the second phase begins. The algorithm checks whether an occurrence of x_l occurs before by scanning the letters of x_l from right to left. Any shift occuring during this phase must be compatible with the period $p(x)$ of x (or in some cases with $q(x)$) and must eliminate a possible mismatch. After such a shift a prefix of x matches a factor of t. This prefix is kept in memory via a variable s for use in the next execution of the first phase concerning x_r. The following property is an invariant of the main "whole" loop of the algorithm in figure 3.4.

$$x[r] = t[k+r], \quad 1 \le r \le s.$$

```
k ← 0 ;        s ← 0 ;
while  k + | x | ≤ | t |  do
    {      i ← max (| x |, s) + 1 ;
```
```
           while  i ≤ | x |  and  x [ i ] = t [ k+i ]  do  i ← i+1 ;
           if  i ≤ | x |  then
```
2
```
               { k ← k + max (i - | x_l |, s - p(x) + 1) ;   s ← 0 ; }
           else
               { j ← | x_l | ;
```
3
```
                 while j > s  and  x [ j ] = t [ k+j ]  do  j ← j-1 ;
                 if  j = s  then  { output  k ;  j ← 1}
```
4
```
                 k ← k + min  (⌈ j / p(x) ⌉ . p(x), q(x)) ;
```
5
```
                 s ← | x | - min (⌈ j / p(x) ⌉ . p(x) , q(x)) ;
           }
    }
```

Figure 3.4 Searching t for x.

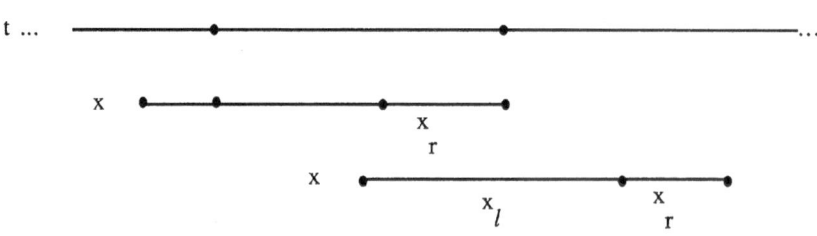

Figure 3.5

The correctness of the algorithm is based on several properties that we prove now. We shall not prove here formally the correction of the algorithm. In the following lemmas (x_l , x_r) is any critical factorization of the fixed word x.

LEMMA 3.1 *Let v be a prefix of x_r and z be a suffix of v. If vz is a suffix of $x_l v$ then p(x) divides | z |.*

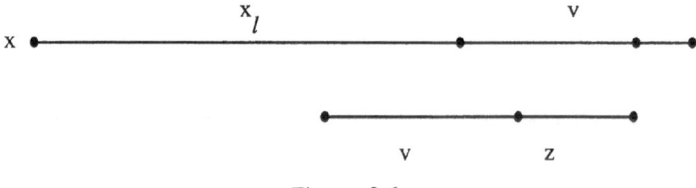

Figure 3.6

<u>Proof</u> Let w be the prefix of v of length | z |. Since p(x) is the period of x, w may be written (uu ')^e u with | uu' | = p(x) and 0 ≤ | u | < p(x). One can see that | u | is a repetition at the critical factorization (x_l , x_r). The only possibility is then | u | = 0 which proves that p(x) divides | w | = | z |. □

Lemma 3.1 has the following consequence on the analysis of the algorithm.

Assume that a mismatch occurs in a letter a of x_r which immediately follows the word v as in lemma 3.1. Any shift to the right less than $|v|$ places and compatible with the occurrence of v in t is a multiple of $p(x)$ by lemma 3.1. So, the corresponding occurrence of v in x is also followed by letter a and the mismatch remains. Thus, to have any chance to recognize x in t, we must shift more than $|v|$ places to the right. This corresponds to the expression "$i - |x_l|$" at line 2 of the algorithm. The next lemma explains in a same manner the other expression "$s - p(x) + 1$" at line 2 again.

LEMMA 3.2 *Let w and y be such that w and yw are prefixes of x and $|w| \geq p(x)$. Then $p(x)$ divides $|y|$.*

Proof The word y can be written $(uu')^l u$ with $|uu'| = p(x)$ and $0 \leq |u| < p(x)$. Since w is a prefix of x of length greater then $p(x)$, uu' is a prefix of w. Since yw is a prefix of x, $(uu')^{l+1}u$ is a prefix of yw, hence u'u is also a prefix of w. So, uu' = u'u which implies that uu' is a non trivial power of a word (see Lothaire [19]). This gives a contradiction with the minimality of $p(x)$ and shows that the only possibility is $|u| = 0$. Thus $p(x)$ divides $|y|$. $\square$

Consider the case where aux_r is a suffix of x, bux_r occurs in t and a mismatch occurs between letters a and b during the second phase of the algorithm (line 3). Then the next lemma shows that any worthy shift must rule out the mismatch by sliding x beyond letter b. This idea gives the expression at line 4. Note that when an occurrence of x is found, in which situation k is output, instruction "$i \leftarrow 1$" works as if a mismatch has been encountered on the first letter of x and this leads to the smallest valid shift.

LEMMA 3.3 *If $x_r z$ is a suffix of x then $p(x)$ divides $|z|$.*

Proof If z is a suffix of x_r, the conclusion holds from lemma 3.1 with $v = x_r$. In the other case, one can write $z = ux_r$. Observing that $|x_r u|$ is a repetition at (x_l, x_r) the result follows as in the proof of lemma 3.1. $\square$

The time complexity of the algorithm is proportional to the number of letter comparisons between x and t (lines 1 and 3). This number is bounded by $3|t|$.

LEMMA 3.4 *The maximum number of letter comparisons during a run of the algorithm on words x and t is $3|t|$.*

Proof Each comparison at line 1 strictly increases the value of k+i. Hence, their number is at most $|t| - |x_l|$, since expression k+i has initial value $|x_l| + 1$ and terminal value $|t|$ in the worst case.

At most $|x|$ comparisons are done during one execution of the "while" instruction at line 3 with $k = k_0$. Any execution of this "while" leads to a shift of length greater than $p(x)$ (note that $q(u) \geq p(x)$). Thus, if $p(x) \geq |x|/2$ the next values of k+j are all greater than $k_0 + |x|/2$. If $p(x) < |x|/2$, one may see that the next values of k+j are all greater than $k_0 + |x| - p(x)$ since the shift done at line 2 is greater than $s - p(x)$. Again the next values of k+j are greater $k + |x|/2$. Thus a letter of t is considered at most twice during the "while" instructions at line 3.

The total number of comparisons is thus less than $3|t|$. $\square$

4. REFERENCES

[1]   A.V. AHO & M.J. CORASIK, Efficient string matching : An aid to bibliographic search, Comm. ACM, 18 (1975) 333-340.

[2]   A.V. AHO, J.E. HOPCROFT & J.D. ULLMAN, The design and analysis of computer algorithms, Addison-Wesley, Reading, Mass., 1974.

[3]   A. APOSTOLICO & F.P. PREPARATA, Optimal off-line detection of repetitions in a string, Theoret. Comput. Sci. 22 (1983) 297-315.

[4]   D.R. BEAN, A. EHRENFEUCHT & G.F. McNULTY, Avoidable patterns in strings of symbols, Pacific J. Math., 85 (1979) 261-294.

[5]   A. BLUMER, J. BLUMER, A. EHRENFEUCHT, D. HAUSSLER, M.T. CHEN & J. SEIFERAS, The smallest automaton recognizing the subwords of a text, Theoret. Comput. Sci., 40, 1 (1985) 31-56.

[6]   K.S. BOOTH, Lexicographically least circular substrings, 10, 4, 5 (1980) 240-242.

[7]   R.S. BOYER & J.S. MOORE, A fast string searching algorithm, Comm. ACM, 20 (1977) 762-772.

[8]   Y. CESARI & M. VINCENT, Une caractérisation des mots périodiques, C.R. Acad. Sc., t. 286, série A (1978) 1175.

[9]   M. CROCHEMORE, Transducers and repetitions, Theoret. Comput. Sci., 45 (1986) 63-86.

[10]   M. CROCHEMORE, Longest common factor of two words, CAAP' 87, Theory and Practice of Software development, Vol. 1, Springer Lecture Notes in Comput. Sci., 1987.

[11]   J.P. DUVAL, Mots de Lyndon et périodicité, RAIRO Informatique Théorique/ Theoretical Informatics, 14, 2 (1980) 181-191.

[12]   J.P. DUVAL, Factorizing Words over an Ordered Alphabet, Journal of Algorithms, 4 (1983) 363-381.

[13]   Z. GALIL, On improving the most case running time of the Boyer-Moore string matching algorithm, Comm. ACM, 22, 9 (1979) 505-508.

[14]   Z. GALIL, String Matching in real time, J. Assoc. Comput.Mach., 28, 1 (1981) 134-149.

[15]   Z. GALIL & J. SEIFERAS, Time Space Optimal String Matching, J. Comput. and System Sci., 26 (1983) 280-294.

[16]   L.J. GUIBAS & A.M. ODLYSKO, A new proof of the linearity of the Boyer-Moore string searching algorithm, Proc. 18 th Annual IEEE Symposium on Foundations of Computer Science (1977) 189-195.

[17]   R.M. KARP, R.E. MILLER & A.L. ROSENBERG, Rapid identification of repeated patterns in strings, trees, and arrays, ACM Symposium on Theory of Computing, Vol. 4, ACM, New York, 1972, pp. 125-136.

[18]   D.E. KNUTH, J.H. MORRIS & V.R. PRATT, Fast pattern matching in strings, SIAM J. Comput., 6, 2 (1977) 323-350.

[19]   LOTHAIRE, Combinatorics on Words, Addison-Wesley, Reading, Mass., 1982.

[20]   G.M. LANDAU & U. VISHKIN, Efficient string matching with k differences, Technical Report 186, Courant Institute of Mathematical Sciences, New York University (1985).

[21]   E.M. McCREIGHT, A space-economical suffix tree construction algorithms, J. Assoc. Comput. Mach., 28, 2 (1976) 262-272.

[22]   R.L. RIVEST, On the worst-case behavior of string-searching algorithms, SIAM J. Comput., 6, 4 (1977) 669-674.

[23]   G. ROZENBERG & A. SALOMAA, The Mathematical Theory of L-systems, Academic Press, 1980.

[24]   W. RYTTER, A correct preprocessing algorithm for Boyer-Moore string-searching, SIAM J. Comput., 9, 3 (1980) 509-512.

[25]   Y. SHILOACH, Fast canonization of circular strings, J. of Algorithms, 2 (1981) 107-121.

[26]   A.O. SLISENKO, Determination in real time of all the periodicities in a word, Soviet. Math. Dokl., 21, 2 (1980) 392-395.

[27]   K. THOMSON, Regular expression search algorithm, Comm. ACM, 11 (1968), 419-422.

[28]   E. UKKONEN, Finding Approximate Patterns in Strings, J. of Algorithms, 6 (1985) 132-137.

[29]   U. VISHKIN, Optimal parallel pattern matching in strings, Proc. 12th ICALP, LNCS 194, Springer-Verlag (1985) 497-508.

[30]   P. WEINER, Linear pattern matching algorithms, IEEE Symposium on Switching and Automata Theory, Vol. 14, IEEE, New York, 1972, pp. 1-11.

IMAGE ANALYSIS PROBLEMS IN ASTRONOMY

F. Murtagh*

Space Telescope — European Coordinating Facility
European Southern Observatory
Karl–Schwarzschild–Straße 2
D–8046 Garching–bei–München, FRG

1. INTRODUCTION

The topics dealt with in this paper are as follows. Firstly, to set the scene, we briefly overview the major astronomical image processing systems, and make reference to current software engineering problems in this area. Secondly, we survey a range of pattern recognition problems, which all have classification as their central objective. These pattern recognition problems are: object searching and classification in photometry; the classifying of galaxies on the basis of their morphological shapes; and the classification of stellar spectra. Finally, we review some of the specific problems expected for Hubble Space Telescope image data.

2. ASTRONOMICAL IMAGE PROCESSING

2.1 Image Processing Systems

Current observational, optical astronomy begins at the telescope, often situated in an isolated region of particularly good visibility. The positioning of an "observatory" above the earth's atmosphere (the Hubble Space Telescope, currently due for launch in late 1988) represents a further major step forward in data capture. Image data from ground–based instruments is collected in observing runs, and then usually studied at the astronomer's home institute. Major image processing software systems are available for aiding this analysis. Papers in Scarsi et al. (1987) may be referred to for image processing techniques in general use.

Observing time is often the scarcest resource, among all resources related to the human astronomer, the analysis and processing computers available, or the detecting instrumentation. Hence, astronomical databases are also assuming ever greater importance (Ochsenbein, 1986; Murtagh and Heck, 1987b).

Among the more well established astronomical image processing systems are the following:

Affiliated to Astrophysics Div., Space Science Dept., European Space Agency.

• AIPS (Astronomical Image Processing System) from the National Radio Astronomy Observatory (NRAO), Charlottesville, Virginia, which was developed initially for radio astronomy data from the VLA (Very Large Array) observatory, New Mexico.

• IRAF (Image Reduction and Analysis Facility), from Kitt Peak National Observatory (KPNO), Tucson, Arizona, and the National Optical Astronomy Observatories (NOAO), Tucson, Arizona.

• GIPSY (Groningen Image Processing System) from the Astronomy Department, Groningen University, Holland, was originally designed for radio astronomy.

• STARLINK is not a homogeneous system but rather a wide–ranging collection of software (including graphics, database management, and applications) maintained at the Rutherford Appleton Laboratories, Oxfordshire, England. It is accessible over the astronomical research network of the same name.

• IHAP (Interactive Handling and Analysis Package) developed at the European Southern Observatory, Munich. MIDAS (Munich Image Data Analysis System) is the current system developed and distributed by the European Southern Observatory. Its focus is also on optical astronomy.

All the systems mentioned above have special characteristics making them suitable for particular areas of application. In the following, we look briefly at aspects of MIDAS.

In its current version, MIDAS runs under the VAX VMS operating system, uses Gould DeAnza image displays, several graphics monitors, different hardcopy devices, and a high quality image recorder. MIDAS consists of about 50,000 lines of FORTRAN code and has, to date, involved about 15 man years of work. The typical work station consists of an alphanumeric terminal, an interactive graphics monitor and an image display with joystick.

A portable version of MIDAS is currently nearing completion. It will be C– and Unix-based, and will allow the straightforward implementation of MIDAS on a wide range of smaller (and less expensive) hardware configurations.

MIDAS has a command driven monitor, tailored along the lines of DEC's DCL command language, with the general syntax:

```
command/qualifier parameter1 parameter2 ... ! comments
```

Additional control statements are available for looping, branching, etc., which yield a programming language for astronomical applications. The advanced user can define complicated procedures, but interaction is also sufficiently simple to allow access to data and algorithms to users with little previous computer experience. New applications can be included as new commands with little effort.

The IRAF system (mentioned above) on the other hand has a forms–based menu interface. This lends itself to handling a large number of parameters with little inconvenience to the user.

The data structures in MIDAS were designed for astronomical applications. The two most important high level data structures available to the user are images and tables:

• Images are raster type data used to store pictures or spectra collected by photographic or digital detectors. Pixels in an image can be referred to by pixel number, as in any image processing system, or by users' coordinates (using wavelengths, spatial positions, etc.).

• Tables are collections of n-tuples which represent the properties of objects in a structured form. They are used to store results of the analysis of images, and to keep catalogues and small databases inside the image processing environment. This data structure is looked at in the next section.

At present, MIDAS provides a comprehensive set of basic algorithms operating on images and tables, and controlling the devices associated with the user work station. Several application packages are also available in the areas of spectroscopy and photometry, and they are used to reduce and analyze data originating in different astronomical instruments. A suite of about 300 commands is currently available to perform these operations.

MIDAS software is available free of charge to the astronomical community (— it is currently used by about 30 institutes in Europe and America).

2.2 The MIDAS Table Subsystem

Most image processing systems allow for connections with database management systems (DBMSs); a feature of MIDAS is that some of the capabilities of DBMSs are provided in MIDAS itself. The primary, and original, objective of MIDAS is for the display and processing of digitised, astronomical images. From the user point of view, system integration should be maintained. For this reason, the intermediate *table* data structure is used, which is closely linked to the image processing routines.

The MIDAS table subsystem (Grosbøl and Ponz, 1985) grew out of 2–dimensional photometry: scanning digitised images for objects (e.g. stars, galaxies, or other objects) led to lists of coordinates, and associated flux or intensity values. Both during an image *reduction* session, and as an end–product, tables are needed for storing data ranging from stellar magnitudes, through galaxy isophotal diameters, to spectral line intensities. Most image processing systems use tables in some form, whether visible or otherwise to the user. An efficient table system is therefore highly desirable.

It must also interface effectively with graphics. This arises from the need of the astronomer to compare various sets of results, and from the central rôle played in this fundamentally experimental science by the plotting of data.

MIDAS tables allow for various data types, default display formats, reference to columns by label or by sequence number, missing value handling, automatic handling of row/column dimensions, vaious operations such as sorting, deleting, arithmetic calculations, editing, a range of plotting and display I/O options, and statistical methods.

The table subsystem provides an effective and satisfactory link between, on the one hand, image processing and database material and, on the other hand, pattern recognition and other analysis procedures. Additional tables capabilities envisaged are special data types to facilitate processing of astronomical coordinates (e.g. right ascension and declination), or to allow error estimates to be more conveniently stored with associated measurements.

2.3 Current Problems

In the past, portability has not been a great problem (more than 80% of astronomical institutes in Europe were estimated in 1986 to have been VMS-based). The current tendency however is to have astronomical systems available on UNIX, and hence on relatively inexpensive configurations.

Considerable attention has also been paid to problems of having software from one system, or developed in another hardware environment, more widely available. A result of this has been the adopting of device–independent plotting and display interfaces.

A more long–term effort involves the porting of enduser–written applications software. Much (and, it could be argued, in any sophisticated domain the best) applications software is written by the individual astronomer for his/her particular tasks. General approaches to implementing such software in different systems — including the use of standard I/O interfaces, interface emulation, physical replacement of calling sequences, and restriction of applications code to callable subroutines — are overviewed by Adorf et al., (1986). A considerable challenge, for instance, is provided by the IRAF–MIDAS porting of applications software: the structures of these two systems have been developed separately, and IRAF's use of the SPP preprocessor (*Subset Preprocessor*, which produces Fortran IV code) in applications software is a new development against a general backdrop of VAX–11 Fortran or Fortran 77 usage.

Although at present not a full DBMS, the MIDAS table system goes some of the way towards being a *scientific DBMS*. As Shoshani and Wong (1985), among others, have pointed out, commercial relational DBMSs are often unsuitable for scientific and statistical applications. The current evolution of the MIDAS table system will take it further in the SSDBMS (scientific and statistical DBMS) direction.

3. CLASSIFICATION IN ASTRONOMY

Classification is a very general problem. Here, we look at some of the major contexts in which classification is carried out in astronomy, and where work is on–going. These contexts are: photometry; galaxy classification; and ultraviolet spectral classification. This overview relates primarily to optical astronomy. Ancillary reading, from a data analysis perspective, includes Murtagh and Heck (1987a) and Di Gesù et al. (1984, 1986).

3.1 Photometry

Digitised astronomical image frames are widely used in observational astronomy. They provide such obvious advantages as automated, objective study of the area of sky, or objects of interest. They also allow such study to be carried out at the astronomer's convenience, rather than instantaneously at the observatory. Objects studied include

• stars: usually of large central brightness; of intensity distribution which is roughly gaussian in shape; usually radially symmmetric; sometimes giving rise to diffraction spikes and intensity saturation effects;

• galaxies: usually extended in shape; of a wide range of morphological shapes;

• other objects of interest: quasars; clusters of stars or galaxies; visual corroboration of radio sources, etc.;

• a range of unwanted "objects" such as cosmic ray hits, often occassioning strong intensities in single, isolated pixels; instrumental effects, including, in widely used CCDs (charge coupled devices) pixels reporting clearly an incorrect situation (hot/cold pixels, charge overflow columns), left–over effects of a previous bright image, and a varying image background.

Additionally, problem–areas include dust lanes obscuring galaxies, stars superimposed on galaxies, and very crowded stellar fields which are difficult to resolve.

Photometric packages encompass object detection and recognition, determining the PSF (point spread function, — the intensity distribution of a star), and deriving measurements of interest. The PSF (also known as impulse response or system degradation) is the output for an idealized point source (or mathematical delta function) input. As a rule, galaxy photometry packages determine parameters for the objects studied, which allows operations to be carried out in parameter space. The objective in stellar photometry packages is often the deriving of very accurate estimates of certain values.

One package, initially designed for galaxy analysis, is INVENTORY (MIDAS, 1987; application in West and Kruszewski, 1981). It was designed for the fast detection of objects in large plates. It has been used for Schmidt, CCD, and electronographic plates. For blended objects, it produces a fast but not optimal solution.

It is a modular system and, not unlike other packages, commands are available for the following:

• Preparing a list of objects. A local sky background is defined, and an object detected if the average intensity over 9 central pixels is greater than the background by a user–defined threshold. A user–defined separation threshold defines distinct objects.

• Refining the centroids of the objects; and determining parameters associated with them.

• Finally, classifying the objects into stars, galaxies, defects, and unclassified (using, in INVENTORY, a non–parametric, seed–initiated, iterative algorithm).

INVENTORY does not provide the extraction precision of other packages, but is very suitable for the quick scanning of large images. It was originally designed for detecting faint distant clusters of galaxies. INVENTORY is written in VAX–11 Fortran, it uses MIDAS command language, and an upgrade by the author is currently taking place.

Another example of a parameter–extracting system is COSMOS (*Coordinates, Size, Magnitude, Orientation and Shape*; Stobie, 1980, 1986; applications in MacGillivray and Stobie, 1984; MacGillivray et al., 1976) It was developed and is maintained at the Royal Observatory Edinburgh (ROE). It has been used for large photographic plates, including non–astronomical applications such as aerial photographs and radar pictures.

An example of a stellar photometry package is DAOPHOT (Stetson, 1984; 1987a; King, 1986). It is a modular package with commands to allow: synthetic aperture photometry — magnitudes at varying radial distances from the centroids of the objects are determined; and defining a PSF, from sampled bright stars in the frame, where the PSF is taken as a Gaussian plus a table of residuals (hence, allowing a very general distributional form for the PSF). The analysis of a crowded field, involving blended stars, is carried out by multiple, least squares PSF fitting.

3.1.1 Classification and Related Algorithms in Use in Photometry

It rapidly becomes clear that the problems of object recognition and the consequent extraction of information are strongly associated with many classification problems. In the following, some of the more sophisticated algorithms which have been used to date will be overviewed.

3.1.1.1 Connectivity Analysis

The processing of large images, containing many hundreds or thousands of objects, can require extensive computational time. COSMOS uses an algorithm due to Lutz (1980) and improved in Thanisch *et al.* (1984) for determining connected components in a thresholded (binary) image "on the fly". The connected components are simply the pixels which are above the threshold in value, and which are adjacent. The raster scanning of images allows connected parts of an object to be built up in one pass, given that the necessary bookkeeping is carried out.

3.1.1.2 Clustering

Clustering, for automatically classifying detected objects in parameter space, is used in INVENTORY and COSMOS, among other packages. The primary objective is classification into classes corresponding to stars, galaxies, and plate faults.

In FOCAS (*Faint Object Classification and Analysis System*, Jarvis and Tyson, 1981, Tyson, 1984), objects are built up as contiguous, thresholded pixels. From these, parameters are calculated. Functions of the moments are used to determine shape, size, elongation, peak and average intensities, effective radius, and magnitude. In parameter space, "seeds" for the classes of "galaxy", "star" and "other" are defined manually. A variance–based iterative procedure is then used on part of the data set, allowing the quadratic decision surfaces arrived at to be used for a subsequent assignment stage.

If not saturated or crowded, and if not displaying diffraction spikes, then stars are usually of large central brightness, of steep intensity gradient, and symmetric in shape; galaxies are usually more extended. Using a variety of single or pairs of parameters has allowed objects to be discriminated: Kurtz (1983) lists 17 studies using relatively simple star/galaxy separation methods along these lines.

INVENTORY (up to May 1987) determines the values of, in all, 20 parameters for each object (local sky background, average of central nine pixels, isophotal magnitude where the isophote may be user-specified, intensity gradient related paramaters, and so on). Most of these parameters are subsequently used for classification purposes. The two parameters *central magnitude* and *relative gradient* are used for selecting seeds (stars having high values for both of these parameters, galaxies having low values; using both parameters simultaneously helps to cater for faint objects). The seeds are defined for less faint objects, and iterative partitioning is used for the remainder of the data (essentially a k–means approach).

A parametric appraoch has also been used for this problem. A Bayesian classifier used by Valdes (1982) is compared with the non–parametric approach in Tyson (1984). The former wins in accuracy but looses out in computational efficiency. Similarly in the Guide Star Selection System for the Hubble Space Telescope (GSSS, 1985), a Bayesian approach is used. This system, nearing completion, involves the building of a catalogue of 20 million guide star candidates in order to provide sufficient pointing information for the Space Telescope (Jenkner, 1983). The method used for distinguishing stars from other objects is based on multivariate normal discriminant analysis.

3.1.1.3 Matching

The optimal matching of sets of objects (defined by two-dimesional pixel or world coordinates) is a useful tool for appraising the outputs produced by different packages; or it may provide a utility for defining a transformation between different subframes.

An approach to the matching of patterns of 2–dimensional points using geometric data structures (Delauney triangulations and maximal cliques) is described in Ogawa (1986). Another approach is adopted in the PAIR program of Lauberts (1986): it constructs a histogram of all pairwise distances between points in the first list and points in the second list. Under the assumption that the second list of coordinates is a small perturbation of the first list, a peak in the histogram of these distances should allow homologous — paired off — points to be determined. Note that this approach is insensitive to uniform translation, and is fast (compared to other approaches looked at in this section).

For a solution to the very general problem of matching point patterns based purely on a visually intuitive notion of *configuration*, it appears that a fine-tuned analysis of the triangles which can be formed from triplets of points is necessary. Properties of triangles, derived from ratios of the side lengths, can be used to provide a characterisation of point shape which is invariant to coordinate translation, rotation, rescaling, inversion (flipping), and to small random perturbation or distortion. Groth (1986) describes such an algorithm, with care being taken to reduce substantially the number of triangle comparisons to be carried out (which have worst case $O(n^6)$ complexity). Stetson (1987b) has also developed a similar algorithm: it uses magnitudes associated with objects to selectively carry out the comparison of triangles. Reported studies using the triangles–based approach have been limited to 20–30 objects in the two lists of objects.

3.2 Galaxy Morphology Classification

The Hertzsprung–Russell diagram summarizes in a two–dimensional diagram many characteristics of stars. The galaxy equivalent of this has often been pursued, but has proved more difficult to bring about. In the 1930s Hubble published his "tuning fork"

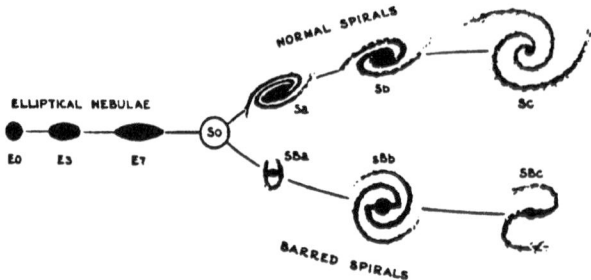

Fig. 1. The Hubble "tuning fork" classification of galaxies.

classification of galaxies (Figure 1) and more complex morphological sequences have been proposed since then (see, e.g., Sandage, 1975; Mihalas and Binney, 1981, chapter 5; for background material on galaxies, see Takase *et al.*, 1984, or Audouze and Israël, 1985, pp. 332 *et seq.*). The use of quantitative data on galaxies in order to define classifications is a burgeoning research topic, and if successful will complement the dominant current approach of visually deciding the morphological class of a galaxy.

The inherent difficulty of characterising spirals (especially when not face–on) has meant that often work focusses on whether the galaxy under study is elliptical or spiral, using the observed ellipticity as a criterion. The inherent noisiness of the images (especially for faint objects) has additionally meant that the parameters measured ought to be made as robust as is computationally feasible.

3.2.1 Galaxy Parameters

Two approaches to investigating morphological shapes of galaxies have made use of galaxy radial profiles and more general parameters.

Known theoretical properties of galaxy profiles motivate the former approach: ellipticals and spirals may be approximately modelled as "bulges" and/or "disks" by fitting, respectively, $r^{1/4}$ or exponential laws. Noisiness and faintness of data require attention to robustness in measurement. It may also be supposed that the the galaxy is face–on, optically–thin and axisymmetric. The luminosity profile along the major axis of the object, determined at discrete intervals, can be carried out by the fitting of elliptical contours, followed by the integrating of light in elliptical annuli (Lefèvre *et al.*, 1986). Alternatively, circles of given radius (Watanabe *et al.*, 1982) or isophotal contours (Thonnat, 1985; see also Thonnat and Berthod, 1984) have been used.

Specific morphology–related parameters may be derived instead of using the profile. A measure of galaxy magnitude can be provided by the integrated magnitude within some limiting surface brightness (Takase *et al.*, 1984; Lefèvre *et al.*, 1986). A measure of the extent of the galaxy can be provided by the logarithmic diameter at such a user–specified limiting surface brightness (Okamura, 1985). It may be interesting to fit to galaxies under consideration model bulges and disks using, respectively, $r^{\frac{1}{4}}$ or exponential laws (Thonnat, 1985), in order to define further parameters related to the degree of fit. Some catering for the asymmetry of spirals may be carried out by decomposing the object into octants: at successive radii, the set of eight values gives an indication of the regularity or irregularity of shape (i.e. elliptical or spiral). The taking of a Fourier transform of the intensity may also usefully indicate aspects of the spiral structure.

From the foregoing, the difficulty of mapping the objects (galaxies) into parameter space may be appreciated, especially when one is working near the noise limit of the data gathering equipment. Galaxy morphology classification is, today, predominantly accomplished by visually examining each object individually. A real challenge exists for automated techniques (image processing in general, and classification in particular) to make inroads into this labour–intensive process.

3.2.2 Multivariate, Rule–Based and Fuzzy Classification

The classification of galaxies, using luminosity profiles, has been carried out in Murtagh and Lauberts (1986). While successful, a two way classification only (into ellipticals and spirals) was attained. Using a set of parameters to characterize the galaxies studied, Okamura (1985) and Takase et al. (1984) used Principal Components Analysis to simplify and to interpret the parameters.

In order to try to more closely emulate the human (visual) classification process, rule–based expert systems have also been used. Thonnat and Berthod (1984) and Thonnat (1985) specify rules, which relate together morphological galaxy types with parameters. The latter are chosen in a similar way to other studies discussed above. This forward chaining expert system approach was a prototype (with approximately 100 rules). A similar type rule–based system was also used by Bernat and McGraw (198x) for general object classification.

One advantage of such a system over multivariate methods is that the knowledge initially built into the system (in particular, the rule–base) can be very conveniently communicated to (and subsequently modified by) the non–mathematical user. On the other hand, the validation of complex expert systems may fall behind the coherence of mathematical frameworks offered by multivariate methods.

To handle imprecise data (e.g. as arises in γ and X–ray data) fuzzy set theory has also been employed. Possibility theory has been used for determining shapes of galaxies as a preliminary to their morphological classification (see Di Gesù and Maccarone, 1987, and references therein).

3.3 Spectral Classification

Stellar spectral classification is based on comparison with standard stars. The MK system (named after Morgan and Keenan, who proposed the system in 1943) is two-dimensional: spectra have a value in *spectral sequence* and a *luminosity class*. The fomer is the sequence O, B, A, F, G, K, M (celebrated in the mnemonic *"Oh, Be A Fine Girl, Kiss Me"*) and is associated with decreasing temperature. Figure 2 shows some sample spectra from these classes. Note that the shape of the spectrum continuum varies from "downward" to "upward" in going from O to M.

The luminosity classes are V, IV, III, II, I, and are associated with dwarfs, V (which are commonest and faintest); giants, IV, III, and II; and supergiants, I (which are

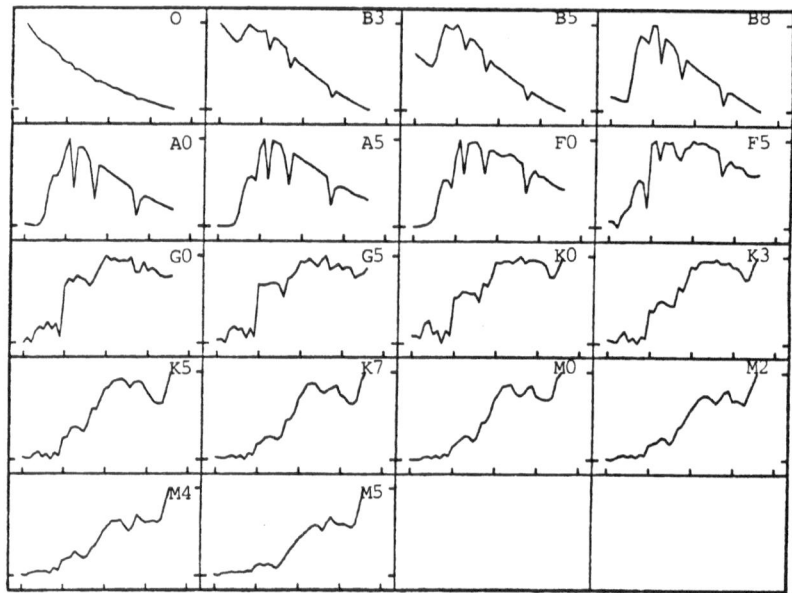

Fig. 2. Sample stellar spectra.

brightest and largest). Ignoring further refinement in spectral sequence, our Sun is a
G V star.

Stellar spectral class is two–dimensional. Noise is never absent from measurements
and the most appropriate spectral resolution is usually not at the control of the
investigator. Two dominant approaches have been used as the basis for automating
spectral classification (Kurtz, 1984): either to measure spectral parameters (e.g. the
role played by various lines, related to the physics of the object studied; or continuum
shape); or alternatively Kurtz (1982) used the spectra themselves (following suitable
correction and calibration). Most studies have addressed less than the full $O-M$
range.

Principal Components Analysis (PCA) has an established place in the tools available
to the investigator to assess the parameters used: it has been employed by Deeming
(1963), and later in such studies as Christian (1982) and Whitney (1983a, 1983b).
Kurtz (1982) used PCA, Multiple Discriminant Analysis (MDA), and a weighted
Euclidean distance between given spectra and standard spectra for classification purposes. The results were partially successful, -- the author indicates where he found
systematic misclassification, and how the resolution of the spectra relates to the performance of the methods used.

The IUE (International Ultraviolet Explorer) satellite has been a privileged source
for spectral data in the ultraviolet. The publication of an extensive set of IUE reference spectra is on–going, and such data have also been subjected to multivariate,
confirmatory analyses. Using carefully chosen lines in the spectra, with adjustment

for interstellar reddening, PCA, MDA and clustering methods were used (see Heck, 1987, which also indicates many earlier references; and Murtagh and Heck, 1987a, chapter 6). One of the principal aims was to assess the similarity with classification arrived at using data measured in the visual wavelength range. It was not expected that differing wavelength ranges should necessarily give rise to the same interrelationships between stars (since in a different wavelength range, a star could present quite different characteristics). It was assessed (on the basis of a fairly small sample) that about 94% of stars had in fact similar interrelationships in the visual and in the ultraviolet (Jaschek and Jaschek, 1984).

4. THE HUBBLE SPACE TELESCOPE AND BEYOND

The role of astronomical databases and archives will necessarily increase in the future. This requires both the infrastructure to handle this — to be looked at below; and additionally the software superstructure to suitably process this data. In the ambit of the latter, we can include multivariate data analysis and expert system analysis approaches. A comprehensive range of algorithms for the former currently exist in MIDAS, and appraisal of the latter area for astronomical research problems is proceeding rapidly.

With regard to increasing storage needs for astronomical data, an oversubscription rate of about a factor of 10, for example, is expected for observing proposals for the Space Telescope. Hence archival research will be in heavy demand. With this in view, a Data Management Facility is currently being set up. It comprises two parts — an archive which will use write–once, read–many–times (WORM) optical disks; and a catalogue which is currently being established as a commercial, relational database. The archive will store all data in the form of one- or two–dimensional images. The catalogue, implemented on an IDM–500 database machine, is to provide an index to the latter and to permit interactive querying.

As far as Hubble Space Telescope images are concerned, the lack of atmospheric conditions implies an altered PSF (including colour dependence and non–monotonic decrease in radial value). Extrapolating from the work of Bendinelli et al. (1985) and Bragaglia et al. (1987), the photometry of blended objects may therefore require extra capabilities in analysis software.

In conclusion, this article has argued that for astronomy,

• applications,

• image processing and pattern recognition, and

• databases

represent three separate but converging fields. The convergence of technologies may not always be smooth (cf. the problem of interfacing major astronomical image processing systems), but it is clear that there are always lessons to be learnt. Disciplines far removed from computational astronomy have much to gain from this experience also.

Acknowledgements

Discussions with many colleagues contributed to the material presented here. I am grateful to H.–M. Adorf for Figure 2 (see Adorf, H.–M., "Classification of low-resolution stellar spectra via template matching", in [6], pp. 61–69).

REFERENCES

Adorf, H.–M., Baade, D., and Banse, K., 1986, "Preparing analysis of Hubble Space Telescope data in Europe", in Di Gesù *et al.* (eds.), *Data Analysis in Astronomy. II*, Plenum Press, New York, 165:170.

Audouze, J. and Israël, G., 1985, *The Cambridge Atlas of Astronomy*, Cambridge University Press, Cambridge.

Banse, K., Crane, P., Ounnas, Ch. and Ponz, D., 1983, "MIDAS — ESO's interactive image processing system based on VAX/VMS". Proceedings of the Digital Equipment Computer Users Society (Zurich, Switzerland, August/September 1983), 87:91.

Bendinelli, O, Di Iorio, A., Parmeggiani, G., and Zavatti, F., 1985, "Some clues to Hubble Space Telescope image analysis", *Astronomy and Astrophysics* **153** 265:268.

Bernat, A.P. and McGraw, J.T., 1986, "An intelligent object recognizer and classification system for astronomical use", Preprint, Steward Observatory.

Bragaglia, A., Ferraro, F.R., Fusi Pecci, F., Buonanno, R., Corsi, C.E., Ferraro, I., and Iannicola, G., 1987, "ST–FOC observations of globular clusters: simulations", *Journal of Astrophysics and Astronomy* **8** 57:68.

Christian, C.A., 1982, "Identification of field stars contaminating the color–magnitude diagram of the open cluster Be 21", *The Astrophysical Journal Supplement Series* **49** 555:592.

Deeming, T.J., 1963, "Stellar spectral classification", *Monthly Notices of the Royal Astronomical Society* **127** 493:516.

Di Gesù, V., Scarsi, L., Crane, P., Friedman, J.H. and Levialdi, S., 1984, *Data Analysis in Astronomy*, Plenum Press, New York.

Di Gesù, V., Scarsi, L., Crane, P., Friedman, J.H. and Levialdi, S., 1986, *Data Analysis in Astronomy. II*, Plenum Press, New York.

Di Gesù and Maccarone, M.C., 1987, "Possibility functions and computational geometry: tools for the analysis of astronomical images", *Newsletter of Working Group for Modern Astronomical Methodology*, No. 4; and *Bulletin d'Information du Centre de Données de Strasbourg*, No. 32.

Groth, A.J., 1986, "A pattern–matching algorithm for two–dimensional coordinate lists", *The Astronomical Journal* **91** 1244:1248.

Grosbøl, P. and Ponz, D., 1985, The MIDAS table file system. In Proceedings of International Course on Data Handling in Astronomy. G. Sedmak (ed.), Trieste: Trieste Observatory.

GSSS, 1985, Preliminary documentation, Guide Star Selection System Group, Space Telescope Science Institute, Baltimore.

Heck, A., 1987, "UV stellar spectral classification", in Kondo, Y., (ed.), *Scientific Accomplishments of the IUE*, D. Reidel, Dordrecht, pp. 121:137.

Jaschek, M. and Jaschek, C., 1984, "Classification of ultraviolet spectra", in Garrison, R.F., (ed.), *The MK Process and Stellar Classification*, David Dunlap Observatory, Toronto, pp. 290:304.

Jarvis, J.F. and Tyson, J.A., 1981, "FOCAS: faint object classification and analysis system", *The Astronomical Journal* **86** 476:495.

Jenkner, H., 1983, "Astronomical and statistical algorithms used in the Space Telescope Guide Star Selection System", in Rolfe, E.J., (ed.), *Statistical Methods in Astronomy*, European Space Agency SP–201, pp. 65:68.

King, I.R., 1986, "Cluster photometry: present state of the art and future developments", in Di Gesù, V. *et al.*, (eds.), *Data Analysis in Astronomy II*, Plenum Press, New York, pp. 17:30.

Kurtz, M.J., 1982, "Automatic spectral classification", PhD thesis, Dartmouth College, New Hampshire.

Kurtz, M.J., 1983, "Classification methods: an introductory survey", in *Statistical Methods in Astronomy*, European Space Agency SP–201, pp. 47:58.

Kurtz, M.J., 1984, "Progress in automation techniques for MK classification", in Garrison, R.F., (ed.), *The MK Process and Stellar Classification*, David Dunlap Observatory, Toronto, pp. 136:152.

Lauberts, A., 1986, description of suite of galaxy reduction programs, European Southern Observatory.

Lefèvre, O., Bijaoui, A., Mathez, G., Picat, J.P. and Lelièvre, G., 1986, "Electronographic BV photometry of three distant clusters of galaxies", *Astronomy and Astrophysics* **154** 92-99.

Lutz, R.K., 1980, "An algorithm for the real time analysis of digitised images", *The Computer Journal* **23** 262:269.

MacGillivray, H.T., Martin, R., Pratt, N.M., Reddish, V.C., Seddon, H., Alexander, L.W.G., Walker, G.S. and Williams, P.R., 1976, "A method for the automatic separation of the images of galaxies and stars from measurements made with the COSMOS machines", *Monthly Notices of the Royal Astronomical Society* **176** 265:274.

MacGillivray, H.T. and Stobie, R.S., 1984, "New results with the COSMOS machine", *Vistas in Astronomy* **27** 433:475.

MIDAS, 1987, *MIDAS Manual*, European Southern Observatory, Munich. Chapter 11, "Object searching and analysing commands".

Mihalas, D. and Binney, J., 1981, *Galactic Astronomy*, W.H. Freeman and Co., San Francisco.

Murtagh, F. and Lauberts, A., 1986, "A curve matching problem in astronomy", *Pattern Recognition Letters* **4** 465:469.

Murtagh, F. and Heck, A., 1987a, *Multivariate Data Analysis*, D. Reidel, Dordrecht.

Murtagh, F. and Heck, A. (eds.), 1987b, *Astronomy from Large Databases: Scientific Objectives and Methodological Approaches*, European Southern Observatory, Garching (forthcoming).

Ochsenbein, F., 1986, "Data storage and retrieval in astronomy", in V. Di Gesù *et al.* (eds.), *Data Analysis in Astronomy. II*, pp. 305:313.

Ogawa, H., 1986, "Labelled point pattern matching by Delauney triangulation and maximal cliques", *Pattern Recognition* **19** 35:40.

Okamura, S., 1985, "Global structure of Virgo cluster galaxies", in Richter, O.G. and Binggeli, B., (eds.), *ESO Workshop on The Virgo Cluster of Galaxies*, ESO Conference and Workshop Proceedings No. 20, pp. 201:215,

Russo, G., Richmond, A. and Albrecht, R., 1986, "The European scientific data archive for the Hubble Space Telescope", in Di Gesù, V. *et al.*, (eds.), *Data Analysis in Astronomy. II*, pp. 193:200.

Sandage, A., 1975, "Classification and stellar content of galaxies obtained from direct photography", in Sandage, A., Sandage, M. and Kristian, J., (eds.), *Galaxies and the Universe*, University of Chicago Press, Chicago, pp. 1:35.

Scarsi, L., Di Gesù, V. and Crane, P., 1987, *Selected Topics on Data Analysis in Astronomy*, World Scientific Publishing Co., Singapore.

Shoshani, A. and Wong, H.K.T., 1985, Statistical and scientific database issues, *IEEE Transactions on Software Engineering*, **SE-11**, 1040:1047.

Stetson, P.B., 1984, *DAOPHOT User's Guide*, Dominion Astrophysical Observatory, British Columbia.

Stetson, P.B., 1987a, "DAOPHOT: a computer program for crowded-field stellar photometry", *Publications of the Astronomical Society of the Pacific* (forthcoming).

Stetson, P.B., 1987b, private communication, software description of point matching program.

Stobie, R.S., 1980, "Application of moments to the analysis of panoramic astronomical photographs", *Applications of Digital Image Processing to Astronomy*, SPIE Vol. 264, The International Society for Optical Engineering, 208:212.

Stobie, R.S., 1986, "The COSMOS image analyser", *Pattern Recognition Letters* **4** 317:324.

Takase, B., Kodaira, K. and Okamura, S., 1984, *An Atlas of Selected Galaxies*, University of Tokyo Press, Tokyo.

Thanisch, P., McNally, B.V. and Robin, A., 1984, "Linear time algorithm for finding a picture's connected components", *Image and Vision Computing* **2** 191:197.

Thonnat, M. and Berthod, M., 1984, "Automatic classification of galaxies into morphological types", Proc. 7th International Conference on Pattern Recognition, IEEE, 844:846.

Thonnat, M., 1985, "Automatic morphological description of galaxies and classification by an expert system", INRIA Rapport de Recherche, No. 387.

Tyson, J.A., 1984, "Galaxy counts", in Capaccioli, M., (ed.), *Astronomy with Schmidt Type Telescopes*, D. Reidel, Dordrecht, 489:498.

Tyson, J.A. and Jarvis, J.F., 1979, "Evolution of galaxies: automated faint object counts to 24th magnitude", *The Astrophysical Journal* **230** L153:L156.

Valdes, F., 1982, "Resolution classifier", *Instrumentation in Astronomy IV*, SPIE Vol. 331, The International Society for Optical Engineering, 465:471.

Watanabe, M., Kodaira, K. and Okamura, S., 1982, "Digital surface photometry of galaxies toward a quantitative classifier. I. 20 galaxies in the Virgo cluster", *Astrophysical Journal Supplement Series* **50** 1:22.

West, R.M. and Kruszewski, A., 1981, "Distant clusters of galaxies in the southern hemisphere", *Irish Astronomical Journal* **15** 25:35.

Whitney, C.A., 1983a, "Principal components analysis of spectral data. I. Methodology for spectral classification", *Astronomy and Astrophysics Supplement Series* **51** 443:461.

Whitney, C.A., 1983b, "Principal components analysis of spectral data. II. Error analysis and applications to interstellar reddening, luminosity classification of M supergiants, and the analysis of VV Cephei stars", *Astronomy and Astrophysics Supplement Series* **51** 463:478.

A PANEL ON: PATTERN RECOGNITION AND IMAGE PROCESSING WITH OR

WITHOUT INTELLIGENCE?

>   Panelists: Tonis Kasvand (Canada), Piero Mussion
>   (Italy), Armando Roy (Spain), and
>   George Stamon (France)
>   Conductor: Stefano Levialdi (Italy)

Introduction

It is difficult to summarize the different views as expressed by the panelists on a tricky subject, i.e. what help (if any) can Artificial Intelligence provide to pattern recognition and image processing as scientific disciplines.

A short introduction was given so as to define what is AI today, briefly: 1) the introduction of **knowledge** for the representation of useful information and know-how, for dealing with incomplete and uncertain information, for having open/growing information bases and for enabling the use of relevant information in a specific domain; 2) the definition of **inference** mechanisms (or engines) that deduce/induce the correctness of a working hypothesis and 3) a sophisticated level of **interaction** with powerful HELP for the user, GUIDE to enable smooth navigation in the system and WHY for explaining the partial/total conclusions of the system after a specific conclusion has been reached.

Within the realm of AI, pattern recognition is achieved in a paradigm which may be expressed as **shape-from** (shading, motion, stereo, texture, contour, projections, etc) or **shape-to** (infer processes of growth, disambiguate interpretations, etc).

A suspicion that AI is a different way of naming things and not a new way of thinking was also mentioned. For instance, we may group PR&IP techniques on the left column and AI&CV (Computer Vision) techniques on the right column, as shown below.

PR & IP

Pre-processing
Feature selection
Clustering

AI & CV

Image construction
Rule evaluation
Model matching

Classification   Interpretation
Control of an interactive loop   Explicit WHY

Another way of looking at this problem is to quote the kee starting points
of classical PR & IP approaches:

PR & IP approach

Ad-hoc models (of the objects, of the world, of exceptions)
Significant features
Space dimensionality reduction
Classfication by simple computation

AI approach

Inference of shape from boundary, shade, primitive model, etc
Iteration with reinforcement
Improved models
Use of heuristics
Use of physical environment & constraints
Consistency proofs

The questions for the panelists

At this point a number of questions (given to the panelists only a day
before) where the backbone of the panel. These questions were
subdivided into three groups: the first group touched on the TOOLS, the
second one on the RESEARCH STRATEGIES and the third one on the
choice to build systems that worked like those in nature or behaved in the
same way, i.e. SIMULATION versus EMULATION.

The questions to the panelists were the following ones subdivided into
three main topics: tools, strategies and replication of results/functions.

TOOLS

Which are the classical tools? / Which are the AI tools?
Do we have "more" tools or "better" tools?

RESEARCH STRATEGIES

Can we "integrate" both approaches?
Is the choice of the problem the reason for the "good" solution?
Can we generalize good results?
Which is the direction to go?

EMULATION OR SIMULATION

Should we emulate man (or biological systems) or simulate them?
How may we profit from Perception and Cognition studies?
Is the field at a turning point by having more computational power (parallel machines, parallel algorithms, high resolution monitors, networks, etc)?

Armando Roy suggested that the existing tools (linguistic, algorithmic and technological) developed by the AI community were not directly applicable to PR and IP so that the experts in these two areas should try to adapt existing tools to the specific problems posed by the pictorial world. This conference has included presentations on recent expert systems for image analysis (in specialized fields) and that is the way to go, profit from the AI approach and then tune the systems to particular applications.

Regarding strategies, a large experimental basis is needed since it is difficult to use the same approach and algorithms to different problems, all directions should be explored and tested on a wide variety of cases.

Finally, regarding emulation of human capabilities, he showed the figures below to point at the distance we are from the "technology of nature". Nevertheless some simple biological vision systems are less powerful than some computing systems so that we may start emulating the performance of primitive vision systems.

	N° of elements	Speed	Connectivity	Product
Man	10^{10}	10^3	10^4	10^{17}
Machine	10^6	10^7	10^1	10^{14}

It has been noted that the Product column is not very meaningful since it is the result of multiplying non-homogeneous quantities.

George Stamon has remarked that the main requirement of an efficient image analysis system is to have a good data representation with convenient data structures so that all the relevant computation may be efficiently performed. It may be that different problems, in the practical world, require different data structures but if these are well engineered the problem is half solved.

On the problem of research strategy he feels all roads must be pursued and common sense more than AI techniques should be used so as to have a really good feeling for how the problem must be solved.

Finally a caution regarding simulation of living beings: we are still far from success in any field and must go a long way before we may generalize results, i.e. have intelligent programs.

Tonis Kasvand remarks that since the vision problem in itself is badly stated and ill defined it is impossible to discuss tools for a task being so vague; yet some systems seem to be working well on a limited set of problems so that many more practical tasks should be faced to learn from this experience. The strategy of research is connected to the clever choice of problems in a scale of growing complexity so that some conclusions may be drawn. AI systems have only reached a level of amateurs and he is pessimistic regarding their upgrading to a professional level in a short future. In this respect it is difficult to immagine an artificial system simulating (internally and externally in terms of behaviour) a natural counterpart.

Piero Mussio, stating he represents the user's point of view, recalled the definition of tools given by Prof. Serra, i.e. no universal tools exist for PR & IP, each problem is solved with special purpose tools and, for best results, a combination of methods is employed like the statistical, morphological and syntactic approaches. The AI tools, on the other hand, offer the possibility of managing knowledge bases, data driven and event driven strategies but have misleading words like "heuristic", "engine" and "abduction" which sometimes may confuse issues and purpose.

Regarding the research strategy, more should be done to understand how an expert understands an analog image and to evaluate AI tools using natural intelligence so as to decide on the relative merits.

Finally, in connection with biological systems, they are still widely unknown so that models must be formulated by both biologists and computer scientists in close cooperation.

We had a very lively discussion, with personal confessions from the panelists and good questions from the audience: I do hope not to have misunderstood the ideas nor distorted the views of the speakers. The problem was a very debatable one so that no conclusion could be reached except that being presently far from a general solution to the problem of recognition and understanding of images (independetly form the pattern), a good mixture of the available techniques should be used for each particular problem, enabling a working system to be built: one that does the job in a reasonable, economical way.

KNOWLEDGE BASED APPROACHES

A KNOWLEDGE BASED APPROACH TO INDUSTRIAL SCENES ANALYSIS:

SHADOWS AND REFLEXES DETECTION

M. Adimari +, S. Masciangelo +,
L. Borghesi +, and G. Vernazza *

+ ELSAG S.p.A, SRC Research Department
Via Puccini 2, 16154 Genova - Italy
* Biophysical and Electronic Engineering Department
University of Genova, Via All'Opera Pia, 16145 Italy

ABSTRACT

This paper describes a framework for detecting regions of special interest, like reflexes and shadows in industrial scenes. A new approach involving Artificial Intelligence techniques is tested; in fact procedural approaches exhibit several drawbacks such as lack of flexibility and of transparency in knowledge representation because knowledge is buried in code. A Knowledge-based system is proposed that is strictly interfaced with several low level image processing tools.
A brief description of the image processing modules is presented together with details concerning knowledge representation and the control strategy adopted. For a better understanding of the system performance and man-machine interface, a detailed analysis session is reported.

1. INTRODUCTION

A great effort has been devoted to image analysis and interpretation, involving Pattern Recognition methods and more recently Artificial Intelligence techniques; many and different algorithms have been developed for image processing, so that a large amount of tools is now available.
These algorithms can be classified into three levels (low, intermediate, high) as shown in Fig.1, each corresponding to a specific processing stage, from image filtering up to the description of an image in terms of known primitives (interpretation).
However this classification is only valid for algorithms, and cannot be adopted for the knowledge which justifies the employment of particular techniques.
In fact knowledge, in its most general sense, is used in every processing stage and can be classified into:
- procedural knowledge;
- visual knowledge;
- task world knowledge.

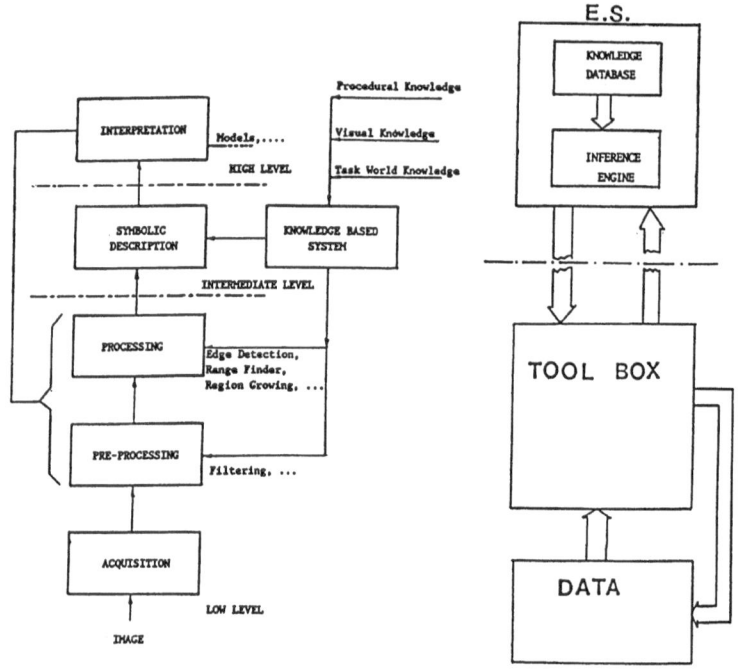

Figure 1.   Figure 2.

The first refers to the determination of algorithms and operators, to the selection and setting of related parameters, combination of different algorithms; the second concerns physical
phenomena generating an image, like different orientation in
object surfaces and illumination conditions; the third is "high
level" knowledge about the image domain considered.
    A procedural approach to these problems exhibits several
drawbacks, the most important of which are:

    - lack of flexibility;
    - lack of clear and modular representation of knowledge.

Both drawbacks are due to the fact that knowledge is embedded in
code; therefore application of procedural systems to
environments different from the one they were designed for,
requires these systems to be redesigned.
A very important aspect to be pointed out is that several low
level algorithms are based on "domain independent knowledge":
for example, filtering operators, such as smoothers or edge
detectors, can be designed on the basis of well defined
mathematics and segmentation algorithms often use rules like:
"merge neighbouring regions with similar properties", and so on.
So we can define these algorithms as "general" although their
applications are not general since parameter setting depends on
a particular environment and on the final goal of the image
processing.
For these reasons, we propose a system in which the whole a
priori knowledge is well separate from the basic algorithms,

102

and it is embedded in the expert system's knowledge database so that it can be clearly and homogeneously represented by rules. The expert system acts as supervisor of a "tool box", that is a set of low level algorithms (Fig.2).
Main tasks of such an expert system are:

- planning the use of each tool;
- selection of parameters in accordance with the a priori knowledge about the image and sequence of analysis steps;
- modifying parameters (and algorithms) according to previous and intermediate results;
- synthesis of final results.

The variety of attainable goals leads to a large number of proposed systems. For istance the followings have been recently proposed in literature as expert systems for image processing :

- consultation system for image processing [1];
- automatic generation of image processing programs [2];
- rule-based segmentation system [3];
- system for automatic location of address block on mail pieces [4].

The first application [1] concerns man-machine interface improvement in interactive image processing systems; the operator of such a system is usually requested to select a command from a large number of possibilities and to specify appropriate arguments. The consultation system uses manuals as knowledge sources to help the user to select commands and parameters.
In the second situation [2] the system combines several existing subroutines or program modules to build up consistent image processing packages: the user has only to write an abstract program specification without any knowledge on implementation details.
Nazif-Levine's system [3] uses production-rules to implement suitable heuristics to split/merge regions and lines into meaningful ones, using the concepts of Gestalt psychology [5].
Finally in [4] an expert system is used as a tool manager which supervises a set of specialized tools. The system manager performs benefit/cost estimation and selects parameters for each tool and evaluates results and performance.
The present work adopts an approach similar to Wang-Srihari's system [4], though in a completely different environment.
Our objective is the detection of regions of interest in industrial scenes with 3-D mechanical parts. The target regions include the background, shadows, reflexes and so on.
At first, some low level processing modules were developed, each to be used for a specific task, like histogram analysis, edge detection and image segmentation. The design was oriented toward a flexible structure based on a large number of parameters which can tune the appropriate tool according to the given application. Subsequently a statistical model of the regions of interest was developed and some human expertise on the selected environment was acquired. Finally such expertise was translated into a knowledge database format (knowledge engineering phase) in order to create an expert system which could accomplish certain goals.

2. SYSTEM ARCHITECTURE

Fig.3 shows the general architecture of the system.

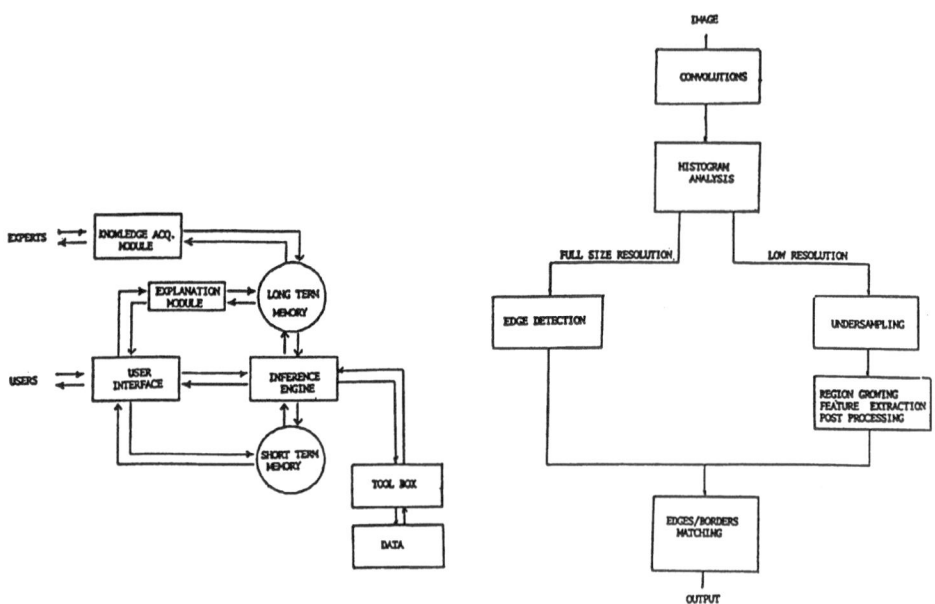

Figure 3.   Figure 4.

The knowledge acquisition module is used off line during the development of the knowledge database.
The core of the system is the inference engine which interfaces the long term memory containing the knowledge about the domain, the short term memory containing facts deduced during the analysis session, the user through a carefully designed man-machine interface module and the low level image processing modules with the data structures [6].
For a better understanding of the system as a whole, it is useful to introduce a brief description of the low level archi-tecture. Fig.4 summarizes the principal processing modules and shows how they interact each other from a functional point of view.
In the preprocessing stage preliminary filterings are performed in order to obtain data matrices, like intensity gradient and Laplacian, useful in the following processing stages.
In image processing, histograms are important sources of infor-mation. The histogram analysis module is designed to extract information about the image from the intensity and gradient histograms. The goal of the analysis is to link the primary modes of histograms to pictorial entities (like background, shadows, objects, etc.). The algorithm extracts the degree of multimodality of the histogram and, for each meaningful mode, estimates a number of parameters like mean, variance, peak value, and weighting factor. The basic idea, suggested by Bhat-tacharyya's distribution function analysis method [6], is to characterize a histogram as a superimposition of different Gaus-sian components and then to detect each of such components by

the evaluation of related mean, variance and weighting factor (m, σ, w).

At this point the diagram of fig. 4 is divided into two branches: edge detection on a full-size scale and segmentation at low resolution.

This approach depends on the available resources: edge detection, usually involving great computation time, is based on convolutions performed by a fast 2-D hardware convolver. On the other hand the segmentation module, which is used many times during an analysis session, must allow efficient computation: this is obtained by averaging and subsampling the image to perform segmentation on a reduced scale.

The edge detection module is based on Marr and Hildreth's theory [7]: we introduced some implementative improvements in order to enhance results in this specific context.

The original image is convolved with Laplacian of Gaussian function and zero crossing points in the resulting image define boundaries. The main drawback of this approach is the presence of a large amount of noise contours. To select meaningful contours from those due to noise we imposed a hysteresis threshold mechanism on gradient [8]; the meaningful edges are picked out whether they have at least one zero crossing with gradient value above a prefixed threshold T1; besides the gradient value of each point in the same edge must be greater than a second threshold T2, where T1 > T2.

Segmentation is performed using a region growing algorithm: the algorithm is highly flexible since it allows one to choose the aggregation law and the data on which to work.

A feature extraction algorithm supplies statistical (mean, standard deviation, gradient value mean and so on) and geometric parameters (area, perimeter, elongatedness) for each detected region.

A post processing module, based on a priori knowledge, improves the results by merging adjacent and homogeneous regions or deleting the noise ones.

The final module combines evidence from different sources like the edge map and the partitioned image in order to extract the final segmented image; the main problem consists in solving the ambiguity between edges and borders of segmented regions (obtained at low resolution) by matching them appropriately. Actually there is no ambiguity because edges are considered (and really they are) much more accurate from a localization point of view. This problem is encountered whenever one has to cope with a pyramidal data structure.

The architecture of this section of the system is highly modular so that it is quite easy to insert new tools, for example, a texture analysis module. Furthermore it is evident that many of the modules are completely independent and they could be implemented in parallel, on a multiprocessor computer. At present there are only two resolution levels: a higher degree of multiresolution could improve the overall performance of the system.

3. CONTROL STRATEGY

This section illustrates how human expertise on industrial scene environment was acquired and how the Expert System's knowledge database was built up.

At first a statistical model of industrial scenes was developed in order to select which feature, or subset of features, could be useful in recognizing each region of interest in the image.

REGION	GRAY LEVEL VALUES	GRADIENT VALUES	VARIANCE
Shadow	VL-L	VL-L	VL
Shade	L-M	M-H	M-H
Background	M-H	VL-L	VL
Internal Reflexes	VH	M	M-H
External Reflexes	VH	L	L-M
Object	L-H	L-H	L-H

Figure 5.

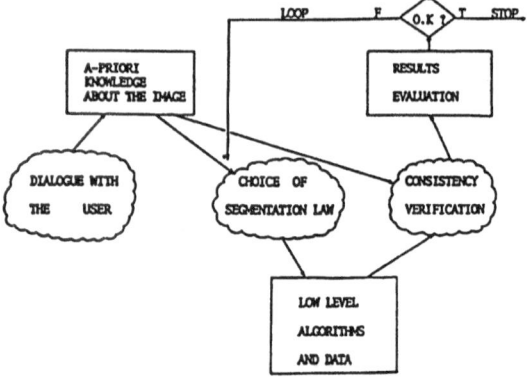

Figure 6.

Figure 5 gives an example of the format in which the model can be synthesized: the regions of interest are described in terms of primary statistical features, using a nominal scale. It is also interesting to examine the range of values for each region: for example the shadow is characterized by the lowest values in the gray level histogram with some overlap only over object's range so that the intensity feature can be considered discriminant for that region. Geometric features like area, perimeter, elongatedness are considered too.

The definition of a control strategy for the system was the major problem. The most appropriate approach is to repeat trial-and-error experiments by modifying parameters and algorithms until the objective is achieved. The difficulty of this approach lies in the lack of robust criteria for segmentation evaluation; a number of heuristic rules strictly dependent on a priori knowledge about the scene were used to validate the results.

Figure 6 illustrates how the expert system works.

At the beginning of the session the expert system starts a dialogue with the user in order to obtain some a priori knowledge about the image (e. g. illumination conditions, geometric properties of the object, reflectivity of object surface). Then the expert system selects the best segmentation law according to this a priori knowledge and to the target region the user wants to detect.

The low level processing system performs the segmentation and extracts all the necessary features from the partitioned image; finally the expert system checks on the consistency of the results with the a priori knowledge using some extracted features such as number of detected regions, their areas and so on. If the results are validated the session ends, otherwise a new segmentation is performed after modifying the parameters according to the kind of errors that occurred in the previous run.

The user can choose from among several objectives: detection of a single region of interest (background, shadow, etc.), detection of a subset of regions, complete analysis of the scene; in this case a consistency check of the scene as a whole is performed by the system.

4. KNOWLEDGE REPRESENTATION

It is well known that the core of an Expert System is the inference engine.
In this application we used CIP, an Expert System Shell developed by ELSAG S.p.A, Genoa [9].
It works in two different ways:

- forward chaining;
- backward chaining.

The forward chaining strategy recalls typical production rules systems. This strategy is suited to expressing procedural knowledge, and is mainly adopted for this application. Knowledge is represented by facts, rules and metarules.
Facts are the basic elements of representation: they allow one to state whether a relationship between two domain entities is true or false; the format for facts is a triple of this kind:

< entity > < relationship > < entity >.

Rules and metarules are IF-THEN statements: rules express declarative knowledge about a given domain and they are used for the backward chaining strategy only.
Metarules, used in the forward chaining strategy, can assume the following formats:

-IF <condition> THEN <facts>;
-IF <condition> PERFORM <procedure identifier>;
-IF <condition> ASK <fact/entity>;
-IF <condition> VERIFY <facts>.

The left hand side (LHS) of a metarule is a list of facts linked by the AND logical operator: the metarule is active only if all the facts on its LHS match the facts in short term memory. If more than one metarule are active at the same time, the conflict is resolved by following the order in which the metarules are stored in the database: the first active metarule is performed, then it is marked and cannot be performed anymore. Of course, each metarule can be executed only once so that no backtracking is allowed at the present.
The right hand side (RHS) of a metarules expresses the action performed by the inference engine: it can infer a new fact in the short term memory, activate a low level processing module, gather new evidence by asking the user for additional information, switch the strategy from forward to backward.
In the present work a relevant problem has been revealed in interfacing the inference engine with image processing modules.
A software shell, which has to call the procedures and to manage their parameters, was implemented so that only a small amount of numerical values were treated directly by the expert system in order to improve the overall performance and to increase the knowledge database trasparency.
The possibility of asking explanations for the inferred facts or for the questions the user has to answer represents one of the most remarkable features of the system.

5. EXPERIMENTAL RESULTS

The system has been implemented on a DEC VAX 11/780 computer in VAX/VMS Pascal.
The hardware architecture is completed by a convolver [10] and a VDS 701 graphic system both linked to the host computer via DMA.
This section describes a typical analysis session, from the image input to the final results.
The original image is shown in Fig. 7: it represents a hollow cylinder illuminated by a directional light source.
This 3D scene is quite representative of industrial environment for robotic purpose (workpiece recognition).
Zero crossing points are shown in fig. 8, while figure 9 illustrates the edge map obtained after application of double threshold mechanism.
In fig. 10 the result of the search for background can be seen: two regions are aggregated and this fact meets the expectations of the expert system because the object is hollow.
In fig. 11 there is an intermediate result of the search for the shadow; in this case at least two regions grow completely inside the object; the validation heuristics of the system perceive that and select a new aggregation law in order to obtain more correct results [Fig. 12].
Detection of shade and reflexes follow; in search for shade [fig. 13] two regions overlap the object almost completely; this fact is pointed out by the global validation rules which remove the wrong items from the final result. As a side effect also the remaining small shade regions disappear. Then after global consistency checking, the final segmentation result is shown [fig. 14].
It should be pointed out that meaningful contours (object versus shadow) are really refined; on the contrary, wherever contours are difficult to define accurately even by a human observer, the effects of undersampling are clearly evident (background versus shade). The image considered is particularly critical because of low contrast between object and shadow; it is displayed to illustrate the reasoning capabilities of the system.
The results of about 15 experiments show that the system is "conservative", that is, it rarely takes wrong decisions even if sometimes this leads to unclassify some regions (shade in the present example).

Figure 7.   Figure 8.

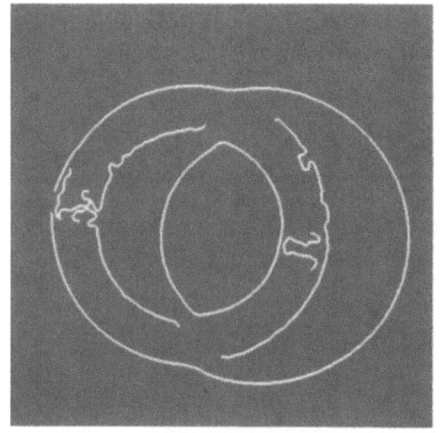

Figure 9.

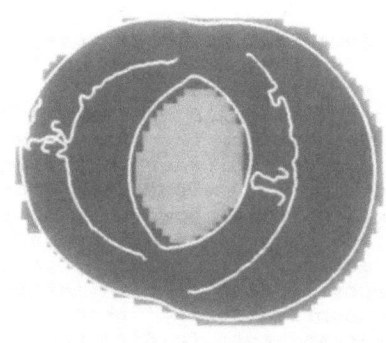

Figure 10.

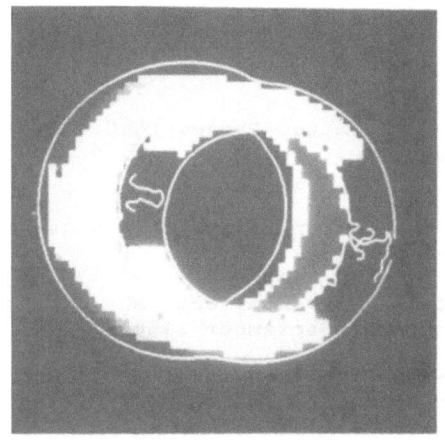

Figure 11.

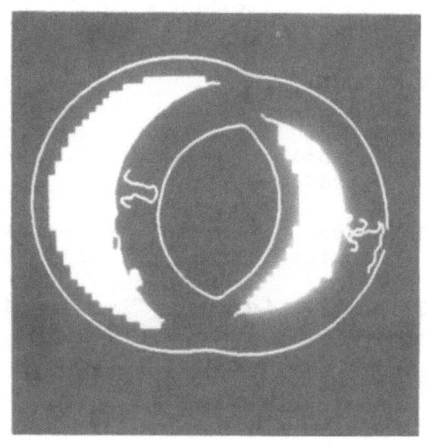

Figure 12.

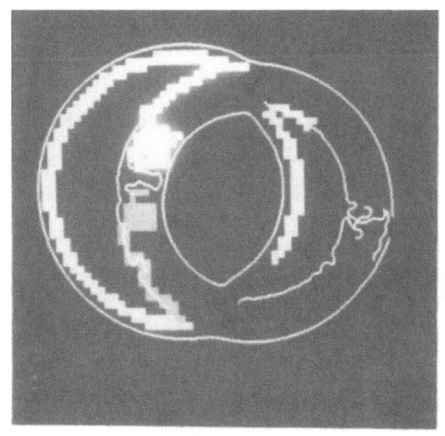

Figure 13.

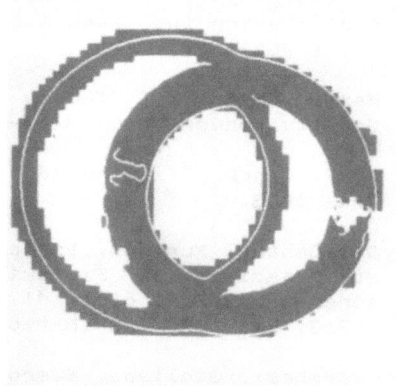

Figure 14.

6. CONCLUSIONS

The foremost aim of the paper has been to propose a new architectural approach and to describe an "open system"; the shadows and reflexes detection is an example of application. At the present the system is to be regarded as a research and development tool.

The system could be improved; we have seen that no backtracking is allowed by the expert system shell: such a limitation could be removed in order to implement a more powerful trial-and-error mechanism. Besides this facility would allow the development of complex (and powerful) consistency rules on the whole image. On the other hand the number of available tools can be increased by adding, for istance, texture analysis modules (based both on Markov or Fractal mathematics).

New applications, based on the same system architecture, can be foreseen, even in different fields (e. g., processing of biomedical and aerial images).

The system can be viewed not only as a "stand alone tool" oriented toward region detection, but also as the first stage of a larger and more complete scene analyzer for which it could work as a "hypothesis generator".

REFERENCES

[1] Sueda, "An Expert System for Image Processing", Image Technology and Information Display, vol. 17, no. 9, 19-22 (1985).

[2] Sakane, Tamura, "Automatic Generation of Image Processing Program", Proc. of CVPR, 189-192 (1985).

[3] Nazif, Levine, "Low Level Image Segmentation: an Expert System", IEEE Transactions on PAMI (September 1984).

[4] Wang, Srihari, "Object Recognition in Structured and Random Environments: Locating Address Block on Mail Pieces", Vision and Signal Understanding (1986).

[5] Kofka, "Principles of Gestalt Psychology", Hartcourt, New York, Brace & World (1935).

[6] Bhattacharyya, "A Simple Method of Resolution of a Distribution into Gaussian Components", Biometrics 23, 115-135 (1967).

[7] Marr, Hildreth, "Theory of Edge Detection", Proc. Royal Society London, B207, 187-217 (1980).

[8] J. F. Canny, "Finding Edges and Lines in Images", MIT Report June 1983.

[9] Del Canto, Fusconi, Piano, Roncarolo, "Uno strumento per la realizzazione di Sistemi Esperti e sua applicazione alla generazione dei cicli di lavorazione ", Relazione Interna ELSAG, Rel. SRC/125 (Marzo 1986).

[10] Borghesi, Giuliano, Musso, Cabiati, Ottonello, "Programmable Modified Systolic Array for Fast One-and-Two Dimensional Convolutions", Journal Opt. Soc. Am. Vol. 3, N.9, Sept.1986.

AN APPROACH TO RANDOM IMAGES ANALYSIS

V. Di Gesu'*,+, and M.C. Maccarone+

* Dipartimento di Matematica
  Univ. di Palermo, Italy
+ Istituto di Fisica Cosmica ed
  Informatica del CNR, Italy

ABSTRACT

The study of **random images** requires new approaches, as a matter of fact that their meaning is strongly dependent from the context of the research field and the classical techniques for shape analysis are not sufficient. Aim of the paper is the definition of an "open" **knowledge based system** for the analysis of problems dealing with such kind of image data.

Keywords: random images, expert system, fuzzy sets.

1.INTRODUCTION

Images whose "on" pixels are spatially spread and embedded in a noisy background are named **random images** (RI). Such kind of data derives from many research fields (biomedicine [1], X and **gamma** ray astronomy [2]) or from detectors with poor resolution [3].

The phases of a complete RI analysis follow the standard scheme (preprocessing and segmentation, shape classification, structural analysis, interpretation), also if the meaning of each step and the techniques applied may be different, depending from the application. In effect:

- **density** of points is often used instead of **intensity** of pixels;
- **probability** is combined with **possibility**;
- **proximity** measures are introduced in order to describe the topological features of the shape.

Perceptual properties may be recovered by considering RIs as a **random graph**. C.T.Zahn has shown the gestaltical properties of the Minimum Spanning Tree [4]; K-nearest neighbours are useful to detect "compactness" and to

estimate density [5]. Computational geometry [6] is useful
to define contours, concavities and connectivities [7] in
the phase of the structural analysis.

The steps of the analysis are influenced by the
human-knowledge about the underlying physical model, so that
some of them may be revisited during the process.

Knowledge Based Systems (KBS) have been widely used in many
research fields [8,9,10]. This approach seems to be
applicable, also, to the analysis of **RIs**. However in this
case the knowledge representation complexity is increased by
the existence of several techniques that may be applied
alternatively for the analysis of the same data. On the
other hand the **human knowledge** is not easily formalizable,
because mathematical and probabilistic reasoning must be
joined to the "heuristic" one. **Fuzzy logic** [11,12] supplies
a good basis for the representation of such knowledge and
must be added to the control structure of the **KBS**.

In the paper is presented the design of an open **KBS** for the
analysis of **random images RIMA**. In Section 2 an example of
the analysis scheme for such kind of data is shown. In
Section 3 is described the structure of **RIMA**. Section 4 is
dedicated to the present state of the implementation.
Section 5 is dedicated to final remarks.

2. PROBLEMS EXEMPLIFICATION

The images coming from X and gamma ray astronomy, as well as
the degraded characters, can be considered random images
(see Fig.1). We will use them to test the performance of the
RIMA system with regards to two different problems: galaxies
classification and character recognition. The analysis
follows the same phases:

* preprocessing
* features extraction and segmentation
* classification
* structural analysis
* interpretation

also if the methods applied may be different for each case
as well as the relations between different phases. For
example the classification of galaxies may be performed by
using probabilistic and fuzzy methods together combined [1],
while in the case of character recognition the
classification may be realized via syntactic analysis [13]
or graph matching, strongly related to the segmentation
procedure. In the comparison of galaxies with the physical
model the interpretation is very influenced by the human
knowledge, well represented via a fuzzy approach, while in
the case of characters this step corresponds to the semantic
analysis for which the contextual information is relevant.

The phases above described are not independent in the sense
that the result of each of them may guide the analysis
strategy. So in the **galaxies classification** the

interpretation may bring to revisit the **preprocessing** or the
segmentation phase; in the case of **character recognition**,
the **shape analysis** may reveal inconsistencies in the
segmentation or in the **classification**.

3. OVERALL STRUCTURE OF THE OPEN KBS

The main features that are required for a **KBS** dedicated to
the analysis of **RIs** may be summarized as follows:

Adaptivity. The **KBS** must be able to choose the strategy in
accord to the input data and to dinamically
change the verity values of the decision rules.

Generality. The .KBS is devoted to the solution of a wide
class of homogeneous problems.

Supervised. The **KBS** must provide for explicative session
(log file) which allows the user to change the
strategy path depending from the information
contained in it.

Generality feature is achieved if a set of inference
engines, $\mathbb{E}$, and a set of knowledge basis, $\mathbb{K}$, are
available. **Adaptivity** of the **KBS** depends from the set of
relations $\mathbb{R}$ that are settled between the inference engines.

Formally the **KBS** may be defined as follow:

$$KBS = \langle \mathbb{E}, \mathbb{K}, \mathbb{R}, L, D \rangle$$

where:

$$\mathbb{E} = \{E_1, E_2, \ldots, E_n\}$$

$$\mathbb{K} = \{K_1, K_2, \ldots, K_n\}$$
$$K_i \neq K_j, \; i \neq j$$

$$D: \mathbb{R} \rightarrow \mathbb{E} \times \mathbb{E}$$

$$L: \mathbb{R} \rightarrow \{0, 1, -1\}$$

From this point of view the **KBS** may be seen as a labelled
directed graph. The nodes are the inference engines, the
arcs are the set of relations and they define a sort of
hierarchy among the elements of $\mathbb{E}$. The orientation of an
arc from E_i to E_j means that E_j is activated starting from a
set of hypothesis inducted by E_i. The labels are related to
the navigation of the knowledges in the **KBS**:

$L(E_i, E_j) = 1$ means that E_i does not need the knowledge base
of E_j to perform inferences.

$L(E_i, E_j) = 0$ means that E_i and E_j could share some
knowledges.

$L(E_i, E_j) = -1$ means that some knowledges of E_j may be added
to the knowledge base of E_i.

In addition the following rules:

$$K_i \cap K_j = \emptyset \text{ if } L(E_i,E_j) = 1$$

$$K_i \cap K_j \neq \emptyset \text{ \& } K_i \neq K_j \text{ if } L(E_i,E_j) = 0$$

allow to establish opportune correspondances between the sets $\mathbb{E}$ and $\mathbb{K}$, in order to preserve the **KBS** consistency. Figs.2a and 2b show an example of **KBS** graph and the **intersection graph** built by using the rules above stated.

The proposed structure takes in account the human reasoning model that also modifies and adapts its strategy in relation to the dynamic change of the knowledge base and to avoid redundance in the inference process.

The decision rules in the heuristic search are built in the framework of the fuzzy logic [14]: **AND-OR** trees are built by means of the fuzzy connectivities **min, max, minmax**. The resolution principle is also stated in the framework of the fuzzy logic:

$$\frac{\begin{array}{ll} {\sim}P{\vee}Q & \max((1-f(P)),f(Q)) \\ {\sim}Q{\vee}R \Rightarrow & \max((1-f(Q)),f(R)) \end{array}}{{\sim}P{\vee}R \qquad \max((1-f(P)),f(R))}$$

Here P, Q, R **are predicates and f** is an appropriate **membership-function**, form of which depends from the specific problem or knowledge that must be represented.

4. RIMA DESIGN

As we showed in section 2 the problems that are present in the analysis of **RIs** belong to the data analysis domain, for which are necessary different types of knowledge sources (facts, rules, procedures, prototypes,...) to be activated whenever they contribute to the solution. These different knowledges contribute jointly to solve the problem using different inference mechanisms as appropriate [15].

At the present state of the design the **OPS5** system [16] seems to be the more appropriate for the implementation of **RIMA**. In fact it is a production system with essentially data-driven control strategy (forward chaining), constituted by:

- a working memory, containing the representation of the state of the problem;
- a set of production rules "condiction->action", which operate on the data in the working memory;
- an inference engine which defines a control strategy and chooses the rules to be applied.

Moreover **OPS5** may be used in backward way, by undoing the effects derived from the application of some rules, modifying the state of the system and restarting from the previous level.

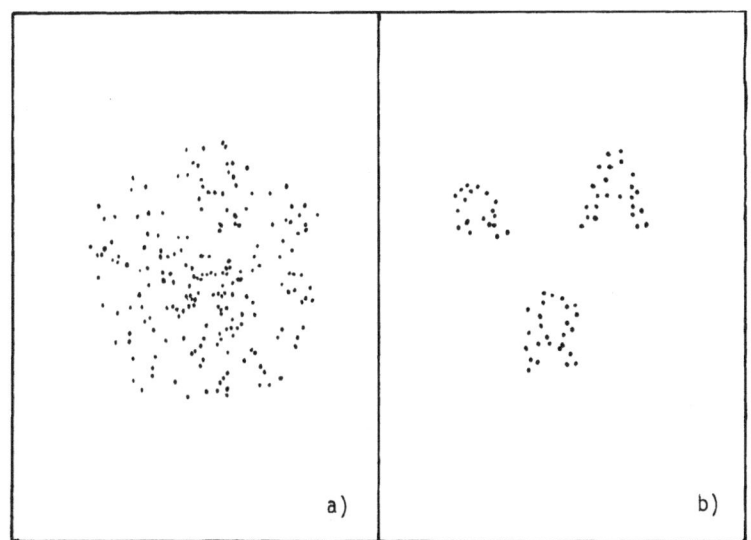

Fig. 1. a)- X-ray object THICO as detected by the EXOSAT
satellite; b)- example of degraded characters.

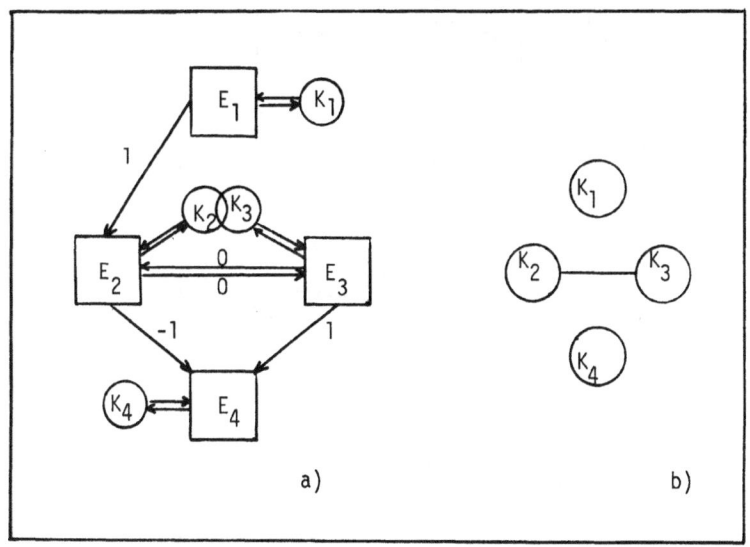

Fig. 2. a)- Example of KBS-graph; b)- Intersection graph
of the K-set.

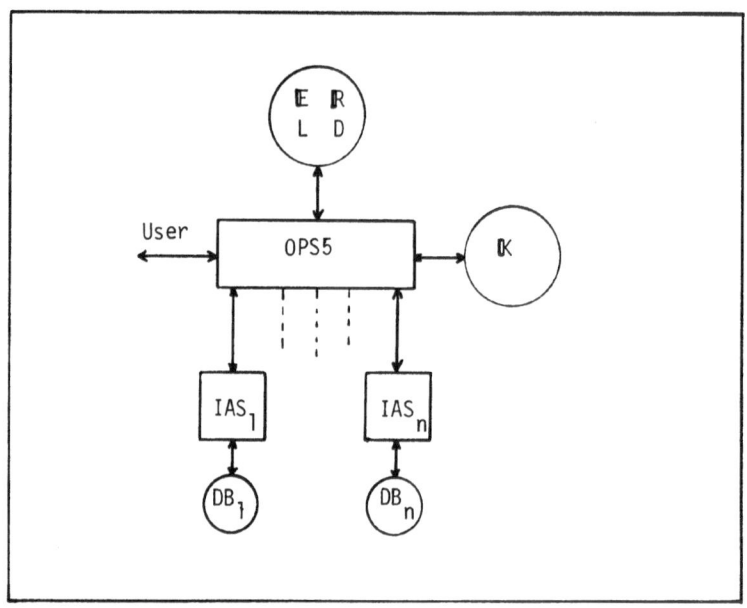

Fig. 3. General architecture of the RIMA system.

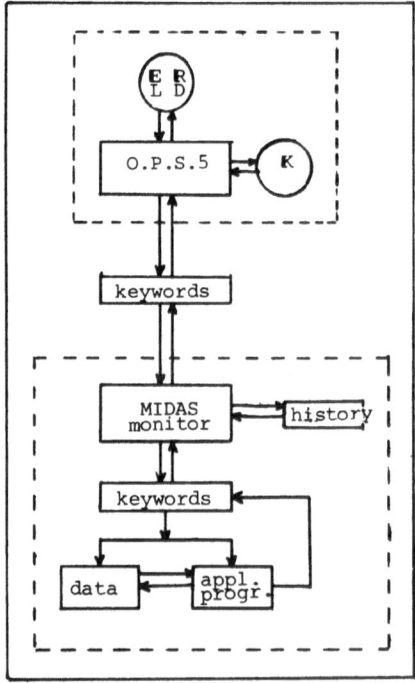

Fig. 4. Interaction between OPS5
and MIDAS systems.

The forward chaining allows to dialog with external software modules, that may constitute the environment monitoring of the physical system. This feature is useful to include in RIMA methods, data bases (catalogs, bibliographies, history,...) and models developed in existing Image Analysis systems (IAS) (see Fig.3).

Several IASs have been developed in the last years [17]. For the first release of RIMA the MIDAS Image Processing System [18], developed by the Image Processing Group of ESO-Munich, has been chosen. MIDAS consists of a control program (monitor) and an unrestricted set of application programs built around a common data structure. The application programs interact via key-words which allow easy communication of data, parameters and global conditions. The monitor realizes also the interface between the applications and the host operating system. MIDAS satisfies the following requirements:

* handling of relational data base;
* management of multidimensional data-sets;
* easy navigation of data among the applications;
* opening to new user applications (modularity);
* device independence for graphics.

At the present time it works on DECVMS operating system and its implementation under UNIX is in the process. MIDAS constitutes then the data base (methods, numerical and nonnumerical information) on which the knowledge basis are built and the OPS5 shell communicates with MIDAS-monitor, by exchanging key-words, see Fig.4.

5. CONCLUSIONS

The paper summarizes the design ideas to realize an open system for the analysis of RIs. Two main features are required: generality in order to handle classes of homogeneous problems, and distributed intelligence in order to select an appropriate inference engine for a specific goal. Knowledge is also represented by frames to consider prototypes; fuzzy logic is useful to represent no-formalizable human knowledge.

REFERENCES

[1] V.Di Gesu', M.C.Maccarone, "Features Selection and Possibility Theory", Patt.Rec., Vol.19, No.1, pp.63-72, 1986.

[2] G.A.De 'Biase, V.Di Gesu', B.Sacco, "Detection of Diffuse Clusters in Noise Background", Patt. Rec.Letters, Vol. 4, pp.39-44, 1986.

[3] V.Di Gesu', S.Mazzola, "Random Sampling of an Image and its Restoration", Proc. II Int.Conf.APRDT, Calcutta, Jan.1986, in press.

[4] C.T.Zahn, "Graph Theoretical Methods for Detecting and Describing Gestalt Clusters", IEEE Trans.on Compt., Vol.C-20, No.1, pp.68-86, 1971.

[5] G.Zeng, R.C.Dubes, "A Test for Spatial Randomness based on K-NN Distances", Patt.Rec.Lett., Vol.3, No.2, pp.85-91, 1985.

[6] G.T.Toussaint, "Computational Geometric Problems in Pattern Recognition", in Pattern Recognition Theory and Applications, J.Kittler, K.S.Fu, L.F.Pau (eds.), pp.73-91, 1982.

[7] V.Di Gesu', M.C.Maccarone,"Description of Fuzzy Images by Convex Hull Techniques", Proc. 8th Int.Conf.on Patt.Rec.IAPR, Vol.2, pp.1276-1278, 1986.

[8] B.G.Buchanan, "Expert Systems: Working Systems and the Research Literature", Knowledge Systems Lab., Stanford, Rep.No.KSL-85-37, 1985.

[9] J.M.Chassery, C.Garbay, "Expert Systems, Image Processing and Image Interpretation", Proc.8-th, Int.Conf. on Patt.Rec. IAPR, Vol.1, pp.175-177, 1986.

[10] M.Thonnat, "Automatic Morphological Description of Galaxies and Classification by an Expert System", Rapport de Recherche INRIA, No.387, March 1985.

[11] L.A.Zadeh, K.S.Fu, K.Tanata, M.Shimura, "Fuzzy Sets and their Application to Cognitive and Decision Processes", Academic Press, 1975.

[12] C.V.Negoita, "Expert Systems and Fuzzy Systems", Benjamin Cummings, 1985.

[13] M.G.Thomason, "Syntactic Methods in Pattern Recognition", in Pattern Recognition Theory and Applications, J.Kittler, K.S.Fu, L.F.Pau (eds.), pp.119-137, 1982.

[14] A.Kandel, "Fuzzy Mathematical Techniques with Applications", Addison Wesley, 1986.

[15] "Knowledge Systems Laboratory 1985", Dept.Comp.Science, Dept.Medicine, Stanford University, 1985.

[16] L.Brownston, R.Farrell. E.Kant, N.Martin, "Programming Expert Systems in OPS5", Addison Wesley, 1985.

[17] D.Wells, "Data Analysis Systems", in Selected Topics on Data Analysis in Astronomy, V.Di Gesu', L.Scarsi, P.Crane, (eds.), pp.123-151, 1987.

[18] "MIDAS User's Guide", Image Processing Group, ESO-Munich, F.R.G., 1986.

DESCRIPTION-BASED IMAGE INTERPRETATION AS A TOOL

FOR HEURISTIC INFERENCE: A GEOLOGICAL APPLICATION

Anna Della Ventura*, Piero Mussio**
and Raimondo Schettini*

(*)   Istituto di Fisica Cosmica e Tecnologie
      Relative, CNR, via Ampere 56, Milano

(**) Dip. di Fisica, Sez. di Fisica Medica
     Universita' degli Studi di Milano
     Via Viotti 5, Milano

INTRODUCTION

The interpretation of gravimetric surveys combines with other methodologies such as the interpretation of seismic and magnetic surveys, on-the-spot exploration, etc., to create a geological model of the region in question.   This will provide a great deal of information on below-ground tectonics.
The development of specially quick, reliable techniques has resulted in the availability of a great mass of surveys result, whose interpretation makes a considerable call on resources.
There are two possible approaches to the interpretation of a gravimetric survey: the first tends to associate, through analytical methods, the gravity measured with the causes that generated it; the second, which is a qualitative approach, involves recognition by an expert of characteristics shapes in bi-dimensional representation of the signal.

The analytical approach is not, at the present time, much used in practical applications, since the operator associating the field source (geological structure) with the gravity measured is not invertible.[13] The problem is therefore not "properly posed", and attempts at automatic inversion based on digital analysis methods, such as for example the method of the least squares, required the introduction of constraints to limit the space of solutions.[12]

The qualitative approach requires that the digital image be further processed by means of the so-called "contouring" procedure, which enables the expert to regulate the identification of characteristic configurations taken on by the signal processed that can be correlated with the presence of possible geological structures on the scene.

In the process of interpretation, the expert has recourse to knowledge regarding the image, the experiment, the discipline, and any other data sources other than his own

experience. Such knowledge, since it cannot be directly
deduced from the image being interpreted, is called "external
knowledge".

This paper describes the developement and verification of
an automatic instrument, known as Geological Automatic
Assistant (GAA) that, based on the same principles as visual
interpretation and on appropriate coding of knowledge related
to the experiment and discipline, will automate all routine
procedures and make, whenever it is possible, the process of
interpretation both objective and repeatible.[3] Indeed, the
definition and implementation of an automatic instrument does
involve the formalization of criteria of interpretation and
the organization, in suitable structures, of the external
knowledge.

In this work, the expert exploits heuristic methods, which
are anyhow time expensive.

The aim of the experiment is to improve the performance of
the human expert as a decisor by a progressive automatization
of the clerical tasks he now executes as well as those
heuristic decisions which can be descibed as Pattern Directed
Information Systems. The definition of GAA and his
implementation follows a general image interpretation based
on structural-attributed descriptions of image features.[4]
This kind of approach combines the syntactic and the
parametric approaches, since the image primitives are
elements of an alphabet to which some semantic parameters
describing their numerical or morphological characteristics
are related.[9]

After an examination of the data to be analyzed, presented
in Section 2, the subsequent sections describe the
development of the experiment with the identification of the
structures in the image of geological interest, and the
definition of their descriptions and of the recognition
algorithms.

Examples drawn from application of the instrument to one
of the sample images by which it was set help the exposition.
We aknowledge the AGIP Company for supplying data and
specific skills.

2. THE AVAILABLE DATA

The gravity acceleration values measured by gravimetry in
the region in question, and which were the subject of the
survey, cannot be compared with each other, inasmuch as they
were measured at stations located at various heights above
sea level and under different topological conditions - for
example, proximity of mountains. In order to make this
measurements match, corrections are applied to the data
recorded that depend only on the position in the scene at the
measuring stations.[5] The result of these corrections is a
phisical quantity called "Boguer Anomaly".

We define an anomalous mass or causative body a certain
volume of ground whose density is different from that which
would be given by zero Boguer anomaly. If a Boguer anomaly
other than zero occurs, this therefore supplies information
on the difference in density between the anomalous mass and
the matrix containing it.

Since the measurements are not uniformly distributed over
the scene, the Boguer anomaly values are given, by means of
mathematical methods [5], in the nodes of a rectangular grid
with a 250-m. sweep. The values at the nodes represent a

weighted average of the anomalies measured in an area all round them of radius R.

3. THE METHODOLOGICAL APPROACH

In interpreting a digital image, an observer needs to determine whether any set of pixels in the image can be related to an object in the underlying scene: the sets of pixels that are "similar" with respect to some characteristic, group together to form clusters that may be shadows of sets of "similar" regions belonging to objects in the scene.
The relationship between regions in the scene and numbers in the matrix representing the digital image is generally unknown and complex: equal numbers are sometimes associated with different physical entities and vice-versa.
A human interpreter resorts not only to contextual information in the image, but even to his own knowledge of the discipline and the experiment to identify and describe these structures and to interpret them.
A system that has to mimic the human interpreter must be able to associate the structures of the image with entities of the real scene exploiting the external knowledge suitably coded.

The formal definition of "description" provides a tool for identifying the structures that the observer is looking for and their meaning, during his interpretation. Once the descriptions of the structures sought have been defined, rules for recognizing them and tools for managing the application of those rules can be deduced.

In our approach, descriptions are derived through recognition of the structures present in a digital image and significant for the experiment and through computation of their properties.[8]
A "structure" is a set of pixels that satisfies some relations. Thus, for instance, a "blob" is defined as a structure composed of all the pixels connected to a given one.
A "feature" of a structure is a characterizing subset of the structure itself. Features are in general denoted by names and are the sets of pixels that should correspond to parts of the structure which have a meaning for the interpreter.
For instance a geologist sees outlets and inlets of each region - named "geological outline" - as significant features, that he names "thrusts".
A structure is characterized by a set of variables named "attributes", each of which may assume a value, named "property", in a well defined set.
Features, attributes and properties provide the description of a structure. The description is a triple:

$$N:P1 \diamond P2 \qquad where:$$

N   is the name of the description and denotes
    the entity being described
P1 is the set of global descriptors
P2 is the set of the names of descriptions
    which concern with the features of the entity.

A descriptor is a couple $< a_i , v_i >$ where a_i is the name of an attribute and v_i is the associated property.
The names of the set P2 are in their turn names of descriptions which must be defined.
P2 is void at the lowest level of description, that is the level at which the structures are no more decomposed. Also P1 may be empty, at every level except the lowest one. The set

of descriptions related to the same entity forms a multilayer hierarchical scheme, called Description Scheme (DS).

In our approach, the image interpretation is based on four types of DSs: "Intensional", "Discriminant", "Measured" and "A Posteriori".

For each kind of object which might be observed in the real scene, an Intensional DS (IDS) is defined by the experimenter. This DS collects all the features relevant to the current experiment and their descriptors, and can be viewed as a combination of the descriptions of the known instances of objects in the class, described according to models specific to the discipline to which the experiment refers.

From every IDS one or more Discriminant DS (DDS) are derived. A DDS collects a minimal set of features and related descriptors, sufficient for assigning an unknown structure to the class identified by the DDS's name. It is therefore a minimal discriminating set of descriptors and of names of features derived from the corresponding sets in the IDS.

The a priori knowledge of the patterns observable in a given class of images, and available before interpretation, must be organized into the aforementioned two kinds of DS. As the human interpreter uses this a priori knowledge to extract the interpretation of a single unknown image, the automated system is designed to exploit these DSs. The GAA in fact performs the image analysis by trying to build the "actual" copies of the corresponding DDS which are organized into a Measured DS (MDS).

This means that, for each entity of a given class, the system identifies in the input image, the candidate structures at the different synthesis levels, and measures their attributes by trying to follow the corresponding DDS.

The system achieves an interpretation by evaluating the MDS. The structures, once interpreted, are described according to the end user requirements, resulting in an a posteriori DS which constitutes the interpretation output. The A Posteriori DS (APDS) is defined as one that collects all the properties significant to the experiment including those measured after the interpretation has assigned a structure to a meaningful class.

In the following exposition of the design and implementation of GAA, some examples of DSs concerning the structures of interest will be shown.

4. THE EXPERIMENT

To define the GAA, it was necessary, from the very first stages of the project, to have a close interaction with the disciplinary expert that made it possible to define which structures of the image are considered important, and which logic and heuristic transitions he performs to arrive at their recognition. GAA is therefore created by the co-ordinated work of the data analyst and the disciplinary expert who, in addition to supplying the specifications of the problem, ensures the significance of the proposed instrument by confirming its results[2].

The images analyzed by the expert, which have led to identification of some of the relevant structures and their descriptions, are, as previously stated, obtained through the contouring of the Boguer anomaly images. This procedure, which consists in interpolating the data with a third-degree polyneumial with variable coefficients and in representing the surfaces thus obtained by means of level curves, is normally used because it enables the geologist to refer back easily to a three-dimensional model of the quantity measured.

A not uncertain assignment of specific configurations assumed by the processed signal to the anomalous mass distribution that generated it can happen only with integration of data from another source. However, the expert usually associates with the structures of the image the actual names of the geological structures with which they are likely to be connected.

Similarly, GAA has been designed to recognize the structures of the image that have, for the reasons stated, an indicative meaning, since their recognition is an indication of the presence, in the scene, of the geological structure with the same name. The process of visual interpretation begins, for example, to identify structures of the faults represented by thin, compact strips of isoanomalies.

The digital equivalents of the structures visually identified by the expert can be obtained by means of a simple slicing operation, in which pixels are grouped together in equivalence classes having the same amplitude T.[10] Let us call this structures isoanomalies, which is in line with the nomenclature used in gravimetry.

An initial segmentation of the image can be obtained, by applying, in parallel to all the pixels of the image, the algorythm described in I.

$$I \qquad f(x,y) \sim_T f(x',y') \quad \Longleftrightarrow \quad (f(x,y)/T) = (f(x',y')/T)$$

Between the isoanomalies produced by the segmentation, we identify those whose slope properties are such that they can be made to correspond to the presence on the scene of the geological structure "Fault".

To this end, a suitable descriptor, shown in II, which associates with each component an isotrope gradient measurement, is applied in parallel to all the pixels of the image.

$$II \qquad GRAD\ (P) = MAX\ |P - V(i)|$$

in which $V(i)$ indicates the eight near to P taken as a whole.

A subsequent evaluation of the slope values associated with each isoanomaly makes it possible to pinpoint the "Fault" structure.

The fault represent transitional isoanomalies in the image; in fact, these are structures that we may define stress structures, in which the quantity measured varies greatly. The faults therefore separate regions that are geologically different from each other: these regions of the image, which have presumably had the same tectonic evolution, have been given the name of "geological outlines".

Identification of the faults thus makes possible a new segmentation; analysis of the contours of the regions thus pinpointed provides the disciplinary expert with a considerable amount of observation: their primary forms and their deformations can be correlated with substantial geological movements.

The contours of the geological outlines, (fig 1), are described at different levels of syntheticity, starting with a topological description based on local properties of the pixel common to all the types of image dealt with.[5,8]

TAB I

IDS OF THE ENTITY GEOLOGICAL OUTLINE

GEOLOGICAL OUTLINE: <TYPE: SIGNIFICANT, NOT SIGNIFICANT,
 SIGNIFICANT AND FULLY INCLUDED, NUCLEUS> <AREA: {min, max}>
 ·PERIMETER:{min, max}> <MEDIUM SLOPE: {min, max}> <NUMBER OF
 PIXELS ON THE BORDER: [0, max]> <BOUGUER'S ISOANOMALY
 RANGE:{1 ,max]> <FAULT DELIMITATING (BOUGER'S ISOANOMALY
 VALUE): [0, max]> <EXISTENCE OF INCLUDED FLATS: {yes, no}>
 <NUMBER OF INNER CONTOURS: [0, max]> ◇ CONTOUR, FLAT
CONTOUR: <LENGTH: [min,max]>   <NUMBER   OF   GENERALIZED
 CONCAVITIES: [0, max]> <NUMBER OF GENERALIZED CONVEXITIES:
 [0, max]> <NUMBER OF FLEXES: [0, max]> <NUMBER OF EXTREMES OF
 CONTOUR NOT CLASSIFIED: [0, max]> ◇ GENERALIZED CONCAVITY,
 GENERALIZED CONVEXITY, FLEX
GENERALIZED   CONVEXITY: <TYPE: reginal   positive   push,
 residual positive push> <LENGTH:[min,max]> <BASE (DISTANCE
 BETWEEN EXTREME POINTS OF THE STRUCTURE):   [min,max]>
 <HEIGTH (DISTANCE BETWEEN MIDDLE POINT OF THE BASE AND
 MIDDLE POINT OF THE STRUCTURE): [min,max]> <NUMBER OF NOISY
 CONVEXITIES: [0, max]> <NUMBER OF CONVEXITIES: [1, max]>
 <NUMBER OF STEP-WISE-SEGMENTS:[0, max]> <NUMBER OF
 SEGMENTS:[0,   max]>   ◇ |NOISY   CONVEXITY,   CONVEXITY,
 STEP-WISE-SEGMENT, SEGMENT|
GENERALIZED   CONCAVITY: <TYPE: reginal   negative   push,
 residual negative push> <LENGTH:[min,max]> <BASE:[min,max]>
 <HEIGTH: [min,max]> <NUMBER OF NOISY CONCAVITIES : [0, max]>
 <NUMBER   OF   CONCAVITIES: [1,   max]>   <NUMBER   OF
 STEP-WISE-SEGMENTS:[0, max]> <NUMBER OF SEGMENTS:[0, max]> ◇
 |NOISY CONCAVITY, CONCAVITY, STEP-WISE-SEGMENT, SEGMENT|
FLEX: <IN-BETWEEN: GENERALIZED   CONCAVITY,   GENERALIZED
 CONVEXITY> ◇ STEP-WISE-SEGMENT
NOISY CONCAVITY: ◇ CONCAVITY, LOCAL CONVEXITY, CONCAVITY
NOISY CONVEXSITY: ◇ CONVEXITY, LOCAL CONCAVITY, CONVEXITY
CONVEXITY: ◇ PV,(PV)*
CONCAVITY: ◇ PC,(PC)*
STEP-WISE-SEGMENT: ◇ (PC,PV)*
SEGMENT: ◇ PC,PVV
LOCAL CONVEXITY: <IN-BETWEEN: CONCAVITY, CONCAVITY> ◇ PV
LOCAL CONCAVITY: <IN-BETWEEN: CONVEXITY, CONVEXITY> ◇ PC
PV ◇F|I|N|S|C|A|W|J
PC ◇ G|P|O|E|D|B|Z|K
PVV ◇ L|T|M|H
FLAT: <AREA: [1, max]> <PERIMETER: [1, max]> <SLOPE: 0> <MAIN
 GROWTH DIRECTION: 0> <BARYCENTRIC COORDINATES:[1...m,1...n]>
 <BOUGUER'S ISOANOMALY VALUE: [0, max]> ◇

TAB II

DDS OF THE ENTITY SIGNIFICANT GEOLOGICAL OUTLINE

SIGNIFICANT   GEOLOGICAL   OUTLINE: <AREA: [min,   max]>
 <PERIMETER: [min, max]> <NUMBER OF PIXELS ON THE BORDER:
 [min, max]> <EXSISTENCE OF INCLUDED FLATS: {yes, no}> <NUMBER
 OF INNER CONTOURS: [0, max]> ◇ CONTOUR, FLAT
CONTOUR: ◇ GENERALIZED CONCAVITY, GENERALIZED CONVEXITY,
 FLEX
GENERALIZED   CONVEXITY: ◇ |NOISY   CONVEXITY,   CONVEXITY,
 STEP-WISE-SEGMENT, SEGMENT|

GENERALIZED   CONCAVITY: ◇ |NOISY   CONVEXITY,   CONCAVITY,
 STEP-WISE-SEGMENT, SEGMENT|
FLEX: <IN-BETWEEN: GENERALIZED   CONCAVITY,   GENERALIZED
 CONVEXITY> ◇ STEP-WISE-SEGMENT
NOISY CONCAVITY: ◇ CONCAVITY, LOCAL CONVEXITY, CONCAVITY
NOISY CONVEXSITY: ◇ CONVEXITY, LOCAL CONCAVITY, CONVEXITY
CONVEXITY: ◇ PV,(PV)*
CONCAVITY: ◇ PC,(PC)*
STEP-WISE-SEGMENT: ◇ (PC,PV)*
SEGMENT: ◇ PC,PVV
LOCAL CONVEXITY: <IN-BETWEEN: CONCAVITY, CONCAVITY> ◇ PV
LOCAL CONCAVITY: <IN-BETWEEN: CONVEXITY, CONVEXITY> ◇ PC
PV ◇F|I|N|S|C|A|W|J
PC ◇ G|P|O|E|D|B|Z|K
PVV ◇ L|T|M|H
FLAT: <AREA: [min, max]> <PERIMETER: [min, max]>

 mxn is the image dimension.
 max is a variable assuming values scene-depending

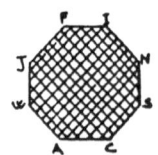 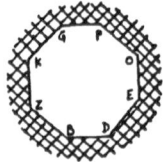 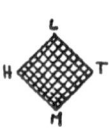

124

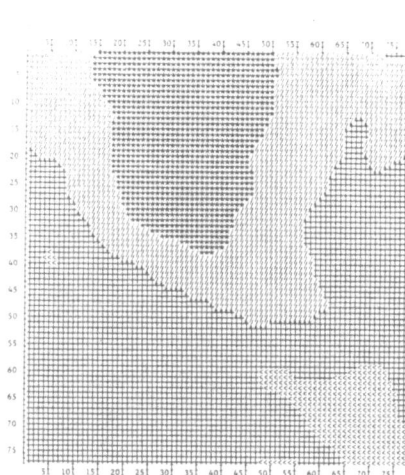

Fig.1

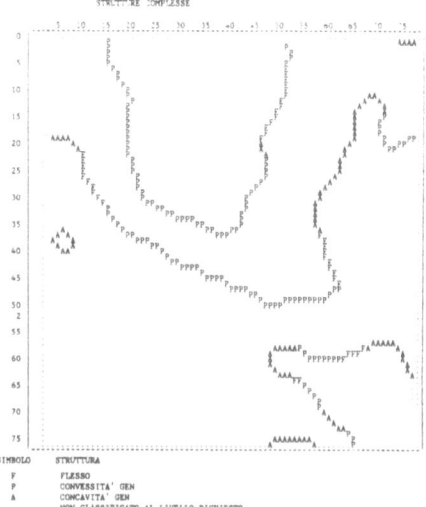

Fig.2

The morphological description of the contours at last level of syntheticity is shown in fig.2.

The presence of structures at a certain level of their description and verification of the correctness of their relationships makes it possible to recognize a structure at a higher level. Evaluation of the attributes present in the discriminant DS and associateds with the structures identified enables structures at the top level of the hyerarchy to be interpreted.

Table I gives the IDS related to the structure "geological outline", while the DDS that identifies a "significant geological outline" is given in Table II.

The morphological attributes that occur in the description of the contour exclude geometrical verifications, inasmuch as identification of morphological structures (e.g. concavities, convexities, step-wise segment, segment) takes place on the basis of the presence of code sequences that describe the local properties of the pixel.[1]

Once the GAA has interpreted some structures of the image as clues of the presence of geological structures of interest in the scene, it is possible to measure their morphometric properties in order to distinguish phenomena connected with substantial (regional) geological movements from those connected with irrelevant (residual) geological movements.

The property considered discriminant, in accord with the experts, is the ratio between the length of the structure and the distance between its extremities. The geometrical verification and possible approximation with a stepped side of structures that are residual in character take place at the first description level (e.g. concavities, convexities, stepped side, side).

Since, in interpretation of a gravimetric survey, the "regional" or "residual" character of the structures interpreted is determined by the aims of the analysis (minimum dimension of the geological structure that it is intended to identify), it is not reasonable to define rigid threshold values to distinguish the structures; however,

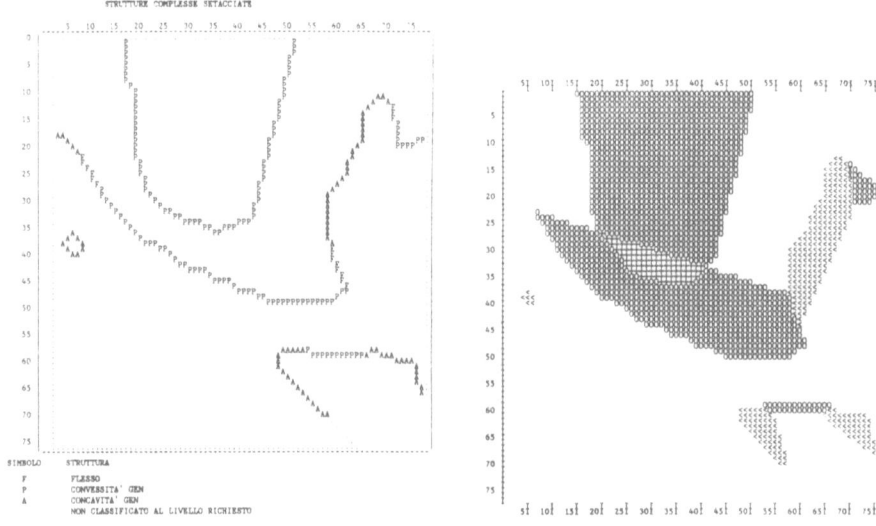

STRUTTURE COMPLESSE SETACCIATE

SIMBOLO   STRUTTURA

F   FLESSO
P   CONVESSITA' GEN
A   CONCAVITA' GEN
           NON CLASSIFICATO AL LIVELLO RICHIESTO

Fig.3   Fig.4

appropriate values are obtained from the external knowledge
base.   In the example that follows, the value below which a
structure is considered "residual" has been estimated 1.5.
This threshold value makes it possible to eliminate those
structures that can be correlated with movements in the
tectonics that are not particularly intense and therefore not
significant at this level analysis.   The morphological
description of the contours at last level of syntheticity
after removing of residual structures is shown in fig.3.
Comparison of the structures in fig.2 and fig.3   makes it
possible to observe the "residual" structures disappeared.
As already said, deformations in the contours described above
yeld information on the possible tectonic evolution of the
regions under study.   In particular, it has been noted that
convexities represented "wedges" in the adjacent regions.
These structures, called gravimetric "Thrusts", are of great
importance to interpretation.
The morphological description adopted, while containing all
the information relating to gravimetric thrusts, has been
found to be unattractive for the disciplinary expert,
inasmuch as it cannot be directly compared with the results
of his interpretative process.   The convexities in the
geological outline have therefore been extruded.[11] and
represented as individual entities in the image (fig.4).
Moreover, once identified and extruded, the thrusts are
described in metric terms.
At the end of the interpretation, GAA. supplies a description
of the image (the "a posteriori" description), which lists
the structures of interest that have been recognized,
together with the properties that the expert has indicated to
be significant.   This description groups together a
sub-assembly of the properties present in the intensional
description, including some that cannot be measured until
after the interpretation has estabilished a relationship
between the structures of the image and those of the scene.

126

5. CONCLUSIONS

The formal definition of description, as a hierarchical scheme in which, at different structural levels, some semantic parameters relate to morphological or numerical characteristics, provided the base on which different types of descriptions useful in interpretative process were identified.
Following this general approach, an automated instrument has been designed to help the interpretation of gravimetric surveys by which structures in the digital images are associated to possible below-ground geological structures on the scene.
The application of these tools to sample images enabled the GAA to b completed in its preliminary form: the following step will be the definition of other more complex geological structures, and the validation of GAA against new uninterpreted data.

6. REFERENCES

(1) S.Bianchi, A.Della Ventura, M.Dell'Oca, P.Mussio, A.Rampini: An APL pattern directed module for bidimensional data analysis, APL Quode Quad, vol 12, N 1, Sept.1982.
(2) U.Cugini , D.Merelli , M.Dell'Oca , P.Mussio: A computer aided system for interactive definition of digital image processing,(S. Levialdi ed.) ,Pitman, 1983.
(3) A. Della Ventura, A.Maggioni, P.Mussio, A.Pawlina: Knowledge acquisition for automatic interpretation of radar images, Image analysis and processing, (V.Cantoni et al.Eds.),Plenum Press, 1985, p.189.
(4) K.S.Fu: A syntactic-semantic approach to pictorial pattern anaylis, in Pictorial Data Analysis, R.M. Haralick ed, Springer Verlag; 1983
(5) F.S.Grant, G.F.West: International series in the earth sciences, Mc Graw-Hill book Company, N.Y. 1965.
(6) Hayes-Roth, Watermann, Lenat: Principles of Pattern Directed Inference Systems, in Pattern Directed Inference Systems, Ac Press, 1978.
(7) K.E. Iverson: Elementary Analysis; APL Press, Pleasantville, N.Y. 976 1980.
(8) D.Merelli, P.Mussio, M.Padula: An approach to the definition, description and extraction of structures in binary digitized images , in Computer Vision Graphics and Image Processing, Ac. Press, 1984.
(9) P.Mussio: Design of a pattern-directed system for the interpretation of digitized astronomical images, Proc. of the inter. Workshop on data Analysis in Astronomy, Erice, 1984.
(10) R.Schettini: Studio di fattiblita' di un Assistente automatico per l'interpretazione di immagini gravimentriche digitali, R.I. CNR, 1986.
(11) L.Shapiro: A structural model of shape, PAMI, vol II, N 2, march 1980.
(12) A. Simonotti, C. Zenucchini: Bidimensional gravity interpretation by damped linear-square methods using inequality constrains, Mathematical Geophysics Forth Seminar, 6-8 feb. 1986, Free Univesity Berlin.
(13) Tikhonov, Arsenine: Methodes de resolution de problems malposes, editions MIR, Moscow, 1974.

THE FUR PROJECT: UNDERSTANDING FUNCTIONAL REASONING

M. DiManzo[†], E. Trucco[*], F. Giunchiglia[‡ *], and F. Ricci[*]

[†]Inst.of Comp.Sci.,University of Ancona,via Brecce Bianche, 60100 Ancona, Italy
[‡]Philosophy Dept., University of Stanford, 94305 Stanford, CA
[*]DIST−Univ.of Genova,via all'Opera Pia 11/a, 16145 Genova, Italy

INTRODUCTION

The FUR (FUnctional Reasoning) project aims to develop a computational model for the representation and use of functional knowledge. Reasoning in terms of function seems to be very common: in everyday language, for instance, objects are often referred to in terms of the function they provide (e.g. a "washing mashine"). In many cases a relation exists between function and shape, i.e. function can be inferred from structure and vice versa. Tools like hammers, screwdrivers or spanners are well−known examples.

The main goal of FUR is to model the integrated use of functional and structural information about the world. Intelligent robots could exploit functional reasoning to achieve a greater flexibility in many generic tasks, e.g. choosing the right tool for a given operation. Examples of target problems for FUR are therefore function identification in a given structure ("what is this tool for?"), function−driven object recognition ("Is this a chair?"), functional improvisation ("what else can I use as hammer if no hammer is at hand?").

This document reviews briefly some intial thoughts, the work plan and the development state of the project.

SHAPE VS. FUNCTION

"Suppose you find yourself in a supermarket. It is particularly hot and you begin to feel tired. So you look around for a chair, but of course there is none. Eventually, you see a case. You decide to sit on it to rest a little."

Everything sounds obvious in this story. However, the decision to sit on a case is an example of our ability to recognize in an object other possible functions than those related to its main design goal. After all, chairs are meant to sit on and cases to contain stuff. Nevertheless we have no difficulty to use cases as chairs if necessary, even if their shapes can hardly be related [1] . In computer vision complex objects are usually represented according to some structural decomposition criteria (a well−known example is ACRONYM [2]), and the most common approach to recognition is to try to match shapes or other geometric properties against stored models [3,4]

Facing high−level vision tasks, however, this geometry−based approach appears to suffer from two main drawbacks. On the one hand, the object representation techniques currently employed either specialize in a particular class of shapes or ignore geometric details in favor of more abstract information (see [5] for a recent detailed discussion). On the other hand the description of all the possible variations of an object shape is quite an expensive computational task. If the size of the model data base cannot grow indiscriminately, a shape−based system is likely to consider only a limited number of different instances for each given objects.

To avoid the explosion of the data base, one could think to describe an object through its function. The starting point is the observation that the shape of many man−made objects is closely related to the task they are designed to accomplish [6,7]. Tools like hammers and screwdrivers are a plain example [8]. According to this hypothesis, a chair would not be defined by a particular shape or set of shapes but through a set of functional constraints. Therefore, one could think that a good definition for a chair could be "something on which you can sit on without falling back". Unfortunately, things are not so easy.

Following the destiny of shape, function alone is not sufficient to avoid ambiguities. From a functional point of view, the chair definition above individuates all objects on which you can sit without falling back, and this comprehends chairs, winding staircaises and many other structures. It individuates chairs unambiguously only if chairs are the sole objects satisfying the functional constraints in the given domain.

Deducing function from shape turns out to be not at all straigthforward. For some objects a relation between shape and function makes little or no sense: what is the "function" of a tree? One could observe that this relation is evident for those objects whose shape has been designed to satisfy a given function. This is the case of tools and, in general, of many man−made objects. But even some classes of man−made objects show similar shapes but are built to perform rather different functions (e.g. tables and stools). Moreover, a model−based system should be capable of producing a picture of the known objects. If objects are described through functional constraints, than a level must exist in the descriptors where function is eventually associated with a primitive shape or to a complex structure.

This document reviews briefly some initial thoughts, work plans and the development state of the project. The next section gather some functional reasoning tasks that we would like to model in FUR. Section 3 is an overview of the project, and section 4 reports about its development to date. Section 5 and 6 sketches briefly the theory of our approach to functional reasoning. Finally, section 7 is about the first FUR prototype.

TARGET PROBLEMS

It is interesting to notice that facing some problems require manipulation of 3−D shapes, while others tasks can be performed basing on abstract functional descriptions. Another interesting fact is that a number of functions can be expressed in terms of geometric relations (see section 5 and 6). This suggests a possible hyerarchy of levels for functional reasoning:

GEOMETRIC REASONING (in terms of 3−D models)

FUNCTION MODELING (in terms of geometric properties)

OBJECT REPRESENTATION (in terms of primitive functions)

TASKS AND PLANS REPRESENTATION (function−driven planning)

This section sketches some sample problems that FUR is expected to face in the future. The classification proposed is quite informal. In the following, "object" means the functional descriptor of an object, that's its integrated functional−structural representation; "structure" refers to a 3−D shape model, e.g. an octree.

DEDUCING FUNCTION FROM SHAPE:
"What is this structure for?" (GENERAL FUNCTION IDENTIFICATION)

"Is this object graspable?" (PRIMITIVE FUNCTION RECOGNITION)

DEDUCING SHAPE FROM FUNCTION
"Which structure can I use for grasping nuts?" (FUNCTION IDENTIFICATION IN A SET OF GIVEN STRUCTURES)

"Is this structure a screwdriver?" (FUNCTION−DRIVEN IDENTIFICATION)

"Which of these structures is a screwdriver?" (FUNCT.−DRIVEN CLASSIFICATION)

"What else can I use as screwdriver?" (FUNCTIONAL IMPROVISATION)

ABSTRACT FUNCTIONAL REASONING
"Which known objects are designed for containing liquids?" (FUNCTION−DRIVEN CLASSIFICATION)

"Which objects can be of use in a (given) plan?"

"What is the most plausible plan involving these objects?" (PLAN RECOGNITION)

PROJECT OVERVIEW

The final output of FUR is expected to be twofold. First, a theory of functional reasoning will be delivered. Such theory should consist in

1) a representation of objects integrating functional and structural knowledge. In other words, a language to describe both the structure of objects and their function. In FUR, objects are represented in terms of semantic functional descriptors (SFD). SFDs are semantic networks composed by primitive functions or functional experts, which embed eventually the formal model for the shape−function relation;

2) a functional representation of tasks in the same formalism;

3) knowledge about the above mentioned set of problems (see section 2).

Second, a software system implementing the theory will be developed. This system should be able to:

1) let users define easily their own environment, that's a set of objects, their 3−D structures and functional properties. This corresponds in FUR to instantiate new SFDs using a catalogue of predefined functional experts;

2) have the system solve a number of problems, such as choosing the right tool for, say, grasping a nut. This exploits functional reasoning on a user−defined environment.

The next sections reports briefly about work done on both aspects to date (theory of functional reasoning − implementation).

STATE−OF−THE−ART: GENERALITIES

The FUR theory is being developed under two main assumptions:

131

1) the functional goal of most man−made objects can be expressed in terms of some very common, basic functionalities; such functions (called primitive functions) are described in turn by frame−like structures.

2) most man−made objects can be hierarchycally decomposed in subparts. Each SFD suggests a hierarchycal decomposition for the object it describes, associating each subpart to a basic function.

Primitive functions are implemented by functional experts, which collect information about the conditions under which a function is possible. As detailed later, these conditions can be expressed through sets of geometric constraints for a number of common functions. In this sense, one typical activity of the FUR system is to decide whether something can accomplish a given primitive (maybe primitive) function, e.g. to decide whether something is graspable by a robot hand, or to find all graspable objects in a scene, or again the grasping points in an object.

An initial phase was devoted to the detailed analysis of some actions and the development of a first list of functional primitives [9−11]. A language for SFDs was defined, largely inspired by the Conceptual Dependency Theory formalism [12]. The main system task considered was function−driven object recognition. A first prototype was implemented which could identify simple objects using SFDs.

Consideration of more general tasks led to a critical revision of the theory. The study of the shape−function relation introduced a number of restrictions on the kind of functions represented (see next section). Current research is devoted to:

1) the definition and implementation of a set of functional primitive experts meeting the restrictions stated in the next section;

2) the definition and implementation of a computational environment for FUR;

3) the implementation of geometric reasoning algorithms working on octree representations of solid objects [13−15].

The expected output of this phase is a second prototype capable of basic functional reasoning.

RESTRICTIONS

Objects are involved in plans and relate to actions in different ways. Sometimes the conditions necessary for the feasibility of an action regard only their structures. The action of sitting on something needs e.g. a support, whose surface is constrained by the human body. Much in the same way, structural constraints let us grasp (with one hand) pencils, telephone receivers and hockey stocks, but not computer terminals unless they have special handles.

In other cases an action is declared possible on the basis of properties which are not geometric. To hit something means first of all to apply a great force in a short time. One can hit a nail with a hammer but also with a spanner or a shoe heel. In general, the action "hit" establishes rather loose structural constraints between actor and subject, surely much looser than those imposed by the sitting human body to the chair, or by the hand to a handle. The necessary conditions for "hit" to be performed with success are met by everything graspable and hard enough, to which a proper speed can be impressed. The crucial condition here seems more dynamic in nature than geometric.

For the time being, the class of primitive functions modeled by FUR comprehends only functions whose feasibility conditions can be expressed through relations between 3−D structures. Functional experts define functions as related to the geometric constraints that a shape must satisfy to be associated with the represented function. E.g., the action of GRASPing something establishes a structural interaction between the actor and the object, which can be modeled in terms of structural constraints. It is therefore possible to consider such constraints as a feasibility model for the GRASP function.

THE FUR THEORY: OVERVIEW

This section gives a brief sketch of the main ideas underlying the FUR theory. It is meant just to give the flavour of our approach to functional reasoning.

A number of PRIMITIVE FUNCTIONS have been identified. Objet functions are described composing such primitive functions. E.g., from the functional point of view a cup must contain liquids and be graspable.

Functions relate often to shape in the sense introduced in section 5. The shape — function relation can be modeled at two distinct levels:

Primitive Functions Descriptors (Functional Experts):
embeds knowledge about the conditions of a given primitive function.

Object Functional Descriptors:
organize multiple primitive functions in a complex functional descriptions, which specify both structural and functional properties of objects.

The primitive functions considered are listed below. They undergo the restrictions stated in section 5.

SUPPORT
GRASP
ENTER
CONTAIN
HANG
CUT
PIERCE
STOP
PLUG
EQUILIBRIUM

Primitive functional experts are implemented in terms of structural relations and geometric constraints. These can again be expressed by the same FUR formalism employed for object description. Network arcs in the primitive functional experts are composed by low — level geometric experts functions and predicates. They all work currently on octree — based object models and implement geometric reasoning algorithms such as e.g. extracting an object subtree from the environment, computing some integral properties of objects or extracting concavities).

Object descriptors or SFDs integrates structural and functional knowledge. They describe the hyerarchical decomposition which can be considered prototypical (e.g., chair = seat + seat_back + legs) and associate each subpart to a function. Therefore, this decomposition is both structural and functonal in nature. However, no prototypical SHAPE is defined. Subparts are classified and identified according to their function, which in turn is described by functional experts. In other words, the seat of a chair is not mandatorily supported by cylindrical or parallelepipoidal legs or the like, but by "anything which can succesfully accomplish the action of supporting it". While supporting a flexible and powerful approach to object recognition and classification, this choice et FUR handle very different instances within the same object class (say a rococo and a bauhaus chair).

Functional reasoning can take place at different levels, depending on the particular task. E.g. object recognition requires manipulation of SFDs; the detection of a primitive function in a 3 — D structure is performed using primitive functional experts; and functional improvisation involves both of them.

A FUR representation of tasks and its use for modeling complex functional reasoning is ongoing.

Uncertainty in the visual recognition process is due to several factors. Broadly speaking we can individuate two qualitatively different factor classes: the noise introduced by early processing and the incompleteness of high−level models. Their difference in nature makes a simple binary answer with an uncertainty coefficient ambiguous. An example will make the point clearer. Let's imagine an observer looking at a picture representing a chair with a broken leg. He would probably classify it as a chair where neither functional nor shape constraints are completely satisfied. A corresponding binary answer with uncertainty could be something like "yes−0.60". Now suppose the same observer is looking at a flawless chair in a very old and spoiled picture. This time he would probably say that he's seen "something like a chair". Again, the corresponding binary answer should be something like "yes−0.60". Apart from the particular numbers, there is nothing in these answers which let us discriminate between the two cases; however, the chair in the spoiled picture is somehow similar to the 3−D reconstruction of a low−level vision system, while the broken chair is a "perfect" description not corrupted by noise.

The problem is therefore to identify the **reason** of hypothesis refusals and, in general, of any system answer. From the point of view of functional experts, this implies that object descriptors must be able to individuate lacking or redundant object subparts and possibly reconstruct details spoiled by noise. Basing on such knowledge, a chair descriptor could classify the broken chair as an imperfect chair and recognize the broken leg as the uncertainty cause.

IMPLEMENTATION STATE

An implementation of the presented approach as been realized. Object descriptors are implemented as networks of active entities called actors. There are two actor classes: expert−actors (EXAs) and ego−actors (EGAs). The EXAs correspond to the functional experts in the semantic descriptor. They can exchange messages only with their close neighbours, that's actors of the second class. EXAs implements the functional constraints and associate each accepted structure with an uncertainty coefficient, due to imperfect input data or ambiguity arisen in the constraint satifaction analysis. EGAs correspond to the semantic roles in a semantic descriptor. They are directly connected to EXAs and are responsible for the formation and evaluation of compound hypotheses from local ones. EGAs implements knowledge about the object substructures and evaluate their consistency using local uncertainty coefficients.

So, in the case of searching an object (say a hammer) in a simple scene, the recognition activity of an object descriptor network can be depicted as a message exchange among actors, each of which has generated a set of data base parts after a search. The network tries to converge to a global set of mutually compatible hypotheses. Such activity stretches troughout the network grouping more and more complex entities in the data base until the so−called target ego−actor is reached, that is the EGA referring to the overall object.

A first system prototype has been implemented on a DEC GPX Workstation running ULTRIX. Its experimental task was to classify 3−D structures basing on SFDs. A running example has been concerned with the recognition of a structure as a chair. A scheduler organizes the activity of a network related to the CHAIR functional semantic descriptorriptor. Input data are a volumetric representation of a quasi−standard chair which makes the SUPPORT search generate several hypotheses: since the chair might be found in any stable position with respect to the ground, all the plane surfaces are considered possible seat surfaces. The network rejects actually the wrong seat surface hypotheses for the structural constraints due to the connections among experts, accepting eventually the input data as a possible chair. A graphic interface monitors the analysis running, showing the experts active in each time slot and the data base structures under analysis.

CONCLUSIONS

We have presented in this paper the FUR project, aiming to develop a computational model for the representation and use of functional knowledge. The theory and implementation have been briefly sketched, along with a first implemented prototype. Hints about the current development state of the FUR project are interspersed in the paper.

ACKNOWLEDGEMENT

This work has been supported in the framework of the ESPRIT P419 Project "Human Movements Understanding"

REFERENCES

1. J. Connell, *Learning Shape Descriptions: Generating and Generalizing Models of Visual Objects*, MIT Department of Electrical Engineering and Computer Science, 1985. M.S. Thesis.

2. R. Brooks, Symbolic Reasoning among 3−D Models and 2−D Images, *Artificial Intelligence XVII*, (1981), .

3. R. Nevatia and T. Binford, Description and Recognition of Curved Objects, *Artificial Intelligence VIII*, 1 (1977), .

4. L. Shapiro, J. Moriarty, R. Haralick and P. Mulgaonkar, Matching Three− Dimensional Objects Using a Relational Paradigm, *Pattern Recognition XVII*, 4 (1984), .

5. P. Besl and R. Jain, Three−Dimensional Object Recognition, *Computing Surveys XVII*, 1 (1985) .

6. F. Ingrand and J.−C. Latombe, Functional Reasoning for Automatic Fixture Design, *Proc. 13th Annual Meeting "Man or Machine: A Choice of Intelligence"* , (1984), .

7. P. Winston, T. Binford, B. Katz and M. Lowry, Learning Phyical Descriptions from Functional Definitions, Examples, and Precedents, *Robotics Research*, Cambridge, 1984, 117−135.

8. M. Brady, P. Agre, D. Braunegg and J. Connell, The Mechanic's Mate, *ECAI 84: Advances in Artificial Intelligence*, Amsterdam, 1984.

9. M. DiManzo, G. Adorni, F. Ricci and F. Giunchiglia, Building Functinal Descriptions, *Proceeding 5th ROVISEC*, Amsterdam, October 1985.

10. M. DiManzo, G. Adorni, F. Ricci, A. Batistoni and C. Ferrari, Qualitative Theories for Functional Description of Objects, *Esprit Report P419−TK4− WP2−DI1*, , May 1986.

11. M. DiManzo, F. Ricci, A. Batistoni and C. Ferrari, A Framework for Object Functional Descriptions, *Proc. 2nd Intern. Conf. on Artif. Intell.*, Marseille, December 1986.

12. R. Schank, *Conceptual Information Processing*, North−Holland , Amsterdam, 1975.

13. C. Jackins and S. L. Tanimoto, Oct−trees and their Use in Representing 3−D Objects, *Computer Graphics and Image Processing XIV*, (1980), 249−270.

14. D. Meagher, Geometrical Modeling using Octree Encoding, *Computer Graphics and Image Processing XIX*, (1982), 129−147.

15. H. Chen and T. Huang, Octrees: Construction, Representation, and Manipulation, *Proc. SPIE Conf. on Intelligent Robots and Computer Vision SPIE 579*, (1985), .

A KNOWLEDGE BASED APPROACH

FOR IMAGE UNDERSTANDING

S.Losito+,   G.Pasquariello+,   G.Sylos-Labini++,
and A.Tavoletti+

(+)Istituto Elaborazione Segnali e Immagini
Bari Italy
(++)Piano Spaziale Nazionale Roma Italy

1. INTRODUCTION

Many works in the field of image processing stress the
utility of using Artificial Intelligence tools to obtain
enhanced performances in the scene understanding task.
Feigenbaum (1977) states that "..the power of an expert
system derives from the knowledge it possesses not from the
particular formalism and inference schemas it employs...".
In our opinion this assertion can be extended to all complex
cognitive problems (such as natural language and automatic
image understanding) in which human reasoning capabilities
are required.   "From this point of view a theory for the
vision problem resolution must necessarily contain elements
of a more general theory of thinking (Minsky, 1974)". In
order to translate these philosophical statements into
action, we have to understand what kind of knowledge is
useful for the vision problem resolution and how   to
represent it. Looking at the image understanding problem as
a perception problem it is possible to individuate two
different knowledge sources.   The first one descends from
the perceptual grouping laws of visible entities; the other
derives directly from an explicit description of the objects
to be recognized in the image describing a real scene.
Normally these two knowledges are used in a hierarchical
fashion with a special emphasis on one source at the
expenses of the other.   Moreover in many of these
applications a low level processing attached to the image
extracts features and a high level step tries to match the
previously selected features with the model descriptions. A
drawback of this approach is that an intelligent matching
can be applied to an essentially non intelligent image
partition.
The approach to the image understanding problem we
propose is basically knowledge driven, with a distinction on
the areas in which knowledge is applied. In fact while the
image processor part applies procedural knowledge, in order
to produce a partition of the input image on a non semantic

base, the object recognition process uses descriptive knowledge to "bind" the image to the internal model.

This approach extends the flexibility of the imaging system downward with respect to different sensors and image acquisition systems and upward with respect to different scenes (outdoor, indoor) and objects.

Following this schema the proposed vision system is pervaded by this twofold knowledge structure: one in the grouping process and during the match between the image and the internal world representation (bottom-up) and the other in the descriptive models, in order to introduce the internal object models into the observed scene (top-down).

The two described different knowledge structures enforce two distinct, but cooperating, levels in the image understanding process. This simmetry provide both robustness to the understanding process and economy in the complexity of the representation.

2. RELATED WORKS

Starting from the late 70's there is a growing attention to meet image processing needs using the tools offered by the artificial intelligence methodologies.

The ACRONYM System (Brooks, 1983), developped at the Stanford Artificial Intelligence Lab, is an example of a model based system which incorporates a detailed 3-D geometric description of the object target of the recognition process. The identification strategy generates hypotheses based on edges extracted from gray-levels image, then verifies the consistency of the positions of the edges and vertices with the geometrical constraints obtained from the models. So far results have been reported on the recognition of airport photographic images.

Nagao and Matsuyama (1980) have presented a system for the segmentation and classification of high resolution colour aerial photographs. In this system the extraction of complex geographical and cultural structures (such as roads, rivers, forests, residential and agricultural areas) is based on rules that use multispectral, shape and adjacency properties to explicity characterize each goal region in a knowlegde based description.

Another interpretation system , VISIONS, has been proposed by Hanson and Riseman (1978). In this system the knowlegde is represented in a hierarchical structure with layers, such as objects, volumes, surfaces and regions. At each level of the hierarchy a hypothesis-and-test paradigm is used to construct a scene description from the image data.

More recently McKeown et al. (1985) have proposed the SPAM, a rule based system for the interpretation of airport scenes from aerial photos and map data base. In this system existing maps and domain specific knowledge are used for the coordination and the control of the segmentation process: objects are hypothesized and verified via consistency rules starting from fragments generated in the low level process.

In all the works reported above a specific domain knowlegde is used in the high level step in order to generate hypotheses and to drive the low-level processing, but essentially no-explicity knowlegde based low level tools are employed.

A totally different approach is offered by Nazif (1984) and Levine (1985) : in their works for the low level image segmentation an expert system is proposed , based only on the general knowledge of the laws on the perceptual grouping.

Our idea of a production system as a first step in the image understanding process is strongly influenced by the Levine's experiment, with the principal modification that the strategy rules (metarules in the inference engine and focus of attention mechanism for the region of interst selection) are directly driven by the object hypotheses formulated in the model analysis step.

3. SYSTEM OVERVIEW

The overall architecture of our system is represented in fig. 1.The fundamental parts of this schema are: a production subsystem (sect. 3.1) dedicated to the image segmentation using a set of rules based on elementary characteristics of the image elements; a frame based subsystem (sect. 3.2) containing the explicit description of the objects goals of the recognition task. Both these subsystems are completely open to the user by means of user interfaces enabling, for the former one, the input of rules and strategies for the segmentation step and, for the latter, the definition of object attributes and relations between them. Between these two modules a supervisor (sect. 3.3) acts as a message router throughout the Short Term Memory containing the status of the processed image, the production subsystem and the frame data base.

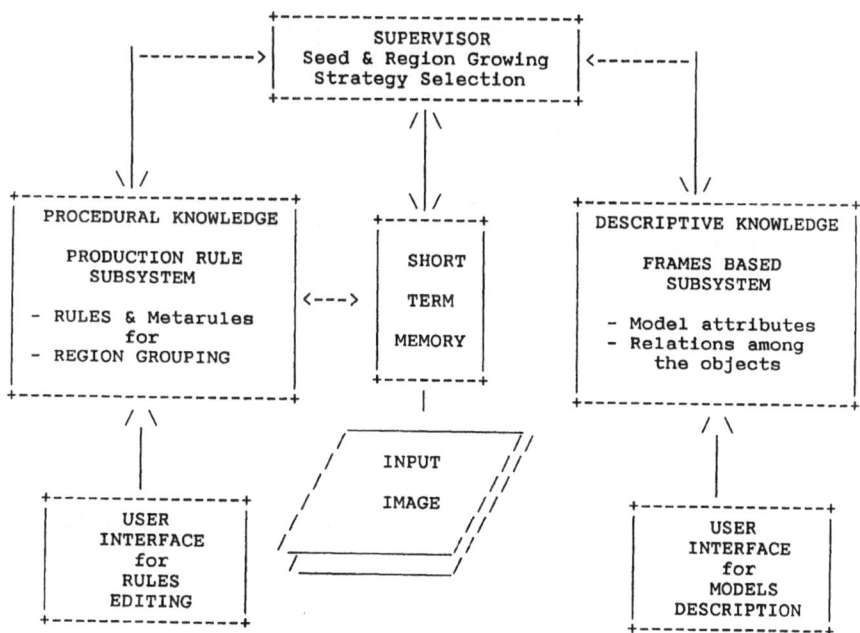

Fig.1. Overall system architecture

The Supervisor, on the basis of the STM status and model base descriptions, selects a guess model, sends to the production subsystem the appropriate rules and strategies to use, then analyzes the grown region in the image and compares it with the hypothesized model. If the region matches the guess model requirements, it is bound to the object described by that model; otherwise the initial hypothesis is rejected. It's important to point out that in this way the hypothesis that a region belongs to a specific class is used to drive the growing of that region.

This result is obtained considering the perceptive attribute of a model (such as homogeneus, not homogeneus, textured etc...) not as object of "a posteriori" verification, but as principal engine for a region growing strategy. This approach permits the use of different segmentation strategies for each different model attribute. From this point of view our rule based segmentation subsystem recalls the Low-Level Vision Expert proposed by Matsuyama (1980), but, in our case, specialized low level processors can be obtained selecting and ordering the proper rules derived from the general laws of the Ghestalt visual perceptive grouping, such as similarity, proximity, uniformity, continuity and so on. At last it must be noted that an extensive use of logical qualifiers, such as very low, low, high, very high and any possible logical combination of them, is made. It allows an easier interaction between the user and system and more generality in describing models and in establishing the rule predicates.

3.1 The Rule Based Subsystem for Image Segmentation

The knowledge based segmentator can be defined as a production system partially data driven and partially driven by the specific domain rules fed by the model description section of the system.

Following these statements, the segmented image is seen as a Formal System (FS), where the composition laws are constructed by the production system as "theorems" defined by the model characteristics of objects to be "proved". The "atoms" of this FS are the pixels. A very low-level algorithm produces, by merge operations on pixels, a list of regions oversegmenting the image. It must be pointed out that the aim of this operation is only to produce a structured representation of the input image.

In this representation the image is described by a list of elementary regions. The following properties are associated with each region:

```
REGION
        ------- label of the region
        ------- mean of the region
        ------- variance of the region
        ------- dimension of R (in number of windows)
        ------- Minimum Bounding Rectangle(MBR) to R
                (Xmin, Xmax, Ymin, Ymax)
        ------- Adjacency-list
where

adjacency-list ::= ( (adj-region-label . number-of-adj)+)
```

A parameter of interest for the segmentation process is the number of adjacencies between regions. This feature is expressed in term of the length of the two regions boundary (in number of boundary pixels) normalized for each region to the length of the total boundary.

As already noted one of the key issues of the rule based segmentator is to be free from numerical constraints on data, reducing numerical comparisons among quantitatives estimators to predicate calculus expressions (large, not very small etc.). This need is addressed following the approach described by Nazif (1984).

The oversegmented image is sent to the rule based section of the program. In this section it is appropriate to define two areas: the Short Term Memory (STM) and the Long Term Memory (LTM).

The STM actually is the list of regions scanned and modified by the system using the rules contained in the LTM. In this schema all the knowledge of the system about the segmentation is buried in the LTM. An user interface supports the user in rules definition, editing and verification. The rules contained in the the LTM have the following structure:

```
            IF < CONDITION >  THEN  < ACTION >
    where
        < CONDITION > ::= and/ or <test-1>...<test-n>
          < ACTION > ::= always MERGE action
          < test-k > ::= (<Logical-qualifier>
                         (<operator> < CURRENT-REGION >
                                     < TEST-REGION>    ))
 <Logical-qualifier> ::= boolean combination of
                         very-low/low/high/very-high
          <operator> ::= a function that returns the value
                         of a specific  region slot or the
                         difference  between  two  of them.
```

The <CURRENT-REGION> is selected by the supervisor. The user is allowed to select metarules that choose a <TEST-REGION> in the adjacency-list of <CURRENT-REGION>. These metarules can privilege the lowest difference in mean values or the highest adjacency value.

The set of segmentation rules performs, when firing, the only action of merging the test region into the current one.

3.2 The Frame Based Subsystem

The frame based subsystem can be seen as a data base containing the structured description of known objects. This module contains the data base handling facilities to allow the user to input, verify and edit the desired model. For each object the user is guided to fill the model characteristics through its spectral and positional properties and relative positions among the different objects of a real scene. Each object is represented by a frame with the following slots:

```
OBJECT
       ------- NAME
       ------- INTENSITY-VALUE
       ------- HORIZONTAL-POSITION
```

```
------- VERTICAL-POSITION
------- GROWING-STRATEGY
------- TOPOLOGICAL-RELATIONS
```

For the intensity and positions slots the value assigned by the user can be a logical qualifier or any possible combination of them.

This first implementation of the system is devoted to black and white images analysis : for color images the INTENSITY-VALUE slot will be the triplet of the values Red,Green and Blue.

The GROWING-STRATEGY slot value refers to the degree of uniformity of the objects in the image: it is a value associated to the perceptive characteristic of the object. The user has the option of selecting predefined growing strategies such as: very-uniform, not-very-uniform, not-uniform, noisy. At each of them is associated a corresponding set of rules in the production subsystem LTM. If not satisfied by these strategies, the user can define a new one using an appropriate rule list.

Through the last slot value, the user can link the position of an object with another. The syntax for TOPOLOGICAL-RELATIONS slot value is :

(<relation-1 object-name-1>....<relation-n object-name-n>) where the relation-k key can be one of the following: ABOVE, BELOW, NEAR, INSIDE and the object-name-k key refers to an existing object in the data base.

These topological relations for an object can substitute the absolute (referred to the image) values of the position slots. If an object is defined only by its relations with other objects the supervisor classifies it as a second order object and starts the recognition step for it only after binding the referenced objects to the image.

3.3 Supervisor

The Supervisor subsystem acts as a message router between the procedural knowledge base of the production subsystem, the descriptive knowledge base of the frame based subsystem and the Short Term Memory containing the image status. The operations performed by the Supervisor can be described as follows:

- Starting from information supplied by the descriptive KB, it establishes some restrictive and exclusive requirements on a region to be candidate as center of growth or "seed" of a class. The seed constraints are fixed choosing a subinterval of each model definition in the space of the spectral and positional model values, so as to discriminate from the other models. The existence of such discriminating conditions are guaranteed by the requirement that different objects are represented with not completely overlapping descriptions.

- The supervisor scans the STM looking for an object seed and transfers to the production subsystem the model information about the ad hoc growing strategy for that seed.Then two alternatives are possible: the region grows to his limit and then it is bound to the object; the region does not grow, the seed is rejected and the process restarts with a new seed.

- Every time a new region is bound to an object, the

supervisor verifies if the condition for a second order object growth are satisfied. If that is the case, the seed for the second order model· is considered.

- Once all seeds have been bound, the supervisor starts a process of fuzzy labelling of the yet unbound regions in the STM. For each of these regions (Reg) and for each model (Mod) a score is computed by means of the following function:

$$SCORE = MIN \{ MATCH**p, ADJ**(1-p), APRIORI\}$$

where:

MATCH(Reg,Mod) = degree of matching between the region and the model characteristics (1 = Reg completely contained in the model description; 0 = no correspondence at all).

ADJ(Reg,Mod) = function which weighs the context information: 1 if some region adjacent to Reg is bound to Mod; 0.5 if any of the adjacent regions is unbound; 0 if all adjacent regions are bound to models other than Mod.

APRIORI(Mod) = model dependent score referring to the a priori occurrence probability of Mod in the scene.

p = weigth factor dependent on the region size.

The region is bound to the model which maximizes the value of SCORE(Mod).

4. RESULTS

The vision system described above has been implemented on a VAX 11/780 computer in Common Lisp; the very low level algorithm is implemented in FORTRAN. Fig. 2 shows the results of the production system applied to a test image with a high noise pattern. In this case no specific knowledge is necessary: the only information used by the supervisor is about the region growing strategy to be applied (i.e. "noisy" object).

An outdoor scene image has been selected (fig. 3A) as test of the overall process performance. The logical description of the objects to be recognized is reported in fig. 4: in table A) the used object description is shown; in the same table the seed constraints obtained by the supervisor are also shown. The objects "Shadow-of-Church" and "Shadow-of-Wall" are defined as second order objects.

The result of the overall process is shown in fig. 3B: as it is possible to see, the binding of the models to the image is satisfactorily accomplished.

The present results must be considered preliminary. The future evolution of the work will be devoted to an enhancement of the descriptive characterization of the models, enabling the frame subsystem to handle more attributes for an object and improving the supervisor capabilities concerning the seed selection conflict resolution and the hypotheses verification strategy.

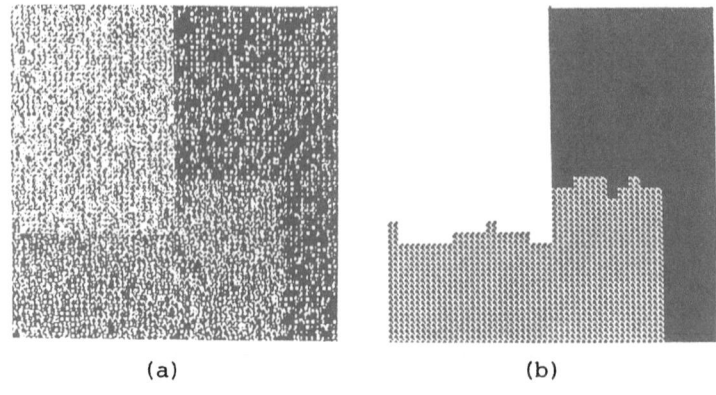

(a)　　　　　　　　　　　(b)

Fig. 2: Test image

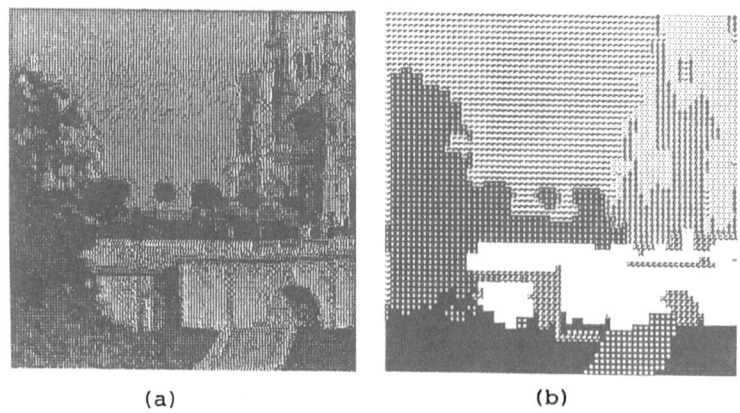

(a)　　　　　　　　　　　(b)

Fig. 3. Outdoor scene application

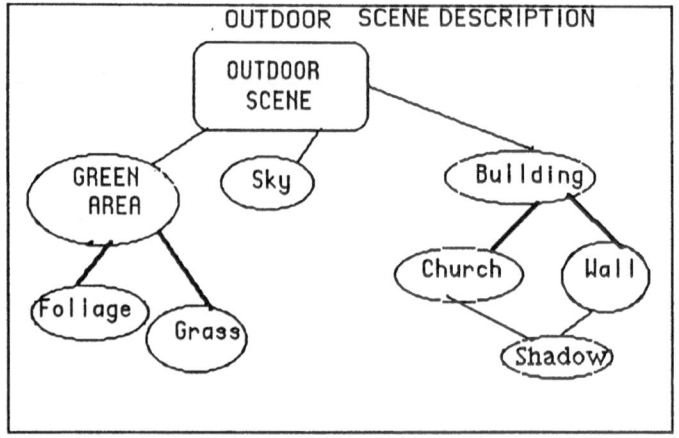

Fig 4. Outdoor scene logical description

Table A. Model description for outdoor scene

```
OBJECT:NAME......SKY             SEED_CONSTRAINTS
      :INTENSITY.not low,not high   .not low,not high
      :HOR-POS...                .
      :VER-POS...not low         .very-high
      :GROW-STR..homogeneus      .
      :TOP-REL...                .
OBJECT:NAME......GRASS           SEED_CONSTRAINTS
      :INTENSITY.low             .very-low
      :HOR-POS...                .
      :VER-POS...low             .very-low
      :GROW-STR..homogeneus      .
      :TOP-REL...                .
OBJECT:NAME......FOLIAGE         SEED_CONSTRAINTS
      :INTENSITY.low             .very-low
      :HOR-POS...low             .very-low
      :VER-POS...not low,not high   .not low,not high
      :GROW-STR..noisy           .
      :TOP-REL...((above GRASS)(below SKY))
OBJECT:NAME......CHURCH          SEED_CONSTRAINTS
      :INTENSITY.high            .very high
      :HOR-POS...not low         .very-high
      :VER-POS...not low         .very-high
      :GROW-STR..not homogeneus  .
      :TOP-REL...                .
OBJECT:NAME......WALL            SEED_CONSTRAINTS
      :INTENSITY.high            .very-high
      :HOR-POS...                .
      :VER-POS...not high        .low
      :GROW-STR..not homogeneus  .
      :TOP-REL...((above GRASS)(below CHURCH))
OBJECT:NAME......SHADOW-OF-CHURCH   SEED_CONSTRAINTS
      :INTENSITY.low             .very-low
      :HOR-POS...                .
      :VER-POS...                .
      :GROW-STR..homogeneus      .
      :TOP-REL...(INSIDE CHURCH) .
OBJECT:NAME......SHADOW-OF-WALL  SEED_CONSTRAINTS
      :INTENSITY.low             .very-low
      :HOR-POS...                .
      :VER-POS...                .
      :GROW-STR..homogeneus      .
      :TOP-REL...(INSIDE WALL)   .
```

5. REFERENCES

Brooks R. A., "Model-based three-dimensional interpretations of two-dimensional images", IEEE Trans. Pattern Anal. Mach. Intell., vol. PAMI-5, No.2, March 1983, pp. 140-150.

Feigenbaum E. A., "The art of Artificial Intelligence: themes and case studies of knowledge engineering", in IJCAI 5, pp. 1014-1029, 1977.

Hanson A. R. and Riseman E. M., "VISIONS: a computer system for interpreting scenes", New York: Academic, 1978, pp. 303-333.

Levine M. D. and Nazif A. M., "Rule-based image segmentation: a dynamic control strategy approach", Computer Vision Graphics Image Process. 6,1985, pp. 104-126.

Matsuyama T. and Nagao M., "A structural analysis of complex aerial photographs", New York: Plenum, 1980.

Minsky M., "A framework for representing knowledge", M.I.T. A.I. Memo No. 306, June 1974.

McKeown D. M. jr., Harvey W. A. jr. and McDermott J., "Rule-based interpretation of aerial imagery", IEEE Trans. Pattern Anal. Mach. Intell., vol. PAMI-7, No.5, Sept. 1985, pp. 570-585.

Nazif A. M. and Levine M. D., "Low level image segmentation: an expert system", IEEE Trans. Pattern Anal. Mach. Intell.,vol. PAMI-6, No.5, Sept. 1984, pp. 555-577.

A DYNAMIC PROGRAMMING APPROACH

TO KNOWLEDGE BASED CONTOUR SEGMENTATION

F. Mangili, and G. Viano

ELSAG spa , SRC Research Department
Via Puccini,2 I-16154 Genova (Italy)

ABSTRACT - Contour segmentation problem is investigated in the context
of an object recognition system dealing with overlapping parts. A
knowledge based approach is adopted, in order to get stable results.
Contours are described by curvature function. Angles are detected in a
preliminary phase. Dynamic programming allows a fast approximation with
straight lines and circle arcs. A simple rule based system is developed
in order to choose the best approximation. Some results are discussed.

1 INTRODUCTION

Recognition of 2-D overlapping parts requires an object
representation scheme that cannot be based on global features [1]. In
fact, the values of these features, computed for only partially visible
objects, are not correlated to the values for the whole shape. A
suitable representation has to relay on features that can be computed
with local support [2].

A particular approach is to describe shape using contours [3]. We
follow this approach describing contours in term of geometric primitives
of three types: angles, straight lines and circle arcs. Angles are by
all means local features; the other two kinds of primitives have some
topological and geometric properties that are not modified by occlusion.
Furthermore, the considered primitives have some peculiar
characteristics:

- they describe intrinsic features of the shape and are invariant to
  rotation and scaling.

- they allow a compact description, because regular contours can be
  usually represented by a few number of them.

Primitives have to be extracted by a segmentation process whose
main feature should be stability. In fact segmentation must be as much
insensitive as possible to shape rotation and scaling and to the
presence of a small amount of noise; moreover, partial visibility should
not modify the results.

We have developed a quite complex segmentation algorithm that

allows to deal with regular contours on which curvature changes
smoothly. In fact avoiding shapes with irregular contours, for which
curvature cannot be a good descriptor, the main problem is how stability
can be obtained when there are not sharp changes of curvature, for
example along a curve that can be approximated with more consecutive
circle arcs with different radii.

Easier situations, such as contours with only single straight lines
or circle arcs between angles, could be solved also with more
straightforward techniques; but the "knowledge based" nature of our
method allows to deal with these simple cases by "cutting off" reduced
versions specially drawn for particular applications.

In Section 2 we give some reasons for the use of curvature for
segmentation and show as curvature function is computed for digitized
contours. Next section deals with angles; they are detected through a
morphological filtering and examining peaks of curvature. Section 4
describes how segments between two consecutive angles are approximated
with straight lines and circle arcs. This task is performed by a
constant piecewise approximation of the curvature. A set of possible
partitions is computed, examining optimal approximations with increasing
number of constant pieces. Some criteria for choosing the
"OK-partition" are implemented in a simple rule based system. This
method, supported by a fuzzy logic, allows to introduce heuristic and
domain dependent knowledge. Section 5 briefly sketches how the symbolic
primitives are obtained. Finally, some results are presented in
Section 6.

2 CONTOUR DESCRIPTION USING CURVATURE

Curvature analysis is often used in Pattern Recognition
applications. Much of the initial effort for the use of curvature has
come from the areas of physiology and psychology. In fact, experiments
in these fields point out the high informative content of the contour
portions corresponding to critical points such as maxima, minima and
discontinuity points of curvature function [4]. Curvature can be used
as the fundamental descriptor for geometric primitives such as straight
lines and circle arcs [5]. Furthermore, shape representations based on
curvature are independent of rotation and have a known dependence of
scaling [6].

Computing Curvature

Curvature has to be computed for contours described by chains of
edge pixels. The following relation defines curvature as a function of
a parameter t:

$$c(t) = \frac{x'(t)y''(t) - x''(t)y'(t)}{[x'(t)^2 + y'(t)^2]^{3/2}}$$

where x(t) and y(t) are the coordinates of the curve points, and x', y',
x", y" are the corresponding first and second derivatives. In computing
curvature for a digitized curve, the problem of derivative estimation
arises, due to the quantization noise [5]. We adopted a gaussian filter
for noise reduction It allows to estimate the derivatives convolving the
function with the corresponding derivatives of the filter function. The
function obtained after convolution can be thought of as an average
curvature computed over a window whose size depends on the smoothing
filter function.

148

Curvature is equal to zero for straight lines and constant for circle arcs, positive or negative due to the direction in which the contour is ran over, and angles correspond to discontinuities of curvature (Dirac deltas). Examining an image contour, we have pieces of roughly constant curvature for contour portions we can approximate with straight segments and circle arcs, and peaks corresponding to angles. As the area of a peak is proportional to the tangential angle change undergone in rounding the corner, obtuse angles will determine small amplitude peaks, while acute angles very evident and high peaks (see Fig. 1). These considerations suggest not to use a gaussian filter with very narrow bandwidth, to avoid the risk to lose some information about wide angles. Due to the choice of primitives that correspond to opposed characteristics of the same function (curvature discontinuities for angles and constant values for straight segments and circle arcs), it is not efficient to use a single filter that works uniformly on the whole contour.

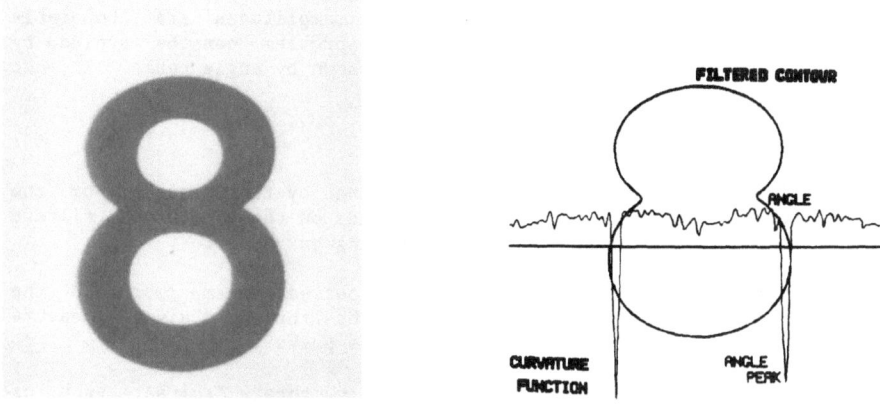

Fig. 1 - Computed curvature

Morphological Filtering

Curvature function presents a characteristic "peak noise", due to the derivative operations performed on quantized data (see Fig. 1). Then a filtering process that gets rid of these noisy peaks but keeps all the essential information about the angles is needed. The filter we use is a morphological one [7]; it has two most important goals:

- performing an edge preserving smoothing on the curvature segments corresponding to contour portions between angles.

- allowing the angle detection.

The filtering operation is performed on the absolute values of the curvature function. It consists of two phases: the first "erosion-expansion" phase deals with external peaks (i.e. peaks whose central absolute value is a local maximum); the second "expansion-erosion" phase deals with internal peaks (i.e. peaks whose central absolute value is a local minimum). The first phase is composed of two steps: in the first the curvature is "eroded", by replacing the value of each point with the minimum value over a window centered on the point. In the second step, the resulting function is "expanded", by replacing the value of every point with the maximum value over the window. The second phase is quite similar to the first; only the order between the two steps is reversed.

149

The result of this process is the supression of peaks having basis size less than the filter window and the "cutting" of the larger ones at a curvature value where the basis size becomes less than the filter size. The smoothing effect consists in peak elimination, while smooth changes of the curvature function are preserved.

3 ANGLE DETECTION

Angles produce external peaks in the curvature function that are cut off by morphological filter. Angle localization is performed examining the difference between the curvature function and the filtered one. Peaks in the difference function can derive both from angles and from noise.

As angles correspond to high peaks of curvature, a straightforward method to determine them could be to put a threshold on the amplitude. But wide angles often determine peaks whose amplitudes are comparable with that of some noisy peaks. This problem can be avoided by exploiting the structural characteristics shown by angle peaks. In fact an "angle-peak" must have:

- maximum value over a threshold.

- basis size in a range of values determined by the dimension of the filters. In fact, the peak size depends on the size of the filters used for edge detection and for computing curvature.

- a symmetric structure and a difference between the maximum and the two extrema values over a threshold; this threshold gives a measure of the discontinuity thas has caused the peaks.

These three characteristics are substantially three "rules"; each of them involves a threshold, but the values of them are not critical, as the important thing is they must be satisfied all together, for determining an angle.

4 CONTOUR APPROXIMATION BETWEEN ANGLES

After angle detection, the contour portions between consecutive angles are segmented and approximated. As the involved primitives correspond to constant values of curvature, the goal is to perform an approximation of each portion with a constant piecewise function. This means to determine a partition of the filtered curvature in constant pieces. In the following, we establish some requirements for the final partition:

1. stability : it must be robust against these phenomena:

    - variations of scaling and rotation of the examined shape in the scene

    - small amount of noise along the contour

    - small deformations due to the imaging camera system.

2. conciseness : it must describe every segment with a number of primitives as small as possible.

3. naturalness : it must determine break points on the contour portions close to the natural ones for a human observer.

We will call a partition with these characteristics an "OK-partition".

The strategy we use to obtain the OK-partition consists in the following steps:

- a preliminary approximation of the curvature function with short constant pieces.

- the determination of the "optimal" partitions, (i.e. such that the error on them is minimized) with a fixed number of constant pieces, obtained using dynamic programming.

- the choice of the OK-partition among the optimal partitions.

Let us examine in some detail the realization of each step and the good characteristics of this approach.

Preliminary Constant Piecewise Approximation

The preliminary approximation is obtained directly from the morphological filtering. As a consequence of this process, the curvature function already contains many short constant pieces corresponding to the erased peaks. The remaining portions of the function are replaced by constant segments, whose values are computed as average values over a fixed window (see Fig. 2).

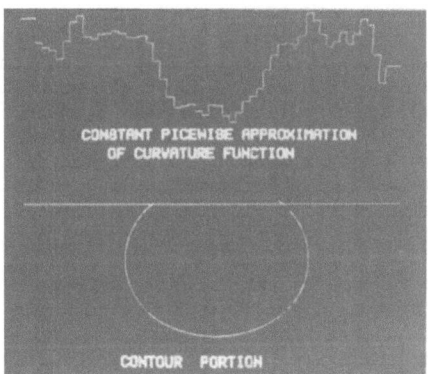

Fig. 2 - Constant picewise approximation of curvature function

Optimal Approximation Using Dynamic Programming

The goal of our procedure is to determine the optimal partitions with fixed number of elements for each curvature segment. A "partition" is a constant piecewise approximation that can be obtained only merging the preliminary short pieces. That is, we impose that all elements of it must have start and end points coincident with some start and end points of the initial segments.

With N starting elements, there will be k^N possible partitions with k elements. A partition is "optimal" if it minimizes the squared

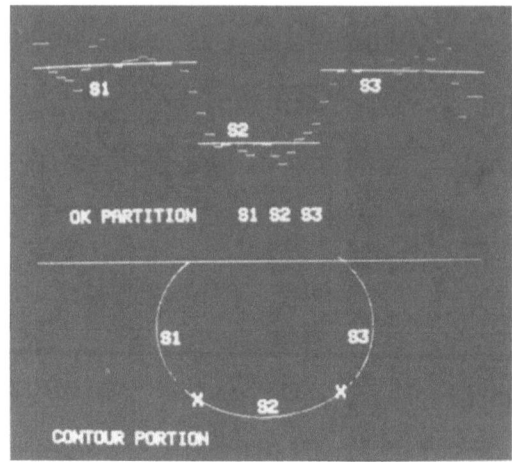

Fig. 3 - Optimal partition

The system uses a fuzzy logic [10]; each partition has associated a fuzzy variable, which in the following will be referred as "identification-likelihood", whose starting value is zero; its value is then incremented according to the answers of the partition to the rules. Each rule, using some suitable functions, gives to the partition a positive or a negative score that is added to the current value of the partition fuzzy variable. We have a stop condition when the value of the variable overcomes the value 1; as this happens, we choose as OK-partition the partition under examination or the immediately previous one, depending on the behaviour of the error (if the error change between them is low, the partition with less elements is preferred).

The fuzzy logic has been adopted as it seems suitable when we have to integrate many different information, and to translate qualitative concepts, such as those involved in the definition of OK-partition, into quantitative terms.

Let us illustrate some of the rules. There is a rule of the first type that examines the normalized squared error value for every partition and gives to it a score according to a decreasing function of the error. Another rule involving the single partition gives negative scores to partitions which have elements with non homogeneous length. It allows to avoid too close break points in the contour segmentation.
As the error is a decreasing function of the number of elements in the partition, there is a rule of the second type that deals with the error decreasing speed; it has been implemented in a way similar to the previous one, by using a function of the ratio between the errors corresponding to two successive partitions.
We have also a rule of the third type to take into account the error stability that is present near the OK-partition; it uses a function of $(E_1 \cdot E_3)/E_2^2$, where E_1, E_2 and E_3 are the errors corresponding to three successive partitions. A metarule, i.e. a rule working on other rules, is implemented in order to give more strength to partitions that have given positive answers to many rules, in comparison with other partitions with similar identification likelihood value, but determined by a less number of rules.

error in the approximation of the original function with a fixed number of constant segments. The optimal partition with only one element corresponds to the average value calculated on the whole segment.

The optimal partitions could be obtained by computing explicitly the errors for all possible partitions, but this process is too time consuming. To obtain the sequence of optimal partitions we can use the dynamic programming [8,9]. In fact, dynamic programming determines an increase of memory, but allows to reduce the number of operations for a partition with k elements from k^N to $k \times N$. The minimum error E_k^N to divide N segments into k parts is obtained with the following relation:

$$E_k^N = \min_{1 < j < N-k+1} [E_{k-1}^{N-j} + \epsilon_{N-j,N}]$$

where $\epsilon_{i,j}$, computed for $i, j \leq N$ and $i \leq j$, are the squared errors calculated by approximating the segments between i and j with their average value. The computation of $\epsilon_{i,j}$ is done for all possible choices of i and j and the results are stored in a triangular matrix. The E_m^l are the errors obtained by approximating l segments with m elements in an optimal way. Their computation with $1 \leq l \leq k$ and $1 \leq m \leq N+1-k$ can be performed in an iterative way. In fact, for $k = 1$ we have $E_1^l = \epsilon_{1,j}$; then the computation of the optimal partitions with increasing values of k is performed with successive iterations of the relation (2), with suitable indexes. For each step, only the storing of the values of E_m^l corresponding to the previous step, is required.

Experimental evidence suggests to look for the OK-partition among the optimal ones. This means that the procedure "steps through" the desired solution; the problem is the identification of the right solution in a limited set of possibilities.

From Optimal Partitions to the Ok-partition: a Rule Based System

It would be possible to compute the optimal partitions with k elements for 1 < k <N and to choose the OK-partition only after a complete computation. But, since our algorithm performs iteratively the determination of the error for partitions with increasing number of elements, we have developed a set of rules that allow to establish when the OK-partition has been obtained and to stop the process. Dealing with regular contours, it is often possible to obtain a good approximation with a small number of elements, and then the desired partition can be obtained rather quickly (see Fig. 3).

A simple rule based system has been developed to perform the choice of the OK-partition. The use of a rule based system determines remarkable advantages [10]. First of all, it allows to introduce knowledge about the problem in an explicit way; this provides flexibility and expandibility to the system itself. Furthermore, the system has a structure convenient for learning; in fact, according to the considered domain, the examination of the answers to the rules can allow to understand the real importance of each rule and can suggest the addition of some new rules or the removal of some not very useful ones.

In our system there are three kinds of rules:

- rules that involve only the current partition.

- rules that compare a partition with the immediately previous one.

- rules that compare a partition with some previous ones.

153

5 SYMBOLIC PRIMITIVES

The last phase of the segmentation process is to compute the parameters of the primitives used for approximation. Each OK-partition determines as many primitives as its elements; elements whose curvature value is very close to zero give straight segments and the other ones give circle arcs. Both for the straight segments and for the arcs, the parameters are computed with simple algorithms of linear regression and minimization, performed on the functions $x(t)$ and $y(t)$ obtained after the first filtering operation. For each angle the amplitude is computed, and for arcs and angles also a parameter giving the orientation is estimated, using the corresponding curvature sign.

6 RESULTS

The process feature we have been mainly interested to test has been stability. We have chosen a test domain composed of nine black shapes with regular contours The segmentation process has been performed on 108 images. acquired with a solid-state TV camera. Each of them contains a single shape; there are twelve images for every shape, with different rotations and scaling factors.

The process shows a good stability with respect to rotation and scaling, as long as the shape dimension in the image is not comparable with the filter window size. About the primitives, we have obtained good results (about 100%) of stability) for angles and "simple" portions of contours, i.e. portions on which the curvature value is constant or changes abruptly. "Complex" contour portions , i.e. portions on which the curvature changes in a smooth way, are approximated with a stability of about 70%, that is, we obtain the same partition in about eight or nine cases over twelve; in the remaining cases the resulting partition is not very different: usually no more than an element is added or missing. Fig. 4 shows results for two scenes.

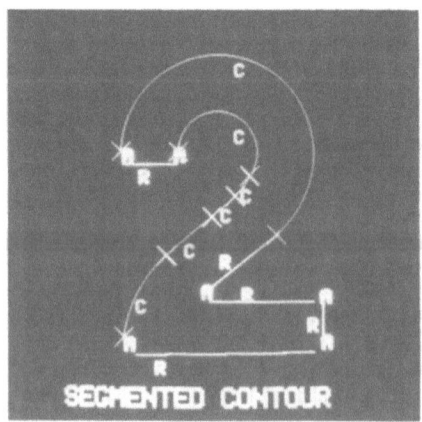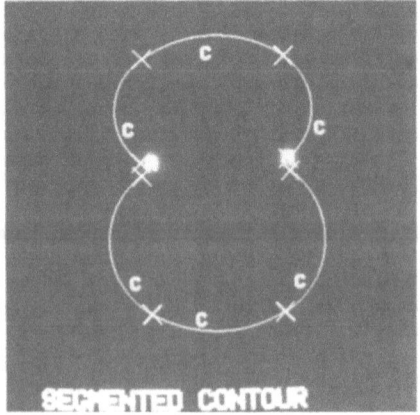

Fig. 4 - Segmentation examples

The system has been tested on scenes of overlapping shapes with good result. In fact contour portion between overlap points are segmented in the same way as in the situation without overlap. Fig. 5 and Fig. 6 show examples involving two of the test shapes and an industrial part.

154

As the system is intended to obtain stable results when smooth changes in curvature are present, it may result redundant when only simple contours are involved. Reduced versions can be easily adopted to deal with simple applicative contexts. At present the system runs on a VAX-11/780 computer and takes about one minute for processing a 512x512 pixel image containing a single shape.

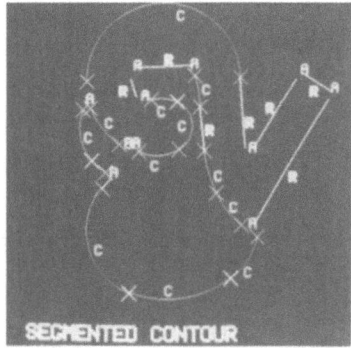

Fig. 5 - Overlapping shapes

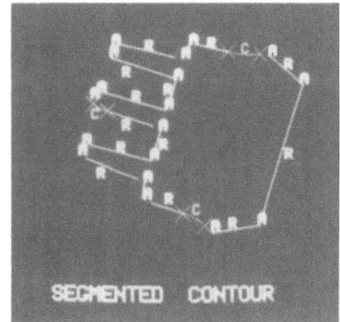

Fig. 6 - Industrial part

REFERENCES

[1] Berman S., Parikh P., Lee C.S.G., "Computer Recognition of two Overlapping Parts Using a Single Camera", Computer 3, (1985).

[2] Hakalahti H., Harwood D., Davis L.S., "Two-Dimensional Object Recognition by Matching Local Properties of Contour Points", Pattern Recognition Letters 2, (1984).

[3] Ayache N., Faugeras O.D., "HYPER: A New Approach for the Recognition and Positioning of Two-Dimensional Objects", IEEE Transactions on Pattern Analysis and Machine Intelligence, Vol. PAMI-8, No. 1, (1986).

[4] Bennet J.R., Mac Donald J.S., "On the Measurement of Curvature in a

Quantized Environment", IEEE Transactions on Computers, Vol. C-24, No. 8, (1975).

[5] Freeman H., "Lines,Curves and the Characterization of Shape", Information Processing 80, (1980).

[6] Pavlidis T., Horowitz S.L., "Segmentation of Plane Curves", IEEE Transactions on Computers, Vol. C-23, No. 8, (1974).

[7] Serra J., "Image Analysis and Mathematical Morphology", ACADEMIC Press, INC. (1982).

[8] Horowitz E., Sahni S., "Fundamental of Computer Algorithms", PITMAN, (1978).

[9] Bellman R.E., Dreyfus S.E., "Applied Dynamic Programming",Princeton University Press, (1962).

[10] Buchanan B.G., Shortliffe E.H., "Rule-Based Expert Systems", Addison-Wesley Pubblishing Company, (1984).

LEARNING FROM EXAMPLES IN COMPUTERVISION:

PRELIMINARY STATEMENTS

V. Cantoni, and L. Lombardi

Dipartimento di Informatica e Sistemistica

Universita' di Pavia, Italy

1. Introduction

Maybe learning from examples is the field where inductive learning is more helpful and the results of the studies
seem clearer.

The learning systems usually need an accurate
description of a set of experimental examples and try to take
out a simple description of the situation that we are
studying, where all descriptors, that do not concern with the
current problem, are neglected.

Let us now remember an important logical concept:
induction never warrants the truth of the results. Indeed the
number of examples may be as large as we want (this is true
for nearly all situations, except for the simplest cases) and
so we may get a potentially infinite number of hypotheses
that imply our set of instances.

2. The problem of clustering

Let us show a general description of the methodology we
must follow in inductive inference (this part is taken from
Michalski [1]).

We start from:

1) "Observational statements" (that is experimental facts),
    F, that make up the particular knowledge about objects
    and situations of the problem we are studying.

2) "A tentative inductive assertion" (which may be null).
    That is a hypothesis we want to develop.

3) "Background knowledge" where we store the constraints
    that the observational statements must meet and the
    preference criterion characterizing the inductive
    assertion.

From this basis we take out

+   An inductive assertion (that is a hypothesis), H, that
     implies tautologically or weakly (that is it is true
     only probabilistically) our observations and our
     background knowledge.

     We say that a hypothesis H implies tautologically facts
F, when F derive logically from H, that is, when the
expression H ==> F is not in odds with all possible
interpretations.
     Formally we can write: H !> F or F !< H
     We use the symbols !> and <! to mean the specialization
of a concept by the first one and the generalization by the
other one. The symbol ==> is the logical implication and ::>
the implication linking a concept description with a concept
name.

     Let us suppose that the proposition H ==> F is true. If
we know that H is true therefore F must be necessarily true.
Then we say that the deductive inference is truth-preserving.
     We may remember the famous Aristotle's syllogism:

     All men are mortal   (a)

     Socrates is a man   (b)

Thus

     Socrates is mortal   (c)

(And we know that really he died).
     We know a set of true assertions (a,b) and we get a true
statement.

     On the contrary when we want to take out the hypothesis
H from the facts F (that is we are talking about inductive
inference) we cannot preserve the truth: that is if one of
the facts in F is false then the hypothesis H must be false
(inductive learning is falsity-preserving).
     In our life we see that the sun rises each day (let us
remember Hume) and we take out the rule "the Sun rises every
morning", but really we cannot be sure that next morning it
will rise again, we can only hope it.
     In this example F is:

     On January,1st the sun rose

     On January,2nd the sun rose

     . . . . . .

and H is the previous rule.
     It is sufficient that tomorrow the sun does not rise so

that the rule is false. (In the same way we might not be sure of (a))

Really the generalization methods that try to get rules from sets of examples (inductive inference) never can warrant the correctness of the results. This is one of the reason why it is important to know some counterexamples and specially the ones we call "near missing". They are negative instances that differ only a little from the positive ones.

Now we must remember a last aspect that implies some problems: the hypothesis that we may draw out from a finite set of facts are infinite (or at least a number difficult to handle), that is way the background knowledge must offer us a method to cut the number of the chances.

In this connection we may distinguish two main kinds of inductive learning: learning from examples (concept acquisition) (this is the part that will be examined more closely because it is the most interesting for the purpose of this work) and learning from observation (descriptive generalization).

In the first situation the observed instances are descriptions of some objects that a teacher (that is an expert) classified into one or more classes. The rules, we take out by induction, may be considered as recognition rules (that is, they are definitions of concepts).

For instance, we may define motor-vehicle as:

"whatever vehicle with an engine and two or three wheels"

In contrast in learning from observation we look for a general description that characterize a set of instances.

We know that motor scooters and motor-bikes are typical motor-vehicles, then we may say:

"most motor-vehicles have two wheels"

We must note that while concept acquisition gives a set of rules (or hypothesis) to classify objects from the knowledge of their properties, descriptive generalization give us one (or more) description(s) of objects belonging to the class we want to study.

In formal way we may write:

$$\forall i \in I \ (Ei \implies Di) \qquad (1)$$

and

$$\forall i,j \in I \ (Di \implies \sim Ej), \ if \ j \neq i \qquad (2)$$

The expression (1) is known as completeness condition and it ensures us that whatever positive example need to be recognized in this way by the extracted rules; the expression (2) is called consistency condition and tell us that a counterexample will never be recognized as a positive one (we want to be sure that if an event belongs to the description then it is a positive example).

We need both these conditions for accepting an inductive
assertion as a rule to recognize an object.

The completeness and consistency concepts we defined
formally allow us to explain better the distinction between
characteristic descriptions and discriminant descriptions. In
the first situation we want to describe an object by a
logical expression that satisfies the completeness condition.
We want to find out the properties that are common to all
examples of the class we are studying, obviously the possible
descriptions are countless: so in the practical applications
we use only the one called "maximal conjunctive
generalization", that is the one which uses the greatest
number of descriptors, or, in other words, the one that
describes the smallest set including all objects of the class
in consideration.

Let us now stop studying the discriminant descriptions:
(we will look for the "minimal discriminant description":
that is the description having the minimum number of
descriptors) the simplest situation we can deal with is
single concept acquisition: in this case we need to
distinguish all examples between two different classes which
classify all the situations (we classify all instances as
positive and negative examples, this is possible if these
previous ones are available, if it is not we can get only a
characteristic description).

Problems rise if we want to classify more concepts at
the same time (in this situation we cannot talk about
counterexamples). However we may report this case to the
last one: we may classify the set of all observations of the
other classes as counterexamples for each class we study.

When we study multiple concepts, it may be easier to
work if the set of rules is linearly-ordered; then
expressions become simpler.

For instance the set of unordered rules:

    D1 ::> K1

    ~D1 & D2 ::> K2

    ~D1 & ~D2 & D3 ::> K3

is equivalent to the set of linearly-ordered rules:

    D1 ::> K1

    D2 ::> K2

    D3 ::> K3

3. Vision applications

In vision applications of machine learning we may
distinguish two fundamental points of view: learning of
method for helping system to get informations and their use
for taking out salient features.

After image acquisition we must work out data again to
get the description of the scene (this first step may be

developed by users, they will give the system a complete description of the image, so system will work only on the second step); this operation may be done by basic operations, now widely employed: edge followers, shape descriptors, Fourier transform, texture analyzers.

Of course we have to supply system with these tools, but not all of them are always useful in any situation: here we have a first step of learning by rules. The system have to receive informations for dealing correctly with all possible situations that we can submit it, so that we use only most helpful operators (doing so next step is simpler and faster).

Let us study now a simple problem: recognition of "nearly two-dimensional" mechanical pieces (flat pieces with only one steady position). First of all we must give the system a data-base in which all representations of objects are stored in term of descriptors of the scene, the system, working out this information, selects discriminant characteristics of each object, which are used to recognize objects in complex scenes.

In theory the system should work correctly, in practice we may notice several problems.

Every times we change the set of objects (or simply their definitions) we must recalculate the discriminant descriptions; if this updating often takes place, the data processing time becomes a critical parameter.

With respect to data correctness we may think that informations in our data-base are faultless because they were taken under optimum conditions without pressing constraints of time. We cannot say the same about data of objects we want to recognize.

Then it is helpful a careful analysis of likelihoods that a descriptor may be badly measured and in case we must reject most unreliable descriptors, even · if they can be judged very helpful. At last we must mention another possible source of errors: the presence in the image of more objects at the same time and possibility of occlusions, shading, etc.

4. A simple example

Let us think of a simple example in which inductive learning may be very helpful: we must teach a robot to handle the pieces that make up a common electrical switch. Its parts are made of metal and plastic; it is very important that the robot is able to distinguish easily, quickly and confidently between these two chances.

We are dealing with a situation in which we must take out a discriminant description from our informations, a characteristic description it is not helpful at all. We ourselves might give this description to the system (then we should speak about learning from instruction); let us suppose that we do not want (or we are not able) to do this.

We have another course that may be followed: we show the system a lot of real pieces and classify each of them. If we have a system like INDUCE [2,3,4,5], we get a description of piece in term of a set of descriptors (our experimental objects are shown in the next figure):

a) circle(Pi):
   number of circle shapes in the piece Pi
   domain: integer numbers

Figure 1

b) shape(Pi):
   shape of piece Pi
   domain: (rectangular, spring, "very long", irregular)

c) angle(Pi):
   number of angles in the piece Pi
   domain: integer numbers

d) colour(Pi):
   colour of piece Pi
   domain: (black, non-black)

e) area(Si,j):
   area of surface Si of piece Pj
   domain: real numbers

The preference criterion should include rules that allow the system to calculate the cost we must pay to get a particular description or to value the confidence of some measures: for example in our situation colour should probably be more affected by noise.

In our case the discriminant description might be:

plastic piece: a piece with more than two circular shapes

or formally:

circle(B)>2

Of course also:

plastic piece: a black piece with more than two circular shapes

circle(B)>2 & colour(B)=black

is a valid discriminant description, but it is not a "minimal discriminant description" and so it should be less helpful for our purpose.

We must note that the first definition describes a greater set of possible objects (we must say "possible" because the two sets are made up exactly by the same real objects).

5. Conclusions

In general criteria like the one just shown may help to analyze images; however very anomalous situations bring the system to error; it is really true that optical illusions have similar causes: images that are very different from usual contexts do not bring human brain to look for correct interpretation, but to choose the most ordinary interpretation.

Considering what was said, a vision system should be endowed with hardware to get and store a high resolution image and software to recognize simple features of the image. Then we need a knowledge base to help the system to use hardware and software in the best way (learning by being told). Finally we need a data-base in which we store the recognition rules that may be modified (using heuristic metarules) by system itself. So that the system may use informations in the best way (learning from examples and from observations).

Only now some packages, applied to computer-vision, begin to be available (we know for instance ASTRA [6]). This work is a theoretical survey about all what was developed up to now. We want to direct our studies in this field, for using this type of knowledge to improve object recognition in known contests.

ACKNOWLEDGEMENTS

We must thank ELSAG of Genova for the contributions which gave us in our work and particularly Doctor Musso.

References

.1   R. S. Michalski "A Theory and Methodology of Inductive Learning" in Machine Learning 1980

.2   J. B. Larson, R. S. Michalski "Inductive inference of VL decision rules Proceedings of the Workshop on Pattern Directed Inference Systems, SIGART Newsletter 63, June 1976

.3   T. G. Dietterich "Description of inductive program INDUCE 1", Technical Report (Internal), Department of Computer Science, University of Illinois, Urbana, October 1978

.4   R. S. Michalski "Pattern Recognition as rule-guided inductive inference", IEEE Transactions on Pattern Analysis and Machine Intelligence, Vol. PAMI-2, n. 4, 1980

.5   W. A. Hoff, R. S. Michalski, R. E. Stepp "INDUCE 2 - a program for learning structural descriptions from examples", Technical Report 82-5, Intelligent Systems Group, October 1982

.6   R. S. Michalski, R. E. Stepp "Conceptual Clustering: Inventing Goal-Oriented Classifications of Structured Objects" in Machine Learning II 1986

AUTOMATIC TRAINING IN STATISTICAL PATTERN RECOGNITION

Nils Lid Hjort and Torfinn Taxt

Norwegian Computing Center
Oslo, Norway

0. SUMMARY

The traditional way of constructing a statistical classification procedure
involves estimation of class densities from a set of feature vectors from each
class. In order to be effective these training sets often must be rather large,
for example of the order of 100 from each class. This training stage is sometimes very costly and can involve many hours of tedious labelling and editing
work. The present paper proposes a way of greatly reducing or almost avoiding
this bottleneck stage, by automatically updating class descriptions via exploitation of the unclassified feature vectors. We treat the multivariate normal
case in particular, but mention generalisations suitable for non-normal cases,
including completely nonparametric updating methods. Application of the normal-
based updating methods to two symbol recognition tasks is discussed.

1. INTRODUCTION

The statistical classification problem involves a framework with K classes
$1,...,K$ and a vector X of some chosen feature components that can be computed
for each object to be recognised. X vectors, when they come from class k, follow a statistical distribution with density $f_k(x)$. The a priori or prior probabilities for the K classes are $\pi(1), ..., \pi(K)$.

The optimal classification rule, in the idealised case where the class
densities and the prior probabilities are known in advance, is the following:
For a new, observed X evaluated on an object of unknown class, compute the
posterior probabilites

(1.1) $$P(k|X) = \pi(k)f_k(X)/\sum_{m=1}^{K}\pi(m)f_m(X),$$

and assign class label k to X, where k maximises the above expression. "Optimal"
refers to the fact that this rule minimises the long-term proportion of misclassifications. It is often convenient to include a "doubt" option as well.
If every misclassification is equally serious, and the cost of handling such a
doubt case is a threshold c times the cost of a misclassification, then decision "doubt" is taken in cases where each of the K posterior probabilities are
less than 1-c. See Hjort (1986, Ch. 1), for example.

Of course the class densities are never (completely) known in practice,
and the rule described above takes the role of an idealised procedure for later
approximations to imitate. The classical approach has been to build some parametric, semi-parametric, or nonparametric model for the class distributions,

obtaining an estimate $\hat{f}_k$ based on a training set from class k, say $X_1^{(k)},\ldots,$ $X_{n(k)}^{(k)}$, and then mimic the ideal rule by inserting $\hat{f}_k$'s for f_k's, i.e., using

(1.2) $\quad \hat{P}(k|X) = \pi(k)\hat{f}_k(X) / \sum_{m=1}^{K} \pi(m)\hat{f}_m(X), \quad k = 1,\ldots,K.$

In cases where the $\pi(k)$'s are unknown these must be estimated too.

In order to be effective it is usually required to have moderate to large training sets, depending upon the circumstances in the situation at hand: class separation, feature space dimension, and the chosen model for class densities (parametric, semi-parametric, or nonparametric). In difficult cases (meaning moderate or low class separations) training set sizes of the order of $n(k) =$ 100 from each class may be needed. This usually entails many hours of tedious label assigning and editing work, however, and may be practically impossible in some applications. We have been specifically concerned with the problem of recognising printed and handwritten symbols on maps, where the training stage also would involve digitisation.

It is the aim of our statistical updating techniques to reduce this training phase to a minimum, exploiting instead the potential of unclassified vectors to reveal the separate class structures. They aim, in other words, at "unsupervised learning" after a perhaps very brief introductory session of "supervised learning".

A typical application of the methods we describe in this paper could involve the following steps: Start out with an initial and perhaps very small training set from each class. Next obtain a file of unclassified feature vectors (of which there are as many as there are objects to classify), and carry out one of the updating procedures described below, changing the initial, crude estimates $\hat{f}_k$ to more realiable ones, say $\tilde{f}_k$. Finally classify the unknown objects using (1.2) but with $\tilde{f}_k$ replacing $\hat{f}_k$. (When carried out in this way some rather intensive computations need to be performed for each new set of objects. A less cpu-consuming variant that also should perform well is to declare the learning phase finished at some suitable stage, say after initial training on five from each class and further, automatic training on a couple of thousand objects, and then apply the resulting classifier on all forthcoming objects.)

The reason why such updating methods are possible at all is that these feature vectors with unconfirmed class labels carry some statistical information about the K underlying class densities, in that such vectors X come from the mixture distribution

(1.3) $\qquad\qquad f(x) = \pi(1)\, f_1(x) + \ldots + \pi(K)\, f_K(x).$

After observing a perhaps large number of such vectors $X_1,\ldots, X_M$, therefore, some statistical technique can be employed to estimate the unknown parameters involved in (1.3), say the mean vectors and the covariance matrices of the K classes (in addition to the prior probabilities when these are unknown too).

A naive proposal is to classify a new, incoming vector according to the current classification rule, and then treat it as really coming from that class. This way, trusting the classification procedure, class parameters can be estimated using new data, even sequentially. Procedures like this are called decision directed, but they are generally biased and not recommended. Among remedies that have been proposed in the literature is the probabilistic teacher, which randomly assigns class labels to incoming vectors according to the computed posterior probabilites, and then update parameter estimates, again on the assumption that the labels are correctly assigned. This method also has shortcomings, however, and other methods, including the quasi-Bayes and the probabilistic editor approaches, seem to have greater potential. This potential is largely unexplored, however. See Titterington, Smith, and Makov (1985, Ch. 6) for comments about some particular situations and for clues to generalisations.

The approach used in the present paper seems to be the statistically soundest one: natural models for the class densities entail a statistical model also for the unclassified vectors, and parameters can be estimated using well understood principles, like that of maximum likelihood. By specifically tailoring the parameter estimates to the set of vectors that is going to be classified this updating approach has a second advantage over the traditional one, and may actually outperform it. So, hopefully, more is won with less money. Let the computer work for us!

Section 2 describes this general statistical approach in more detail, and treats the multivariate Gaussian (or normal) case in particular. In Section 3 two experiments are discussed in which the normal-based updating methods performed quite well. Section 4 mentions some amendments that can be implemented in cases where class separation is low or moderate and the class densitites are non-normal.

2. THEORY OF AUTOMATIC UPDATING OF CLASS DENSITY ESTIMATES

2.1 General considerations

It is important to realise that the problem is much more delicate than the "classical" one of estimating parameters from classified vectors. The statistical information contained in the additional, labels-unknown sample $X_1, \ldots, X_M$ above is less than what it would have been if the labels (or true classes) were checked as well. The problem of obtaining reliable estimates for the parameters of a mixture distribution is well known to be fraud with difficulties for a number of reasons, and an always trustworthy, consistent method does not seem to have been constructed in the literature. Several methods are discussed in the recent book by Titterington, Smith, and Makov (1985); see also Hjort (1986, Chapter 7 and Section 3.2.E).

The somewhat pessimistic remarks made here concern chiefly the situation where parameter estimates must be extracted from the sample $X_1, \ldots, X_M$ of labels-unknown vectors alone. The problem is made much more tractable and "safer" when some feature vectors with known class labels are available as well. This situation is in force for the pattern recognition applications we have in mind, where an initial and perhaps very modest training phase produces some vectors $X_1^{(k)}, \ldots, X_{n(k)}^{(k)}$ from class k, $k = 1, \ldots, K$. In such a situation the total likelihood for the data becomes

$$(2.1) \qquad L_{total} = \prod_{k=1}^{K} \prod_{j=1}^{n(k)} f_k(X_j^{(k)}) \prod_{i=1}^{M} [\pi(1)f_1(X_i) + \ldots + \pi(K)f_K(X_i)].$$

Several approaches are now possible for obtaining parameter estimates. The most appealing one is the method of maximum likelihood, and proceeds by finding the parameter values that maximise expression (2.1), or equivalently, its logarithm. This is a non-trivial numerical problem, as the log-likelihood surface can have several local maxima. A satisfactory solution is provided by the so-called Expectation-Maximisation (EM) technique, which in the present case essentially involves taking partial derivatives of the log-likelihood and use these as "iteration equations", with suitable starting values obtained from the initial labels-known training set.

2.2 The multivariate normal case

Hjort (1986, Chapter 7) develops the updating methods more carefully, and considers both parametric, semi-parametric, and nonparametric models for the class densitites. We shall be content here, in view of space pressure, with a brief description of how the multivariate normal case is handled.

The model in this case specifies that X vectors from class k follows a $N_d(\mu_k, \Sigma_k)$ distribution. The maximum likelihood estimates for the parameters here, using the initial training set only, are

(2.3) $\hat{\mu}_k = \sum_{j=1}^{n(k)} X_j^{(k)} / n(k)$, $\hat{\Sigma}_k = \sum_{j=1}^{n(k)} (X_j^{(k)} - \hat{\mu}_k)(X_j^{(k)} - \hat{\mu}_k)' / n(k)$.

The task is to automatically update these using new, incoming data $X_1, ..., X_M$ whose class labels are unknown.

The maximum likelihood estimates $\tilde{\mu}_k, \tilde{\Sigma}_k$, based on the full-data likelihood (2.1), are arrived at in the following iterative manner: after having obtained t-th generation parameter values $\mu_k(t)$, $\Sigma_k(t)$, compute current posterior probabilites

(2.4) $P^{(t)}(k|X_i) = \pi(k)N_d(\mu_k(t), \Sigma_k(t))(X_i)/\text{sum}$,

and then find $\mu_k(t+1)$ and $\Sigma_k(t+1)$ using certain explicit formulae, involving (2.4), and too lengthy to record here. (These equations can be seen as natural generalisations of known ones for the case where only unclassified vectors are used to obtain parameter estimates for the mixture, cf. Duda and Hart (1973, p. 200).) Start values are $\mu_k(0) = \hat{\mu}_k$ and $\Sigma_k(0) = \hat{\Sigma}_k$. Stopping criteria for this iteration procedure can be based on the increase of the log-likelihood and/or changes in parameter values.

2.3 Some practical and theoretical issues

Although the description above of the normal-based updating procedure is "complete" and can be followed directly, there are several points to be raised of general practical and theoretical importance. Should one start directly with the "full model", or is it sometimes advantageous to update in a preliminary, less ambitious model, say one with equal covariance matrices? What model should one use? Will the multi-normal density suffice, or is a more sophisticated description necessary? What constitutes an efficient stopping criterion for the EM iterations? How small can we allow the initial training set to be? Are there potential dangers involved with a particular updating algorithm, and how can warnings be given? How small error rates can be expected? How should outliers be handled?

We do not intend to provide complete answers to any of these admittedly difficult questions in this paper, and refer instead to the fuller discussion in Hjort (1986), where also further remarks are offered on several of the issues involved.

3. TWO APPLICATIONS TO SYMBOL RECOGNITION

3.1 General description of the experiments

The theory described in this paper has been applied to automatic recognition of two different sets of symbols. The first set consisted of printed, uppercase letters from the English alphabet. The second set consisted of handwritten numerals. Both training sets and test sets were made available, these being completely disjoint. There were about 150 of each letter in both training and test set, and about 400 from each numeral class in both training and test set. There was little variation within each symbol class of the printed letters by visual inspection; the noise still present is the unavoidable one that comes from the scanning and vectorising process. The handwritten numerals were much more noisy, and any chosen feature vector method must face the possibility of confusing some of the classes.

Two different feature measurement methods were used in the experiments reported on here. The grid method (Ullman, 1973) consists of dividing the smallest box enclosing the symbol candidate into subrectangles and then measuring the length of the skeletonised symbol within each of these; also linear combinations of these can be used to reduce dimension. These features depend upon the orientation of the symbols. A Fourier descriptor method leading to certain amplitudes to compute on each symbol, and essentially taken from Zahn and Roskies (1972), was used to get a rotation independent feature vector.

Only experiments carried out with the normal-based updating methods are discussed in the present paper. We intend to implement and test some of the more sophisticated updating methods for non-normal situations later on. We expect better classification results for the difficult case of handwritten, possibly rotated symbols with such extended methods.

The procedures described in this paper have been implemented in VAX Fortran, and run on a VAX 11/780 minicomputer. The software is part of the GEOREC software system developed by the Norwegian Computing Centre for SysScan Ltd, Kongsberg, Norway.

3.2 Recognition of printed letters

To minimise the human workload, the number of manually classified symbols for each class should be as low as possible. Hence, we started out with only a single symbol from each class being manually labelled. This turned out to be sufficent for excellent estimation of mean values and covariance matrices, as compared to the values obtained by the traditional and more laborious method, where the full training set is manually labelled. Accordingly it was not necessary to study cases with more than one from each class in the initial training set.

A special updating procedure was used for this optimistic, but successful case where the initial manual training consists of finding only one symbol from each class. The normal updating system passed through three initial sub-models corresponding to covariance matrices being respectively proportional to the identy; a common diagonal matrix; and individual diagonal; before ending up in the full normal description (see Section 2; we have developed model choice criteria of the Akaike (1974) and Schwarz (1978) type).

The updating machinery performed excellently in this situation. The grid feature extraction method was used with a five-dimensional feature vector. Similar results were obtained with higher feature vector dimensions, and indeed also with the rotation-invariant Fourier feature extraction method. The classification based on updating in fact exactly matched that of the traditional, "full training" approach with 150 manually labelled symbols from each class.

The mean parameter estimates based on updating gave almost identical results to the classical "full training" approach. Estimates for the covariance matrices for the various classes also agreed satisfactorily, as witnessed by the example below, for class 'A':

```
64.1 -19.8 -20.7  -3.8   5.1         64.5 -20.0 -20.8  -3.8   5.2
      32.5 -33.4   5.0  -4.9               32.8 -33.6   5.0  -4.9
            80.6  -6.0  10.2                     81.2  -6.0  10.3
                   9.2  -5.3                             9.3  -5.3
                        16.4                                  16.5
```

The left hand side displays the estimate based on updating, having started with only one label-known symbol from each class. On the right hand side is the estimate based on "complete training", with 150 manually labelled 'A' symbols. (A similar degree of agreement was observed also for the other symbol classes.)

The classification accuracy achieved was impressive. A single real error was made using the grid method with five features (an 'O' mistaken for a 'G'), when the 26 x 150 test set was classified (and 0.6% of them were classified as outliers and 0.0% of them were assigned the "doubt" label). The orientation-independent Fourier amplitudes method made no real errors at all (but had outlier rate of 1.7% and doubt rate of 0.0%, in this particular example.)

3.3 Recognition of handwritten numerals

The mean parameters for those of the feature components that have an approximately normal distribution were estimated quite well using the unsupervised training approach with as little as three manually labelled samples of each symbol class. However, the mean parameters for components with a bimodal distribution were unsatisfactorily estimated. Unfortunelately, previous experience had shown us that such components were necessary to achieve classification results with success rate above 98%.

Nevertheless, a preliminary classification was carried out using the grid feature extraction method. Our partially unsupervised training method resulted in 93.8% correct classification and 5.2% real errors (among a total of 4614 symbols in the test set), after having started with three manually labelled symbols from each class. Increasing the number of manually labelled symbols from each class to 20 gave better results; 96.4 % correct classification and 2.3 % real errors. Still, however, we are confident that the percentage of errors can be lowered, without more extensive training, by extending the updating machinery to cope with typically ocurring non-normal phenomena, like bimodality. Some suggestions are briefly discussed in the following section, and will be implemented and tested in the near future.

4. UPDATING IN NON-NORMAL CASES

The paper has so far only discussed updating methods based on the multinormal model for class densities. Our experience indicates that these can perform very well even in non-normal cases, provided the interclass distances (as discussed e.g. in Hjort, 1986, Ch. 10) are moderately large. In more difficult classification problems, however, non-normality may cause problems. Our study of handwritten, rotated numerals is a case in point: several of the feature components we have used in order to get more than 98% and 99% success rate have two-peaked class distributions.

This section briefly describes some updating methods outside those based on normal class densities. We should also mention that there is a potential gain in sticking with the normal distribution but with robustly estimated parameters instead of those described in Section 2. A simple alternative is to use multivariate t-distributions instead of the normal.

4.1 Discrete-times-normal models

One reason for seeing bimodal class distributions in applications with handwritten symbols is that they are drawn in two similar, but principally different ways. This suggests that the data structure sometimes can be captured by including one or more discrete feature components in the vector. Imagine X = (A, Y), where A is discrete (taking on a low number of possible values, say one/zero indicating presence/absence of a certain characteristic in the simplest case) and Y being continuous. A simple and often effective statistical model for such data is the discrete-times-normal one, in which Y is normal for given A-value, i.e.

$$f_k(x) = f_k(a, y) = g_k(a) \, N_d(\mu_k(a), \Sigma_k(a))(y).$$

Such a density description can be updated in a partially unsupervised way; details will be made available elsewhere.

4.2 Mixture of two normals: tackling bimodality

Having bi-modality (two-peaked-ness) specifically in mind, a second proposal is to model each class as a mixture of two normal distributions; see Section 3.2.E of Hjort (1986). Once an updated class description of the above type is established, classification can be carried out in the usual way. This approach should be successful provided estimation methods can be found that are both effective (i.e. statistically reliable) and efficient (i.e. acceptable w.r.t. cpu-time).

In the simplest cases the class distribution in question is clearly divided into two sub-clusters, and these can be determined by clustering methods. In many cases the sub-classes are far more difficult to identify, however, and more sophisticated machinery may be needed.

A natural starting point for updating the mixture model is the updated class descriptions obtained for the traditional normal model. One may write down iterative "updating equations" for the full, ambitious model, and search for appropriate maxima of the log-likelihood function. This is somewhat involved but can be accomplished, by an EM-iteration within each M-step of the "outer" EM-iteration. It may be necessary to carry out the "preliminary updating" in special cases of the model first. Model choice criteria must be devised.

4.3 The updating, third order corrected normal

A more flexible model than the normal is the "third order corrected normal" discussed in Sections 5.4 and 7.3 of Hjort (1986). An updating procedure is also described there. Serious study of its properties is required before its degree of success can be assessed.

4.4 Updating the k-nearest-neighbour rule

The solutions considered in 4.1 and 4.2 are parametric, while that of 4.3 above may be descibed as semiparametric. Totally nonparametric methods can also be updated, like the k-nearest-neighbour rule discussed in Hjort (op. cit., Sections 5.5 and 7.3). The updated k-NN rule proposed there requires separate implementation of rapid k-NN finding search algorithms.

4.5 Updating kernel estimates and histograms

Section 6 in Hjort and Taxt (1987) outlines also how the kernel density estimation approach to classification can be updated. Along similar lines also histogram-based classification methods can be updated; see also Hall and Titterington (1985). These are perhaps the most convenient nonparametric rules. Their force is that they are conceptually and computationally easy (this remark applies to classification of future objects, once the computationally somewhat involved updating job has been accomplished), while having the potential to outperform say the normality-based procedures. They may on the other hand lose to more sophisticated semiparametric and nonparametric updating methods. Matters involving complexity and cpu-time should be explored, in relation to statistical efficiency, before a "full implementation" of either the kernel method or the histogram method can be recommended.

5. CONCLUSIONS AND FUTURE WORK

The theoretical basis for updating class descriptions via unclassified vectors, of which there are as many as there are objects to recognise, is discussed in Hjort (1986; especially Ch. 7). A particular set of such algorithms, tailored to work best in cases where feature vectors are close to normally distributed within each class, has been discussed in Section 2, and tests of their performance and usefulness were discussed in Section 3.

The normal updating system has proved effective already in several types of applications, as demonstrated in Section 3. The methods have the potential of reducing and almost avoiding the sometimes costly training phase without sacrificing classification accuracy, by exploiting unclassified vectors to reveal the underlying structure of class distributions. The procedures are admittedly burdensome, but still acceptable, w.r.t. cpu-use. Computational shortcuts and clever approximations can possibly be invented, but the issue of cpu-time has not been a serious problem in our applications (recognising symbols on maps, for example).

Both experience and theoretical considerations suggest that the normality-based algorithms will have difficulties when the class distributions are markedly non-normal, however. If the classes are very well separated (as measured by the generalised Mahalanobis distance, for example, see Hjort, 1986, Chapter 10), then non-normality does not matter much. In many often-occurring cases the situation is less favourable for normality-based updating, however.

This problem is not caused by any fundamental shortcoming of the updating approach itself, and can be solved by appropriate statistical models for the class densities, and corresponding, more general updating algorithms. Some suggestions for further work along these lines have been offered in Section 4.

We finally point out that the automatic updating approach is certainly valuable also in applications outside the field of pattern recognition; see Hjort (1985) for methods of estimating also "spatial parameters" in models for analysis of remotely sensed data. Similar-spirited methods are currently explored at the Norwegian Computing Centre for use in analysis of well log data.

6. REFERENCES

Akaike, H. (1974). A new look at statistical model identification. IEEE Trans. on Autom. Control AC-19, 716-723.

Duda, R.O. and Hart, P.E. (1973). "Pattern Recognition and Scene Analysis." Wiley, New York.

Hall, P. and Titterington, D.M. (1985). The use of uncategorised data to improve the performance of a nonparametric estimator of a mixture density. J. Royal Statistical Society Series B 47, 155-163.

Hjort, N.L. (1985). Estimating parameters in neighbourhood based classifiers for remotely sensed data, using unclassified vectors. Technical Report, Department of Statistics, Stanford University.

Hjort, N.L. (1986). Notes on the Theory of Statistical Symbol Recognition. Report No. 778, Norwegian Computing Centre, Oslo.

Hjort, N.L. and Taxt, T. (1987). Updating methods: Automatic training in statistical symbol recognition. Technical Report, Norwegian Computing Centre, Oslo.

Schwarz, G. (1978). Estimating the dimension of a model. Annals of Statistics 6, 461-464.

Titterington, D.M., Smith, A.F.M., and Makov, U.E. (1985). "Statistical Analysis of Finite Mixture Distributions." Wiley, Chichester.

Ullman, J.R. (1973). "Pattern Recognition Techniques." Butterworth, London.

Zahn, C.T. and Roskies, R.Z. (1972). Fourier descriptors for plane closed curves. IEEE Trans. on Computers C-21, 261-281.

ACKNOWLEDGEMENTS

We are greatful to Knut Bråten and Rune Solberg for taking part in the system design and implementation of the computer system referred to in this paper. We are also indebted to Sysscan Ltd., Kongsberg, Norway and the Royal Norwegian Council for Scientific and Industrial Research for partly financing the research.

A NEW KNOWLEDGE DRIVEN, OMNIFONT, MULTILINE OCR PROCESS

Jose Paster and Evelina Zemelman

Corporate Engineering and Technology Division

Pitney Bowes, Inc. - Stamford, CT, USA

This paper describes a new knowledge–driven, multiline, omnifont, feature extraction OCR process. A new representation of the alphabet knowledge allows usage of a limited set of rules when building a character description, disregarding particularities of font style and size, and directs the recognition process to the creation of a specific limited set of hypotheses, and their consequent test and verification. The process, although tested mostly with printed characters, was designed with the possibility of extension to handprinted and handwritten characters.

The known problems encountered in existing OCR systems usually correspond to the areas of line finding and segmentation, as well as font recognition. Line finding and segmentation techniques accomplished by projecting pixels in vertical and horizontal directions within the scanning area and analyzing the result, impose limitations for segmentation of italics, script and writing styles with kerning.

The existing systems are based mostly on "template" matching, the "template" being a digitized geometric model of a character or an algebraic expression for a tree or string of features. The best match is achieved by evaluating "Euclidean" or "Manhalanobis" (statistical) distances between the unknown character and the "templates". Due to geometric limitations of the templates and sensitivity of features, most of the OCR systems are statistical [1], [2]. The "deterministic" systems [3] define often artificial and unrelated to human intuition primitives that lack font generality and are conceptually limited.

None of the existing systems use features with enough abstraction to make recognition truly omnifont, and usually require large training sets and dictionaries that include neither script, no handprinted samples [2].

Unlike the methods mentioned above, the process presented here deals with those problems in a different way, and offers solutions to most of them:

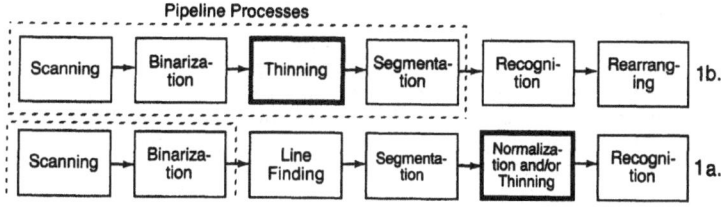

Fig. 1 The Recognition Process Steps
1a. Conventional process   1b. The process presented in this paper

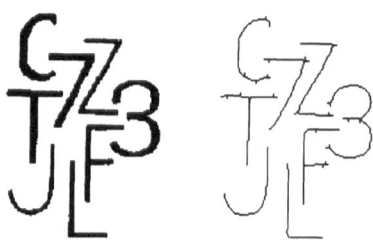
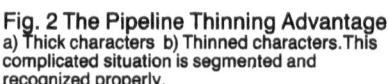

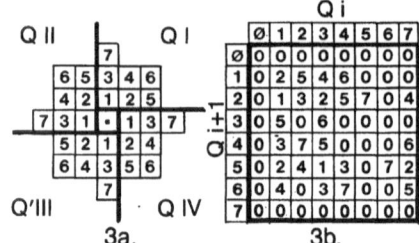

Fig. 2 The Pipeline Thinning Advantage
a) Thick characters b) Thinned characters. This complicated situation is segmented and recognized properly.

Fig 3
The weighted neighborhood distance measuring matrix (3a), with non-conmutative arithmetic (3b), where 0 means outside the distance range.

Qi	Ø	1	2	3	4	5	6	7
Ø	0	0	0	0	0	0	0	0
1	0	2	5	4	6	0	0	0
2	0	1	3	2	5	7	0	4
3	0	5	0	6	0	0	0	0
4	0	3	7	5	0	0	0	6
5	0	2	4	1	3	0	7	2
6	0	4	0	3	7	0	0	5
7	0	0	0	0	0	0	0	0

- The segmentation is done on pipe line configuration which is made easier by the preceding pipeline thinning. The process can handle multiple lines simultaneously, and is limited only by the size of the scanner and allocated memory. A rearrangement procedure reconstructs processed characters into a meaningful sequence.
- The process uses a knowledge representation based on topological properties of the plain graphs, which allows to direct recognition to a limited subset of the possible characters. According to "hypothesize–and–test" approach [4], [5] we hypothesize that a given graph is a legitimate character, and then test this hypothesis [6]. This knowledge–based representation also helps to identify connected characters.
- It is a feature extraction process with features both intuitive and general, the process is truly omnifont. Special thinning and smoothing filters applied to original characters help to uncover their basic abstract shapes.
- The use of psychological Gestalt laws of continuity and proximity [7] allows to connect broken and isolated parts of a character, and to complete characters with missing portions.

PROCESS STEPS: PRERECOGNITION

The process steps are described in Fig. 1. Up to thinning the process is conventional. We do the thinning before segmentation using "Weighted Convolution Thinning Technique" [8], [9], to take advantage of the consequences of thinning, i.e. background growth and corresponding increase of space between characters (Fig. 2). We preserve the top left corner of a graph, what is very useful when metric information is required to distinguish between characters like 5 and S. Thinning reduces characters to plane graphs (one pixel thick strings) as shown on Fig. 2 [12].

The pipeline processing of this stream of pixels results in segmentation which is successfully accomplished even on italics and other situations with unusual kerning (Fig. 2). A segment is regarded as a set of connected links where each link is a set of colinear pixels. The pixels and links critical for graph identification are assigned arbitrary numbers. Links are disconnected if the distance between them exceeds the range of the weighted neighborhood distance matrix (Fig. 3a). If the voids between links are smaller than the matrix range (we call those soft–breaks on the graph), the links are connected. To avoid unnecessary connections leading to small bounded regions the connectivity is decided upon examination of the non–commutative table (Fig. 3b). When the stream of pixels is analyzed after thinning, the neighborhood distance is always evaluated for the pixels of value 1 (black); at the same time a buffer is being filled with current information and with look ahead connections (Fig. 4).

In operational terms the result of the pipeline processing is a flow of segments, where each is a non–ordered array of links and vertices identified by the assigned numbers.

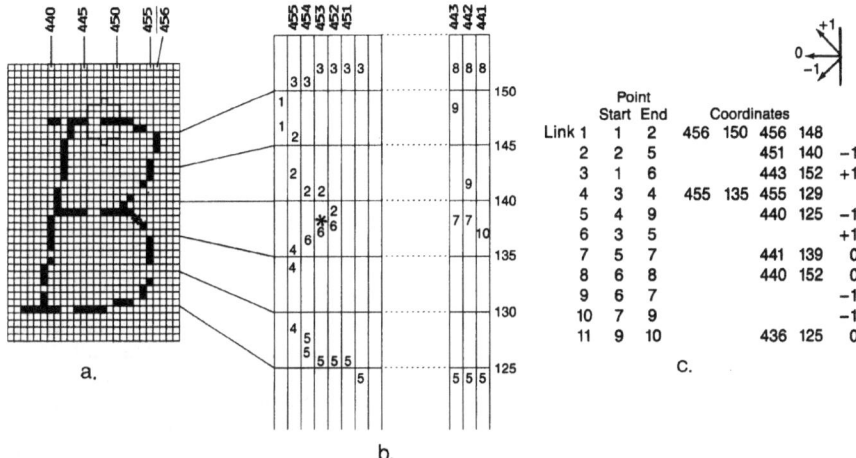

Fig. 4 Scanning (a), Look-ahead buffer (b), and Output (c)

Point

Link	Start	End	Coordinates			
1	1	2	456	150	456	148
2	2	5			451	140 −1
3	1	6			443	152 +1
4	3	4	455	135	455	129
5	4	9			440	125 −1
6	3	5				+1
7	5	7			441	139 0
8	6	8			440	152 0
9	6	7				−1
10	7	9				−1
11	9	10			436	125 0

c.

a) Binary image being scanned. Scanning direction top-down, right-left. The distance neighborhood matrix is shown in heavy line. The look-ahead buffer is being filled at every scan, see ∗. The output (c) is represented on Fig. 5.

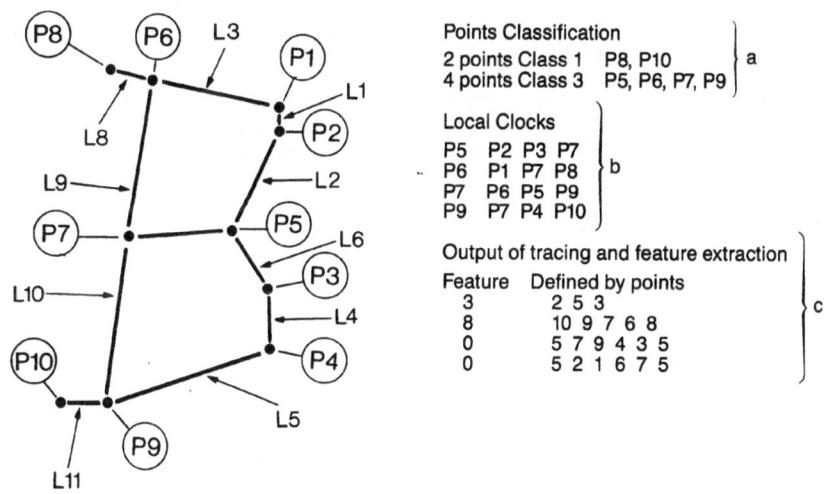

Points Classification
2 points Class 1   P8, P10
4 points Class 3   P5, P6, P7, P9   } a

Local Clocks

P5	P2	P3	P7
P6	P1	P7	P8
P7	P6	P5	P9
P9	P7	P4	P10

} b

Output of tracing and feature extraction

Feature	Defined by points
3	2 5 3
8	10 9 7 6 8
0	5 7 9 4 3 5
0	5 2 1 6 7 5

} c

Fig. 5 Steps in Recognition

a) Points Classification. If graph is a character, since it belongs to class (2,4) can only be 1) B, 2) &, 3) Ø, 4) æ

b) The local clock enables the tracing, ordering of links, and extraction of features c).

c) Since graph has only 4 features can be only 1) or 4). Geoarithmetic differentiates 1) from 4) and recognizes B

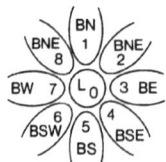

Fig. 6

Nine geofeatures
represented
by one nibble.

The links of the individual segments
are ordered into continuous strings by
a tracing routine that first classifies
the vertices as class 1, 2, 3 or 4, and
then defines the local orientation
clock for each of class 3 or 4 vertices
(Fig. 5).

The tracing procedure starts on any existing vertix class 1, and continues
until it finds another vertix class 1 (it could be the same one closing up the
path). Every time the tracing passes through the vertices class 3 or 4, where
more than one path is possible, the local orientation is used to decide on the
one to choose. When all the vertices class 1 have been used as starting points
of the traces, we check if there are any paths left which include vertices class
3 or 4 and have not been traced yet. If any such paths are detected, they
are traced starting from any point class 3 or 4. The procedure is continued
until all the possible paths (strings) are traced.

During the tracing procedure concavities and convexities are detected.
Concavities and close loops are the only features of interest, we call them
geofeatures. Nine geofeatures are used (Fig. 6), the close features are called
lagoons, and eight open features – bays: BN, BNE, BE, BSE, BS, BSW, BW, BNW.

In case of a perfect graphical representation of the characters we would
reject all the convexities. Unfortunately, global concavities found in real life
binarized and thinned characters often include small convexities, and viseversa.
Therefore, a set of special smoothing filters is applied before starting the
recognition process itself (Fig. 7).

At this point the only parts of the strings preserved are Bays and Lagoons.
The result of the operations described is an abstract numerical representation
of what could be called The Paradigmatic Symbols [10] of the Alphabet, that
constitutes the basic knowledge used in our recognition process [6].

The graphical representations of the paradigmatic symbols are topologically
classified according to the number of vertices class 1 and 3, and, within each
of those subsets, characters are differentiated by the number, nature and
relative location of the geofeatures. What it means is the following:
-    If a graph has n1 vertices class 1 and n3 vertices class 3, and we hypothesize
that the graph is a character, then it shall be a character within a corresponding
cell of the Topological Table (Fig. 8).
-    Furthermore, if f is the number of geofeatures in the graph, then our
hypothesis is reduced to even smaller set of characters.
-    And, finally, if geofeatures are located in a specific disposition, then the
graph is a specific character.

RECOGNITION PROCESS

Operationally, the process of recognition is conducted in the following way:

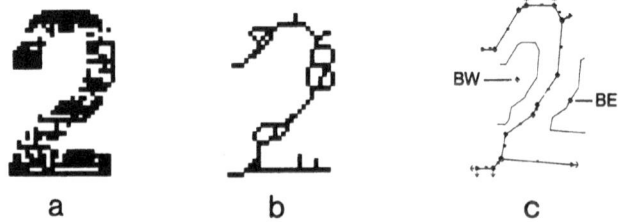

a   b   c

Fig. 7 The Results of Prerecognition on the PB Process

a, b, and c are done on pipeline. The original a, is abstracted into
the two bays of c.

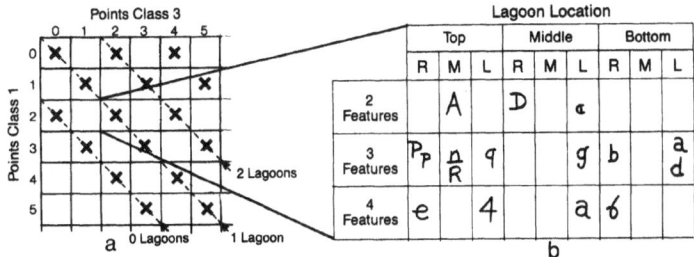

Fig. 8 Topologic Classification of English Characters and Symbols
a) There are characters only in the boxes **x**   b) Possible characters on cell (2.2)

1. As the result of prerecognition, we have information about the number of points class 1 and 3 and number of geofeatures. If those numbers point to a possible character from a particular cell of the Topological Table, then a hypothesis is created, and a next step is its verification. Otherwise, such information indicates a possibility of connected characters.

2. To verify a hypothesis made on the step 1, we examine the relationship of geofeatures, and since this relationship is not topological anymore but geometrical, we define a geoarithmetic, i.e. arithmetic for geofeatures.

GEOARITHMETIC: DEFINITIONS

A thinned plane graph divides the inside of a convex poligone (the envelope of the graph) into subplanes associated with geofeatures. The features are located with respect to each other in two general intuitive ordered locations 'above/below' and (or) 'next-to' represented by the operators '/' and '+', respectively, the order being defined by the selected scanning directions.

The direction of scanning is assumed up-down and right-left. According to that, EAST has priority over all the other features in the 'next-to' direction, so that the EAST oriented bays (BE, BNE, BSE) shall always be placed before the WEST oriented bays (BW, BNW, BSW) in a character description. Moreover, NORTH has priority over SOUTH in the direction 'above/below'.

When both operators are used in an algebraic string, the priorities between "+" and "/" are based on human spatial intuition, as in the example of Fig. 9, where F1, F2, F3 are features identifying the subplanes. Both descriptions are acceptable, but in order to avoid ambiguity and select one description consistently we use additional rules related to the angular resolution of the simple cells of the cortex [11] and to the location of the features in respect to bisectors of the quadrants. However, once the definition is accepted and used with consistency, it is as valid as any other definition proposed by someone with different spatial intuition.

Based on the above, there are only 11 patterns describing the configurations around a single point class 3 or 4 (Table I).

	Table I		BASIC PATTERNS
PATTERN #	DESCRIPTION		POINT CLASS
1	F1 + F2/F3		Class 3
2	F1/F2 + F3		Class 3
3	(F1 + F2)/F3		Class 3
4	F1/(F2 + F3)		Class 3
5	(F1 + F2)/(F3 + F4)		Class 4
6	F1/F2 + F3/F4		Class 4
7	F1 + F2/F3 + F4		Class 4
8	(F1 + F2 + F3)/F4		Class 4
9	F1/(F2 + F3 + F4)		Class 4
10	F1/F2/F3 + F4		Class 4
11	F1 + F2/F3/F4		Class 4

$$\frac{F_1}{F_2} + F_3$$

$$\text{or } (F_1 + F_3) / F_2$$

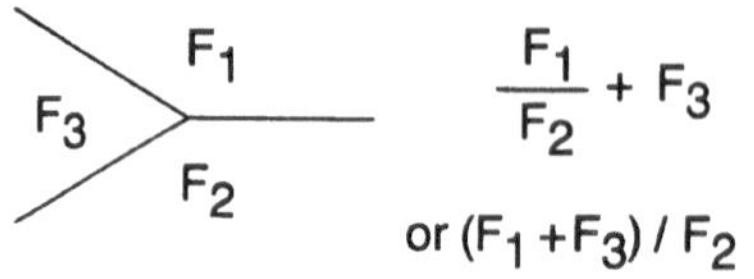

Fig. 9

Two possible algebraic descriptions for same spatial configuration

It is possible that one of the features clustered around class 3 or class 4 point is "absent". It happens when the links outlining a feature form a convex border. In that case the same pattern is used but the absent feature is eliminated.

GEOARITHMETIC: USE   FOR RECOGNITION

A graph representing a character may include more than one point class 3 or 4. In such a case the recognition procedure builds partial descriptions of a graph around each of those points using the basic expressions (Table I), and then performs an addition of the partial descriptions. During addition individual relationships of the points class 3 and 4 and their relative are preserved. The processing order of the points class 3 and 4 is based on the priority rules determined in the scanning directions (Fig. 5). The addition of the partial descriptions on Fig. 5 produces the result BE+ L1 / L2 + BW. This corresponds to the "Capital B" or "Number 8" entries in the dictionary. However, since the graph has two points class 1 and four points class 3, it belongs to the topological classification cell (4,3), and the answer can only be "CAPITAL B".

The characters without points class 3 or 4, spatially built in above/below or next-to relationship, are described with the same "+" and "/" operators, but the selection follows special rules. For example, the character W can be described as (BN + BS + BN), and the number 3 can be represented by a string BSW/BS/BNW.

CONNECTED CHARACTERS

When fitting a graph into the topological table we face two posibilities: a graph either belongs to a cell with meaningful characters, or does not. In the first case the regular processing continues until the final description is made. If the description does not coincide with any in the dictionary, then the graph is either incomplete character or a set of connected characters, and the procedure dealing with connected characters is immediately invoked. In the second case the graph is always presumed to be a set of connected characters.

The connections between characters are always characterized by a sequence BN/L1/L2/.../Ln/BS with the possibility of any element missing. Therefore, a procedure to hypothesize, break, and test for meaningfulness is applied. Eventually, the characters are disconnected or assumed to be incomplete for case 1, or disconnected or declared meaningless for case 2 (Fig. 10).

INCOMPLETE AND BROKEN CHARACTERS

The isolated graphs could be incomplete characters as mentioned above, or parts of the broken characters. In both cases a procedure using the Continuity and Closeness Gestalt laws is applied. In a case of incomplete characters we hypothesize that a point (points) class 1 can be connected to close the small gaps. In a case of broken characters we hypothesize that the closest bays can be connected. In both cases we then test the validity of our hypothesis. Eventually, the graph is either completed or reconstructed into a meaningful character or declared unrecognized (Fig. 11).

MULTIFONT

Since the recognition process works with thinned characters, the thickness

178

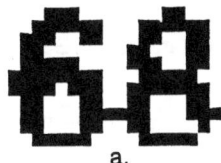
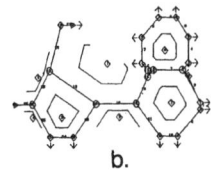

a.   b.

Fig. 10 Connected Characters

a) original   b) after prerecognition
2 points class 1, 6 points class 3. On cell (6,2)
there are no characters. Therefore the connected characters
algorithm is invoked.

of strokes is irrelevant. Furthermore, since we use topologic characteristics describing the graphs (vertices, lagoons, bays), the geometric details and style become almost irrelevant also. In consequence the process is multifont. Fig. 12 shows multifont letters A, all represented by a kernel L/S and some additional features. The letters were recognized as A, despite of the additional features.

CONCLUSION

The process has been implemented in software (hardware: PDP-11/34, with Image Processor IP64000, VAX-11/750) and tested with 8000 characters from real mail envelopes. The tested set included characters printed with different printing techniques and   quality, and handprinted characters. The results are very satisfactory: of the 8000 characters 16% were annotated as unrecognizable by a human observer. Our process recognized properly 91% of the humanly recognized characters. Some results are presented on Figs. 13. It is worth to note, that the results have been obtained without any contextual recognition enhancements.

The process is made fairly insensitive to noise by using filters as on Fig. 7, or, if the noise is an isolated speck, by eliminating it using the technique for broken characters.

The timing characteristics of the recognition are not available at the present, because the simulation has been done only to prove the validity of the concepts.

By design the process is implementable in hardware up to the segmentation. The remaining steps could be implemented in a parallel distributed process, either completely in software or in a hybrid system.

We believe the vailidity and soundness of the concepts have been proven, as well as the advantages of the "hypothesize and test" phylosophy of the process. The results indicate also the validity of this approach for the handprinted unconstrained characters and for the connected characters. It could be also applicable to script.

ACKNOWLEDGEMENTS: The contributions of H. L. Parker and D. H. Wilson during the initial phases of this work, and R. McClellan during the testing are much appreciated. The continuous support of Dr. R. A. Connell is recognized.

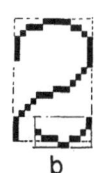
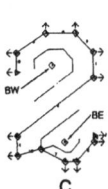

a   b   c

Fig. 11 Example of Broken Characters

a) Original   b) Thinned and segmented into two segments
c) Properly connected using the Proximity Gestalt Law. 2=BW/BE

A **AAA⊖AA** Basic Character **L/BS**

АААААА *АААА*А **A** BNE+L/BS+BNW

АА BNE+L/BS+BNW/BW

Fig. 12 The Omnifont Characteristics of the PB Process
Observe the kernel of letter A, **L/BS**. All the characters of this figure
are recognized by the process.

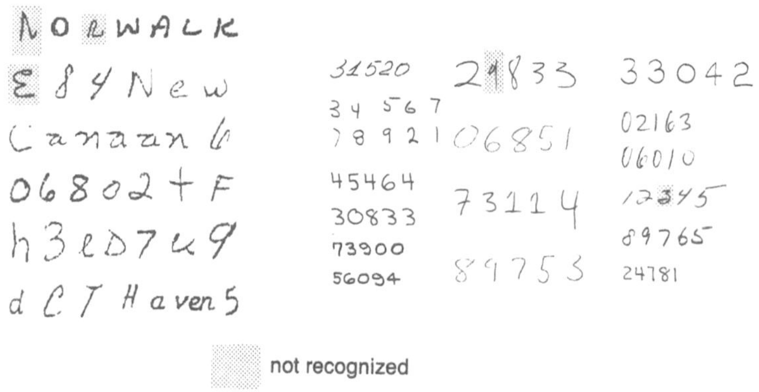

NO2WALK
E84New
Canaan 6
06802+F
h3e07u9
d C T Haven5

31520	2٩833	33042
34 567		02163
78921	06851	06010
45464	73114	12345
30833		89765
73900		
56094	89755	24781

not recognized

Fig. 13 Some Handwritten Characters Recognized with the PB Process

REFERENCES

[1] J. R. Ullmann "Advances in Character Recognition" in "Applications of Pattern Recog " (K. S. Fu, ed.), pp. 197-236, CRC Press, Inc., Florida, 1982.
[2] C. Y. Suen "Character Recognition by Computer and Applications" in "Handbook of Patters Recognition and Image Processing" (Y. T. Young, K. S. Fu, ed.), pp. 569-586, Academic Press, Inc., 1986.
[3] M. Shridhar, A. Badreldin "High Accuracy Character Recognition Algorithm using Fourier and Topological Descriptors", Patters Recognition, Vol. 17, #5, 1984, p. 515.
[4] R. M. K. Sinha "A Knowledge Based Script Reader", Proceedings of 7th Int. Conf. on Pattern Recognition, Montreal, Canada, 198, p. 763.
[5] R. Rozinovic, S. N. Srihari "Knowledge Based Cursive Script Interpretation", Proceedings of 7th Int. Conf. on Pattern Recognition, Montreal, Canada, 198, p. 774.
[6] J. Pastor, E. Zemelman "Knowledge Representation of Alphanumerics in Terms of Graph Theory, and its Application to a New Omnifont OCR Process", II Inter Sympo on Knowledge Engineering, Madrid, Spain, 1987.
[7] K. Koffka "Principles of Gestalt Psychology", p. 110, Harcourt, Brace & World, NY 1935.
[8] J. Pastor "Image Thinning Process", US Patent 4539704, Sept. 3, 1985
[9] J. Pastor, H. L. Parker, D. H. Wilson "Real Time Character Thinning Process", US Patent 4574357, March 4, 1986
[10] Watanabe, Satosi "Pattern Recognition, Human and Mechanical", p. 443, John Willey & Sons, Inc. NY, 1985.
[11] J. P. Frisby "Seeing, Illusion, Brain and Mind", p. 111, Oxford University Press, 1980.
[12] Kahans S., Pavlidis T. and Baird H., IEEE; PAM 1-9, 2, 274, 1987 use also thinned stroke characters as primitives.

BASIC PATTERN RECOGNITION TOOLS

OPTIMAL CONVEX SET INCLUDED IN A BINARY FIGURE

Jean-Marc Chassery

Equipe de Reconnaissance des Formes et de Microscopie
Quantitative Laboratories T1M3
Cermo BP 68 38402 Saint Martin d' Heres Cedex
France

Abstract

The interpretation of a binary figure requires geometric methods. The access to interpretation can be obtained either by an approximation process or by a decomposition process.
For example when using the concept of convexity, we use the convex hull notion to obtain an approximative representation of the initial figure.
In this paper we present a new method oriented to the search for a convex set included in an initial figure.

Introduction

In image Analysis the concept of shape is currently matched with descriptive notions permitting the parametrized representation of binary figures.
Such a point of view induces classical notions of perimeter, area, curvature, convexity and topological quantifiers or morphological quantifiers.
These shape descriptors are well adapted in particular situations of figure identification or figure caracterization. It is principally in the domain of Pattern Recognition that we identify a figure resulting from a pre-identified set of patterns. As in case of identification of figures derived from a same pattern, such parameters are sufficient.

Nevertheless, there exist some situations for which the analysis of a complex figure or the analysis of a binary architecture requires the introduction of some caracteristic elements to obtain a representation of these complex figures.
The notion of caracteristic elements is related to the use of a reference set or a basis in which the figure has to be described.
In shape analysis a reference can be used either by introducing a geometric constraint (for example the convexity), or by considering a morphological constraint (for example, the introduction of a set of elementary figures such as morphological masks).

183

The use of a notion of reference can be organized in two classes :
- decomposition methods ;
- approximative caracterization.

The first part of this paper introduces different shape understanding methods currently encountered in the litterature. This part will be oriented to decomposition methods and approximative caracterization methods.

The second part will focus on the presentation of a new specific method designed to obtain an optimal convex set included in a binary figure. The presentation will be essentially in an algorithmic form completed by proofs and illustrations.

I. FIGURE ANALYSIS
I.1. Representation of elements

The analysis of figures requires the introduction of a representation mode supported by an algorithmic process in order to access the shape description.

Two representation modes are currently used in reference to spacial environment of the image support. They are the contours and the regions. For example, a polygonal figure can be seen in terms of a subset of space or in terms of delimited regions.

It is generally easy to pass from one representation mode to the other and the justification of the definition of such two modes is related to problems of complexity. Our discussion in this paper will concentrate on these two modes, but we will also mention other existing modes not directly associated with concept of spacial distribution.

Consider, for example, the representation by Quadtrees or Hough Transform or skeletization. Such modes are associated to numerical considerations (coding, compression, minimal representation of figures, ...).

These modes are not directly correlated to visual indices for figure caracterization.

With the concept of visual indices we must to refer the notions of globality and deformation. When a figure is examined, we analyze it from a global level to a more local level. the notion of globality and deformation are transposed in terms of compacity, circularity and regularity. Very often, the geometric concept of convexity is introduced to measure the globality in terms of a convex hull.

In order to transpose the visual interpretation into a computer vision environment we define two classes of methods :
- decomposition methods ;
- approximative caracterization methods.

Although these two classes use the same geometrical environment and tools, they differ in the following point. Decomposition methods associate a set of figures to an initial figure. That set of figures is constructed by geometric or morphologic constraints.

Conversely, approximative caracterization methods associate an unique figure based on an optimality criterion.

I.2. Decomposition or Approximative Caracterisation

Let a shape figure be represented by its contour or by the notion of connected component. There exist different methods to access its decomposition.

184

- The Decomposition in convex elements is currently used in Computationnal Geometry. More specifically we have :
    - the use of Voronoï polygons in a region mode [1] and
    - the use of concave points in a contour mode [2] .

- Decomposition by textural elements is currently used in Mathematical Morphology. In this case, the partition process requires the use of different types of textural elements, each one being caracterized by its geometry [3] .

- The decomposition in rectangles can be performed. The caracteristics of rectangles are used to distinguish the global from the local contribution in the initial figure [4] .

- A contour decomposition can be performed by the definition of caracteristic points in association with some curvature notions [5] . this class of methods is used in shape filtering [6] .

These methods are related to the use of a spacial reference system. But in case of the introduction of a new reference system (Quadtree, skeleton ...), we can refer to methods based on a hierarchical description of figures, a decomposition method based on the concept of medial axis [7] . For example, this decomposition method permits the partitioning of the figure in terms of principal components and components of connexion and local deformations [8] .

In a decomposition environment, the visual interpretation of the figure is performed by successive selections of different elements of the puzzle composed by elementary figures. Such a process requires the use of a descriptive environment in terms of a graph permitting the manipulation of notions of adjacency between the different elements. Each elements is caracterized by its geometric figure.

Concerning the second class of methods, called Approximative Caracterization, we have the following list :

- the search for a minimum area encasing rectangle [9] .

- the search for the optimal ellipse (in the sense of least square approximation) [10] .

- the search for the convex hull [11] .

- the search for the optimal morphological figure included in the initial figure and composed of union of textural elements [3] .

In this class of methods, visual interpretation essentially arises from the analysis of the approaching figure. Moreover, considerations of deformations can be evaluated by a comparison of the initial figure and the approximate figure.
It is in the latter class of methods that we propose a new method to determine an optimal convex set included in a figure.

II. SEARCH FOR AN "OPTIMAL" CONVEX INCLUDED IN A FIGURE
II.1. Preliminary notations

Let A be a binary image dimensionned by NxM and valued with 0 and 1. Pixels valued with 1 compose the object class to be analysed.
We suppose that A contains only one simply connected component (without holes).
For the object class we use the 8-connectivity and the **contour** is

introduced as the set of points of A valued by 1 with a 4-neighbor valued by 0.

The **shrink** of an object is defined as the object obtained after elimination of contour points.

A **level curve** is introduced as the set of object pixels which are localized at the same distance from the background (pixels valued by 0). The level curves are obtained by successive shrinking.

The **convex hull** of an objects is caracterized as the set of pixels of A included in the convex polygonal line constructed from the contour points of an object [12] . In Figure 1 we give an illustration of the convex hull of a connected component.

To present the construction of a convex set included in a connected component, we will first consider an elementary case followed by the general situation.

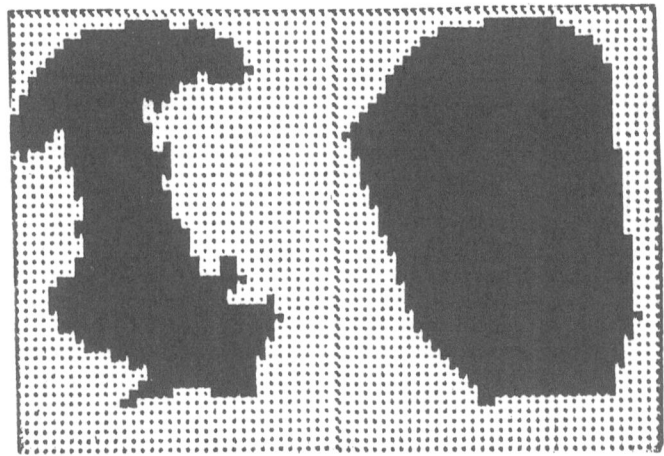

Figure 1.   Initial figure and its convex hull

II.2. Elementary case

In figure 2 we present an elementary case with a non convex object having an unique concavity.

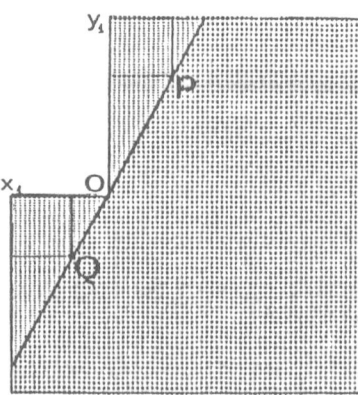

Figure 2.   Polygonal figure to illustrate the elementary case

By use of different notations presented in Figure 2, we can make the following observations.

Let D be the straight line crossing by point O (point of concavity) with slope coefficient noted by **a** and given by $a = y_1/x_1$. This straight line fixes a boundary of a convex polygon included in the initial figure. Moreover, the slope of this line satisfies an optimal criterion concerning the minimum area of the residual part from the initial figure.

The residual area is given by the formula $\frac{1}{2}\left(a\ x_1^2 + y_1^2/a\right)$ and it is minimum for $a = Y_1/x_1$.

It follows that the straight line D contains 2 points P and Q such that d(P, contour) = d(Q, contour) where d corresponds to the metric defined by :

$$d(A,B) = \max\ (|x_A - x_B|,\ |y_A - y_B|)$$

In the case of Figure 2 we have :

$$d(P,\ contour) = Y_1/(a+1)\ et\ d(Q,\ contour) = ax_1/(a+1)$$

Points P and Q are on the same level curve and they are straightened with the point of concavity O.

Points P and Q are unique on the level curve in the sense that the straight line D is tangent to that level curve.

Thus the basic idea of the method can be formulated as the search for points O, P and Q satisfying the conditions :
- **O is an articulation point. It belongs to the contour of the initial figure and it is located at a concavity angle ;**

- **P and Q belong to a same level curve with the condition that the straight line PQ is tangent to that level curve ;**

- **Points O, P and Q are on same straight line.**

In Figure 3 we present an illustration of the result obtained in the elementary case.

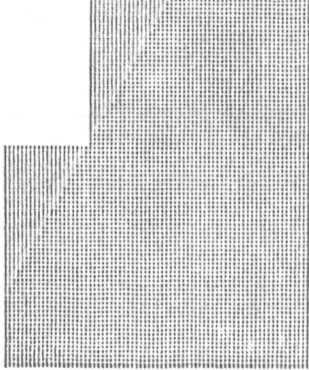

Figure 3.   Illustration of the resulting convex set included in a figure corresponding to the elementary case

To apply the basic idea for finding points O, P and Q, we shall proceed by successive shrinking transformations on the initial figure and at each step we test for the inclusion of the convex hull of the iterated figure in the initial one.

As soon as there exists an iterated figure satisfying the inclusion condition, we determine the points O, P and Q by analysis of the contours of the iterated and the initial figure.

The optimality criterion will be expressed in terms of caracterization of points O, P and Q. In the elementary case such a criterion is equivalent to an area optimization.

II.3. General case of a simply connected component

Let there be a simple connected component defined as the initial figure (Figure 4) on which we shall perform successive iterations.
According to the previously mentionned basic idea, we shall search at each iteration, a point of concavity (noted by O_i) to which we associate corresponding pixels P_i et Q_i.
The process starts with a preliminary convexity test of the initial figure. If the test is negative, then we enter in an iterative phase described as follows.

Begin

    Initialization :
        Iterate figure : = **Initial figure**

    While (**Iterate figure** not convex) do

    begin

        - **Shrink figure** : = shrinking (**Iterate figure**)

        - While (Convex Hull (**Shrink figure**) not included in **Initial figure**) then **Shrink figure** : = shrinking (**Shrink figure**).

        - Research of angle points S_j of the contour polygonal line of the **Shrink figure.**

        - Research of points O_i belonging simultaneously to the convex hull of the **Shrink figure** and to the contour of the **Initial figure.**

        - for each candidate point O_i we determine from the list of angle points S_j the couple P_i, Q_i verifying : "O_i, P_i et Q_i are on a same straight line".

        - Modification of the **Iterate figure** using the half plane generated by the straight line $P_i Q_i$ (this half plane contains the convex hull of the **Shrink figure**)

    End

End

    On Figure 4 we illustrate the different steps of this iterative process.

188

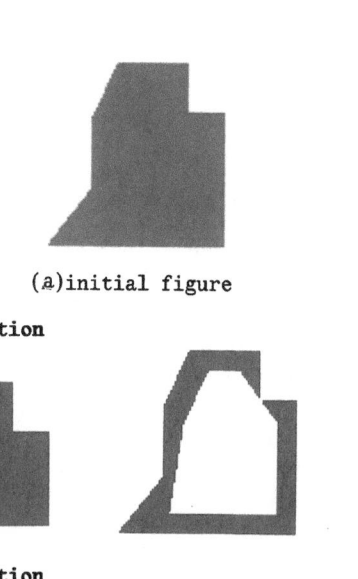

(a) initial figure   (b) initial figure and
                          included convexe

1st iteration

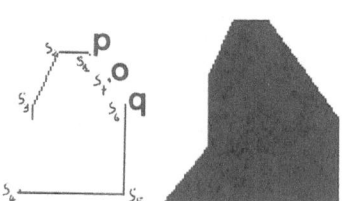

2nd iteration

  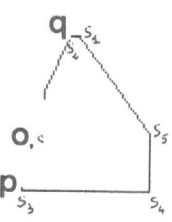 

(c)   shrink figure   convex hull of   Identification of   Result of the
                       the shrink figure   pixels O, P and Q   iteration step

Figure 4.   Illustration of the iterative process
a – Initial figure **b** – Superposition of the initial figure and results
                    c – Details of the 2 iterations related to concavities

II.4. Notion of optimality

The justification of the definition of the optimality criterion in terms
of minimum area is more difficult in the general case than in the
elementary case.
This criterion is oriented to a double caracterization.

- the first caracterization concerns the definition of
  articulation points. These points are associated to concavity
  angles of the contour of the initial figure.

- the second caracterization concerns the definition of contact
  points between a specific level curve and a straight line
  crossing by the articulation point.

Such a caracterization is unique and invariant in translation and
rotation.

In Figure 5 we propose some illustrations of the method.

CONCLUSION

The proposed method enters in the family of methods oriented to the definition of an approximative caracterization of a figure.
There presently exist few methods oriented to the interior decomposition of an object. It is essentially in the domain of Mathematical Morphology that we find the majority of such approaches.
The proposed method constitutes a first element to be compared with more general methods currently named in terms of Split and Merge. The comparison will be oriented to the use of Voronoï Polygons in order to access a local level in a figure decomposition environment.
Therefore, we note the importance of Computationnal Geometry methods in applications to Discrete Geometry for which the proposed method is a specific illustration.

Figure 5. **Illustrations of results of convex set included in a figure**

REFERENCES

1   F.P. PREPARATA, M.J. SHAMOS : Computationnal Geometry, Comp. Science, 1985.

2   B. CHAZELLE, D.P. DOBKIN : Optimal Convex Decomposition, in Computationnal Geometry, TOUSSAINT Ed. North Holland, 1985.

3   J. SERRA : Image Analysis and Mathematical Morphology, Academic Press, 1982.

4   J. SKLANSKY : Measuring concavity on a rectangular mosaïc, IEEE Trans, on Comp. C21, PP 1355-1364, 1972.

5   S.H.Y. HUNG, T. KASVAND : Critical points on a perfectly 8- or 6-connected thin binary line 6th IJCPR, pp 531-533, MUNICH 1982.

6   T. KASVAND, N. OTSU : Regularisation of piecewise linear digitized plane$_{th}$ curves for shape analysis and smooth reconstruction, 6th IJCPR, pp 468-471, MUNICH 1982.

7   C$_{th}$ARCELLI, G. SANNITI DI BAJA : Medial line and figure analysis, 5th ICPR, pp 1016-1018, MIAMI 1980.

8   A. MONTANVERT : Medial line : graph representation and shape description, 8th ICPR, pp 430-432, PARIS 1986.

9   H. FREEMAN, R. SHAPIRA : Determining the Minimum area encasing rectangle for an arbitrary closed curve, Comm. ACM, vol 18, 7, pp 409-413, 1975.

10   T.C. ZAHN, R.Z. ROSKIES : Fourier descriptors for plane closed curves, IEEE Trans. on Comp., C21, pp 269-281, 1972.

11   C.E. KIM : A linear time convex hull algorithm for simple polygons, Dept Comp. Science, Univ. of Maryland, TR 956, 1980.

12   J.M. CHASSERY : Discrète Convexity, CVGIP, vol 21, pp 326-344.

A NEW CONCEPT FOR BINARY IMAGES : THE KERNEL

M. Lamure and J.J. Milan

UA 934 du CNRS
43 boulevard du 11 novembre 1918
69622 Villeurbanne Cedex - France

Introduction

In image processing, skeletonization of binary patterns consists in thinning the pattern until to get a line drawing. The thinned pattern, called the skeleton, must preserve the connectedness and shape of the original one. Many skeletonization algorithms exist such as those of Hilditch (5), Stefanelli and Rosenfeld (6), Chassery (2), ...

In this paper, we propose a closely related concept. By successive thinning of the pattern we get a new one, called the kernel, which is unique, otpimal, i.e. filled with a minimum number of pixels, and from which it is possible to retrieve the original pattern without loss of information. If the shape of patterns is preserved by the kernel, it is not always the case for connectedness. However, if preserved, the kernel becomes a skeleton endowed with all mathematical properties of "euclidean" skeletons.

Our approach is more turned towards data compression with shape preservation and we do not draw our prior attention to connectedness preservation.

The general procedure is as follows(cf. fig. 1)

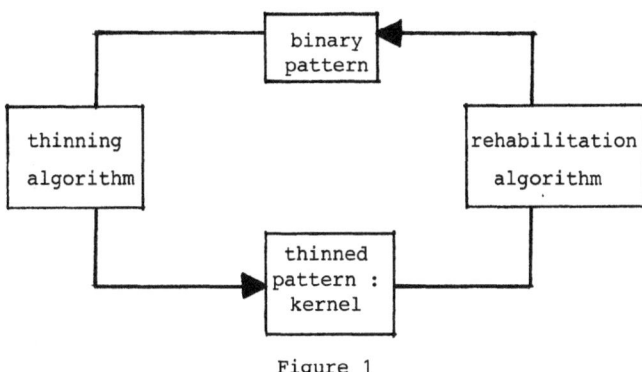

Figure 1

191

The two above algorithms are converse. This is due to the fact **they are based on two mathematical operators** which are the closure and interior functions of the pretopology theory. Pretopology is an extension of topology and was already presented many times in the domain of image analysis (cf. (1), (3), (4)).

Therefore, in this paper, we only present pretopological concepts we need and we subdivide it as follows :
1 - needed pretopological concepts
2 - the kernel
3 - the thinning algorithm
4 - the rehabilitation algorithm
5 - concluding remarks

1 - Needed pretopological concepts
Let E a non empty set. For any x in E, we consider $\mathcal{V}$(x) a family of subsets for x such as :

- $\mathcal{V}$(x) is non empty
- $\forall$ v, v $\in$ $\mathcal{V}$(x), x $\in$ v
- $\forall$ v, v $\in$ $\mathcal{V}$(x), v$\subset$u, u $\in$ $\mathcal{P}$(E)   (v $\subset$ u $\Rightarrow$ u $\in$ $\mathcal{V}$(x))
$\mathcal{V}$(x) is called the family of F-neighbourhoods of x.
Then, if $\mathcal{P}$ is defined as : $\mathcal{P}$ = {$\mathcal{V}$ (x), x $\in$ E }, $\mathcal{P}$ is called a pretopological structure on E and (E, $\mathcal{P}$) is a pretopological space.
Given $\mathcal{P}$ on E, we define :
- a from $\mathcal{P}$(E) into itself such as : a(A) = {x$\in$ E/ $\forall$ v, v$\in$ $\mathcal{V}$(x), v$\cap$A$\neq\Phi$} a is the closure function.
- i from $\mathcal{P}$(E) into itself such as i(a) = {x $\in$ E/ $\exists$v, v$\in$$\mathcal{V}$(x), v$\subset$A} i is the interior function

Example : Let us consider a binary image digitized by means of a square grid. It may be viewed as a finite subset of z^2. We define a pretopological structure on z^2, then on E, as follows : at any pixel x, we assign a subset $\mathcal{B}$(x) of its surroundings (cf. fig. 2).

Then $\mathcal{V}$ (x) = { v $\in$ P(E) / $\mathcal{B}$(x) $\subseteq$ v}   and
           a(A) = {x $\in$ E / $\mathcal{B}$(x) $\cap$ A $\neq$ Φ}
           i(A) = {x $\in$ E / $\mathcal{B}$(x)$\subset$ A}

$\mathcal{B}$(x), the four neighbours $\mathcal{B}$(x), the eight neighbours

Figure 2

The following schemas show the results obtained with a choice of $\mathcal{B}$(x) corresponding to the four neighbours (Cf. fig. 3 to fig. 5)

192

3.a   Original object

3.b   Object transformed one time
         by a

Figure 3

4.a   Object transformed two times
       by a

4.b   Object transformed one time
       by i

Figure 4

Object transformed two times by i

Figure 5

      In this paper, we focus ourselves on the two pretopological struc-
tures, noted $\mathcal{P}_4$ and $\mathcal{P}_8$, defined by means of the four and the eight
neighbours.

2 - The kernel

      Let E be a binary image (in fact a subset of Z^2) and A the set of
the object pixels. For any pixels x in A, we define k(x) the largest
integer such as $a^{k(x)}$ ({x}) is included in A.
Let us put K = {k(x), x $\in$ A}. This set is naturally ordered, we subscript
its elements in a decreasing order, then K = {k_1, k_2, ... k_n}.

Let A_1 defined by A_1 = {x $\in$ A / k(x) = k_1}   and B_1 = A $- a^{k_1}(A_1)$.

$a^{k_1}(A_1) = \bigcup \{a^{k_1}(\{x\}), x \in A_1\}$, then $a^{k_1}(A_1)$ is included in A.

If B_1 is empty, we stop, else we put $k_1 = k'_1$ and we define A_2 and B_2
as follows :
A_2 = {x $\in B_1$/k(x) = k'_2} where k'_2 = Max(K-K_1) and K_1 is the subset
of K obtained when we eliminate all k_i only corresponding to pixels of

193

$a^{k_1}(A_1).$

$B_2 = A-(a^{k'_1}(A_1) \cup a^{k'_2}(A_2)).$

If B_2 is empty, we stop, else we define A_3 and B_3 in the same way that A_2 and B_2.

By this procedure, we build two finite sequences $(A_1, A_2, \ldots A_p)$ and $(B_1, B_2, \ldots B_p)$ until B_p is empty.

Then $N = \bigcup \{A_j, j = 1 \ldots p\}$ is the kernel of A.

Its properties are :
- Given a pretopological structure, N is uniquely defined.
- Card(N) is minimum.

- $A = \bigcup \{a^{k'_j}(A_j), j = 1, \ldots p\}.$

Example

The following schema (cf. fig. 6) shows the kernel from the pattern of fig. 3.a

Figure 6

The two following results allow us to obtain a more convenient way for defining the kernel.

* Given $\mathcal{G}_4$ and $\mathcal{G}_8$, for any subset A of E, $a^k(i^k(A))$ is included in A.
* For any subset A of E, for any pixel x in A, if k(x) is the largest integer such as $a^{k(x)}(\{x\})$ is included in A, if k'(x) is the largest integer such as x is in $i^{k'(x)}(A)$, then $k(x) = k'(x)$.

2 - The thinning algorithm

A is the object of which we want to extract the kernel. Let b(A) be defined by :

b(A) = a-i(A)

b(A) is the set of the border pixels of A. Its elements are coded 1, those of i(A) coded 2, those of i(i(A)) coded 3, and so on.

Then our thinning algorithm can be described as follows :

a - initialisation, k = 1
b - repeat

computation of $i^k(A)$
k = k + 1

until ($i^k(A)$ is empty)

c - l = k - 1
d - repeat

computation of A_1

computation of $a^1(A_1)$

computation of B_1

$1 = 1 - 1$

until $(1 = 0)$

Example

The following schema (Cf. fig. 7) shows how the kernel of the pattern of fig. 3.a is obtained

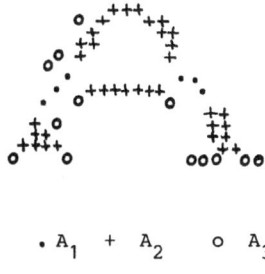

. A_1 + A_2 o A_3

Figure 7

4 - The rehabilitation algorithm

A priori, the sequences $(A_1, ..., A_p)$ with the corresponding orders of interior, noted $k_1, ..., k_p$, are available. Then, the rehabilitation algorithm is very simply described as follows :

a - initialisation, j = 1, A is empty

b - repeat

computation of $a^{k_j}(A_j)$

$A = A \bigcup a^{k_j}(A_j)$

until $(j = p)$

Example

The following schema (Cf. fig. 8) shows the rehabilitation of the pattern of fig. 3.a from its kernel.

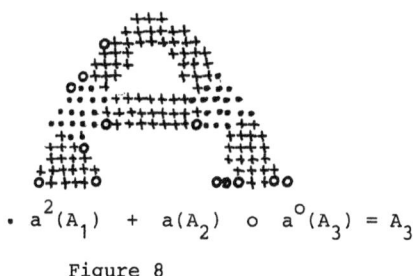

. $a^2(A_1)$ + $a(A_2)$ o $a^0(A_3) = A_3$

Figure 8

5 - Concluding remarks

By means of the concept of kernel, we are able to thin a binary pattern in an unique way and to obtain a result, the smallest as possible, which gives some informations about the shape of the pattern. Moreover, if its connectedness is preserved, the kernel becomes a skeleton on which all mathematical properties are conserved. At least, we can rebuild the original pattern from it without any loss of information.

Bibliography

1 - Arnaud G.M., Lamure M., Terrenoire M., Tounissoux D.
    "Analysis of connectivity in a binary image : a pretopological
    approach". 8th IAPR Conference, pp. 1204-1207, Paris, 1986.

2 - Chassery J.M.
    "Représentation discrète, interprétation numérique et description des
    images : des concepts à l'application".
    Thesis dissertation, University of Grenoble, 1984.

3 - Lamure M., Milan J.J.
    "A system of image analysis based on a pretopological approach"
    Intelligent Autonomous Systems Conference, pp. 340-345, Amsterdam,
    1986.

4 - Lamure M., Milan J.J.
    "An interactive system for image analysis : SAPIN"
    Medical imaging, SPIE Conference, Newport-Beach, 1986.

5 - Naccache N.J., Shinghal R.
    "An investigation into the skeletonization approach of Hilditch"
    P.R., vol. 17, n° 3, pp. 279-284, 1984.

6 - Stefanelli R., Rosenfeld A.
    "Some parallel thinning algorithms for digital pictures"
    J.A.C.M., vol. 18, n° 2, pp. 255-264, 1971.

GRAPH ENVIRONMENT FROM MEDIAL AXIS

FOR SHAPE MANIPULATION

Annick Montanvert

Equipe RFMQ-TIM3
CERMO - BP 68
38402 St. Martin D'Heres Cedex, France

Abstract :
This paper deals with shape description using a representation by shape covering with primitives.
In particular, I will present methods developed from the medial axis transform in discrete space where the primitives are squares.
After a presentation of the methods which define a graph environment to structure these primitives, a method based on the medial line transform will be presented.
The resulting medial line graph can be used for shape manipulation processes such as filtering or decomposition. Each graph node is associeted to a contribution to the original shape.

I. INTRODUCTION

When binary discrete pictures are extracted, after a segmentation process for example, another primordial step still remains : the interpretation of these pictures.
Depending on the application, several kinds of methods can be used.
If the objects to be extracted are known a priori, we speak of pattern recognition. The methods used then are statistical and based on the computation of shape parameters, or they are syntactic and based on representation with binary skeletons (1) .
If the objects are not known, the shapes must be described in order to understand them well enough to obtain a good interpretation.

An automatic process cannot work directly with a global vision of the shape, as it is sometimes say to be the case for the human perception of shape(Gestalt theory) ; it can only proceed by extracting shape characteristics without loss of information.
For elongated shapes with nearly constant width, thinning processes and a coding of the resulting binary skeletons can be used.
For any shape, decomposition into simpler entities is an usually used transformation.

The union of the resulting entities will be equal to the initial shape. Concerning this aspect of figure decomposition, the existing methods can be divided in two classes :

    – First, the search for particular points or configurations on the original shape provides a way to decompose it.
For example, working on the discrete picture and detecting concave points on the contour, the original shape can be decomposed into convex elements (2).
We can also work on a polygonal approximation of the shape with decomposition into Primary Convex Subset (3), or decomposition by graph-theoretic clustering (4). The resulting entities are extracted from relations between the vertices of the approximation.
The entities have properties such as convexity, but they may still be rather complex. We will call these approaches top-down because the entities need to be treated by other processes until an easily interpretable set of entities is found.

    – The second class of methods, which will lead to increasing approaches, is based on an a priori choice of the entities. Such is the case for methods based on mathematical morphology, quadtree or medial axis.
In mathematical morphology, the primary entity, called a structural element, allows tests to confirm or reject the presence of the associated structures in the original shape (5). For quadtree and medial axis, the primary entities are squares. They can be combined to build more complex entities.
For all of the methods, we note that some are partitionning methods, while others provide some overlapping entities ; partitions can also be deduced.
A complementary condition, which is essential for the success of such a method, is the construction of graph coding interrelationships between adjacent or overlapping entities.
This graph is generally generated simultaneously with decomposition processes of the first class.

For the second class of methods, these relations are not directly expressed, but are implicitly present.
In the representation using quadtrees, and more particularly the Quadtree Medial Axis representation (which partially solves the non invariance problem of quadtree under translation) (6), the structure of the quaternary tree provides a way to find adjacent or overlapping squares.
For the Medial Axis representation, we will show several ways to express these relations, devoting special attention to a method based on the medial line transform completed by the construction of the medial line graph. We will then present the bottom up processes of shape manipulation and interpretations from which they can be deduced.

II. MEDIAL AXIS AND GRAPH ENVIRONMENT

    The medial axis notion is issued from the grass fire progagation (7). In discrete space, wavefronts are symbolized by contour lines.
Then at every point of shape we associate its distance $d8$ to the background ; this value is equal to the (radius + 1) of the maximum square included in the shape and centered at this point (the radius is equal to 0 for an isolated point). The medial axis is then obtained by eliminating the points whose square is included in another one. This is performed by using two paths over the picture to build contour lines, and one path to detect local maxima (8).

The multivalued picture of the medial axis provides the low level interpretation of the shape. Each point of the medial axis, which is centered on the shape, codes a square. The union of all these squares is equal to the shape (see fig. 1a).
A major problem of this medial axis transform is its disconnection.

The first relation which appears between medial axis points is one in which associated squares can be adjacent or overlapping. This is an essential property for shape interpretation.
Thus two points can be joined if $d8(A,B) \leqslant rA + rB-1$, where rA and rB are the associated weight of A and B, points of the medial axis A and B.
Rather than test for condition each time, a relation graph can be built (9), which connects centers of the medial axis primitives that overlap or are adjacent. This was obtained using Delauney triangulation ; it provides a structure that can be compared to Quadtree Medial Axis.

An additional characteristic of medial axis in discrete space is that it can be locally thick, since sides of squares always have an odd number of points.
This can be solved by working in a derived space which contains fictitious centers for squares of even size (10).
An algorithm, which needs the storage of only two successive lines of the original shape, builds a relation graph between centers of directly related squares in this new space. Each node codes a square (11) . This transformation is reversible and homotopic (the shape and its core-line have the same number of connected components and holes).
A disadvantage of this method is that the definition of a new network does not allow distances transformed on the original picture to be obtained.

Another approach to the problem was studied. It consists in building a connection on the picture, by adding the lacking points. This defines the notion of medial line. Several algorithms have been proposed (12) ; recent processes are independant of the width of the shape (13,14) (see fig.1b).
This resulting multivalued picture can still have some local faults, such as thickness or undesirable holes.
But, as in the core line tracing algorithm (11) , it provides a structure descriptive representation.
Indeed, from the set of centers, we can obtain centers which are directly related ; the weight of these points provides locally added information about the shape. Shape manipulation methods can then be deduced (15,16).
The principal problem is to parse the medial line correctly since it is not strictly composed of discrete curves.
Thus, it would be interesting to deduce a graph structure from the multivalued picture of the medial line in order to express these relations (17) (see fig.1c).

We shall now consider our construction process of the medial line. It is based on the detection of local maxima (which are grouped into 8-connected points components which share the same label), completed by the detection of double points (which are local configurations of points which will meet more than once during the tracking of a contour line). From these double points, discrete connection paths are built.
This provides an intermedary level for shape interpretation (see Fig.1d):
     - every 8-connected component of local maxima provides a primitive which is a combination of squares,
     - double points and discrete paths define the graph structure between these primitives.

On the resulting graph :

     - a node is a previously defined primitive (we preserve its weight and positions of associated medial axis points), or particular double points, called strict double points, from which several discrete paths are initialized ,

     - an arc is a discrete connection path (on which the weight is strictly increasing) ; they are coded by the adjacency matrix called MAD.

We use the following representation :

     MAD (I,J) = 1, if there is an arc from node I to node J,
     MAD (I,J) = -1, if there is an arc from node J to node I,
     MAD (I,J) = 0, if there is no arc between nodes I and J.

Thus the medial line graph is obtained; it is an oriented planar graph. The associated transformation from the original picture is reversible and homotopic.

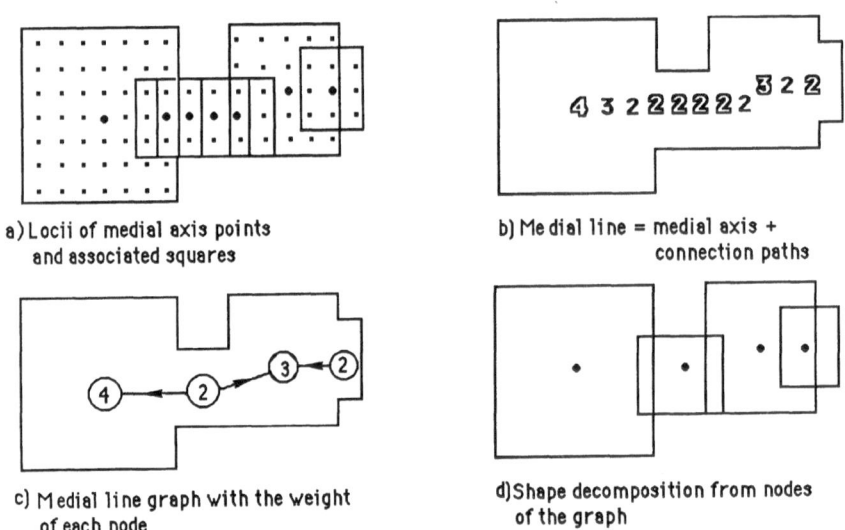

a) Locii of medial axis points and associated squares

b) Medial line = medial axis + connection paths

c) Medial line graph with the weight of each node

d) Shape decomposition from nodes of the graph

Figure 1. Transformation from medial axis to medial line graph.

III. USE OF THE MEDIAL LINE GRAPH

This graph can be compared to the Wakayama's core-line, since its trace is a 'skeleton' of the shape for which information is associated to particular points.

The advantage of such a representation is that two nodes are adjacent on the graph when the associated primitives are directly related. This permits the development of an description process of the initial shape.

It is important to note that :

     If I and J are two nodes such as MAD (I,J) = 1,
     Then weight (J) = weight (I) + lenght (discrete arc (I, J))

Thus the weight of node J is greater than weight of node I, which represents on the shape : the area associated to I is a deformation of the area associated to J.

From this interpretation, a contribution to the shape can be associated to every node ; it will be dependant on its local configuration in the graph :
- I is a principal node if, for every J, MAD (I, J)<0,
- I is a prominence node if there exists only one J such that MAD(I, J) > 0,
- It is a neck node if there exists different J1 and J2 such that MAD (I, J1)>0 and MAD (I, J2)>0.

The tail of an arc can be a prominence node or a neck node, its head can be of any kind.
To principal nodes are associated main areas of the shape, each representing a 'heart' of the shape. These areas are modelled by prominence areas which are adjacent to the principal node, and are themselves modelled by prominence areas, etc... Finally, neck areas (associated to their prominence areas) finish off the structure by connections between more important areas (see fig.2a).
The associated process, starting from principal nodes and adding successively prominence and neck nodes, provides a way of progressively reconstructing a shape from main areas to secondary areas.

From the previous remarks, we note that a prominence node leads to a necknode, or if not, to a principal node. Then at every prominence node, we can associated 'the node that it deforms'. Every neck node is associated to the neck or principal nodes it connects.

This permits us to define different filtering processes.
One classical approach is to remove points of low label of the discrete medial axis ; but this causes important connections established by low label to be lost.
Here, a filtering with connectivity control can be obtained. We must not remove neck nodes, or prominence nodes which are on a connection path from a neck node to a principal node (see fig.2b).
Moreover, filtering can depend on the context. Rather than select a global threshold, we can use a relative threshold between the node to be removed and the node on which it is dependant (see fig.2c).

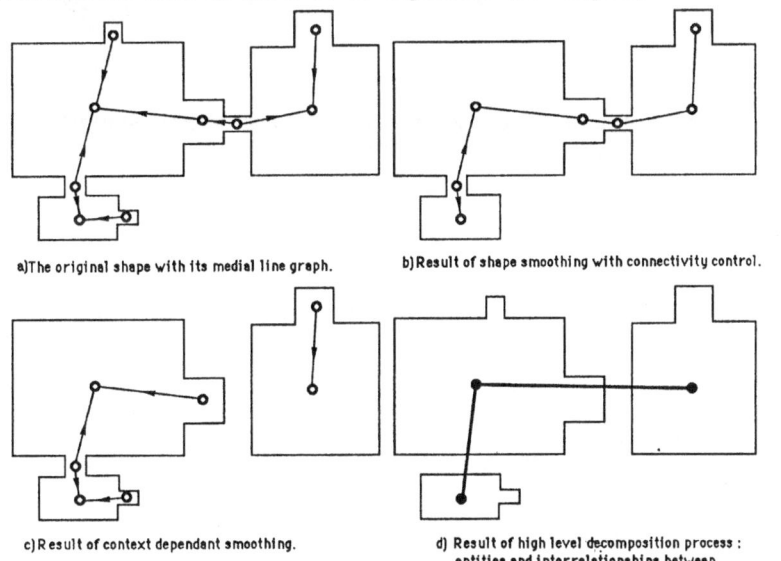

a)The original shape with its medial line graph.

b)Result of shape smoothing with connectivity control.

c)Result of context dependant smoothing.

d) Result of high level decomposition process : entities and interrelationships between these entities.

Figure 2. Shape manipulation from medial line representation.

Thus the medial line graph provides a global description of shape by using principal and neck nodes ; and each of these entities is modelled by its attached prominence nodes.

This can be used for high level decomposition of shape (see fig.2d).

Each neck or principal node completed by its prominence nodes provides a subgraph which is a tree.

Areas associated to these trees are the decomposition entities of a process which is based on detection of necks in the shape, which can then be compared to many other decomposition processes (see fig.3).

This method has been used to quarter an image of complex architecture of submarine sites (18) . All the processes presented here which work on the medial line graph are implemented in a software program which allows us to manipulate the shape ; after each treatment , the associated area can be reconstructed as a result of transformation reversibility.

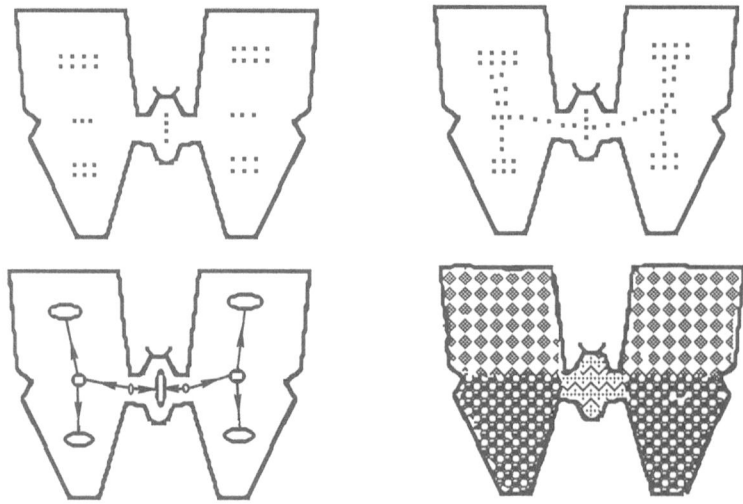

Figure 3. An example of partitionning by medial line graph analysis after its construction from medial axis.

IV. CONCLUSION

The use of the medial line graph as an intermedary representation of discrete pictures makes shape manipulation and the search for shape characteristics easier.

Two essential aspects of this are reversibility and construction of a graph structure.

Indeed,from the graph, each entity can be expressed in relation to its dependant entities ; such a graph environment provides a hierarchical representation of the medial line.

Another advantage of this method is that starting from a low level description of the shape with the medial axis, we progressively understand the shape and deduce an anthropomorphic decomposition process.
We are presently studying complements such as quantitication of contribution of each node on the shape or of configuration of nodes.

REFERENCES

(1) H. Tamura.
     A comparison of line thinning algorithms from digital geometry viewpoint.
     4th ICPR Kyoto. Novembrer 1978. p 715-719.

(2) B. Chazelle ; D.P. Dobkin.
     Optimal convex decompositions. Computational geometry.
     Toussaint ed. North Holland 1985.

(3) T. Pavlidis.
     Structural pattern recognition.
     Springer Verlag. 1977.

(4) L.G. Shap iro ; R.M. Haralick.
     Decomposition of two dimensional shapes by graph-theoretic clustering.
     IEEE Trans. on PAMI 1. p10-20. 1979.

(5) J. Serra.
     Image analysis and mathematical morphology.
     Academic press. 1982.

(6) H. Samet.
     A quadtree medial axis transform.
     Com. ACM 26 n°9. p680-693. 1983.

(7) H. Blum.
     A transformation for extracting new descriptors of shape.
     Sym. on MPSVF. MIT pres Boston 1964. p362-380.

(8) A. Rosenfeld ; J. F. Pfaltz.
     Sequentiel operatons in digital picture processing.
     J. of ACM vol. 13 oct 66. p 471-474.

(9) N. Ahuja ; W. Hoff.
     Augmented medial axis transform.
     7th ICPR Montreal 84. p326-328.

(10) G. Bertrand.
     Skeletons in derived grids.
     7th ICPR. Montreal 84 p326-328.

(11) T. Wakayama.
A core-line tracing algorithm based on maximal square moving.
IEEE Trans on PAMI. vol 4. January 1982 p68-74.

(12) C. Arcelli ; P. Cordella ; S. Levialdi.
From local maxima to connected skeletons.
IEEE Trans on PAMI. vol 2 March 1981. p134-143.

(13) C. Arcelli ; G. Sanniti di Baja.
A width independant fast thinning algoritm.
IEEE Trans on PAMI. vol 7. July 1985. p463-474.

(14) A. Montanvert.
Obtention d'une ligne médiane par connexion de l'axe médian.
5th RFIA. November 1985. Grenoble p777-785.

(15) C.R. Dyer ; S.B. Ho.
Medial axis based shape smoothing
7th ICPR. Montreal 1984. p333-335.

(16) L.P. Cordella ; G.Sanniti di Baja.
Context dependant smoothing of figures represented by their medial
axis transform.
8th ICPR. Paris 1986. p280-282.

(17) A. Montanvert.
Medial line : graph representation and shape description.
8th ICPR Paris 1986. p430-432.

(18) C. Charles ; J.M. Chassery ; A. Montanvert.
Application of image analysis to the determination of the possible
sweeping patterns of a manganese recovery system.
Offshore Technonogy Conference. Houston. May 1987.

WEIGHTED DISTANCE TRANSFORMS: A CHARACTERIZATION

Carlo Arcelli and Gabriella Sanniti di Baja

Istituto di Cibernetica del C.N.R.
80072 Arco Felice, Naples, Italy

INTRODUCTION

In many instances, it is convenient to label the space enclosed within the contour of a single-valued digital figure F. Labeling F by means of its distance transform DT, has been one of the first approaches to give structure to an otherwise amorphous space, and has been useful to reveal some of its features, especially those dependent on shape. In this framework, the set of the local maxima present in the DT plays a crucial role. In fact, the local maxima are necessary to identify the medial axis of F /1/. Moreover, figure decomposition techniques can be derived by suitably grouping the discs associated with the local maxima /2/.

In the last years, an increasing attention has been devoted to distance transforms and their applications (see for instance /3-5/). The problem of computing approximations to the Euclidean DT at a reasonable cost, has also been faced. This goal can be reached by adopting suitable weights so as to measure distance between neighboring pixels. An early example of weighted distance transform can be found in /6/, while some more recent ones are discussed in /7/. In addition to a suitable selection of the weights, also the size of the neighborhood, from which to derive distance information, has a critical role. The approximation becomes better the larger the size is. In /7/, 3x3, 5x5, and 7x7 neighborhoods have been taken into account, and different weights have been proposed. In this paper, we focus our attention on the 3x3 neighborhood, which has already been investigated in /8/, but only as regards the weights 3 and 4.

In general, the structure of a weighted distance transform (WDT, for short) considerably differs from that of the two known weighted distance transforms 4-DT and 8-DT, respectively computed according to the 4-metric and the 8-metric. For instance, the layers cannot be identified as connected sets of equilabeled pixels, and the local maxima not necessarily have label greater or equal to that of the neighboring pixels. Aim of this paper is that of gaining an insight into the structure of the weighted distance transforms. In particular, we suggest a criterion to identify the local maxima in any WDT, as well as a procedure to build the associated discs. A way to identify the pixels belonging to the layers is also presented.

WEIGHTED DISTANCE TRANSFORMS

Let $F = \{1\}$ and $F* = \{0\}$ be the two sets constituting a binary picture, digitized on a square grid. Let F be completely surrounded by F*. Moreover, let u and v be the two integer numbers respectively adopted to weight distance between any two horizontally/vertically and diagonally adjacent pixels. Such adjacent pixels will also be referred to as D-neighbors and I-neighbors, respectively. Finally, let the eight neighbors of any pixel p of F, be indicated by the corresponding Cardinal points in the compass.

The distance transform WDT of F with respect to F* is a replica of F, where each pixel is labeled with its distance from F*, computed according to a given function. In principle, to evaluate the distance of any pixel p from F*, global operations should be performed. However, local distance information can be propagated through neighboring pixels, so that local operations turn out to be sufficient to achieve labeling of F. In practice, the two operations f'(p) and f"(p), as defined below, are applied to every pixel p belonging to F during a forward and a backward raster inspection of the picture, respectively:

$$f'(p) = \min(W+u, NW+v, N+u, NE+v)$$

$$f''(p) = \min(p, E+u, SE+v, S+u, SW+v)$$

Every pixel receives distance information only from its already inspected neighbors. At the end of the forward inspection of the picture, every p in F is assigned a label indicating the length of the shortest path from p towards F*, oriented along any direction among West, North-West, North, and North-East. During the backward inspection, the label of p is confirmed unless a path from p towards F* is found, along any of the remaining directions (East, South-East, South, and South-West), which is shorter than that computed before.

In the resulting multivalued figure, if d and i respectively count the number of D-neighbors and I-neighbors, through which local distance information is propagated from F* towards a pixel p, it is $p = u \cdot d + v \cdot i$. Since d and i can be expressed by means of a linear combination of the 4-distance and of the 8-distance function, the value p is a distance in the mathematical sense. Note that only the numbers which, as p, are linear combinations of the selected weights u and v, are licit distance labels in the WDT.

For every pixel p in the WDT, at least one neighbor exists through which p derives distance information. Such a neighbor is less internal than p in F. In turn, p communicates distance information to any neighbor q more internal than p in F, whenever p is located along one of the shortest paths from q towards F*. In the affirmative, it is q = p + z, where it is either z = u or z = v, depending on whether q is a D-neighbor or an I-neighbor of p. Otherwise, a further pixel exists, from which q derives distance information, and it is q < p + z.

We say that p is a local maximum, if p is not located along the path towards F* starting from any of its neighbors. Formally, p is a local maximum if for every neighbor q there results:

(1)   q < p + z

where z = u   for q ∈ { N,E,S,W } ; z = v   otherwise.

		3	3	3	3			
	3	3	4	6	6	4	3	3
3	4	6	7	8	8	7	6	3
3	6	8	(10)	11	11	9	6	3
3	4	6	7	8	8	7	6	3
	3	3	4	6	6	4	3	3
		3	3	3	3			

Figure 1. Though adjacent to a pixel having greater label, the encircled pixel is a local maximum.

The validity of the previous definition can be guaranteed by proving that the discs, centered on the local maxima of the WDT, are maximal. Namely, the following two properties hold:

i) The union of the discs coincides with F.

ii) Any two discs never overlap completely.

To build the disc associated with any pixel p of the WDT, the reverse distance transformation has to be introduced. To this purpose, let us suppose that all the pixels in the WDT are set equal to zero, except for p. Then, the two local operations $g'(q)$ and $g''(q)$,as defined below, are applied to every pixel q having all the neighbors in the picture, during two inspections of the picture performed in forward and in backward raster fashion, respectively:

$$g'(q) = \max(q, W-u, NW-v, N-u, NE-v)$$

$$g''(q) = \max(q, E-u, SE-v, S-u, SW-v)$$

The shape of the disc depends on the adopted weights and, accordingly, on the selected distance function. Square-shaped and diamond-shaped discs, are obtained by using the 8-distance and the 4-distance function, respectively. Quasi circular discs are obtained by selecting suitable pairs of weights, as the ones suggested in /7/.

As far as the size of the disc is concerned, it increases with the label p of the corresponding center, provided that such a label p is licit in the selected WDT (i.e., two integer numbers d and i exist, such that it is p = u·d + v·i).

It must be noted that if p and p+r, r>1, are two licit labels in the WDT such that no intermediate value p+s, s = 1, r-1, is a licit label, then the disc associated with p+r coincides with the disc associated with any p+s. In particular, it coincides with the disc associated with p+1, and the label p+1 is termed the equivalent label of p+r. As an example, refer to Figure 2, where the weights u=8 and v=11 have been used while applying the reverse distance transformation to pixels labeled 20, 21, and 22. Indeed, no pixel can be labeled either 20, or 21 in the selected WDT, since none of such values can be expressed as a linear combination of the two chosen weights. Since labels 19 and 22 are licit, 20 is the equivalent label of 22.

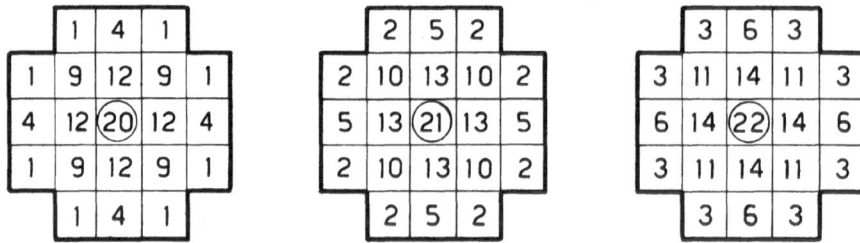

Figure 2. Differently labeled pixels (encircled) are centers of equally sized and shaped discs.

In general, equivalences between labels cannot be disregarded. In fact, suppose that according to condition (1), a pixel p in the WDT is selected as a local maximum. Neither the shape, nor the size of the associated disc changes, if the label of p is substituted by the equivalent label, if any. However, if this is done for every pixel on the WDT, p is no longer guaranteed to be a local maximum, as condition (1) could be no longer satisfied by using equivalent labels. If this was the case, property ii) would no longer be a feature peculiar to the discs associated with the local maxima. An example concerning the WDT computed by using 3 and 4 as weights appears in Figure 3, where for the encircled equivalent labels condition (1) is not satisfied.

a)

			③	3	3	3	3
			3	⑥	6	6	6
3	3	3	4	7	9	9	9
6	6	6	7	8	11	12	12

b)

			①	1	1	1	1
			1	⑤	5	5	5
1	1	1	4	7	9	9	9
5	5	5	7	8	11	12	12

Figure 3. Actual labels (a) and their equivalent labels (b) in the WDT with weights 3 and 4.

To overcome the above problem, condition (1) has to be checked by using the equivalent labels, if any. Under this condition, it can be proved that the discs associated with the local maxima satisfy both properties i) and ii).

From an operative point of view, the WDT is not actually updated with the equivalent labels. Indeed, the equivalences are recorded into a table which, for any WDT, i.e., for any pair of weights, can a priori be built as follows. For any positive value assigned to d (i) , and for i (d) ranging from 0 to d (from 0 to i), all the licit labels are computed and ordered according to their increasing value. Missing values are identified, and equivalences established. When all the integer values in between two successive multiple of u turn out to be licit labels, no further equivalence is searched because each integer number is a licit label since then on. This limits the extent of the table.

The layers on the 4-DT and the 8-DT can be identified as connected sets made of equilabeled pixels. On the contrary, equilabeled pixels are not generally grouped into connected sets in any other WDT, so that an alternative definition of layer is required.

By taking into account both the way in which propagation of distance information occurs, and that the layer index counts how many propagation steps have been already done, we say that a pixel labeled p is located on the j-th layer, if the following condition is verified:

(2) $$u \cdot (j - 1) < p \leqslant u \cdot j$$

According to such a definition, the layers are connected sets whichever weights are adopted to build the WDT. In particular, 8-connected and 4-connected sets of equilabeled pixels are found to be the layers of the 4-DT and the 8-DT respectively, as it was expected. As for the remaining WDT, each layer is likely to include pixels having up to u

different labels. Indeed, pixels belonging to the same layer
have different Euclidean distances from F*, and such
differences are pointed out when using distance functions
approximating the Euclidean distance function. Adjacent
pixels belonging to the same layer are mostly diagonally
connected. However, layers including 4-connected pixels are
also present. In Figure 4, the layers of the WDT, computed by
using 3 and 4 as weights, are outlined.

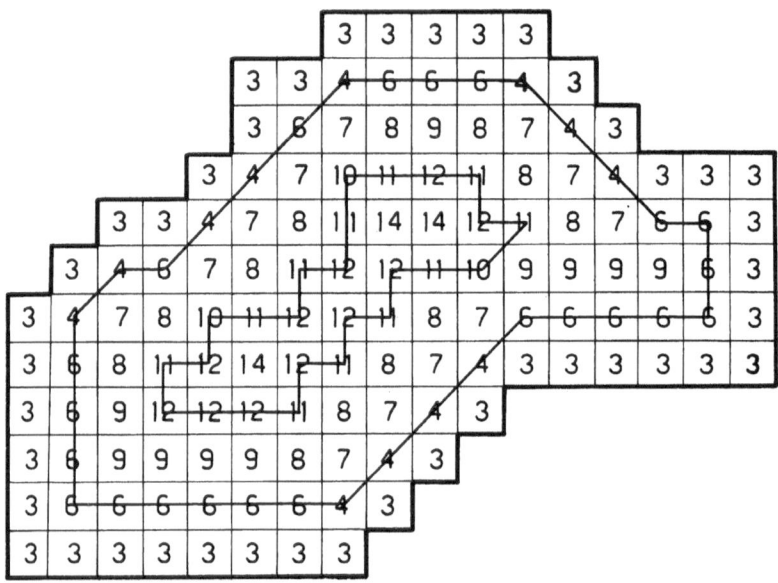

Figure 4. Pixels belonging to the even layers of the WDT
(with weights 3 and 4) are linked to each other.

CONCLUDING REMARKS

Weighted distance transforms have been considered. Any
WDT constitutes a different approximation to the Euclidean
distance transform, and the degree of approximation depends
on the weights selected to measure distance between
neighboring pixels, as well as on the size of the adopted
neighborhood.

Among the WDT evaluated by using 3x3 neighborhoods, the
4-distance transform and the 8-distance transform have widely
been investigated in the past, and criteria to identify their
features have been suggested. Generally, such criteria cannot
always be immediately transferred to any WDT. This is the
case for the criterion to be used for the selection of the
local maxima. By taking into account the way in which the
propagation of distance information occurs, we have suggested
a criterion to find the local maxima, which can be proved to
have general validity.

Furthermore, we have introduced the notion of layer in

the WDT. Differently from the particular cases of both the 4-DT and the 8-DT, the sets made of equilabeled pixels are generally not connected. Thus, the criterion for identifying the layers cannot be based on the equidistance property, which not necessarily characterizes the pixels of the same layer. On the contrary, pixels having different distance from the complement of the figure (i.e., differently labeled pixels) are likely to belong to the same layer. The membership of any pixel p to a given layer depends on the number of steps of minimal weight, necessary to reach p starting from the complement of the figure.

Finally, we note that the suggested criteria can be generalized to the distance transforms discussed in /7/, where local distance information pertaining any pixel p is derived by neighbors located in either a 5x5 or a 7x7 neighborhood, centered on p.

ACKNOWLEDGMENTS

The help of Mr U. Cascini in providing the illustrations is gratefully acknowledged.

REFERENCES

1. J.L.Pfaltz and A.Rosenfeld, "Computer representation of planar regions by their skeleton", Comm. ACM, 10, 119-125, 1967.

2. C.Arcelli and G.Sanniti di Baja,"Shape splitting using maximal neighborhoods", Proc. 6th ICPR, Munich, 1106-1108, 1982.

3. S.Suzuki and K.Abe,"Max-type distance transformation for digitized binary pictures and its applications", Trans. of IECE of Japan, E66, 94-101, 1983.

4. C.Arcelli and G.Sanniti di Baja,"Quenching points in distance labeled pictures", Proc. 7th ICPR, Montreal, 344-346, 1986.

5. T. Matsuyama and T.Y. Phillips, "Digital realization of the labeled Voronoi Diagram and its applications to closed boundary detection", Proc. 7th ICPR, Montreal, 478-480, 1984.

6. J.Hilditch and D.Rutovitz,"Chromosome recognition", Ann. New York Acad. Sci., 157, 339-364, 1969.

7. G.Borgefors,"Distance transformations in digital images", Computer Vision, Graphics, and Image Processing, 34, 344-371, 1986.

8. C. Arcelli, M. Del Sordo and G. Sanniti di Baja, "Maximal neighborhoods in (3,4)-chamfer distance transforms", Proc. 8th International Symposium MECO 86, Taormina,Italy, 108-110, 1986.

DISTANCE TRANSFORMATIONS IN HEXAGONAL GRIDS

Gunilla Borgefors

National Defence Research Institute
Box 1165
S-581 11 Linköping, Sweden

INTRODUCTION

A distance transformation converts a binary image, consisting of feature and non-feature pixels, into a distance image. In this distance image each non-feature pixel has a value that approximates (or is equal to) the distance to the nearest feature pixel. Distance transformation will be denoted DT henceforth. In this paper DTs for the hexagonal pixel grid are derived and presented. A very small example of such a DT is shown i Fig. 1.

In the "normal" square grid the most used DTs have, until recently, been the city block (4-neighbor) distance and the chessboard (8-neighbor) distance. However, pseudo-Euclidean, i.e. more accurate, DTs have been receiving quite a lot of attention recently. Several good DTs have been developed and used for a variety of binary image operations. Examples of such operations are: skeletonizing, segmentation, smoothing and matching.

There are several reasons why the hexagonal image grid is interesting, at least for some applications: The hexagonal grid avoids the 4-neighbor/ 8-neighbor problem. In the hexagonal grid each pixel has six equal neighbors. A second reason is that the hexagonal grid is much closer to the

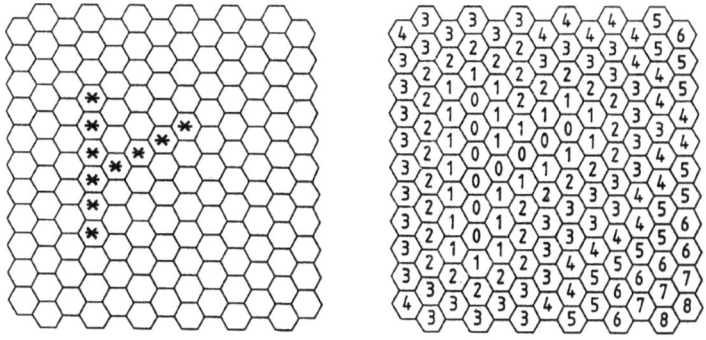

Fig. 1. An example of a distance transformation. In the left, original, image the feature pixels are marked *. In the right image each pixel value is the Euclidean distance to the nearest feature pixel.

"grid" used in the animal kingdom, including human vision. A third reason
is that natural scenes in low resolution look more "natural" presented in a
hexagonal grid, compared to a square grid, if the same number of pixels is
used. Thus, a small but steady trickle of papers using the hexagonal grid
are being published.

DTs in the hexagonal grid should therefore be considered. The simplest
DT possible, equivalent to the city block/chessboard distances, was discussed in [1]. In this paper the geometry of other hexagonal DTs are investigated. The 6-pixel and the 12-pixel neighborhoods are considered. Several
new DTs are derived, among them a very good one, that uses iterated local
operations, integer arithmetic and has a maximum error of about 3%.

Hexagonal grids are usually oriented in one of two ways: either "side
up" or "corner up". One grid is transformed into the other by rotating it
30 degrees. For almost all results in this paper the orientation of the
grid is inconsequential. Where it is not, this is pointed out.

BASIC CONCEPTS

The basic idea for one class of DTs is to approximate global distances
in the image by propagating local distances, i.e. distances between neighboring pixels. This basic idea was first presented in [2] and [3]. Fig. 2
illustrates the local distances in a 12-pixel neighborhood. The propagation
of these local distances can be carried out either sequentially, using two
passes over the image, or in parallel. When the propagation is sequential,
the resulting DTs are often called chamfer distances.

In the binary image, for which the DT is to be computed, the feature
pixels are set to zero and the non-feature pixels to infinity.

The DT algorithms can then be described by masks, where the mask
depicts the neighborhood used. Fig. 2 can be interpreted as the general
12-pixel neighborhood mask for parallel computation. The mask is placed
over each pixel in the image. The local distances in the mask (**a**, **b**) are
added to the image pixels they cover. The new value of the central pixel is
the minimum of the sums. The process is repeated until no value changes.
The number of iterations is proportional to the largest distance value
occuring in the image. Thus the parallel algorithm becomes:

$$v_{i,j}^{m} = \underset{(k,l)\, \in\, \text{mask}}{\text{minimum}} \; (v_{i+k,j+l}^{m-1} + c(k,l)\,), \qquad\qquad (1)$$

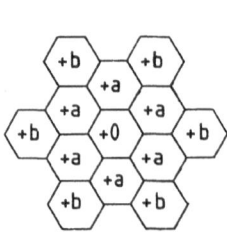

Fig. 2. The local distances in a
12-pixel neighborhood.

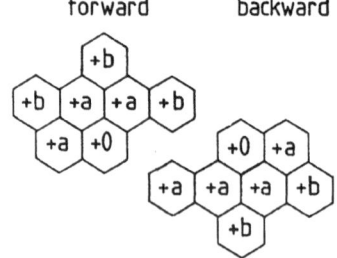

Fig. 3. Sequential masks for the
"corner up" grid.

214

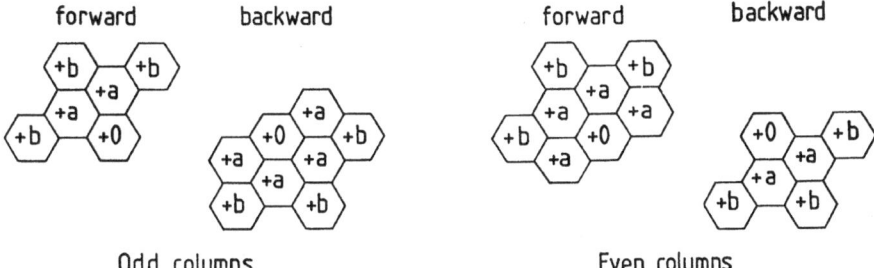

Fig. 4. Sequential masks for the "side up" grid.

where $v_{i,j}^{m}$ is the value of the pixel in position (i,j) at iteration m, (k,l) is the mask-pixel position and $c(k,l)$ is the local distance at that mask-pixel.

When the computation is done sequentially the mask is split in two, the "forward" and the "backward" mask. If the orientation of the grid is "corner up" the masks become symmetrical, Fig. 3. The forward mask moves from left to right and from top to bottom, and the backward mask from right to left and from bottom to top. At each posion the value of the central (zero) pixel is changed to the minimum of the sums, computed as in the parallel case, i.e. as the sum of the mask-pixel values and the image-pixel values. If the orientation of the grid is "side up" the parallel mask must be split asymmetrically, and also differently for every other column, Fig. 4. The masks move as before, but along slightly jagged lines.

OPTIMAL LOCAL DISTANCES

In this section "optimal" local distances are computed. Optimality here means that the maximum difference between the computed distance and the Euclidean distance is minimized in a hexagonal area with radius M (Fig. 7 A). The local distances are real numbers. Other measures of optimality could be used, e.g. minimizing the average difference. Naturally the approximation becomes better the larger the neighborhood is. Here the smallest, 6-pixel neighborhood, and the next, 12-pixel neighborhood have been investigated.

First consider the smallest neighborhood. In this case the only local distance to determine is **a**, see Fig. 2. The **b**-pixels in the masks are ignored. The computed distances are symmetric, so it is enough to consider a 30 degree wedge of the hexagon with radius M pixels, see Fig. 5. The distance to any pixel in column M is **a**M, Fig. 5 again. If the unit distance between the centers of neighboring pixels is set to one, then the Euclidean distance to the y:th pixel in column M, (counting from below and starting with y=0) becomes:

$$\sqrt{\frac{3M^2}{4} + y^2}, \quad y=0, 1, \ldots, \frac{M}{2}, \quad \text{if M even} \quad \text{and}$$

$$(2)$$

$$\sqrt{\frac{3M^2}{4} + (y + \frac{1}{2})^2}, \quad y=0, 1, \ldots, \frac{M-1}{2}, \quad \text{if M odd.}$$

If M is even, then the difference between the DT and the Euclidean distance becomes:

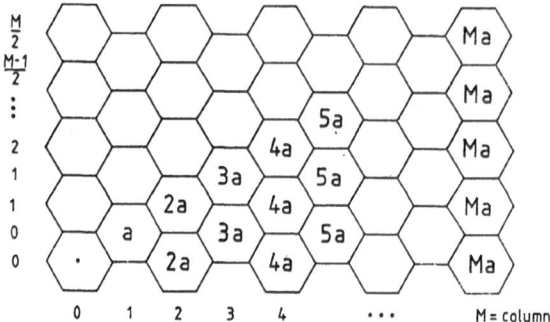

Fig. 5. The distances from a single feature pixel (lower left corner), when the 6-pixel neighborhood is used.

$$\text{diff}(y) = aM - \sqrt{\frac{3M^2}{4} + y^2} \qquad (3)$$

The maximum of diff(y) occurs either for y=0, diff'(y)=0, or for y=M/2:

$$\text{diff}(0) = (a - \frac{\sqrt{3}}{2})\ M,$$
$$\text{diff}'(y) < 0 \quad \text{for all } y>0, \text{ and} \qquad (4)$$
$$\text{diff}(\frac{M}{2}) = (a - 1)\ M.$$

The minimum of the maximum of three expressions (4) occur for diff(0)= -diff(M/2), i.e. for

$$a_{opt} = \frac{1}{2} + \frac{\sqrt{3}}{4} \approx 0.93301. \qquad (5)$$

The maximum difference between the DT, with $a=a_{opt}$, and the Euclidean distance becomes:

$$\text{maxdiff} = (\frac{1}{2} - \frac{\sqrt{3}}{4})M \approx 0.06699\ M. \qquad (6)$$

It can be shown, rather easily, that maxdiff is always equal or less than (6) when M is odd. Thus (5) and (6) are valid for all M. Note that maxdiff is an upper limit rather than a maximum. It need never occur.

Next, we will look at the 12-pixel neighborhood. Here two local distances must be determined, **a** and **b** (see Fig. 2). Using the 6-pixel neighborhood, the distance **b** is approximated by 2**a**. Thus **b** < 2**a** , otherwise the local distance **b** would never be used. Likewise 3**a** < 2**b**, otherwise **a** would never be used for M>2, (two **b**-steps, east + north-east, would be shorter than three **a**-steps east-north-east). If these two inequalities are valid, then the DT values given to different pixels in a 30 degree wedge becomes those that are shown in Fig. 6. The computed values thus become (in the same notation as before):

$$(\frac{M}{2} - y)\ \mathbf{b} + 2\ y\ \mathbf{a}, \quad y=0, 1, \ \ldots\ \frac{M}{2}, \quad \text{M even, and}$$
$$(\frac{M}{2} - (y + \frac{1}{2}))\ \mathbf{b} + 2\ \mathbf{a}\ (y + \frac{1}{2}), \quad y=0, 1, \ \ldots\ \frac{M-1}{2}, \quad \text{M odd.} \qquad (7)$$

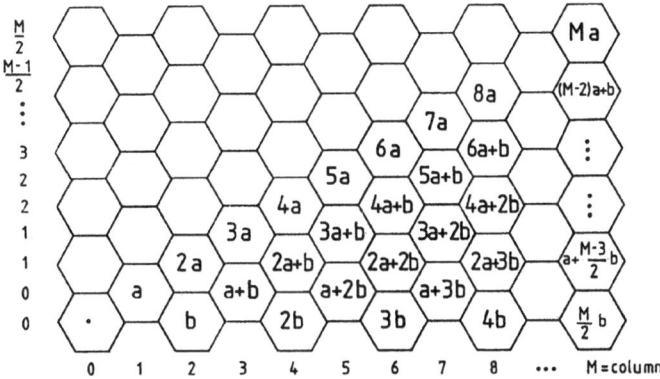

Fig. 6. The distances from a single feature pixel (lower left corner), when the 12-pixel neighborhood is used.

The corresponding Euclidean distances are found in (2). The function diff(y) is defined, as in the previous case, as the difference between the computed distance and the Euclidean distance for pixel number y in the M-column (Fig. 6). As before the maximum difference occurs for y=0, diff'(y)=0 or y=M/2. With M even, these values become:

$$\text{diff}(0) = \frac{M}{2} (b - \sqrt{3}),$$

$$\text{diff}(y_{\text{diff}'(y)=0}) = \frac{M}{2} (b - \sqrt{3} \sqrt{1 - (2a-b)^2}), \tag{8}$$

$$\text{diff}(\frac{M}{2}) = M (a - 1).$$

Minimizing (8) yields the optimal local distances. The computations are not very difficult. The resulting optimal local distances are:

$$a_{opt} = 1 - \frac{\sqrt{3}}{4} (1 - \sqrt{4\sqrt{3} - 6}) \approx 0.98417 \quad \text{and} \tag{9}$$

$$b_{opt} = \frac{\sqrt{3}}{2} (1 + \sqrt{4\sqrt{3} - 6}) \approx 1.70038.$$

With these local distances the maximum difference that can occur becomes:

$$\text{maxdiff} = (\frac{\sqrt{3}}{4} (1 - 4\sqrt{3} - 6))M \approx 0.01583 \, M. \tag{10}$$

As before, it can be shown that maxdiff always is smaller for M odd than for M even.

INTEGER DISTANCE TRANSFORMATIONS

When actually computing the DT of a binary image, it is often impractical to use real numbers. One would like to have good DTs where the local distances are integers.

For the 6-pixel neighborhood the only reasonable value of **a** is **a**=1. This DT was suggested in [1]. A reasonable name for it would be the 6-neighbor distance, as it is the equivalent of the 4-neighbor/8-neighbor distances in the square grid. Inserting **a**=1 in (4) gives the maximum difference for this DT:

217

$$\text{maxdiff} = (1 - \frac{\sqrt{3}}{2}) \ M \approx 0.13397 \ M. \qquad\qquad (11)$$

The values of the 6-neighbor DT are always bigger or equal to the Euclidean distance.

One way of comparing different DTs is to look at their associated "discs". The disc of a DT is the area consisting of all pixels with distance values less than a certain number, counting from a single central pixel. The Euclidean disc in a hexagonal grid with radius 15.5 is shown in Fig. 7 C. This is the ideal that all other DTs try to achieve.

The disc of the 6-neighbor DT is a hexagon. The disc with radius 15.5 is shown in Fig. 7 A. The corners of this hexagon touch the Euclidean circle with the same radius.

Note the possibility of rescaling the distance values of the 6-neighbor DT: If all computed values are multiplied by a_{opt}, (5), then maxdiff decreases to 0.06699 (6), even though all computations but this last multiplication are integer. This rescaling is equal to expanding the hexagonal disc of the DT, so that the hexagon overlaps the Euclidean circle, and thus approximates it better.

In the 12-neighbor case both optimal local distances are multiplied by a suitable factor, and then rounded to the nearest integers. A factor that make both a and b almost integer should yield a good approximation, and thus a good DT. The disadvantage is of course that all distance values get rescaled by a certain scaling factor > 1. Naturally this factor should be as low as possible. If the local distances are re-rescaled, i.e. divided by this scaling factor, their values become $a=1$ and $b=n/m$, where m and n are the integer approximations of a and b respectively. With $a=1$, $b=n/m$ (8) can be used to find the associated maxdiff.

One excellent integer approximation is $a=3$, $b=5$. Using these values maxdiff becomes:

$$\text{maxdiff} = \frac{M}{2} \ (\frac{5}{3} - \sqrt{3} \ )M \approx 0.03269 \ M. \qquad\qquad (12)$$

The 3-5 DT can be both bigger and smaller than the Euclidean distance. The sign of the difference depends on the direction to the nearest feature pixel.

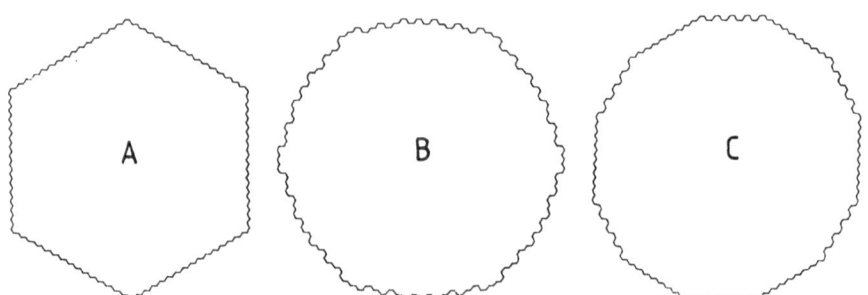

Fig. 7. The digital discs with radius 15.5 for the 6-neighbor DT (A), the 3-5 DT (B) and the Euclidean DT (C).

218

A better approximation is **a**=7, **b**=12, with maxdiff=0.02722M, and still better is **a**=10, **b**=17, maxdiff=0.02386M. However, using these bigger scaling scaling factors is probably not justifiable. The difference from the 3-5 distance is not big enough.

The disc of all 12-neighborhood DTs are dodecagons, i.e. twelve-sided polygons. The disc of the 3-5 distance of radius 15.5 is found in Fig. 7 B. The fact that the disc is a dodecagon is not quite apparent for such a small disc as this. It approximates the Euclidean circle very well.

The local distances and the maximal differences for the most interesting hexagonal DTs are found in the Table. It is also interesting to see how the difference from the Euclidean distance varies in different directions from a feature pixel. In Fig. 8 the difference is drawn as a function of y, where y is defined as in (2). Thus y=0 is the direction towards the middle of the side of the hexagonal area in which the DT is optimized (Fig. 7 A), and y=M/2 is the direction towards the corners. The curves for the different DTs are denoted: 6-neighbor "1", optimal 6-pixel neighborhood "a_{opt}",

3-5 "3-5" and optimal 12-pixel neighborhood "$a_{opt} - b_{opt}$".

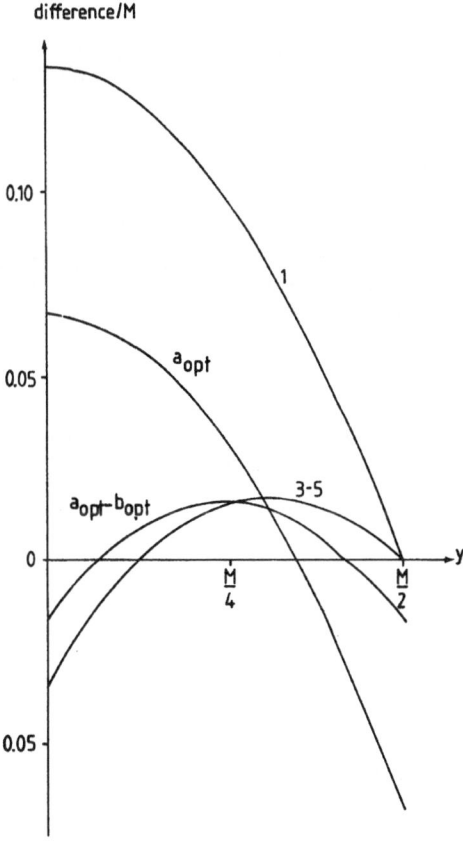

Fig. 8. Difference between computed and Euclidean distance in different directions.

Table. Local distances and maximum differences.

a	b	maxdiff	
opt	-	0.06699	6-neighbor
1	-	0.13397	
opt	opt	0.01582	12-neighbor
3	5	0.03269	
7	12	0.02722	
10	17	0.02386	

CONCLUSION

Pseudo-Euclidean distance transformations are used for many kinds of operations on binary images. Several useful DTs for the hexagonal pixel grid can be found in this paper. The simplest DT sets the distance to all neighbors to one. This 6-neighbor DT is probably good enough for many applications (maximum error about 13%). If a closer approximation to the Euclidean distance is necessary, then the new 3-5 DT presented here is ideal, with a maximum error of about 3%.

The two above DTs use only integer arithmetic. If real numbers can be used, then even closer approximations to the Euclidean distance can be achieved. If the 6-neighbor DT is rescaled, by multiplying all resulting distances by $a_{opt} \approx 0.933$ (5), then the error is reduced from 13% to about 7%. The best approximation using a 12-pixel neighborhood is the $a_{opt} \approx 0.984$-$b_{opt} \approx 1.700$ DT (9), with an error of about 1.6%.

The recommended hexagonal DTs are thus the 6-neighbor DT, the rescaled 6-neighbor DT, and the 3-5 DT. Which one that should be chosen depends on the accuracy necessary and the computing power available.

REFERENCES

1. G Borgefors: Distance Transformations in Arbitrary Dimensions, Computer Vision, Graphics Image Processing 27, pp. 321-345, 1984.
2. A Rosenfeld and J Pfaltz: Sequential Operations in Digital Image Processing, Jour. Assoc. Comp. Mach. 13, pp. 471-494, 1966.
3. A Rosenfeld and J Pfaltz: Distance Functions on Digital Pictures, Pattern Recognition 1 (1), pp. 33-61, 1968.

OPTIMIZATION OF THE GENERALIZED HOUGH TRANSFORM

Makoto Sato and Hidemitsu Ogawa

Department of Computer Science
Tokyo Institute of Technology
Ookayama, Meguro-ku
Tokyo 152, Japan

Introduction

Many image processing problem require curve detection. These include vision directed automation, remote control of vehicles, biomedical applications and so on. The Hough transform[1],[2] is a technique for detecting straight lines within a noisy image and later adapted for the detection of circles, ellipses and other analytically defined shapes. This method has been modified by D.H.Ballard[3] for detecting arbitrary shapes, which is called generalized Hough transform.

The generalized Hough transform is a method for detecting arbitrary curves by exploiting the duality between points on a curve and parameters of that curve. Given an arbitrary shape the transform provides a mapping from the orientation of an edge-element to the set of instances of the shape which could have given rise to that edge-element. This mapping allows all local evidence for a particular instance of the shape to contribute to global decisions about the figure.

In this shape detection schema, the sensitivity of the shape detection depends on the noise and distortion in the edge-element information. So it is necessary to select the description of the edge curve so that the mapping of the generalized Hough transform is stable enough for detecting the shape.

In this paper, we consider the problem to select the location of the local coordinate system for the curve description so that the most stable mapping is obtained. As the criterion of the optimization problem, we use the variance measure of cluster size in the parameter space introduced by S.D.Shapiro [4]. The problem is formalized in the mathematical form and the conditions of the location of the local coordinate system are made clear for the optimal mapping of the generalized Hough transform.

Generalized Hough Transform

Hough techniques for shape detection consist of following basic elements[5]:

a)   a local edge-element detector,

b)   a n-dimensional parameter space, quantized and represented by an accumulator array,

c)   a mapping from the information provided by the detector into the parameter space (and thus the accumulator array),

d)   a voting rule specifying how a particular edge-element affects the values of the accumulator array,

e)   a detection rule, specifying the conditions under which a particular shape has been detected.

Given these basic elements, shapes are found by the following procedure:

a) zero the accumulator array,
b) apply the detector every where in the image,
c) for each edge-element found, apply the mapping to locate cells in the accumulator array. Then apply the voting rule to modify the contents of these cells. (i.e., vote for all possible "causes" of this edge-element),
d) finally, apply the detection rule to the accumulator array (choose the most popular shape).

Assume that the edge curve of an arbitrary shape on a local coordinate system is represented in the parametric form as follows:

$$\begin{cases} x = x_0(t), \\ y = y_0(t), \quad \text{for } t \in [a,b] \end{cases} \tag{1}$$

where $x_0(t)$ and $y_0(t)$ are smooth functions of t with derivatives.

Consider shapes in the image are produced by translating and rotating the above original shape. Then they are given as follows:

$$\begin{pmatrix} x \\ y \end{pmatrix} = \begin{pmatrix} \cos\theta & -\sin\theta \\ \sin\theta & \cos\theta \end{pmatrix} \begin{pmatrix} x_0(t) \\ y_0(t) \end{pmatrix} + \begin{pmatrix} X \\ Y \end{pmatrix} \tag{2}$$

,where (X,Y) is the location of the origin of the shape and θ is the rotation angle about the origin (X,Y). The parameter space in this model is (X,Y,θ). But since our interest is the location of the shapes, we take (X,Y) as the parameter space.

To obtain the explicit expression of the mapping from the edge-element information into the parameter space, we first consider the moving coordinate system, where the origin of the system is on the edge curve and the direction of the x-axis is that of the tangent vector of the curve(, see Fig. 1).

In the moving coordinates system, the vector $(X_0(t),Y_0(t))$ to the origin of the shape is given as follows:

$$\begin{cases} X_0(t) = \dfrac{1}{D}(-x\dot{x} - y\dot{y}) \\[3mm] Y_0(t) = \dfrac{1}{D}(x\dot{y} - y\dot{x}), \end{cases} \tag{3}$$

where $D = (\dot{x}^2 + \dot{y}^2)^{1/2}$.

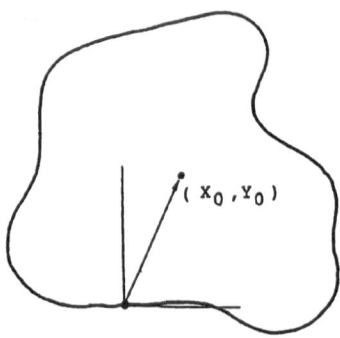

Figure 1. Geometry for generalized Hough transform.

Assume that the edge-element detector provides directional information. Let (u,v) be the location of edge-element and ϕ be the angle of the tangent vector. Taking account of all the possible locations of the origin of the shape, the mapping from (u,v,ϕ) to the parameter space (X,Y) is given as follows:

$$\begin{pmatrix} X(t) \\ Y(t) \end{pmatrix} = \begin{pmatrix} \cos\phi & -\sin\phi \\ \sin\phi & \cos\phi \end{pmatrix} \begin{pmatrix} X_0(t) \\ Y_0(t) \end{pmatrix} + \begin{pmatrix} u \\ v \end{pmatrix}$$

(4)

,where $t \in [a,b]$.

Optimization

It is obvious that the sensitivity of the generalized Hough transform depends on the accuracy of the information from the edge-element detector. Since the estimation errors of edge-element location (u,v) and direction angle ϕ are inevitable, it is necessary to use a mapping stable to the estimation error of the edge-element information.

Since it is free to select the location of the local coordinate system, it is possible to consider the optimization problem to find the location so that the mapping given is optimal in the above sense.

Suppose that additive errors exist in the information of the edge-element detector, that is,

$$\tilde{u} = u + \alpha, \quad \tilde{v} = v + \beta, \quad \text{and} \quad \tilde{\phi} = \phi + \delta$$

(5)

where α, β, and δ are the additive errors of u, v, and ϕ respectively. The errors α, β, and δ are random variables independent each other and satisfy the following equations:

$$E(\alpha) = E(\beta) = 0.$$

(6)

Then the mapping $(\tilde{X}(t),\tilde{Y}(t))$ by the edge-element information $(\tilde{u},\tilde{v},\tilde{\phi})$ is given as

$$\begin{pmatrix} \tilde{X}(t) \\ \tilde{Y}(t) \end{pmatrix} = \begin{pmatrix} \cos\tilde{\phi} & -\sin\tilde{\phi} \\ \sin\tilde{\phi} & \cos\tilde{\phi} \end{pmatrix} \begin{pmatrix} X_0(t) \\ Y_0(t) \end{pmatrix} + \begin{pmatrix} \tilde{u} \\ \tilde{v} \end{pmatrix}$$

$$= \begin{pmatrix} \cos(\phi+\delta) & -\sin(\phi+\delta) \\ \sin(\phi+\delta) & \cos(\phi+\delta) \end{pmatrix} \begin{pmatrix} X_0(t) \\ Y_0(t) \end{pmatrix} + \begin{pmatrix} u \\ v \end{pmatrix} + \begin{pmatrix} \alpha \\ \beta \end{pmatrix}$$

$$= \begin{pmatrix} \cos\delta & -\sin\delta \\ \sin\delta & \cos\delta \end{pmatrix} \begin{pmatrix} X(t) - u \\ Y(t) - v \end{pmatrix} + \begin{pmatrix} u \\ v \end{pmatrix} + \begin{pmatrix} \alpha \\ \beta \end{pmatrix}$$

(7)

From this equation, it is possible to estimate the sum of the variances of $\tilde{X}(t)$ and $\tilde{Y}(t)$, that is,

$$Var[\tilde{X}(t)] + Var[\tilde{Y}(t)]$$
$$= \sigma_\alpha^2 + \sigma_\beta^2 + w_\delta(X_0(t)^2 + Y_0(t)^2),$$

(8)

where $\sigma_\alpha^2, \sigma_\beta^2$, and w_δ are given as

$$\sigma_\alpha^2 = Var[\alpha], \quad \sigma_\beta^2 = Var[\beta], \quad \text{and} \quad w_\delta = 1 - E[\cos\delta]^2 - E[\sin\delta]^2.$$

(9)

From Eq. (3) and Eq. (8) , we have

$$Var[\tilde{X}(t)] + Var[\tilde{Y}(t)]$$
$$= \sigma_\alpha^2 + \sigma_\beta^2 + w_\delta(x(t)^2 + y(t)^2).$$

(10)

If the origin of the local coordinate system is transferred to the position (p,q), then the edge curve is represented as

$$\begin{cases} x_1(t) = x(t) - p, \\ y_1(t) = y(t) - q. \end{cases}$$

(11)

223

In this case, the sum of the variances of $\tilde{X}_1(t)$ and $\tilde{Y}_1(t)$ is given as follows:

$$Var[\tilde{X}_1(t)] + Var[\tilde{Y}_1(t)]$$

$$= \sigma_\alpha^2 + \sigma_\beta^2 + w_\delta(x_1(t)^2 + y_1(t)^2).$$

$$= \sigma_\alpha^2 + \sigma_\beta^2 + w_\delta[(x(t)-p)^2 + (y(t)-q)^2]. \tag{12}$$

It is necessary to find the position $(p^{\cdot}, q^{\cdot})$, which minimizes the above value of Eq. (12) in average of the parameter t. Then the optimization problem is formalized in the following way:

Find $(p^{\cdot}, q^{\cdot})$ such that,

$$J(p^{\cdot}, q^{\cdot}) = \min J(p,q), \tag{13}$$

where

$$J(p,q) = \int_a^b [(x(t)-p)^2 + (y(t)-q)^2]dt. \tag{14}$$

The solution of the problem is easily derived by differentiating $J(p,q)$ with respect to p and q, that is,

$$\frac{\partial J}{\partial p} = 2(b-a)p - 2\int_a^b x(t)dt = 0$$

$$\frac{\partial J}{\partial q} = 2(b-a)q - 2\int_a^b y(t)dt = 0 \tag{15}$$

The optimal position $(p^{\cdot}, q^{\cdot})$ of the origin for the edge curve description is given by the following equations:

$$
\begin{cases}
p^{\cdot} = \dfrac{1}{b-a}\int_a^b x(t)dt. \\[4mm]
q^{\cdot} = \dfrac{1}{b-a}\int_a^b y(t)dt.
\end{cases}
\tag{16}
$$

Suppose that the edge curve is made of uniform wire, then the position $(p^{\cdot}, q^{\cdot})$ in Eq.(16) is just the center of gravity of the curve.

Examples

We consider an analytic example of the generalized Hough transform.

A rational curve is given on the polar coordinate system (r,θ) as follows:

$$r = 1 + a \sin n\theta. \tag{17}$$

The shape of the curve is shown in Fig.2, where $a = -0.5$ and $n = 3$. The curve is represented in x-y coordinate system as follows:

$$
\begin{cases}
x(\theta) = (1 + a \sin n\theta) \cos \theta \\
y(\theta) = (1 + a \sin n\theta) \sin \theta, \quad 0 \le \theta < 2\pi.
\end{cases}
\tag{18}
$$

Since the curve $(x(\theta),y(\theta))$ satisfies the following equation:

$$\int_0^{2\pi} x(\theta)d\theta = \int_0^{2\pi} y(\theta)d\theta = 0. \tag{19}$$

224

the parametric description (18) of the curve is optimal for the generalized Hough transform.

From Eq.(3), the mapping $(X(\theta),Y(\theta))$ of the curve is obtained as follows:

$$\begin{cases} X(\theta) = -na\cos n\theta(1 + a\sin n\theta)/D \\ Y(\theta) = (1 + a\sin n\theta)^2/D \end{cases}$$

(20)

where $D = ((1 + a\sin n\theta)^2 + n^2 a^2 \cos^2 n\theta)^{1/2}$

In Fig.3, the mappings and the voting results of this example are illustrated, where the three cases of the different positions of the origin are represented. The histgrams of the voting results along the x-axis are shown in Fig.4. Fig.4 (a)–(d) are the voting results when the position of the origin (p,q) is $(0,0)$. Fig.4 (e)–(h) are those when the position of the origin (p,q) is $(0.1,0)$. In both cases, the variances of the additive noises are 0.0, 0.05, 0.1 and 0.2, respectively. Clearly, the shape of the histgrams in the case when the origin (p,q) is $(0,0)$ are more sharp and stable than in the case that (p,q) is $(0.1,0)$.

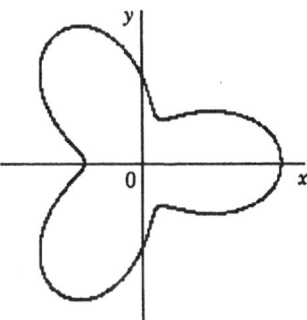

Figure 2. Closed curve $r = 1 + a\sin n\theta$ where $a = -0.5$ and $n = 3$.

Finally, a curve detection experiment results is shown in Fig.5 (a)–(f).

Fig.5 (a) shows the input image, which is a microscope picture of fiber sections. The image is first thresholded to the binary image (Fig.5 (b)) and then skeletonized (Fig.5 (c)). From the original fiber section, the center of gravity is calculated and the optimal mapping is obtained. The voting result is shown in Fig.5 (d). Thresholding it, the candidate points of the center of the section are obtained as in Fig.5 (e). Calculating local average of these points, the centers of the fiber sections are detected as in Fig.5 (f).

Conclusion

An optimization problem of the generalized Hough transform has been considered, where the position of the origin for the edge curve description is optimized. The optimality conditions is derived and some examples are presented.

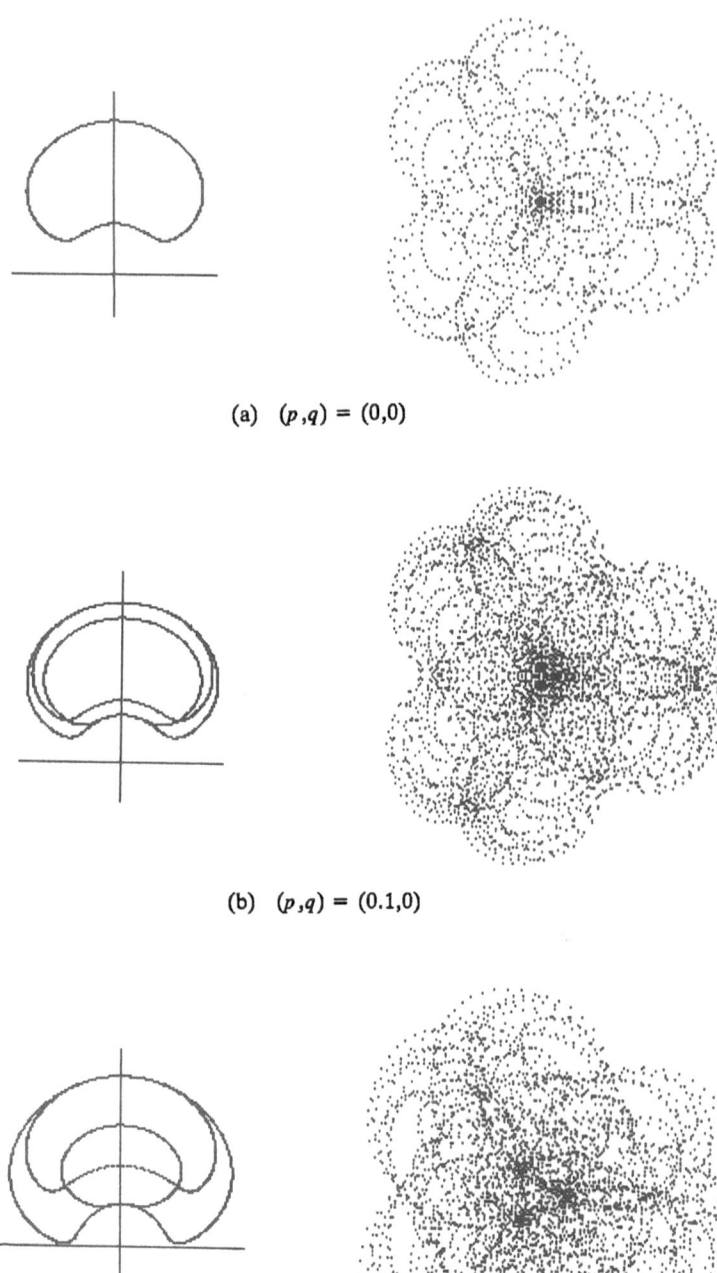

(a) $(p,q) = (0,0)$

(b) $(p,q) = (0.1,0)$

(c) $(p,q) = (0.3,0)$

Figure 3. The mappings and the voting results.
(p,q) is the position of the origin.

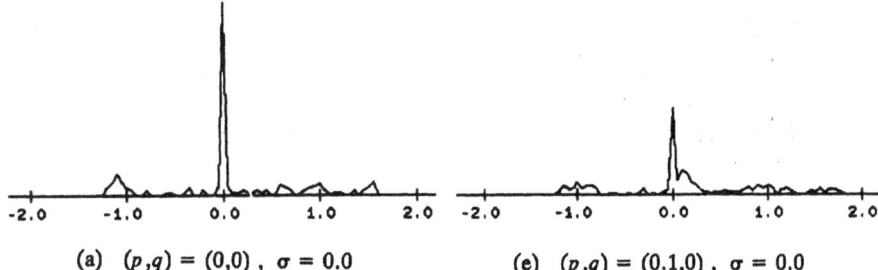

(a) $(p,q) = (0,0)$, $\sigma = 0.0$ (e) $(p,q) = (0.1,0)$, $\sigma = 0.0$

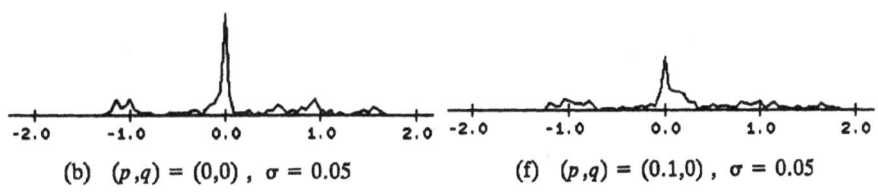

(b) $(p,q) = (0,0)$, $\sigma = 0.05$ (f) $(p,q) = (0.1,0)$, $\sigma = 0.05$

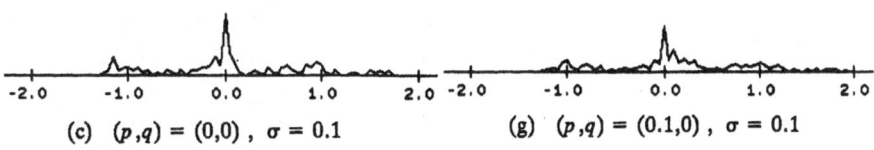

(c) $(p,q) = (0,0)$, $\sigma = 0.1$ (g) $(p,q) = (0.1,0)$, $\sigma = 0.1$

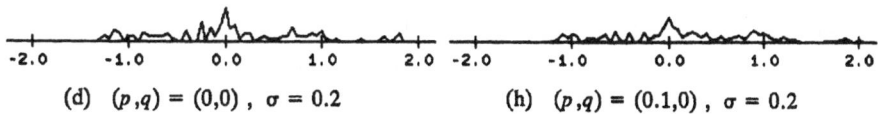

(d) $(p,q) = (0,0)$, $\sigma = 0.2$ (h) $(p,q) = (0.1,0)$, $\sigma = 0.2$

Figure 4. The histgrams of the voting results along the x-axis.

σ is the variance of the additive noises.

227

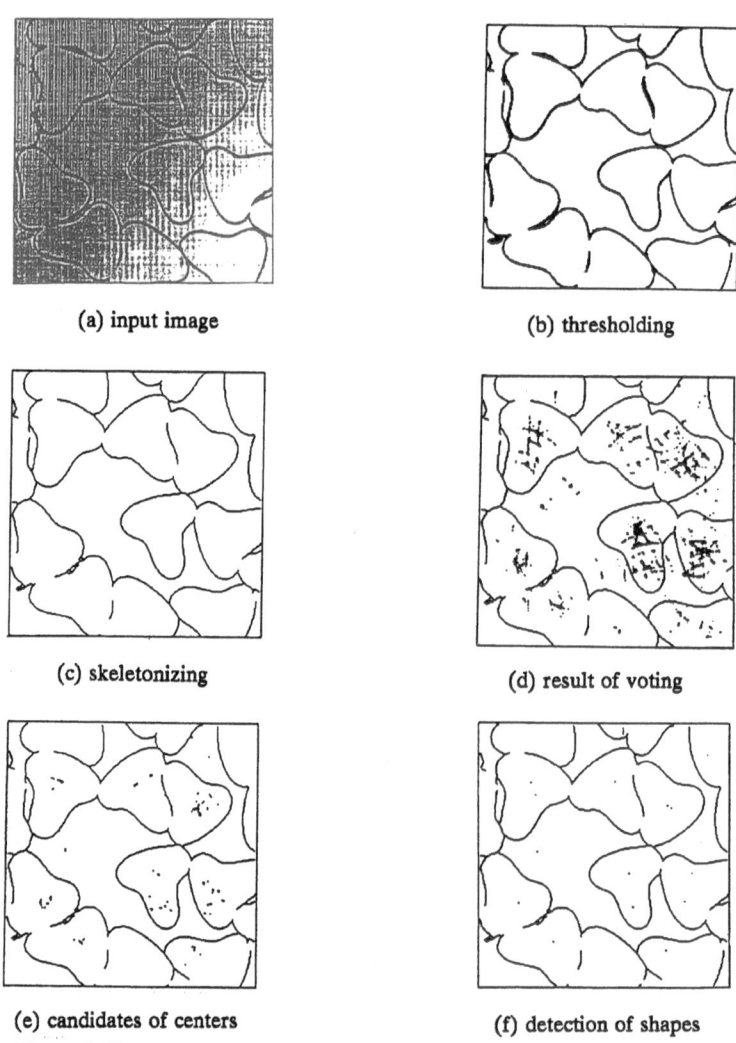

(a) input image

(b) thresholding

(c) skeletonizing

(d) result of voting

(e) candidates of centers

(f) detection of shapes

Figure 5. Example of shape detection by generalized Hough transform.

References

[1] P.V.C.Hough, "Method and means for recognizing complex patterns," U.S.Patent, 3, 069, 654, (1962).

[2] R.O.Duda and P.E.Hart, "Use of the Hough transformation to detect lines and curves in pictures," Comm. ACM, vol. 15, no. 1, pp. 11-15, Jan. (1972).

[3] D.H.Ballard, "Generalizing the Hough transform to detect arbitrary shapes," Pattern Recognition, vol. 13, no.2, pp.111-122, (1981).

[4] S.D. Shapiro, "Transformation for the computer detection of curves in noisy pictures," Computer Graphics and Image processing, vol.4, pp.328-338, (1975).

[5] K.R.Sloan,Jr. and D.H.Ballard, "Experience with the generalized Hough transform ," Proc. of 5th Int. Conf. on Pattern Recognition, pp.174-179, (1980).

SYNAPTIC PATTERNS FOR

STRAIGHT LINE SEGMENT DETECTION

S. Impedovo, M. Castellano, and A. Giannelli

Instituto di Scienze dell'Informazione
University of Bari
Via Amendola 173, 70126 Bari, Italy

ABSTRACT

In this paper the mechanism of the straight line segment detection in the brain of primates is investigated. Specifically, a synaptic pattern model of neuronal small assembly units for segments detection is proposed and some results of its simulation are shown.

INTRODUCTION

Many pattern recognition algorithms have been proposed, but the results have not yielded the same satisfaction in every field of application. For example, whereas in the field of printed character recognition the algorithms proposed have given way to sophisticated commercial machines[1], in the field of hand-written character recognition equal achievements cannot be recorded. Moreover, the human ability to recognize hand-written characters remains unchallenged by machines because the kind of processing which takes place in human recognizing patterns continues to be a mystery. Although it is the opinion of many that several years are needed to understand thoroughly the recognition processes located within the human brain, recent anatomical and neuro-physiological discoveries have demonstrated how simple functions implemented at the level of the visual cortex are similar to those carried out by computers in the pre-processing phase of signals and images. For example, the characteristic of some neurons like the "simple cells" in the visual cortex, to be sensitive to segments of orientated straight lines, illustrate the natural analogy between these cells and the algorithms which use the Freeman code[2] for the description of plane curves.

In this study we have directed our attention to the behaviour of the network of simple cells and, utilizing the Rosen neuronal network model[3], we have studied how the architecture of the network influences the response of the simple cells.

The scheme of the work is as follows:
- In the first section a brief outline of the information pathway through

the primary visual system is presented;
- The architecture of the visual cortex and the nature of signals in neuronal networks are presented in section two;
- The model of the neuronal network is presented in section three;
- Finally, the network architectures for straight line segment detection have been investigated and the experimental results of their simulation are discussed in section four.

SEC. 1   THE PATH OF THE VISUAL SIGNALS FROM THE EYE TO THE BRAIN

It is well known that the phenomenon of image formation on the retina can be investigated by means of geometrical optics [4,5,6].

The image transformation which takes place at the level of the retina when information passes through photoreceptors, bipolar and gangliar cells with the cooperation of horizontal and amacrine cells, is also known [7]. Furthermore, it has been shown [8] that the convolution between the retinal image $I(x,y)$ and the operator $\nabla^2 G$ gives a set of plane curves representing the zero-crossing of the second derivative of the image $I(x,y)$. The operator $\nabla^2 G$ is the Laplacian of Gaussian, that according to the Marr and Hildreth theory [9], represents a filtering taking place at the primary level of the human visual system. Therefore, it can be assumed as a matter of fact, that one component of information transmitted by the optic nerve to the primary visual cortex (or area 17) through the optic tract, optic chiasm and lateral geniculate nucleus, consists of a set of plane curves for each image.

In man, the visual cortex is about 15 cm^2 with a thickness of about 2 mm, located in the occipital lobes, arranged over the cerebral mass but also abundantly bent inwardly. To investigate what happens at the level of the visual cortex, some anatomical and neuro-physiological considerations must be made. Santiago Ramon y Cajal, Rafael Lorente de Nò, David H.Hubel and Torsten N.Wiesel [10] have shown that from the anatomic point of view, the brain consists of a set of neurons, each one having a large number (hundreds or thousands) of dentrites as input channels and one axon as output channel which feed dentrites of another large number of neurons through synapses. In the visual cortex about 10^5 neurons for mm^2 can be detected, and it can be subdivided histologically into 6 main layers, extending from the surface (the first) to the white matter (the last).

It has been shown that input signals to the visual cortex are collected by cells in the lower part of layer IV and layer VI [11]. The cells of the fourth layer have circular symmetrical receptive fields just like the retinal-ganglion and the geniculate cells. The first transformation performed by the visual cortex is the rearrangement of incoming information so that a successive layers of neurons, consisting of simple cells, responds to specifically oriented line segments. These cell are located just above layer IV. The most effective orientation varies from cell to cell and is precisely determined so that a change of about 10 degrees reduces the response. Depending on the single cell, the stimulus can be a bright line on a dark background or the reverse, or it may be a boundary line between a dark and a light area. This would seem to further strengthen the hypothesis of Marr and Poggio that the zero-crossing lines are preferentially transmitted by the optic nerve.

It has been observed that most of the simple cells seem to protrude into the next layer, that of the complex cells, which are sensitive to the translation of a line placed along the optimun direction. The presence of cells which are sensitive to the thickness of the orientated line have also been observed. These cells are called hypercomplex. A more consistent vertical flow in the visual cortex is guaranteed by neurons with a long axon and nucleus within the second, third and fifth layers; these are the pyramidal cells that integrate the bio-electric signals of the different layers and invert the direction of signals spreading from the external layers toward the fourth layer and then toward the output[11]. The output of the visual cortex is thus projected to a number of other destinations: to the neighboring cortical areas, to the lateral geniculate bodies and superior colliculus in order to evaluate more sophisticated components of the perception related to the binocular disparity and chromatic vision. Continuing our examination of the pathway of the propagation of the bio-electric signals, we arrive to the other zones of the cerebral cortex such as the somatic-motorial, and thus proceeding, one has the feeling of no longer knowing where thinking processes like that of recognition have taken place. Therefore it would be opportune to ask ourselves what is the unit of information of the brain and which is the physical support of this unit.

Even if this work primarily concerns the study of neuronal network architecture to recognize straight line segments, indirectly it gives a small contribution to the answers to the above questions. For these reasons in the following section we will discuss the architecture of that part of the brain that seems to be better understood, that is, the visual cortex.

SEC. 2   VISUAL CORTEX ARCHITECTURE AND NEURONAL SIGNALS

It is known that if one plunges a microelectrode into the cortex at a right angle to the surface and records the output from cell after cell (as many as 100 or 200 of them) in successively deeper layers, it can be observed that all the cells along the path of penetration have identical or almost identical orientations (except for cells deep in layer IV, which have no optimal orientation at all) and receptive fields that are piled one on the other. If the microelectrode is plunged in any direction parallel to the surface of the cortex, it can be observed that cells along the path of penetration have a different orientation, and generally the orientation changes about 10 degrees for each 25÷50 μm of penetration; as matter of fact, a full arch of 180° (from 90° to 270°) can be observed in a depth of about 0.4 mm. These observations suggest that the architecture of the visual cortex may be one consisting of slabs having a base of 0.4 x 0.4 mm in size and a height equal to the cortex thickness. Each of these slabs, also called hypercolumn, consists of about 20 x 20 columns, each column including neurons which detect a specific orientation[12]. Furthermore, it has been seen that adjacent columns show adjoining or overlapping receptive fields. Each column includes simple, complex and hypercomplex cells whose responses depend directly on the eye that presents the highest number of connections with them. This eye preference switches from one hypercolumn to the adjacent hypercolumn, passing from a right ocular dominance to left and viceversa.

These observations seem to demonstrate the existence of all circuitery necessary to investigate a small part of the visual world, well confined in a space of 0.4 x 0.8 mm of the visual cortex; this space is known as hypercolumn for ocular dominance or module. To understand the organization of this module of tissue is to understand the organization of all the visual cortex; however, it must be pointed out that about 50.000 neurons may be included in one module.

To define a simplified model of the neuronal network inside a module, further observations must be made about the dialogue which takes place along the direction of the cortex depth, starting from the fourth layer and going through the third and the second and through the fifth and sixth layer. In fact, it has been shown that neurons of the same type show highly complex intra-lamina networks. In this heavily interconnected structure, the diffuse presence of the polyneuronal systems as well as the localized and spatially confined Cajal's "nidi", in addition to other anatomical and physiological considerations, lead to the conception of a cellular system consisting of a few highly connected neurons that function better than a single neuron[13]. Such systems have been called Small Assembly Unit (S.A.U.). Pearson and Shaw have proposed that the most probable number of neurons in the S.A.U. is thirty.

In this paper only the electrical signals due to "action potential" have been taken into account. It has been shown that these signals are impulses[12]. Depending on the kind of synapse, when an impulse arrives at an axon terminal, the neuron next in line is influenced in such a way that its possibility of then generating impulses is modified. In this work two kind of synapses are considered: one excitatory and the other inhibitory. The impulsive neuronal activity is increased by excitatory synapses and decreased by inhibitory synapses.

The exact function implemented by a neuron has not been well defined at the moment, however it has been assumed here, that its function corresponds to the sum of the inhibition and excitation effect of the different synapses which are on the neuron input channel.

As already known, the generic cell receives information from the preceeding cells along the dentrites and sends its résponse through the axon. This response, considered as the impulse activity of a cell, causes a particular sequence of impulsive signals called "spike train", which represents the behaviour of the cell.

SEC. 3   THE SMALL ASSEMBLY UNIT MODEL

Let n be the number of neurons in a S.A.U. and $x_i(t)$ (where i=1,...,n) the function of state of the i-th cell. The meaning of $x_i(t)$ can be obtained by considering it as the number of spikes discharged by the cell i-th up to the instant "t" starting from an initial instant which equals "0". It follows that for each i=1,...,n dx_i/dt represents the frequency of discharged spikes of the neuron i-th and therefore the output that passes through the axon and consequently through the synaptic junctions and which is then trasmitted to the dentrites of all the other cells of the S.A.U..

It is evident that because of the high connection of the system, the output of the i-th neuron will depend on the state of all the others, that is:

$$\forall\ i=1,\ldots,n\ :\ dx_i/dt = f_i(x_1, x_2, \ldots, x_n)\ . \qquad (1)$$

It is fairly obvious to note that for each $i=1,\ldots,n$, the i-th cell will estimate as input signal the d^2x_i/dt^2 and therefore df_i/dt. It must be emphasized that the approximation of the neuronal system with the system in (1) is sustained by experimental results[13] which have demonstrated how the average of the outputs of each neuron belonging to a S.A.U. is repeatable in time (but the same cannot be said for the output of each single neuron) and therefore the neuronal system can be considered a time invariant system.

In the preceeding sections it has been emphasized how the activity of the various synapses is essentially of two types, excitatory and inhibitory, and therefore if one considers the effect that the j-th cell produces on the i-th cell, that is:

$$\frac{\partial}{\partial x_j} \cdot \frac{dx_i}{dt} \quad , \tag{2}$$

it can be easily seen that the (2) will be positive if the junction is excitatory and negative if the junction is inhibitory. If for each $i=1,\ldots,n$ and $j=1,\ldots,n$ u_{ij} represents the junction between the i-th and j-th cells, with u_{ij} equal to $-1, 0$ or $+1$ (where 0 represents the absence of synapse), the system can be defined as:

$$\frac{\partial}{\partial x_j} \cdot \frac{dx_i}{dt} = u_{ij} \quad . \tag{3}$$

Consequently, since

$$df_i = \sum_{j=1}^{n} \frac{\partial}{\partial x_j} f_i(x_1, x_2, \ldots, x_n) dx_j \tag{4}$$

then (1) and (3) combined in (4) gives:

$$df_i = \sum_{j=1}^{n} u_{ij} \, dx_j \quad . \tag{5}$$

The (5) shows how the knowledge of the synaptic pattern permits the understanding of the evolution of a system; in fact, it follows that:

$$f_i = \int \left(\sum_{j=1}^{n} u_{ij} \, dx_j + c_i \right) \tag{6}$$

where c_i is an unknown constant.

In the case of more general functions u_{ij}, than that here defined, details on solutions of the (5) can be found by consulting Rosen[3].

233

It can be assumed that when a small part of the visual field (that which corresponds to a certain number of adjoining neurons of the fourth layer of the visual cortex) is involved by an oriented segment, the neurons of the S.A.U. that analyze this part of the visual field and that are sensitive to the same orientation of the segment, will be excited, thus modifying their spike trains.

In this paper it has been assumed that a S.A.U. consists of 32 simple neurons. This choice is suggested by the fact that 32 is the power of 2 closest to 30 which is exactly the number estimated by Pearson and Shaw[13].

Since no difference in structure has been detected between cells with circular symmetrical fields which are in the fourth layer, and simple cells, we assume that the difference in their behaviour depends on the neuronal network. Therefore, it can also be assumed that the capacity of a S.A.U. to recognize an oriented segment will depend on the synaptic pattern. If we consider the S.A.U. as the support for a psychophysical unit of information, then the information that the S.A.U. records is defined by the synaptic pattern. It must be emphasized however, that for every S.A.U., $2^{32 \times 32 - 32} = 2^{992}$ different synaptic patterns can exist. Consequently, the question becomes, what are the most probable synaptic patterns of a S.A.U. sensitive to oriented segments. In order to identify some synaptic patterns that both maintain the S.A.U. in a state of equilibrium (that is, with a certain frequency of spike trains) in the absence of stimulus, and give a response like those experimentally detected when excited, some considerations must be made:

- It must be noted that a random distribution of an equal number of excitatory and inhibitory junctions throughout all the S.A.U. like that in fig. 1.a, corresponds to the absence of information. The spike train of a S.A.U. with randomly distibuted synaptic junctions has been obtained utilizing the (6), and an histogram of the results is shown in fig. 1.b.

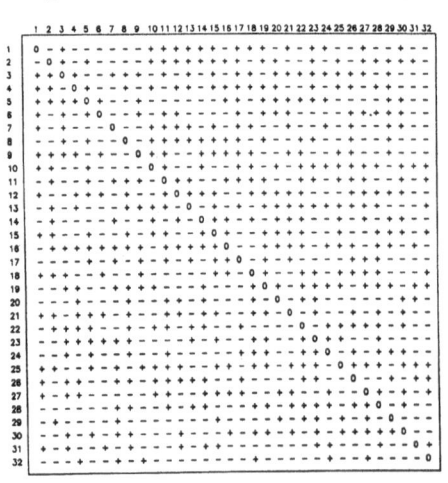

Fig. 1.a

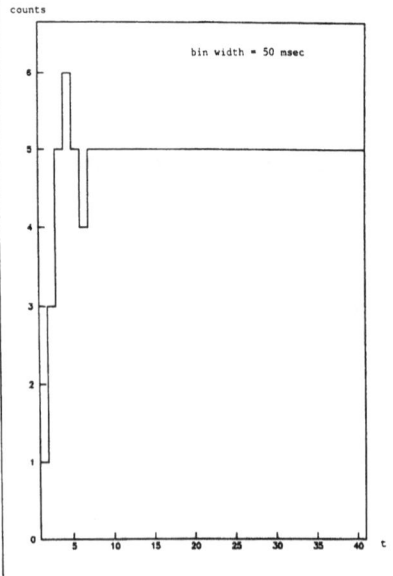

Fig. 1.b →

- Simple considerations of the Information Theory allow us to state that each S.A.U. sensitive to a straight line segment, will exhibit a certain organization degree in the distribution of the synaptics junctions. In other words a certain redundancy must exist in the synaptic pattern of these S.A.U.-s. How much should this redundancy be? To answer this question, some experimental evidence about the sensibility of neurons at the different orientations can be of assistance. A rough analysis seems to confirm that among all the synaptic patterns, only about 9÷18, that is, as many as the orientated segments detected in a 180° angle, are those that have a higher probability of been arranged in these kind of S.A.U.-s. If we then observe that there is no apparent motive why one direction should be perceived better than another, the synaptic pattern relative to one direction should have the same redudancy as that of any other orientation. Likewise, assuming that the probability of confusing one orientation with an adjoining one is constant for all directions, it can be argued that these synaptic patterns must be equidistant in a large number of metrics however defined.

   Some synaptic patterns exhibiting the properties previously stated can be obtained utilizing the isotropic scheme in fig.2 in which the 32 neurons are distributed along a circumference. The synaptic pattern relative to an orientation chosen among 16 pre-established ones, is specified by connecting couples of neurons by chords having the same orientation. The junctions are excitatory when the chords are travelled in one versus and for reasons of stability of the S.A.U., are inhibitory when these chords are travelled in the opposite versus. The fig. 2 shows the set of chords for synaptic patterns of the S.A.U.-s sensitive to segments oriented along the direction 0 and 5 respectively.

   In fig. 3 the synaptic pattern relative to orientation 0 is presented. In this case, the type of synaptic junctions between neurons connected by chords having different orientations than those prefixed, is completely random.

   The function $\bar{f}$, equal to the spatial average of the output f_i of the neurons in the S.A.U. exhibiting the synaptic pattern in fig. 3, is histogrammed in fig. 4.

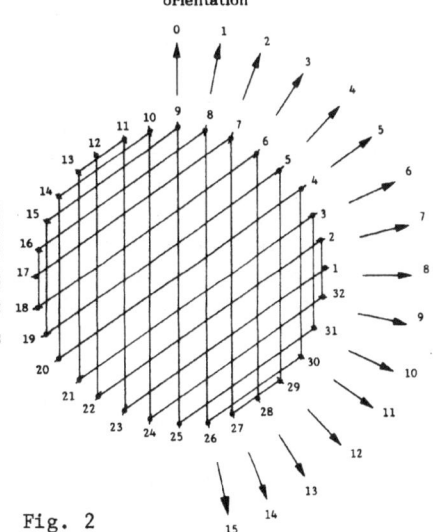

Fig. 2

CONCLUSION

   The synaptic pattern model proposed in this paper is not the only possible model which satisfies the conditions stated in section 4; many other models could be proposed.

   The results reported here however, seem to encourage this type of approach even if we have reason to believe that the process of formation of synaptic patterns is not of a deterministic nature, but rather that of a statistical one. Further investigation in this direction is needed however.

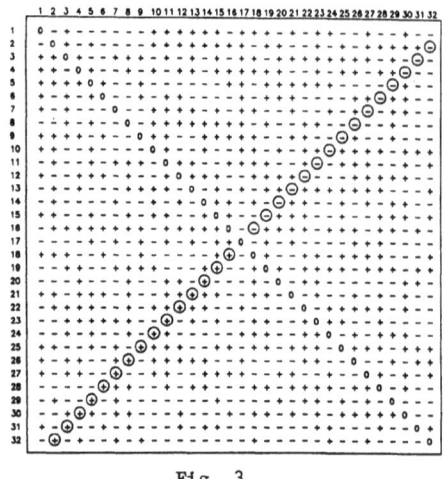

Fig. 3

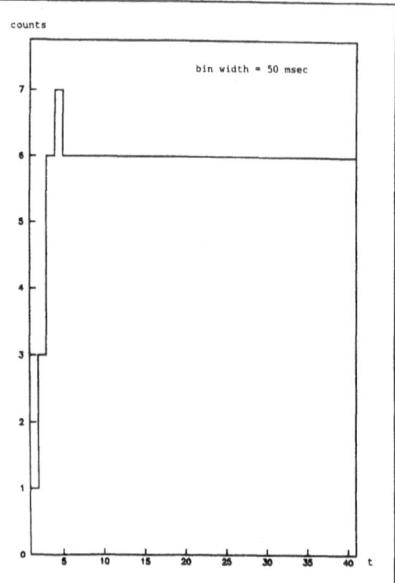

Fig.4 →

REFERENCES

1. C.Y. Suen, "Character Recognition by Computer and Applications" ICPR, Tutorial Text Recognition, Paris (1986).

2. H. Freeman, "On The Encoding of Arbitrary Geometric Configurations" IRE Trans. Electron. Comput., Vol.EC-10, 260÷268 (Feb. 1961).

3. R. Rosen, "Some Comments on Activation and Inhibition" Bull. Math. Biol.,vol. 41, 427÷445 (1979).

4. M. Alonso and E.J. Finn, "Fundamental University Physics" Addison-Wesley Pub. Comp., Vol.II (1967).

5. H.C. Andrews, "Computer Techniques in Image Processing" Academic Press, N.Y. and London (1970).

6. P.N. Slater, "Remote Sensing: Optics and Optical Systems" Addison-Wesley Pub. Comp. (1980).

7. C. Braccini, "A Model of the Early Stages of the Human Visual System: Functional and Topological Transformations Performed in the Peripherical Visual Field" Biological Cyber., Vol. 44, 47÷58 (1982).

8. D. Marr and T. Poggio, "A Computational Theory of Human Stereo Vision" Proc. R. Soc. Lond. B 204, 301÷328 (1979).

9. D. Marr and E.C. Hildreth, "Theory of Edge Detection" Proc. R. Soc., Lond. B 207÷217 (1980).

10. D.H. Hubel and T.N. Wiesel, "Brain Mechanisms of Vision" Science, Vol.29, 130÷144 (1982).

11. R. Pierantoni, "Le Corteccie Visive" Le Scienze, Quaderni, n. 29, 98÷103 (Aprile 1986).

12. D.H. Hubel, "The Brain" Scientific American, vol. 241, n. 3, 39÷47 (September 1979).

13. G.L. Shaw and J.C. Pearson, "Information Processing in the Cortex: the Role of Small Assemblies of Neurons" Studies in the Natural Sciences, vol. 21,1÷30 (1985).

THE MEASUREMENT OF BINOCULAR DISPARITY

Michael R. M. Jenkin and Allan D. Jepson[†]

Department of Computer Science
University of Toronto
Toronto, Ontario, Canada

Current stereopsis algorithms rely on the detection of sophisticated landmarks from bandpass version of the monocular images. The process of extracting these landmarks and determining their inter-ocular correspondence is considered to be one of the hard computational tasks in stereopsis. In this paper we propose that symbolic features should not be extracted in the first stages of processing; rather we propose a technique for measuring the local phase difference between the two images. The local phase difference can be used to measure the relative local disparity between the monocular images. A later level of processing must be used to reduce the "false targets" that may be detected.

1. Introduction

Current computational theories of stereopsis involve three distinct stages (Marr, 1982; Marr and Poggio, 1976, 1977; Grimson 1981). First, the two images of a stereo pair are processed separately to extract monocular features. One common choice of feature is the presence of a zero-crossing in a bandpassed version of the image. Second, the monocular features in one image are matched with corresponding features found in the other image. In practice this second stage cannot be expected to produce only the correct matches, and a third stage must be considered in order to remove the incorrect matches ("false targets"). There are therefore three main issues in the design of such a traditional algorithm for stereopsis, namely i) the choice of image features; ii) the choice of matching criteria; and iii) the way false targets are avoided or eliminated.

In this paper we introduce a different approach. We propose that symbolic features should not be extracted from the monocular images in the first stage of processing. Rather we examine a technique for measuring the local phase

---

[†]also, member CIAR.

difference between the two images. This essentially combines the first two stages of the traditional approach. A third stage is still necessary to remove false targets (which now arise as errors in the relative-phase measurements by roughly an integer multiple of 2π). The benefits of the new approach are discussed below.

There are several factors involved in the choice of suitable image primitives. Of primary importance is that an extracted feature can be expected to correspond to a particular property of a surface in the scene being viewed, and that this surface property is likely to produce the same type of feature in both images. In other words, we wish to use features that can be expected to produce reliable information about *surface structure* once they have been correctly matched. It is also important that matches should occur sufficiently often to provide a fairly dense description of the disparity. While algorithms are available for filling in expected values of disparity given only sparse data (Terzopoulos, 1982), it is obviously preferable to have denser data. Finally, the choice of image features strongly effects the options available for obtaining matches and eliminating false targets. This has been expressed very clearly by Marr (1982, p.127):

> "The basic problem to be overcome in binocular fusion is the elimination or avoidance of false targets, and its difficulty is determined by two factors: the abundance of matchable features in an image and the disparity range over which the matches are sought. If a feature occurs only rarely in an image, the search for a match can cover quite a large disparity range before false targets are encountered, but if the feature is a common one or the criteria for a match are loose, false targets can occur within quite small disparities."

In brief we have, *i*) the features should correspond to surface properties, *ii*) should produce matches that are fairly dense, and *iii*) should have distributions over a typical image such that false matches can be relatively easily avoided or eliminated.

The constraints on the density and on the ease of eliminating false targets are in direct opposition. In particular, the number of possible matches in a given region increases exponentially with the density of a given symbolic feature. Therefore the problem of finding the correct match can be expected to rapidly become more difficult as the density grows.

The approach we take here does not rely on symbolic features to measure the disparity, and therefore we avoid this density/interpretation difficulty trade-off. Instead of using symbolic features, the entire output of various bandpass channels (applied separately to each image of the stereo pair) is used. Operators are constructed to use this output to compute the local phase difference between the two images.

We are not alone in suggesting that the construction of a denser disparity representation should be considered. For example, Mayhew and Frisby (1981) have shown that zero-crossings alone are not enough to account for human stereopsis. They suggest that (at least) peaks and zero-crossings should be extracted as monocular primitives. A similar suggestion has been made by G. Poggio and T. Poggio (1984). However, we are unaware of any other work suggesting that the initial stage of symbolic feature extraction can and should be discarded. This idea is consistent with the general approach of using direct measurement techniques to extract image primitives such as orientation and velocity (see, for example, Fleet and Jepson (1986)).

In this paper we present the basic theory of the measurement approach; the actual use of these measurements in a complete algorithm for stereopsis is beyond the scope of the current paper. However, it is important to note here how these disparity measurements satisfy the constraints discussed above, which suggests that a complete algorithm is possible.

First we note that the local structure of bandpassed versions of an image tend to correspond to properties of objects in the scene, and not on the details of the imaging process. In fact, zero-crossings of such signals were chosen for precisely this reason (see Marr, 1982). Here we are using the entire local structure of the signal to extract relative-phase information. So the case for zero-crossings carries over to the disparity measurements. The disparity measurements can also be expected to be dense since the operators are designed to be insensitive to the particular form in any one image alone. Roughly speaking, only two nontrivial bandpass signals are needed to extract relative-phase. The positions of features, such as zero-crossings, in either image separately is irrelevant.

Finally, we consider the third constraint, namely that false targets can be avoided. The key here is to note that relative-phase information can, of course, only be computed modulo 2π. In particular, false targets arise in a phase measurement approach when an inappropriate integer multiple of 2π is used in the computed relative-phase. The key to avoiding or eliminating many false targets is to note that a good prediction for the distance between possible matches is given by the wavelength associated with the peak frequency of the bandpass filter being used. This is precisely the same information used by Marr and Poggio (1977) (about the expected spacing of zero-crossings) to develop a coarse to fine matching strategy (also see Grimson (1981)). For our purposes here we need only state that a similar strategy could be used on the results of the relative-phase measurements provided by our approach.

We conclude that relative-phase measurements have the potential of doing a significantly better job than the extraction and matching of symbolic features. They should provide denser information with no extra cost in removing false targets. The noise sensitive task of extracting symbolic features at an early stage of processing would be eliminated.

2. The Basic Approach

In this section we present the basic computational approach used to measure local disparity information. To illustrate the proposed technique in its simplest form we consider one-dimensional sinusoidal signals here. We note, however, that only relatively minor modifications of this basic technique will be needed in order to deal with more general two-dimensional images.

In particular, we let $I_l(x)$ and $I_r(x)$ be the left and right "images", where

$$I_l(x) = A \sin(\omega_l x + \theta_l); \quad A, \omega_l > 0, \tag{2.1a}$$

$$I_r(x) = B\sin(\omega_r x + \theta_r); \quad B, \omega_r > 0. \tag{2.1b}$$

Note that, since different perspectives in the left and right views can alter the spatial frequencies of the observed patterns, we do not assume that $\omega_l = \omega_r$. However, we do assume that $\omega_l \approx \omega_r$ (i.e. $|\omega_l - \omega_r|/(\omega_l + \omega_r) \ll 1$), and in practice we ensure this by considering two bandpassed versions of the left and right raw images.

One way to define a local phase difference between two bandpass signals is to consider matching "features" such as peaks and zero-crossings. For I_l and I_r as above, this basically amounts to matching the arguments of the two sinusoids, modulo 2π. Thus, we define the local phase difference to be

$$\phi(x) = \lfloor (\omega_l - \omega_r)x + (\theta_l - \theta_r) \rfloor \in [-\pi, \pi). \tag{2.2}$$

Here $\lfloor \theta \rfloor$ denotes the principal part of the angle θ, which is obtained by adding an integral multiple of 2π to θ so that the result lies in the interval $[-\pi, \pi)$. Note that this mod operation produces a discontinuous function, $\phi(x)$, whenever $\omega_l \neq \omega_r$. Moreover the discontinuities correspond to boundaries of intervals in x where the matching of peaks and zero-crossings between the two images is one-to-one.

The local disparity, $d(x)$, is the distance the images must be shifted with respect to each other so that I_l and I_r agree up to a multiplicative constant (B/A in the above example). This definition leads to the following form for the local disparity,

$$d(x) \equiv \frac{1}{\bar{\omega}} \phi(x), \quad \bar{\omega} \equiv \frac{1}{2}(\omega_l + \omega_r). \tag{2.3}$$

With this definition it follows that the left image shifted to the right by $\frac{1}{2}d(x)$, namely $I_l(x - \frac{1}{2}d(x))$, is a constant times the right image shifted by the same amount to the left, namely $I_r(x + \frac{1}{2}d(x))$. Positive disparity, therefore, corresponds to objects that are *further* from the fixation point of the two eyes.

In order to extract $\phi(x)$ we consider the point-by-point multiplication of the left and right images, that is

$$I_l(x)I_r(x) = AB\sin(\omega_l x + \theta_l)\sin(\omega_r x + \theta_r)$$
$$= \frac{AB}{2}[\cos((\omega_l - \omega_r)x + (\theta_l - \theta_r)) - \cos((\omega_l + \omega_r)x + (\theta_l + \theta_r))].$$

By low-pass filtering this product with a filter, L, having a high frequency cut-off below $\omega_l + \omega_r$, we obtain

$$P(x) \equiv L*(I_l(x)I_r(x)) = \frac{ABK}{2}\cos((\omega_l - \omega_r)x + (\theta_l - \theta_r)) \tag{2.4}$$
$$= \kappa\cos(\phi(x)), \quad \kappa \equiv \frac{ABK}{2}.$$

Here $\phi(x)$ is as in (2.2) and $K = K(\omega_l - \omega_r)$ is the sensitivity of the low-pass filter to the frequency $\omega_l - \omega_r$. This result is encouraging since $P(x)$ depends only on the desired local phase difference and a scale factor involving the product of the amplitudes. However a *local* technique is needed to disambiguate the amplitude and the relative phase information inherent in $P(x)$. The technique must be local since we cannot expect bandpassed images to be well approximated by (2.1a,b) over intervals of roughly the length of one wavelength of $P(x)$, that is, over lengths $\frac{2\pi}{|\omega_l - \omega_r|}$. Therefore it is not possible to estimate the scale factor by examining the behaviour of $P(x)$ over one of its wavelengths. Instead it is necessary to obtain an estimate for $\phi(x)$ based on information available over no more than a few wavelengths of the base frequency $\bar{w}$ (i.e. $\Delta x \approx \frac{2\pi}{\bar{\omega}}$).

A suitable local technique can be obtained by the inclusion of a specific shift in one (or both) of the images before the pointwise product is calculated. For example, we define

$$P(x, s) \equiv L*[I_l(x - \tfrac{1}{2}s) I_r(x + \tfrac{1}{2}s)] = \kappa \cos(\bar{\omega}s - \phi(x)), \qquad (2.5)$$

with $\bar{\omega}$ as in (2.3). Now, to obtain an approximation for ϕ near a given point x_0, the value of s is adjusted so that a particular local feature of $P(x, s)$ occurs at $x = x_0$. Suitable local features of P are, of course, features that are independent of the product AB, and include zero-crossings and local extrema. Here we choose to avoid seeking local extrema of $P(x_0, s)$, which can be fairly sensitive to noise. Instead we consider a zero-crossing of $P(x_0, s)$, say $s_\infty(x_0)$, such that

$$P(x_0, s_\infty(x_0)) = 0, \qquad (2.6a)$$

$$\frac{\partial P}{\partial s}(x_0, s_\infty(x_0)) > 0. \qquad (2.6b)$$

For such a zero-crossing we see from (2.5) and (2.6) that

$$\lfloor \bar{\omega}s_\infty(x_0) - \phi(x_0) \rfloor = -\frac{\pi}{2}.$$

Therefore the local relative phase difference is given by

$$\phi(x_0) = \lfloor \bar{\omega}s_\infty(x_0) + \frac{\pi}{2} \rfloor, \qquad (2.7)$$

which can be computed given $s_\infty(x_0)$ and the mean frequency $\bar{\omega}$. In addition it follows from (2.3) and (2.7) that the local disparity $d(x_0)$ satisfies

$$d(x_0) = s_\infty(x_0) + \frac{\pi}{2\bar{\omega}}. \qquad (2.8)$$

In practice we avoid the dependence of the disparity measurement on the value of $\bar{\omega}$. To do this we introduce a relative phase shift of $\frac{\pi}{2}$ between the two bandpassed images through the use of left and right filters with different phase properties. For example, a sine Gabor filter applied to the right image and a slightly modified cosine Gabor applied to the left image produces the desired phase shift. With such a pair of filters, the function $P(x, s)$ becomes

$$P(x, s) = \kappa \sin(\bar{\omega}s - \phi(x)). \qquad (2.9)$$

A zero-crossing of $P(x, s)$ at which (2.6) is satisfies provides the local disparity value

$$d(x_0) = s_\infty(x_0). \qquad (2.10)$$

That is, the local disparity is given directly by the position of the zero-crossing of $P(x_0, s)$ with positive slope.

3. Phase-sensitive Demons

The general form of the phase measurement scheme discussed above is as follows: The raw images, I_l^0 and I_r^0, are first sent through bandpass filters B_l and B_r, respectively, to produce

$$I_\nu(x) = \int_{-\infty}^{\infty} B_\nu(x-z)\, I_\nu^0(z)\, dz, \ \ \text{for } \nu = l,\, r. \qquad (3.1)$$

The left and right bandpass image are shifted by s, and the phase detector,

$$P(x, s) = \int_{-\infty}^{\infty} L\,(x-z)\,(\, I_l(z-s/2)\, I_r(z+s/2)\,)\, dz, \qquad (3.2)$$

is applied. Here L is a low-pass filter. Finally, zero-crossings of P are detected for each sample point x. We refer to this mechanism as a *phase-sensitive demon*.

We imagine the initial stereo images decomposed into a multiple spatial scale representation, perhaps through the use of a Burt pyramid (Burt, 1981), or the DOLP transform (Crowley and Stern, 1984). Also there is a similar hierarchical scheme that can be used to construct Gabor-like receptive fields (Fleet and Jepson, 1986). The filters B_l and B_r are taken to be members of such a family, and we assume that their output is resampled appropriately. For each spatial scale there is a demon applied at each sample point x and, like all good demons, they are taken to be independent of other demons at different x's or different scales. In particular, we emphasize that the demons are simply collecting *local* disparity information. A globally consistent interpretation of the computed local disparities is the job of either a subsequent level of processing or a network connecting demons to their neighbours. Either way we do not discuss globally consistent interpretations in this paper.

4. A Corrugated Surface

Figure 1 shows the left and right views of a corrugated surface. Such surfaces have been found to pose difficult problems for zero-crossing based stereo algorithms (Mayhew and Frisby 81). In fact, if the standard difference of Gaussians (DOG) filters are used, a simple symmetry argument shows that the exact zero-crossings always occur halfway between the peaks. The matching algorithm can therefore only produce a constant disparity response; the variation in depth is not detected. Moreover, the slope of the zero-crossings in higher frequency channels is extremely shallow. This is exhibited in Figure 2 where the left and right images are plotted after being passed through a DOG filter whose peak frequency is set at twice the fundamental. The shallow slope indicates that the positions of the zero-crossings are extremely sensitive to discretization errors and noise.

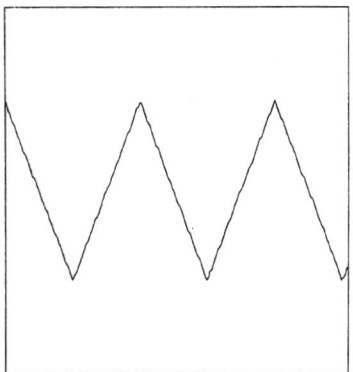

Figure 1. Left and right views of corrugated surface

In contrast, in Figure 3 we present the results from the phase sensitive demons represented in terms of the left camera's view. Figure 3 shows the recovered depth profile marked as crosses, with the actual depth profile given by the solid line. Figure 3a provides disparity values obtained by using bandpass filters with a peak frequency twice that of the fundamental. In Figure 3b the peak frequency is set at four times the fundamental. Both channels detect the variation in depth, with the higher frequency channel obtaining a better approximation near the peaks. Channels based on higher harmonics give even better resolution of the peaks, but do not respond in the linear segments between. Therefore the phase demon which utilizes the full bandpass signal is able to detect structure in the channel that is not encoded simply by the position of the zero-crossings.

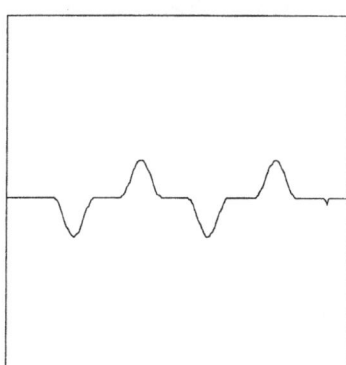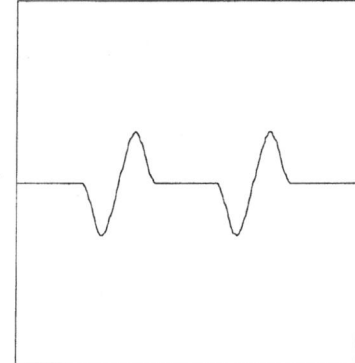

Figure 2. DOG applied to left and right views.

5. Conclusions

By developing the notion of disparity in a bandpass spatial frequency tuned channel, we have constructed a disparity demon which can be tuned for particular disparities. These disparity detectors can be applied independently and are capable of producing dense and robust responses to surfaces that can only be marginally processed using current token-based techniques.

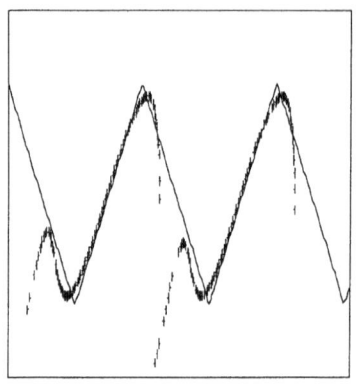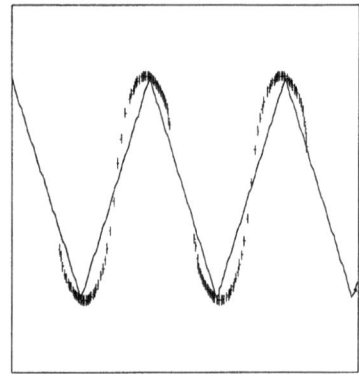

Figure 3. Recovered depth from disparity demons.

References

Burt, P.J. (1981), "Fast filter transforms for image processing", *Comp. Graph. Im. Proc. 16,* pp.20-51.

Crowley, J. L., and Stern, R. M. (1984), "Fast Computation of the Difference of Low-Pass Transform", *IEEE Trans. on P.A.M.I.,* March.

D.J. Fleet and A.D. Jepson (1986), "On the hierarchical construction of orientation and velocity selective filters", submitted to *IEEE Trans. Pattern Analysis and Machine Intel.*

Grimson, W. E. L., (1981), *From Images to Surfaces,* MIT Press, Cambridge.

Marr, D. (1982), *Vision,* Freeman Press.

Marr, D., & Poggio, T. (1976), "Co-operative computation of stereo disparity", *Science Vol 194,* pp.283-7.

Marr, D., & Poggio, T. (1977), "A theory of human stereo vision", A. I. Memo No. 451, MIT.

Mayhew, J., & Frisby, J. (1980), The computation of binocular edges, Perception 9, pp.69-86.

Mayhew, J. E. W., and Frisby, J. P. (1981), "Psychophysical and Computational Studies towards a Theory of Human Stereopsis", *Artificial Intelligence 17,* pp.349-385.

Poggio, G., & Poggio, T. (1984), "The analysis of stereopsis", *Annual Reviews Neurosci.,* pp. 379-412.

Terzopoulos, D., (1983), Multi-level reconstruction of visual surfaces, in "Multiresolution Image Processing and Analysis", A. Rosenfeld, Ed., *Springer-Verlag,* Berin.

POINT PATTERN MATCHING AND CORNER FINDING

FOR LINE DRAWINGS

Hideo Ogawa

Department of Electronics
Fukui University
Fukui, Japan

Abstract

This paper proposes a robust method for matching labeled point patterns. A point pattern is partitioned into a set of triangles using the Delaunay triangulation. For the corresponding triangle pair, a consistency graph is constructed based on the pairwise compatibility between the points in the triangles. The matching is accomplished by locating the largest maximal clique of the consistency graph. A new method for detecting corners is also proposed, based on the local symmetry of a discrete curve. The corners are used as the feature points in point pattern representation and matching of the line drawings.

1. Introduction

Point patterns often appear in processing visual information. For example, objects in an image may be efficiently represented by a small number of labeled points, e.g., spatial feature points. Therefore, many researchers have directed considerable attention to point patterns, especially to the problems of point pattern matching. Ranade and Rosenfeld[1], and Barnard and Thompson[2] discuss a relaxation-based approach. Ogawa also proposes a labeled point pattern matching based on a fuzzy relaxation[3], and also a method[4] based on the Delaunay triangulation and the largest maximal clique. In, Ahuja[5], an overview of dot pattern processing using Voronoi neighborhoods can be found. Bolles[6] discusses how to apply maximal clique techniques to the problems of feature matching. This paper develops the method proposed in Ogawa[4], and discusses an application to the shape matching of line drawings.

Corners can be thought of as the most useful feature points in order to represent the line drawings. The corner detection has been studied by many researchers.[7,8,9,10,11] As a new approach to the corner detection, a method based on the local symmetry of a discrete curve is proposed in this paper. Matuyama et al.[12] have a similar idea.

2. Labeled Point Patterns

A labeled point pattern in this paper is a finite set of spatial points $\{P_i\}$ characterized with the coordinates (X_{pi}, Y_{pi}), labels $\{F_{ij}\}$ and binary relations $\{R^k\}$ on the set $\{P_i\}$. A label F_{ij} denotes a property of the point P_i. In one view, a label may be thought of as a feature extracted at the point. Accordingly, a labeled point pattern M which denotes a prototype or model of some object of interest can be written as

$$M=<\{P_i\},\{R^k\}>=<\{(X_{pi},Y_{pi},F^p_{i1},F^p_{i2},\ldots,F^p_{in},L_i)\},\{R^k\}>,$$
$$i=1,2,\ldots,m(M), \quad k=1,2,\ldots,K. \tag{1}$$

In the above expression, L_i is also a label which is assigned 1 or 0, $L_i=1$ means that the point P_i is an attention point which is a distinguishing point in the sense that some labels $\{F_{ij}\}$ of the point are certain or reliable because they are less sensitive to noise or distortions. This label may be thought of as the knowledge concerning the certainty or reliability of the labels $\{F_{ij}\}$. This kind of knowledge can be obtained statistically or empirically. Binary relation R^k describes some distinguishing relation between two points. For example, as to the line drawings where the points are the corners detected by using the method described in a later section, if a line segment terminates at the points P_i and P_j and the shape is convex, then $(P_i,P_j) \in R^1$, if the shape is concave, then $(P_i,P_j) \in R^2$, and so on.

If more details are required for the description of line segments, the segment can be represented by a series of points or Fourier descriptors instead of the binary relations.

In the labeled point pattern matching discussed in this paper, one of the two labeled point patterns can be thought of as a model of some object of interest, so denoted with the symbol M as mentioned above. The other can be thought of as a decsription of a world in which we are searching for an instance of the object of interest. A world denoted with the symbol W can be described in the same manner as a model M, excepting L_i, as

$$W=<\{Q_j\},\{R^k\}>=<\{(X_{qj},Y_{qj},F^q_{j1},F^q_{j2},\ldots,F^q_{jn})\},\{R^k\}>,$$
$$j=1,2,\ldots,m(W), \quad k=1,2,\ldots K. \tag{2}$$

An example of the representation is shown in Section 5.

3. Matching Method

It is possible to obtain a set of points M^S using the knowledge of attention points and the similarity of labels as follows.

$$M^S=\{P_i \mid L_i=1, \text{ and } \exists Q_j, S(P_i,Q_j) \geq t_1\}, \tag{3}$$

where t_1 is an appropriate threshold and nearly equal to 1. $S(P_i,Q_j)$ denotes the similarity between the labels of points P_i and Q_j.

The partition of a model M is accomplished as a triangulation of the set of points M^S. In this paper, the Delaunay triangulation[5,13] is employed. Figure 1 shows an example of the Voronoi tessellation and the Delaunay triangulation for a set M^S. In this way, a model point pattern can be partitioned into the triangles and the open regions (shown in Fig.2).

Every point $P_i \in M^S$ has the corresponding candidate points $\{Q_j\}$ which are determined in equation (3). Therefore, it is straightforward to construct the triangles $\{T^W_j\}$ (i.e., a set of three-tuples of the points $Q_k \in W$) which are the corresponding candidates for a triangle T^M_i of the model point pattern. But it should be noticed that if Q_i, Q_j ($\in T_j^W$) correspond to P_i, P_j ($\in T_i^M$) respectively and $(P_i, P_j) \in R^k$, then $(Q_i, Q_j) \in R^k$ must be true.

The triangles in each set (i.e., model and world) can be merged so as to preserve the adjacency relation. The procedure is omitted in this paper, but can be found in Ogawa.[4] Consequently, we can obtain the largest set of triangle pairs $\{(T_i^M, T_j^W)\}$ where exists a relational isomorphism from $\{T_i^M\}$ to $\{T_j^W\}$. Hereafter, a polygon means a set of triangles merged.

In general, there are plural polygons $\{G_j^W\}$ which are candidates for the maximal polygon G_i^M. Assume that a G_j^W is selected. This means that a one-to-one correspondence between the triangles in G_i^M and G_j^W (i.e., the largest set of triangle pairs $\{(T_i^M, T_j^W)\}$) is determined.

In order that the matching is invariant under affine transformation, an affine transformation which maps a triangle T_j^W to a triangle T_i^M can be defined for each triangle. Under the affine transformation ψ_i, a set of corresponding candidates $\{Q_m\}$ for a point P_n in T_i^M can be defined as

$$C_n = \{Q_m \mid Q_m \text{ in } R_n^W, \ S(P_n, Q_m) \geq t_2\}, \tag{4}$$

where t_2 is an appropriate threshold and may be allowed relatively small value compared with t_1. The region R_n^W is obtained as

$$R_n^W = \psi_i^{-1}(R_n^M) \tag{5}$$

where the region R_n^M can be determined with the vertices of triangle T_i^M and tolerance θ for the angular displacement of points as shown in Fig.3. In this way, for every P_n in T_i^M, the corresponding candidates $\{Q_m\}$ are obtained. And each Q_m can be transformed to $\overline{Q}_m$ as

$$\psi_i(Q_m) = \overline{Q}_m. \tag{6}$$

Based on the pairwise compatibility of points, a consistency graph can be constructed as follows. Let $\overline{Q}_m$ and $\overline{Q}_j$ be the corresponding candidates for P_n and P_j respectively. When the following conditions are satisfied,

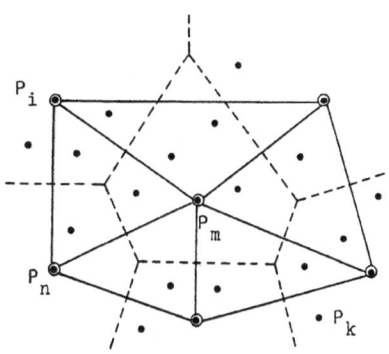

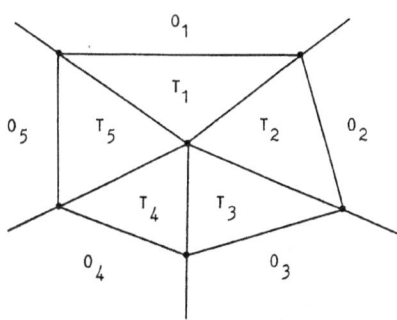

Fig.1. The Delaunay triangulation and Voronoi Tessellation (dotted lines) for a set of points (◉) in M^S, the points (•) belong to M.

Fig.2. The partitioned triangles (T_i) and open regions (O_i).

two nodes corresponding to the point pairs $(P_n, \bar{Q}_m)$ and $(P_i, \bar{Q}_j)$ are introduced into the consistency graph and an arc is defined between the nodes.

Condition 1: the angular difference α between the point pairs (P_n, P_i) and $(\bar{Q}_m, \bar{Q}_j)$ is less than the tolerance θ, as shown in Fig.4.

Condition 2: if $(P_n, P_i) \in R^k$, then $(\bar{Q}_m, \bar{Q}_j) \in R^k$.

The largest maximal clique of the consistency graph represents the largest set of mutually consistent point pairs.

The above process for every triangle T_i^M determines the corresponding point pairs $\{(P_i, Q_j)\}$. This procedure can be extended to the open regions O_i^M. The affine transformation ψ_i which is determined to the region T_i^M adjacent to the open region O_i^M is used for the open region O_i^M.

Finally, as mentioned previously, an appropriate G_j^W must be determined. One method is that the polygon with the largest number of point pairs can be selected as the instance of the model.

4. Corner Detection

The segment $\overline{P_i P_m}$ forms a reflective symmetry with the segment $\overline{P_i P_n}$ at the point P_i in Fig.5. Namely, the segment $\overline{P_m P_n}$ is reflective-symmetric at the point P_i, and half the length of segment $\overline{P_m P_n}$ can be defined as the extent of local symmetry of the curve at the point P_i. Fig.6(a) and (b) show the instances of rotational symmetry of a curve.

In the method proposed in this paper, it is a fundamental idea that the corners locate at the points on a discrete curve where the extents of local symmetry of a curve are local maxima.

The essential idea of finding the reflective symmetry of a segment at the point Q_k is illustrated in Fig.7. The points S_i and T_i are defined as the points which satisfy the following conditions.

Condition 1: They are on the discrete curve.

Condition 2: The distances from Q_k to S_i and to T_i measured along the discrete curve are δ times an integer i, where δ is a predetermined step size. The radius α of a circle centered at T_i represents the tolerance for the displacements of curves. Namely, if there exists the arc $\overline{AB}$ where A and B are the intersections of two circles with radii α and $\overline{Q_k S_i}$ centered at T_i and Q_k respectively, then T_i is considered as a reflective symmetry with S_i. Any straight line which belongs to the sector $CONE_i$ shown in Fig.7 can be thought of as an axis about which T_i forms a reflective symmetry with S_i.

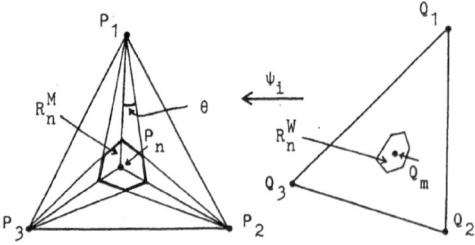

Fig.3. The corresponding triangles

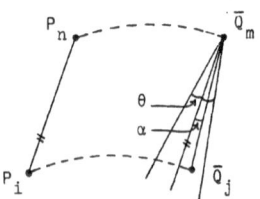

Fig.4. The tolerance θ for the angular difference α.

For the next point pair S_{i+1} and T_{i+1}, the sector $CONE_{i+1}$ can be obtained in the same manner. The intersection $CONE_i \cap CONE_{i+1}$ shows the region of axis about which the segment $\overline{S_i S_{i+1}}$ has a reflective symmetry with the segment $\widehat{T_i T_{i+1}}$. In this way, the index j which satisfies the next condition can be obtained.

Condition: $\overset{1}{\underset{i=1}{\cap}} CONE_i \neq \phi$ (empty set) for $1 \leq l \leq j$, and $\overset{j+1}{\underset{i=1}{\cap}} CONE_i = \phi$.

Then the value $V_k = j\delta$ can be employed as the extent of reflective symmetry of a segment at the point Q_k

The idea of finding the rotational symmetry of a segment at the point Q_k is illustrated in Fig.8. Although the details are omitted in this paper, it is not difficult to compute the value $U_k = j\delta$ (the extent of rotational symmetry of a segment at the point Q_k) on the analogy of reflective symmetry. The extent of local symmetry of a segment at the point Q_k can be defined by

$$L_k = \max\{V_k, U_k\}. \tag{7}$$

The maxima of L_k are the candidates for corners and other critical points such as inflections and middle points of smooth segments. In order to identify corners, the curvature is computed at the candidate points. The curvature κ_i, in this paper, is defined as

$$\kappa_i = \min\{\cos\theta_i, \cos\theta_i^+\}, \tag{8}$$

where θ_i is the angle between two vectors $\overrightarrow{P_i P_j}$ and $\overrightarrow{P_i P_k}$, and θ_i^+ is the angle between two vectors $\overrightarrow{P_i P_{j-1}}$ and $\overrightarrow{P_i P_{k+1}}$ as illustrated in Fig.9. The points P_j and P_k are the adjacent local maxima on both sides of P_i.

The points $\{P_i\}$ which satisfy the next condition are identified as corners.

$$\kappa_i \geq \varepsilon, \tag{9}$$

where ε is a predetermined threshold value. The points which don't satisfy above condition can be regarded as inflections or middle points on smooth segments marked with ▤ in Figs.10,11,12. In addition to the procedure described above, for the line drawings in Figs.10,11,12, intermediate points between the consecutive two points are introduced, in order to make the location of corners exact. The details are omitted here for lack of space.

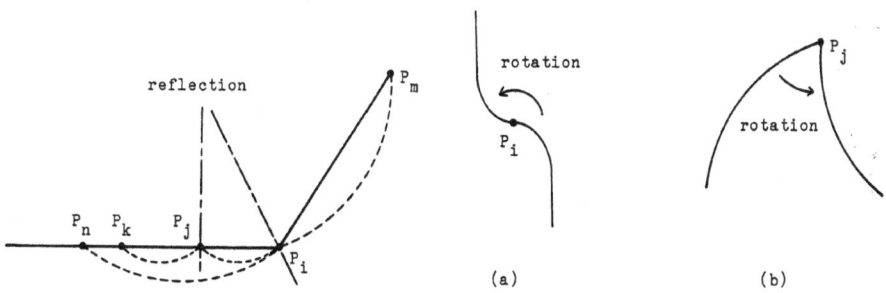

Fig.5. Reflective symmetries of segments at the points P_i and P_j.

(a)　　　　　(b)

Fig.6. Rotational symmetries of segments at the points P_i and P_j.

5. Experiments and Discussions

The proposed matching method was applied to the matching of constellations where the binary relations are not considered. Some of the results were already shown in (4). Through the experiments, the usefulness of the triangulation is shown. Namely, the triangulation allows piecewise linear approximation to a non-linear distortion. It is also a very important merit of the triangulation that the computation cost of clique detection can be greatly reduced. The process of listing all maximal cliques in an undirected graph is known to be an NP-complete problem. Therefore, the graph should be small enough. This requirement can be achieved by the triangulation of a point pattern.

Three experimental results for the different types of discrete curves are presented to show the usefulness of the proposed corner detection method in Figs. 10,11,12. These line drawings are taken from (7),(9),(7) respectively. The detected corners correspond very well to the visual corners. It should be emphasized that the proposed method produces resonable results for the different types of discrete curves, despite of the parameters fixed to the same values as listed on each figure.

Concerning the point pattern representation and matching of line drawings, as an example, an experiment for the Maple leaf is shown where the binary relation "Joined by a line" is introduced. A model of the Maple leaf can be described as

$$M_{leaf} = <\{P_i\}, \{R^1\}>,$$
$$\{P_i\} = \{P_1, P_2, \ldots, P_{17}\},$$
$$P_1 = (50, 90, convex, L=1), \quad P_2 = (58, 70, concave, L=1), \quad etc,$$
$$R^1 = \{(P_1, P_2), (P_2, P_3), \ldots, (P_{17}, P_1)\}.$$

An image of the above model is shown in Fig.13. In the experiment of matching the model in Fig.13 with the object in Fig.12, the correct point pairs were obtained completely. It is shown that the proposed method is tolerant of distortions in the line drawings. The advantages are due to the partition of the line drawings. It is suggested that the method is applicable to the identification of more complicated line drawings distorted or partially occluded by other object.

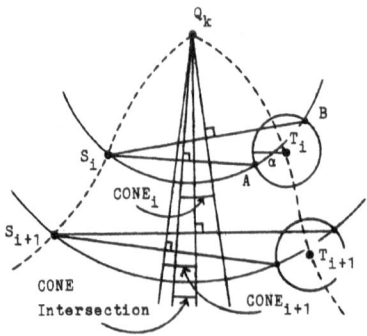

Fig.7. Illustrates the geometry of the computation for reflective symmetry of a segment.

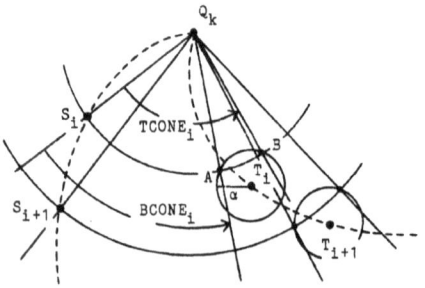

Fig.8. Illustrates the geometry of the computation for rotational symmetry of a segment.

6. Conclusion

   A method for matching labeled point patterns is described. The proposed method is invariant under affine transformation, and allows for distortions and partial occlusion. These advantages are due to the partition of the point pattern. The partition saves the computation cost of matching. The method is applied to the matching of line drawings. It is suggested that the method is applicable to the identification of complicated line drawings.
   A method for detecting corners and other critical points is described. The method is based on the local symmetry of a discrete curve. It is shown experimentally that the proposed method yields resonable results, i.e., the detected corners correspond very well to the visual corners.

Acknowledgement —— This work was supported by a Grant-in -Aid for Scientific Research from the Ministry of Education.

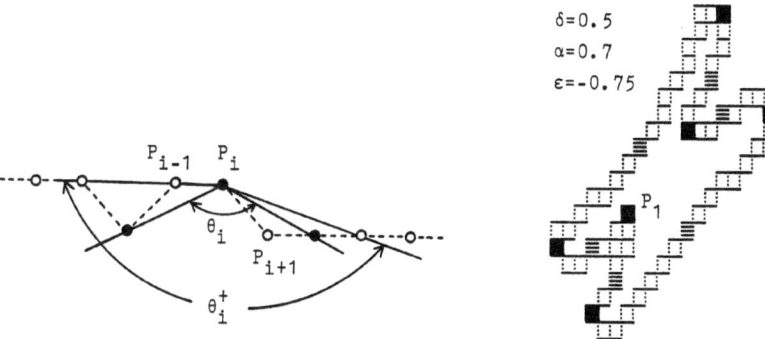

Fig.9. The angles θ_i^- and θ_i^+.

●: maxima of the extent of local symmetry of a curve.

Fig.10. The data set Chromosome.
■: corners, ▤: other critical points.

$\delta=0.5$
$\alpha=0.7$
$\varepsilon=-0.75$

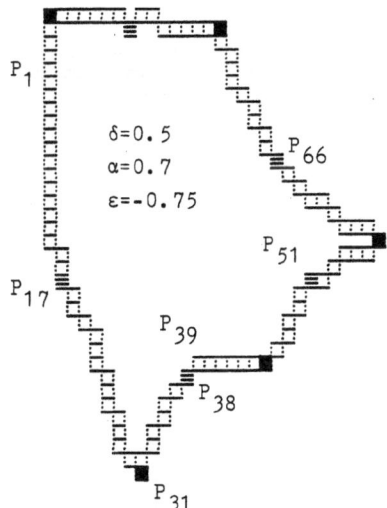

$\delta=0.5$
$\alpha=0.7$
$\varepsilon=-0.75$

Fig.11. The data set Texas.
■: corners, ▤: other critical points.

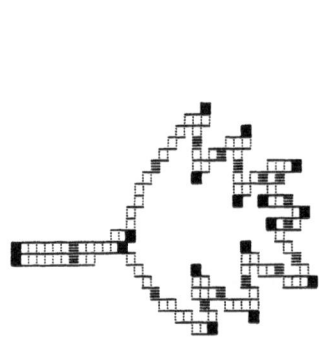

Fig.12. The data set Maple Leaf.
■ : corners, 目: other critical points.

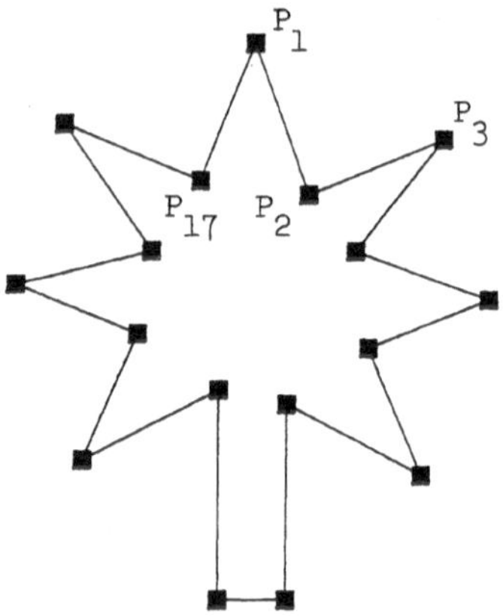

Fig.13. A model point pattern of Maple Leaf.

References

1. S. Ranade and A. Rosenfeld, "Point pattern matching by relaxation," Pattern Recognition 12, 269-275 (1980).
2. S. T. Barnard and W. B. Thompson, "Disparity analysis of images," IEEE Trans. Pattern Anal. Mach. Intell. , PAMI-2, 333-340 (1980).
3. H. Ogawa, "Labeled point pattern matching by fuzzy relaxation," Pattern Recognition 17, 569-573 (1984).
4. H. Ogawa, "Labeled point pattern matching by Delaunay triangulation and maximal cliques," Pattern Recognition 19, 35-40 (1986).
5. N. Ahuja, "Dot pattern processing using Voronoi neighborhoods," IEEE Trans. Pattern Anal. Mach. Intell., PAMI-4, 336-343 (1982).
6. R. C. Bolles, "Robust feature matching through maximal cliques," Proc. Soc. Photo-opt. Instrum Engrs 182, 140-149 (1979).
7. A. Rosenfeld and E. Johnston, "Angle detection on digital curves," IEEE Trans. Computers C-22, 875-878 (1973).
8. A. Rosenfeld and J. S. Weszka, "An improved method of angle detection on digital curves," IEEE Trans. Computers C-24, 940-941 (1975).
9. H. Freeman and L. Davis, "A corner-finding algorithm for chain-coded curves," IEEE Trans. Computers C-26, 297-303 (1977).
10. I. M. Anderson and J. C. Bezdek, "Curvature and tangential deflection of discrete arcs," IEEE Trans. Pattern Anal. Mach. Intell., PAMI-6, 27 -40 (1984).
11. T.Koyama, M. Shiono, H. Sanada and Y. Tezuka, "Corner detection on thinned pattern," Tech. Rep. of the Professional Group on Pattern Recognition and Learning, IECE Japan, PRL80-107, 83-90 (1980).
12. T. Matuyama, H. Yonezawa and M. Nagao, Proc. of 30th national conference, IPS of Japan, 1163-1164 (1985).
13. P. J. Green and R. Sibson, "Computing Dirichlet tessellations in the plane," Comput. J. 21, 168-173 (1977).

PHOTOMETRIC APPROACH TO TRACKING OF MOVING OBJECTS

V. Cantoni(*), L. Carrioli(**), M. Diani(*)
M. Savini(*),and G. Vecchio(***)

(*) Dip. di Informatica e Sistemistica
Pavia Univ., Italy
(**) Ist. di Analisi Numerica CNR, Pavia, Italy
(***) Eledra Systems, Milano, Italy

ABSTRACT

A new approach to object tracking in industrial environments is presented. It is based on
three dimensional information on objects gathered
by means of the stereophotometric technique.
Knowledge about the set of objects we deal with
and constraints on their motion allow a simplified
reasoning to solve the correspondence problem
between objects in two consecutive images.
Finally the performances of such a tracking system
will be discussed.

1. Introduction

In many applications we need to visually identify and
locate objects moving around in a three-dimensional region
of space. When motion is considered, time becomes an
important variable of the entire process and, according to
the speed of the objects involved, we get different time
constraints for the recognition process. For these reasons
it is highly desirable to keep the recognition time within
the limits given by the frame rate; a way to achieve such
performance is to exploit at each frame all the possible
information we gathered at the previous one.
One of the most useful techniques for the recognition of 3-D
objects is based on the stereophotometric method [1] for the
detection of local orientations of a surface, and the
Extended Gaussian Image (EGI) [2] that allows constructing a
signature of the object from these local orientations. The
EGI is obtained by mapping the directions of the local
normals to the surface of an object on the unit sphere. For
this reason the image in the EGI of a plane surface is a
single point on the sphere and the method is particularly
useful when we have objects that are essentially made of
planar surfaces, because the EGI will consist of a set of

253

accumulation points whose distribution corresponds to the peculiar surfaces distribution and orientation of the object, so that it will constitute its "signature".

After an introduction of these methods our paper shows how to exploit them in motion analysis.

2. Preliminary Aspects

The stereophotometric technique allows to compute the local orientation in 3-D space of a surface by observing the apparent brightness of a point and by exploiting knowledge on the nature and the material of a homogeneous object, along with the geometry of the acquisition setting [3].
The apparent brightness of a point in an image depends on several factors, among which the angle of incidence of the light on the surface, the relative position of light source, viewer and surface, and the nature and roughness of the material [5].
The theory of the reflection and diffusion processes [6] allows us to uniquely determine the orientation of the surface from the analysis of three images taken from the same point of view and with three different directions of illumination.
The Extended Gaussian Image is obtained by assigning to each point of the unit sphere a value proportional to the area of the surfaces whose normal is oriented like the radius connecting the center of the sphere to it. Some properties of the Extended Gaussian Image make it an interesting tool for the recognition problems:

a -   an object rotation is reflected in the same rotation of its EGI

b -   the "mass" of the EGI is equal to the area of the object surface.

We have a significant particular case when the object is mainly delimited by planar surfaces. In fact the EGI of a polyhedra is constituted by a number of points on the unit sphere equal to the number of its faces and the height of these points is equal to the respective areas.
The extraction of the local orientations and their statistical distribution (EGI) can be carried out analyzing the three images in raster scan format by using ad-hoc hardware taken from the shelf: essentially ROM containing a three dimensional look-up table and an accumulator for the statistic on RAM.

The problem of motion in image analysis can be seen under different points of view: the problem could require the simple detection of motion or its measurement, either for every point in the image following a differential approach (optical flow) or by a feature-based approach according to [7]. This last case is the problem we address. The motion parameters (linear and angular velocity and acceleration) for rigid objects moving on a "plane surface" under inspection can be determined from the analysis of objects position and orientation in temporal images sequences of the same scene.

3. Stereophotometric Analysis of Motion

In this section we examine the problem of determining the motion parameters of objects taken from a completely known set; in our analysis we consider, as well as in [8], only objects delimited "mainly" by planar surfaces. In fact our approach can be classified among the feature-based ones in which planar surfaces constitute the cues of the identification and consequent motion estimation processes. This solution is justified by the fact that these features are effectively detected by photometric solution. In the literature different issues have been exploited: point and local patterns [9], lines and linear grey level

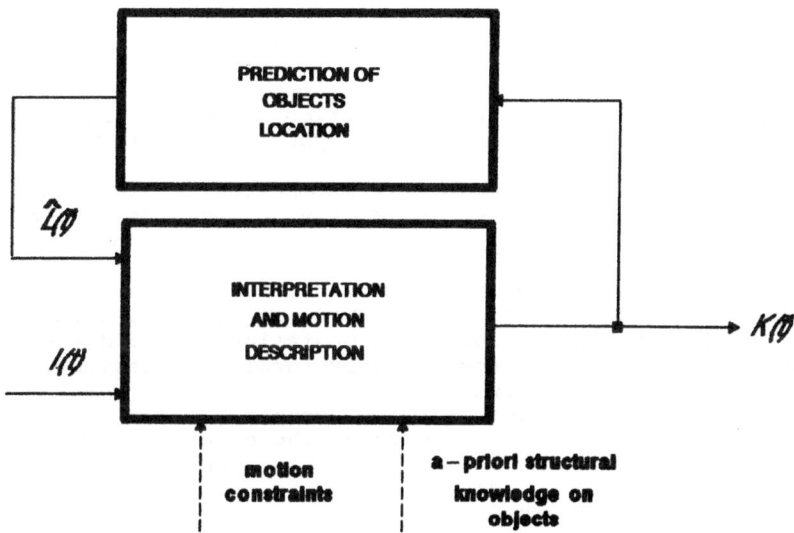

Fig. 1 — Frame processing scheme

discontinuity [10,11], silhouettes or complete shapes [4]. The environment we refer to is composed by a closed world which the objects cannot come out from or enter. This is, for instance, the situation related to the control of motion and surveillance of robot's arms moving objects in a delimited space. When we want to avoid collisions the target of the control system is to prevent objects from approaching to much together. In order to remove this possibility we have introduced the following constraints:

given two objects O_i and O_j, the circumscribed spheres and their radii r_i and r_j, the distance between the centers of the spheres must be greater than $D^-_{ij} = r_i + r_j$ for every t.

Therefor it is necessary to continuously update the forecasting of the future position of the objects based on their actual velocity and on motion constraints.

To satisfy the requirements of motion analysis we must exploit the information gathered from previous frames in the processing of the current one. The global scheme of the process is shown in fig.1.

The segmentation results coming from the EGI and the low level elaboration give in real time information about space orientation, location and apparent extension (then considering the orientation also the true extension) of each surface present in the scene.

The scene and motion interpretation contains information about identity, position, velocity and acceleration of each visible object.

The interpretation and motion description module works on data coming from both the low level processing and the results of the previous interpretation.

The prediction of objects location module outputs a forecasting made at time t of the possible positions of the objects at time t+1.

The advantage of assuming this model for the motion estimator resides in the possibility to avoid performing at each frame a complete recognition process based only on the low level information.

The interpretation module uses the a-priori knowledge on the shape of the objects and their motion constraints along with the results of the previous interpretation to rapidly update the description of the scene and performing an estimation of the new motion parameters.

The interpretation module generates and uses a classification of every surface of the objects in the scene in four classes. This classification is performed after a complete recognition of the scene has been achieved and is updated at every frame.

The discriminating criterion for this classification is based on the actual and future (in the next frame) visibility of a surface: the forecasting is done by considering the actual position and motion parameters of the single surface, the motion constraints and the object geometry.

The four classes are:

A(t) visible surfaces at frame t that cannot disappear at frame t+1

B(t) visible surfaces at frame t that may disappear at frame t+1

C(t) not visible surfaces at frame t that may appear at frame t+1

D(t) not visible surfaces at frame t that cannot appear at frame t+1

The determination of the motion parameters can be reduced to the problem of finding the correspondences among the surfaces in I(t) and those derived form K(t-1) on the base of the first order movement parameters that will be classified in the four groups $\hat{A}(t),\hat{B}(t),\hat{C}(t)$ or $\hat{D}(t)$. The most general procedure for facing this problem can be decomposed in the following steps.

I - for each surface listed in the set $\hat{A}(t)$, we look for the unique correspondent surface that must be present in I(t). Once we have found a couple of correspondent surfaces we have a correspondence between two editions of the same object (O) in two consecutive images. At this point we know which faces of the object O must appear in I(t) and we can look for them. The determination of position and orientation of the object allows us to compute its motion parameters and to classify each of its surfaces in one of the four new classes A(t), B(t), C(t) and D(t).

II - for each surface not yet classified in I(t) we look for the correspondent one among the remaining surfaces in the classes $\hat{B}(t)$ and $\hat{C}(t)$. Also in this case we follow the steps regarding the determination of objects position, orientation and motion and the reclassification of the surfaces.

The procedure described above is pretty general and does not take into consideration the possible problems arising from ambiguities and uncertainities in the search for correspondences. The conflict resolution can require a considerable effort such that it can overload the computation and unsatisfy the time requirements.
For this reason, in order to remove the ambiguities it is necessary to introduce some constraints on the motion parameters and the geometry of the objects the system can deal with.

To avoid the problems in conflict resolution time given by ambiguities it is desirable to have uniqueness in the correspondence between the two editions of a surface in two consecutive images. To have this condition guaranteed we define a "reference point" for every surface and a circular area centered in it with radius r equal to the minimum distance from the reference point to the border of the surface itself.
The displacement of the reference point from an image to the next one will be given by the composition of the actual velocity vector and the acceleration term. The projection on the image plane of the circle in the new position delimits an area in which we will look for the occurrence of the reference point in the new image.
To have a unique occurrence in the search area it is sufficient that the acceleration term does not move the reference point out of the circle projection, that is:

$$|\overline{\Delta v}| = |\overline{v(t+1)} - \overline{v(t)}| < \frac{r\cos(A(t)+|\overline{\Omega}|T)}{T} = fr\cos(A(t)+|\overline{\Omega}|T)$$

where f is the sampling frequency, v is the velocity vector of the reference point in the image plane and A is the angle between the normal to the surface and the observer direction. From this relation we see that when A approaches J/2 rad (the surface normal tends to align to the image plane) the constraint on the surface speed squeezes it to zero.

Therefore we will not take into account for the motion analysis the surfaces tilted of more than an angle A_{max}.

For the definition itself of the class A(t) a surface can be included in it if it is satisfied the following relation

$$|\overline{\Omega}| < \frac{\frac{J}{2}-B}{T} = (\frac{J}{2}-B)\,f$$

where B is equal to the angle between the normal to the surface and the observer direction and $\overline{}$ is the component vector of the angular velocity of the object parallel to the image plane.

If an object has surfaces only in the classes B(t) and/or C(t) and/or D(t) to establish the correspondences we need to match some feature vector describing the shape of the surfaces because we don't have guaranteed any more the visibility of the surface to match with. Since for speed requirements we tend to limit our search for correspondences to the detection of the presence of a surface, we want to set the constraint of having only objects with at least one face in class A(t).

If we suppose that the angular velocity of an object is limited by $|\Omega|_{max}$ one of its faces belongs to class A(t) if the angle between its normal and the direction of

observation is less than $B_{max} = \dfrac{J}{2} - \dfrac{|\Omega|_{max}}{f}$.

Therefore in order to have every object having at least one face belonging to A(t-1) independently from its orientation it is sufficient to set the following condition on its EGI:

- any conic solid angle equal to $2J(1-\cos(B_{max}))$ sterad
- in the EGI space must contain at least one peak.

4. Conclusions

The condition on the EGI of the acceptable objects reduces the size of the class A(t) by considering the maximum value of the angular velocity instead of the actual one. Nevertheless the system has been designed for working at frame rate and achieves this performance by setting constraints on the shape of the tractable objects and their motion parameters.
Possible applications for this tracking scheme can be found in situations where we have objects moving in a close environment in which we want to monitor the situation and record the description of the trajectories followed by the objects.

Bibliography

1. B. K. P. Horn, "Understanding image intensity", Artificial Intelligence 8, 1977, pp. 201-231.

2. B. K. P. Horn, "Extended Gaussian Image", Proc. IEEE Dec.1984, pp.1671-1687.

3. R. J. Woodham, "Photometric method for determining surface orientation from multiple image", Optical Engineer. Jan.,Feb. 1980, vol.19, no.1.

4. R. A. Samy, C. A. Bozzo, "Dynamic Scene Analysis and Video Target Tracking", Proc. 7th Int. Conf. on Pattern Recognition, Montreal, Canada, 1984, pp. 993-995.

5. B. K. P. Horn, R.W. Sjoberg, "Calculating the reflectance map", Applied Optics, vol.18, no.11, June 1979.

6. L. Carrioli, M. Diani, "Segmentazione fotometrica di scene 3-D attraverso l'Immagine Gaussiana Estesa", submitted to Rivista di Informatica, October 1986.

7. H.H. Nagel, "Image sequences - ten (octal) years - from phenomenology towards a theoretical foundation", Proc. 8th Int. Conf. on Pattern Recognition, 1986, pp. 1174-1185.

8. S.Negahdaripour, B. K. P. Horn, "Determining 3-D Motion of Planar Objects from Image Brightness Patterns", Proc 9th Int. Joint Conf. on Artificial Intelligence IJCAI-85 August 1985, pp. 898-901.

9. J. K. Aggarwal, L. S. Davis, W. N. Martin, "Correspondence Processes in Dynamic Scene Analysis", Proc. of the IEEE, vol.69, n.5, May 1981, pp. 562-572.

10.   J.K. Aggarwal, Y. C. Kim, "Computation of Structure and Motion from Stereo Images", NATO ARW on Machine Intelligence and Knowledge Engineering for Robotic Applications, Maratea, Italy, May 12th – 16th 1986.

11.   T.Tsukiyama, T. S. Huang, "Motion Stereo for Navigation of Autonomous Vehicles in Man-Made Environments", Proc. 8th Int. Conf. on Pattern Recognition, Paris, France, October 1986, pp. 165-168.

A FAST ALGORITHM FOR MOMENT INVARIANTS GENERATION

Marwan F. Zakaria, Louis J. Vroomen*, Paul J.A. Zsombor-Murray**, and Jan M.H.M. Van Kessel

McGill Research Centre for Intelligent Machines
Robotic Mechanical Systems Laboratory
McGill University Montrèal, Canada, H3A 2K6

ABSTRACT

Moment invariants have been used as feature descriptors in a variety of object recognition applications. When assuming a continuous image function, moments calculated using a double-integral formulation, are invariant to variations in translation, rotation, and size of the object. However, due to the recursive nature of the calculations and the limited speed of microprocessors, the moments were not computable in real-time. In this paper we present real-time invariant moment computations using the *'Delta Method'*, as a means of scene representation.

INTRODUCTION

Moment invariants have been used as features in object recognition, image classification and scene matching [1-9]. These invariant features extracted from two-dimensional images are invariant under image translation, scaling and rotation. The use of moment invariants was first proposed by Hu [1-2] in 1962, for two-dimensional character recognition. The application of moment invariants to more complex two-dimensional scenes was extended by Sajadi and Hall [3-4], and Hall et al. [7-9]

The concept of moment invariants is based on invariant algebra which deals with the properties of certain classes of algebraic expressions which remain invariant under general linear transformations.

MOMENT INVARIANTS

The two-dimensional moments of order (p+q) of an image, computed from the continuous image intensity function f(x,y), are defined as :

$$m_{pq} = \int_{-\infty}^{\infty}\int_{-\infty}^{\infty} x^p \cdot y^q \cdot f(x,y) \cdot dx \cdot dy \qquad (1)$$

---

* Senior member IEEE
** member IEEE

where $p, q \in \{0, 1, 2, \ldots \}$
The central moments of $f(x,y)$ are defined as :

$$\mu_{pq} = \int\limits_{-\infty}^{\infty}\int\limits_{-\infty}^{\infty} (x - \bar{x})^p . (y - \bar{y})^q . f(x,y) . dx . dy \qquad (2)$$

where $\bar{x} = m_{10}/m_{00}$, $\bar{y} = m_{01}/m_{00}$. The central moments μ_{pq} defined in (2) may easily be shown invariant under translation and can also be expressed in terms of the moments m_{pq} defined in (1).

For a digital image, the double integrals in m_{pq} and μ_{pq} could be approximated by double summations as follows :

$$m_{pq} = \sum_{i=0}^{M} \sum_{j=0}^{N} i^p . j^q . f(i,j) \qquad (1.1)$$

$$\mu_{pq} = \sum_{i=0}^{M} \sum_{j=0}^{N} (i - \bar{i})^p . (j - \bar{j})^q . f(i,j) \qquad (2.1)$$

where $\bar{i} = m_{10}/m_{00}$, $\bar{j} = m_{01}/m_{00}$. The summation limits M and N are the dimensions of the intensity matrix $f(i,j)$ in which i and j are the discrete locations of the image pixels. For many industrial applications images can be represented in black and white. Only size, contour, resemblance and contiguity are important, not the color or the shade of the object. In this case the intensity function $f(i,j)$ would only have the values 0 or 1.

The computations of m_{pq} consist of multiplying the function $f(i,j)$ by a corresponding $i^p j^q$ and integrating the results. In the case of a contiguous image, the function $f(i,j)$ is always 1 inside the boundary of the object in the image, 0 for the background. The double-summations of the nominals of order 3 or less ($i^0 j^0$, $i^0 j^1$, $\ldots\ldots$, $i^3 j^0$) are:

$$m_{00} = \sum_{i=0}^{M} \sum_{j=0}^{N} i^0 . j^0 \quad (1.a) \qquad m_{01} = \sum_{i=0}^{M} \sum_{j=0}^{N} i^0 . j^1 \quad (1.b)$$

$$m_{02} = \sum_{i=0}^{M} \sum_{j=0}^{N} i^0 . j^2 \quad (1.c) \qquad m_{03} = \sum_{i=0}^{M} \sum_{j=0}^{N} i^0 . j^3 \quad (1.d)$$

$$m_{10} = \sum_{i=0}^{M} \sum_{j=0}^{N} i^1 . j^0 \quad (1.e) \qquad m_{11} = \sum_{i=0}^{M} \sum_{j=0}^{N} i^1 . j^1 \quad (1.f)$$

$$m_{12} = \sum_{i=0}^{M} \sum_{j=0}^{N} i^1 . j^2 \quad (1.g) \qquad m_{20} = \sum_{i=0}^{M} \sum_{j=0}^{N} i^2 . j^0 \quad (1.h)$$

$$m_{21} = \sum_{i=0}^{M} \sum_{j=0}^{N} i^2 . j^1 \quad (1.i) \qquad m_{30} = \sum_{i=0}^{M} \sum_{j=0}^{N} i^3 . j^0 \quad (1.j)$$

Computing these double-summations is lengthy due to the recursive nature of the calculations.

262

THE DELTA 'Š' METHOD

Definitions of Variables :
δ : the number of chained pixels in row i. (see Fig. 1)
X_i : the x-coordinate of the first pixel in row i.
Y_i : the y-coordinate of the first pixel in row i.
$m_{pq,i}$: the contribution of row i to the nominals m_{pq}.

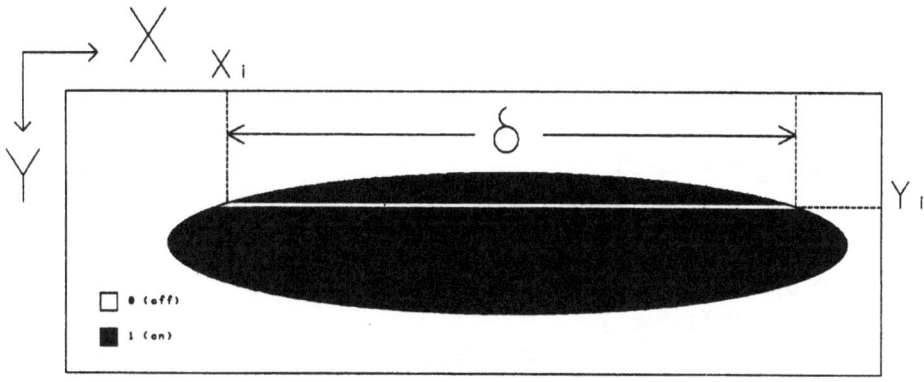

Fig. 1 The *Delta Method*

Each $m_{pq,i}$ can be expressed in terms of X_i, Y_i, and δ.
From (1.a), $m_{00,i} = \delta$
From (1.b), $m_{01,i} = \delta \cdot Y_i$
From (1.c), $m_{02,i} = \delta \cdot Y_i^2$
From (1.d), $m_{03,i} = \delta \cdot Y_i^3$
From (1.e), $m_{10,i} = \delta X_i + (\delta^2 - \delta)/2$
From (1.f), $m_{11,i} = Y_i \cdot [\delta X_i + (\delta^2 - \delta)/2]$
From (1.g), $m_{12,i} = Y_i^2 \cdot [\delta X_i + (\delta^2 - \delta)/2]$
From (1.h), $m_{20,i} = \delta \cdot X_i^2 + (\delta^2 - \delta) \cdot X_i + \delta^3/3 - \delta^2/2 + \delta/6$
From (1.i), $m_{21,i} = Y_i \cdot [\delta \cdot X_i^2 + (\delta^2 - \delta) \cdot X_i + \delta^3/3 - \delta^2/2 + \delta/6]$
From (1.j), $m_{30,i} = \delta \cdot X_i^3 + 3 \cdot (\delta^2 - \delta)/2 \cdot X_i^2 + 3 \cdot [\delta^3/3 - \delta^2/2 + \delta/6] \cdot X_i$
$+ \delta^4/4 - \delta^3/2 + \delta^2/4$

See APPENDIX A for detailed derivations of $m_{pq,i}$.

Let's define the following abbreviations for the sums :

$$S1 = \sum_{n=0}^{\delta-1} n = (\delta^2 - \delta)/2$$

$$S2 = \sum_{n=0}^{\delta-1} n^2 = (\delta^3/3 - \delta^2/2 + \delta/6)$$

$$S3 = \sum_{n=0}^{\delta-1} n^3 = (\delta^4/4 - \delta^3/2 + \delta^2/4)$$

263

The $m_{pq,i}$ gives the contributions to m_{pq} from each line of pixels, in another words:

$$m_{pq} = m_{pq,0} + m_{pq,1} + m_{pq,2} + m_{pq,3} + \ldots\ldots + m_{pq,i}$$
$$+ \ldots\ldots + m_{pq,N}$$

In a simplified format :

$$m_{00,i} = \delta \tag{1.a'}$$
$$m_{01,i} = \delta \cdot Y_i \tag{1.b'}$$
$$m_{02,i} = \delta \cdot Y_i^2 \tag{1.c'}$$
$$m_{03,i} = \delta \cdot Y_i^3 \tag{1.d'}$$
$$m_{10,i} = \delta \cdot X_i + S1 \tag{1.e'}$$
$$m_{11,i} = Y_i \cdot [\delta \cdot X_i + S1] = Y_i \cdot m_{10,i} \tag{1.f'}$$
$$m_{12,i} = Y_i^2 \cdot [\delta \cdot X_i + S1] = Y_i^2 \cdot m_{10,i} \tag{1.g'}$$
$$m_{20,i} = \delta \cdot X_i^2 + 2 \cdot S1 \cdot X_i + S2 \tag{1.h'}$$
$$m_{21,i} = Y_i \cdot m_{20,i} \tag{1.i'}$$
$$m_{30,i} = \delta \cdot X_i^3 + 3 \cdot S1 \cdot X_i^2 + 3 \cdot S2 \cdot X_i + S3 \tag{1.j'}$$

The normalized central moments, denoted by η_{pq} can now be defined as:

$$\eta_{pq} = \frac{\mu_{pq}}{\mu_{00}^{(p+q)/2 + 1}} \tag{3}$$

These are invariant to size change as well as translation.

A set of seven invariant moments (ϕ), invariant to translation, scale change and rotation, has been derived from the normalized central moments. See Hu [1,2] for detailed description:

$$\phi_1 = \eta_{20} + \eta_{02}$$
$$\phi_2 = (\eta_{20} - \eta_{02})^2 + 4 \cdot \eta_{11}^2$$
$$\phi_3 = (\eta_{30} - 3 \cdot \eta_{12})^2 + (3 \cdot \eta_{21} - \eta_{03})^2$$
$$\phi_4 = (\eta_{30} + \eta_{12})^2 + (\eta_{21} + \eta_{03})^2$$
$$\phi_5 = (\eta_{30} - 3 \cdot \eta_{12}) \cdot (\eta_{30} + \eta_{12}) \cdot [(\eta_{30} + \eta_{12})^2 - 3 \cdot (\eta_{21} + \eta_{03})^2]$$
$$\quad + (3 \cdot \eta_{21} - \eta_{03}) \cdot (\eta_{21} + \eta_{03}) \cdot [3 \cdot (\eta_{30} + \eta_{12})^2 - (\eta_{21} + \eta_{03})^2]$$
$$\phi_6 = (\eta_{20} - \eta_{02}) \cdot [(\eta_{30} + \eta_{12})^2 - (\eta_{21} + \eta_{03})^2]$$
$$\quad + 4 \cdot \eta_{11} \cdot (\eta_{30} + \eta_{12}) \cdot (\eta_{21} + \eta_{03})$$
$$\phi_7 = (3 \cdot \eta_{12} - \eta_{30}) \cdot (\eta_{30} + \eta_{12}) \cdot [(\eta_{30} + \eta_{12})^2 - 3 \cdot (\eta_{21} + \eta_{03})^2]$$
$$\quad + (3 \cdot \eta_{21} - \eta_{03}) \cdot (\eta_{21} + \eta_{03}) \cdot [3 \cdot (\eta_{30} + \eta_{12})^2 - (\eta_{21} + \eta_{03})^2]$$

COMPLEXITY ANALYSIS

The advantages of the *delta (δ) method* over the straightforward (S) method, can be shown by a comparison of the time complexity (running time). The space complexity (space occupied by the data required for processing) of both methods proved to be virtually identical.

Time complexity

To compare *worst* case running time for both the δ *method* and the S method, we assume that the image occupies the entire intensity matrix $f(i,j)$.

In the S method a maximum of 10 additions and 20 multiplications is required for each pixel, over the entire image matrix M X N, as each 'on' pixel contributes to the m_{pq}'s (see (1.a)-(1.j)).

In the δ *method* a maximum of N+6 additions and 25 multiplications is required for each *line* of pixels, over the entire image matrix M X N, in order to calculate $m_{pq,i}$ for the corresponding line of pixels i. (see (1.a')-(1.j')).

To calculate the order of time complexity for the intensity matrix $f(i,j)$ of size M X N :

Straightforward method :
    # of additions   = 10 X M X N   (5.1)
    # of multiplications = 20 X M X N   (5.2)

Delta method :
    # of additions   = (N +6) X M   (6.1)
    # of multiplications = 25 X M   (6.2)

For an 80287-8 co-processor the average clock count for a single multiplication, (64 bit real), is 140 cycles, and for a single addition the average is 110 clock cycle [10].

Combining (5.1) and (5.2), (6.1) and (6.2) gives:

Straightforward method :
    Average # of clock counts : 3900 X M X N   (5)

Delta method :
    Average # of clock counts : M X (110N+4160)   (6)

The ratio of (5) over (6) for large N is
    (3900 X N)/(110 X N + 4160) $\approx$ 35

ERROR ANALYSIS

Three factors affect the accuracy of the moment invariants calculations:

1: the size of the intensity matrix.
2: the distance between the object and the digital camera
3: the focal length of the lens used

These factors could vary widely depending on the industrial application. In a typical situation where the intensity matrix is of size 80 x 640, the distance between the object and the digital camera is two meters and using a wide angle lens with a focal length of 8.5 mm, the uncertainty is equal to one pixel, which correspond to less than 0.5% on the object.

10. intel Corporation, *Microsystem Components Handbook (Microprocessors Volume I)*, 1986, Santa Clara, CA, pp. 4.56-4.81.

APPENDIX A

From (1.a), $m_{00} = 1+1+1+1+ \ldots +1 = \delta$

From (1.b), $m_{01} = Y_i+Y_i+Y_i+ \ldots +Y_i = \delta.Y_i$

From (1.c), $m_{02} = Y_i{}^2+Y_i{}^2+Y_i{}^2+ \ldots +Y_i{}^2 = \delta.Y_i{}^2$

From (1.d), $m_{03} = Y_i{}^3+Y_i{}^3+Y_i{}^3+ \ldots +Y_i{}^3 = \delta.Y_i{}^3$

From (1.e), $m_{10} = X_i+(X_i+1)+(X_i+2)+(X_i+3)+ \ldots +(X_i+\delta-1)$

$= \delta.X_i+(0+1+2+3+4+5+ \ldots +\delta-1)$

$= \delta.X_i+(0+\delta-1)/2.\delta$

$= \delta X_i+(\delta^2-\delta)/2$

From (1.f), $m_{11} = X_i.Y_i+(X_i+1).Y_i+(X_i+2).Y_i+ \ldots$

$\qquad +(X_i+\delta-1).Y_i$

$= Y_i.[X_i+(X_i+1)+(X_i+2)+ \ldots +(X_i+\delta-1)]$

The term in brackets equals m_{10} therefore :

$= Y_i.[\delta X_i+(\delta^2-\delta)/2]$

From (1.g), $m_{12} = X_i.Y_i{}^2+(X_i+1).Y_i{}^2+(X_i+2).Y_i{}^2+ \ldots$

$\qquad +(X_i+\delta-1).Y_i{}^2$

$= Y_i{}^2.[X_i+(X_i+1)+(X_i+2)+ \ldots +(X_i+\delta-1)]$

$= Y_i{}^2.[\delta X_i+(\delta^2-\delta)/2]$

From (1.h), $m_{20} = X_i{}^2+(X_i+1)^2+(X_i+2)^2+ \ldots +(X_i+\delta-1)^2$

$= X_i{}^2+X_i{}^2+2.X_i+1+ \ldots +X_i{}^2+2(\delta-1).X_i+(\delta-1)^2$

By grouping terms and factoring out X_i, this becomes :

$= \delta.X_i{}^2+2.(0+1+2+3+4+ \ldots +\delta-1).X_i+$

$\qquad +(0+1+4+9+ \ldots +(\delta-1)^2)$

Using the polynomial theorem :

$$= \delta.X_i{}^2+(\delta^2-\delta).X_i+ \sum_{n=0}^{\delta-1} n^2$$

Where the last term $\displaystyle\sum_{n=0}^{\delta-1} n^2 = \delta^3/3-\delta^2/2+\delta/6$

So the total contribution from (1.h) is:

$= \delta.X_i{}^2+(\delta^2-\delta).X_i+1/3 \ \delta^3-1/2 \ \delta^2+1/6\delta$

From (1.i), $m_{21} = X_i{}^2.Y_i+(X_i+1)^2.Y_i+(X_i+2)^2.Y_i+ \ldots$

$\qquad +(X_i+\delta-1)^2.Y_i$

$= Y_i.[X_i{}^2+(X_i+1)^2+(X_i+2)^2+\ldots +(X_i+\delta-1)^2]$

$= Y_i.[\text{ contribution from one row of } m_{20}]$

$= Y_i.[\delta.X_i{}^2+(\delta^2-\delta).X_i+1/3 \ \delta^3-1/2 \ \delta^2+1/6\delta]$

From (1.j), $m_{30} = X_i{}^3+(X_i+1)^3+(X_i+2)^3+ \ldots +(X_i+\delta-1)^3$

$$= X_i{}^3 + X_i{}^3 + 3.X_i{}^2 + 3X_i + 1 + X_i{}^3 + 3.2.X_i{}^2 + 3.2^2 \cdot X_i$$
$$+ 1.2^3 + \ldots + X_i{}^3 + 3(\delta-1).X_i{}^2 + 3.(\delta-1)^2 X_i$$
$$+ (\delta-1)^3$$

By grouping terms similar to m_{20} this becomes :
$$= \delta.X_i{}^3 + 3.(0+1+2+3+4+ \ldots + \delta-1).X_i{}^2 + 3.(0$$
$$+1+4+9+ \ldots + (\delta-1)^2).Xi + (0+1+8+27+ \ldots$$
$$+ (\delta-1)^3)$$

Using the polynomial theorem :
$$= \delta.X_i{}^3 + 3.(\delta^2-\delta)/2.X_i{}^2 + 3.[\delta^3/3 - \delta^2/2 - \delta/6].X_i$$

$$+ \sum_{n=0}^{\delta-1} n^3$$

Where the last term $\displaystyle\sum_{n=0}^{\delta-1} n^3 = \delta^4/4 - \delta^3/2 + \delta^2/4$

So the total contribution from (1.j) is:
$$= \delta.X_i{}^3 + 3.(\delta^2-\delta)/2.X_i{}^2 + 3.[\delta^3/3 - \delta^2/2 - \delta/6].X_i$$
$$+ \delta^4/4 - \delta^3/2 + \delta^2/4$$

These derivations could be generalized for any $p,q \in \{0,1,2,\ldots\}$ and the contribution of the 2-D moments is as follows:

$$m_{pq,i} = Y_i{}^q . \sum_{n=0}^{p} S(n).\binom{p}{n}.X_i{}^{p-n}$$

Where the term $\displaystyle S(n) = \sum_{t=0}^{\delta-1} t^n$

CONCLUSION

The *delta method*, presented in this paper, only requires the values of X_i, Y_i (the x,y coordinates of the first pixel) and δ (the number of chained pixels) of row i, (i = 0,1,2,3, $\ldots$, M) to calculate all the moment invariants. The straightforward approach uses the x,y coordinates of each pixel of the entire object in the image in order to calculate all the moments as defined in (1). The equations derived using the *delta method* have been implemented in FORTRAN 77 and Turbo PASCAL on a HEWLETT PACKARD *Vectra* (IBM PC/AT compatible) equipped with an 80287 co-processor and a digital camera. The moment invariants, Ú's, were calculated in less than 0.6 second for a matrix of 80 X 640.

The experimental results confirm the invariancy of moment invariants, and the *delta method* exhibits a potential for greatly reducing the amount of data and processing needed to identify an object.

ACKNOWLEDGEMENT

This research was supported by *a contract from IDEAL EQUIPMENT LIMITED*, of Montréal, and grant #A4219 of the Natural Sciences and Engineering Research Council of Canada.

REFERENCES

1. M. K. Hu, Pattern Recognition by Moment Invariants, *Proc. IRE*, 49, September 1961, pp. 1428.

2. M. K. Hu, Visual Pattern Recognition by Moment Invariants, *IRE Trans. Inform. Theory*. Vol. IT-8, February 1962, pp. 179-187.

3. F. A. Sajadi and E. L. Hall, Numerical Computation of Moment Invariants for Scene Analysis, *Proceedings of 1978 IEEE Conference on Pattern Recognition and Image Processing, Chicago. IL. 1978*, pp. 181-187.

4. F. A. Sajadi and E. L. Hall, Three-Dimensional Moment Invariants, *IEEE Trans. Pattern Analysis and Machine Intelligence 2*, No. 2, March 1980, pp. 127-136.

5. E. Elliot, Algebra of Quantics, *Oxford University Press*, 1913.

6. Cho-Huak Teh and Roland T. Chin, On Digital Approximation of Moment Invariants, *Proceedings of 1985 IEEE Conference on Pattern Recognition and Image Processing, San Francisco CA. 1985*, pp. 640-642.

7. E. L. Hall and W. Frei. Invariant Features for Quantitative Scene Analysis. Final Rep., Contract F 08606-72-C-0008, *Image Process. Inst.* Univ. of Southern California, Los Angeles, 1976.

8. E. L. Hall, R. P. Kruger, S. J. Dwyer, R. W. McLaren and G.S. Lodwick, A Survey of Preprocessing and Feature Extraction Techniques of Radio-graphic Images. *IEEE Tran. Comput.* 20, No. 9, pp. 1032-1044.

9. E. L. Hall, *Computer Image Processing and Recognition*. Academic Press,1979.

MODEL GENERATION FROM IMAGES

J. R. Stenstrom and
C. I. Connolly

Aule-Tek, Inc., General Electric Corp. R&D, Schenectady, NY

1. Introduction

With the recent increased interest in model-based matching[1, 2, 3, 4] has appeared a concomitant interest in automated model creation.[5, 6, 7] Ideally, models should be constructed automatically from a scene, so that the process of matching (for example) would require no human intervention. In any case when computer models of an object are needed but are not available, it would be convenient to have an automatic method for constructing the models from image data.

Two prerequisites for constructing computer models of objects are: 1) the models must be geometrically accurate, and 2) they must be topologically sound. What we mean by "geometrically accurate" is that the relative positions of surfaces on the model are not very different from their relative positions on the actual object. The term "topologically sound" means that the model must represent a compact, orientable 2-manifold, which is what we assume an object's surface to be. If the models obtained from individual views are indeed sound, boolean intersection of the models obtained from each view can be performed to obtain a multiple-view model.

These two prerequisites imply that the model creation process should preserve both the geometric integrity and topological information inherent in the original data. This paper describes two methods for model creation which is based upon these prerequisites. Previous work on model generation from multiple range views is greatly enhanced. The new technique will succeed for quite complex objects, and an example is presented. An alternate procedure using multiple intensity views is also introduced. A simple experiment with an intensity based model is also presented.

1.1 Geometric and Topological Considerations

It has been argued many times in the literature[8, 9, 10, 11] that points of high curvature are important features in range images, since they are intrinsic surface features, and are invariant with respect to object or sensor movement.[12] For planar objects, such points denote the boundaries between planes. Even for non-planar objects, such points can indicate a change in surface type. Henceforward, the objects to be modeled will be assumed to be planar. Curvature is zero everywhere on a plane, and becomes infinite across plane boundaries. In range images, this manifests itself as a peak in curvature at plane boundaries. For intensity images, the curvature change usually appears as a sudden change in intensity. To avoid confusion, the term "edgel" will be used to denote an edge pixel, whereas the term "edge" will be used in the traditional geometric sense (i.e., a line segment).

Geometric soundness includes designating any volumes of the viewspace which are not known to be empty as being possibly occupied. Unknown volumes will be handled exactly as occupied volumes until they are reduced by a subsequent view. At the same time, a topologically sound reconstruction of the object surface is achieved when the edge features represent a proper set of polygon boundaries. Occluding boundaries are detected in range or intensity images, since these represent places where the surface normal is orthogonal to the viewing direction.[13, 9] Back-projected faces separate the known empty volumes from the occupied/unknown volumes. A face at the rear is provided, bounding the viewspace volume. Sensed faces, back projected faces, and faces bounding the viewspace volume form the surface of the model.

2. Models from Range Data

2.1 Previous techniques

Previous work[7] has demonstrated the feasibility of using three dimensional edges developed from multiple range views to produce wire-frame representations for an object. The Wesley-Markowsky wire frame fleshing algorithm[14] becomes directly applicable to such a wire frame. The separate view solids thus produced are transformed into a common coordinate system and intersected.

There are several problems that make this procedure difficult to apply. The edges that bound a visible face are connected in the range image. The casting of each intersection vertex into three dimensions guarantees preserving any 1-cycles in three dimensions. Unfortunately, there can be absolutely no guarantee of of the planarity of such edge cycles. The Markowsky-Wesley algorithm identifies possible 2-cycles. Usually, the solids which are so generated are ambiguous, and a tree of possible solid/empty labels of the volumes is the output of the algorithm. While the ambiguity cannot be completely resolved when fleshing wire-frames from engineering drawings, the ambiguity between empty and solid is absent in range data.

The new procedure avoids the planarity issue entirely, and avoids the ambiguity in object representation. It makes direct use of the range image to disambiguate space. The model resulting from the new procedure has strictly planar faces, and is a closed surface.

2.2 A New Algorithm

Step edgels are detected in range images. A roof detector is applied in regions not containing step edgels. Step edgels, because of the relative magnitude of the features, lead to spurious roof edgels in the vicinity of the steps. Near step edgels, the surface is extrapolated beyond the step for the roof detector computations. Typically, there are breaks between roof and step edgels and breaks in what would ideally be a continuous chain of roof edgels. After the conversion of edgels to edges,[15] dangling roof edges are closed by searching for either nearby step edges or other nearby dangling roof edges. Edges that cannot be a part of any cycle are pruned away, and viewplane cycles are formed. The algorithm constructs a cycle tree of out-cycles and in-cycles as in Wesley and Markowsky,[14] where each edge is used in precisely two 1-cycles. Since surface fitting will be used to form the actual object model face within the 1-cycle, it is essential that the basic cycles enclose enough data points for an accurate fit. The area of each cycle is considered. The user specifies a minimum cycle size for a significant feature. Cycles smaller than that minimum cycle size are deleted, since they will provide no meaningful contribution to the model (and in fact may arise from noise in the data[16]). Edges that continue to participate in 1-cycles are collected. The cycle formation algorithm is applied again. A cycle containment tree is produced that properly nests every significant viewplane cycle.

The edges will now form large complete cycles corresponding to major deviations from planarity for polyhedral objects. For general objects, the cycles correspond to regions fully enclosed by the visible boundary as well as surface contours of high curvature.

A single image is assumed to contain one or more objects fully surrounded by a background surface, identifiable by its depth. The outer bounding cycle of that background surface properly contains every other viewplane cycle and is designated as the rear face. Viewplane cycles are converted to three-dimensional faces or holes (if they are coplanar with the rear face). Pixels within the current face but not in a hole are sampled. A least-squares surface fit determines the geometric parameters of the face within the outer edges.

The "holes" in the outer bounding cycle (or rear face) are precisely the occlusion edges of the sensed objects. These may be outermost edges or edges surrounding holes in the sensed objects. To produce a model with holes that properly correspond to those of the sensed object, the cycles surrounding holes in the sensed object must be identified. The faces that correspond to the cycles that bound object holes are identified and collected. At this point all of the three-dimensional faces that correspond to viewplane cycles are fully formed. It remains only to wrap back-projected faces around the object and to close the rear surface.

The first step in the back projection considers each vertex in the range image cycle complex. A line is formed in the view direction that includes the vertex. Cycle, edge, and vertex links permit the identification of all the range-image cycles that include the range vertex in question. Thus, the three-dimensional faces that are encountered along the view direction from the vertex may be readily identified. The intersections of this view direction line and the identified faces are the vertices that will correspond to the range vertex. Each three-dimensional vertex is also paired and linked with the appropriate face. These corresponding three-dimensional vertices are connected pairwise, with three-dimensional edges.

The back projections are done on an edge by edge basis. Each range edge is in precisely two viewplane cycles and thus represents the boundary of exactly two three-dimensional faces. At each endpoint of the the range edge, the front-most face of the two faces is selected.

Ordinarily the same face is front-most at each end of the three-dimensional edge. In this case, forming the back projection is quite easy. The edge in the viewplane together with the view direction determine a plane which may be intersected with the forward and rearward faces, producing two edges (figure 1, left). There are four edges for each back-projected face: these two three-dimensional edges corresponding to the range image edge, and two additional edges extending back along the line of sight, created in the first step of this section. In our system, it is an artifact of the calibration process that two view direction lines are not quite coplanar. By the addition of an edge which triangulates each back-projected face, planarity is achieved.

There is another case for back-projected faces. The same face need not be front-most at both ends of a range edge (see figure 1, right). When it is not, the back projected surface must be split into two triangles meeting at a common three-dimensional vertex. The three-dimensional edges that correspond to the range edge in each of the two faces is found. These edges may be computed knowing the endpoints from the range edge, the two faces, and the correspondence between range vertices and faces. The common vertex for the triangle is the intersection of the two new edges. The three-dimensional vertices in the front and rear face for each endpoint of the range edge, together with this intersection vertex, form the three corners of each triangular back-projected face.

Whether by the simple case or by the two triangle case, *every* back projected face has now been constructed. The final step constructs a valid rear face for the object model. Recall that an outer bounding rear face was created early in the process. During the rest of the procedure, faces may have been created which were coplanar with the rear face. If these faces are arranged in a containment tree, with the rear face at the top level labelled solid, and each interior face labelled in the opposite sense of its ancestor, the tree will represent the final face for the model.

Polyhedral models from multiple views are intersected to refine the model. Figure 2 shows our range camera. The mirrors at the left side direct a stripe of light on an object as shown in figure 3. A three-dimensional plot for one view of a jeep is displayed in figure 4. A three-dimensional plot of the data is shown in the upper right. The original edge cycles are shown in the lower left and the larger cycles in the lower right. Figure 5 illustrates a single view polyhedron. Figure 6 gives a second view polyhedron. The intersection is shown, with hidden lines removed, in figure 7.

3. Intensity Data

3.1 Introduction

In a manner similar to that used for range data, models of the environment can be constructed using multiple intensity views as well. In some ways, this approach is reminiscent of the work of Chien and Aggarwal,[17] but with the emphasis on edge cycles and face-edge-vertex models rather than silhouettes and octrees. It is also related to the work of Wesley and Markowsky,[18] and Baker.[5]

Ideally, objects or portions of an object should produce boundaries (i.e. edge cycles) in the viewplane. In the absence of other information, the properties of the camera used to create the image can be used to generate bounding volumes for the surfaces which gave rise to the features in an image. The most important of these properties are the orientation, focal point

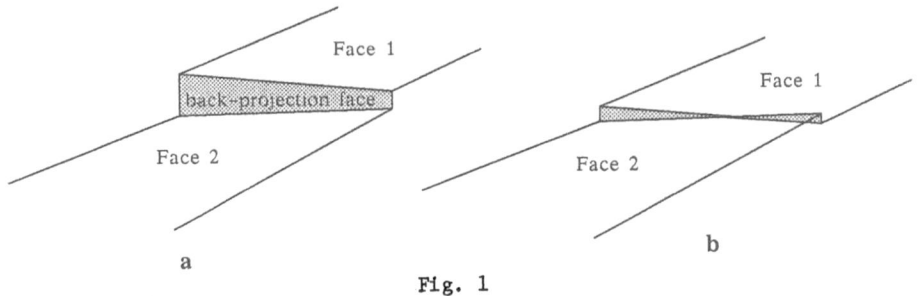

Fig. 1

Fig. 2

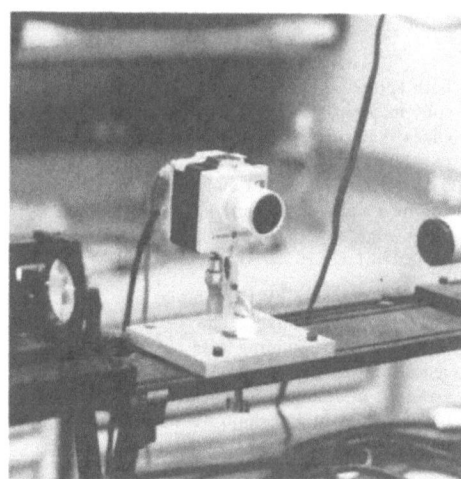

Fig. 3

Fig. 4

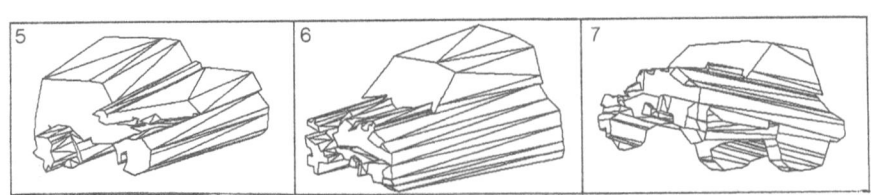

Fig. 5,6, and 7 Geo-Calc SCENE

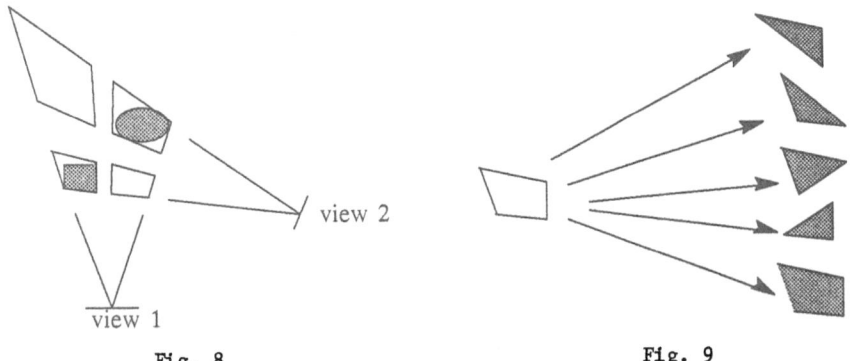

view 2

view 1

Fig. 8   Fig. 9

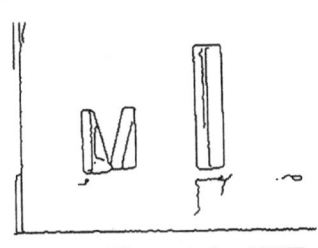

Fig. 10 Geo-Calc SCENE

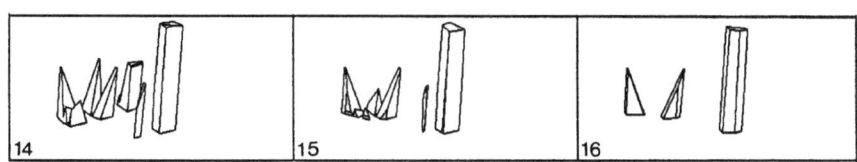

Fig. 11,12 and 13   Geo-Calc SCENE

Fig. 14,15 and 16   Geo-Calc SCENE

Fig. 17

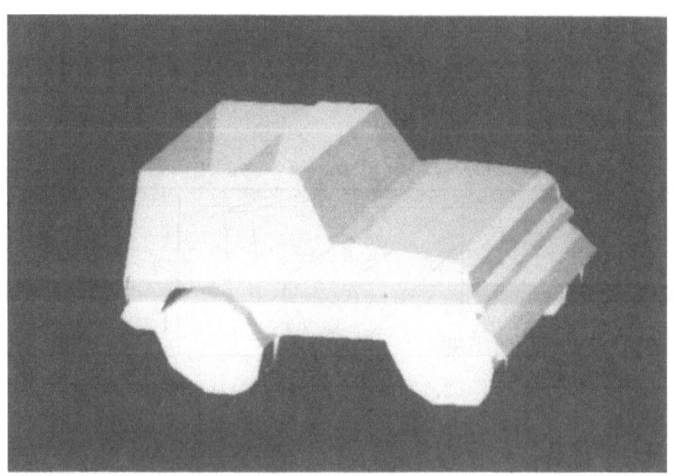

Fig. 18

274

position and focal length of the camera. Using this information, the 1-cycles comprising the features in the image can be extruded to form volumes (2-cycles) in a manner similar to that used in range data. These volumes can be intersected with volumes obtained from other views to obtain the *potential* positions of objects in the scene.

Naturally, the intersection process will usually produce an ambiguous set of volumes. For the purposes of this discussion, the ambiguities will be divided into two types: "shadowing" ambiguities, and "feature" ambiguities. Shadowing ambiguities are caused by the intersection of a volume generated by one set of features in one view with several volumes generated by most (or all) of the features in a second view. Figure 8 illustrates this. Feature ambiguities (see Grimson[19] arise since there can be a few possible surface configurations which explain the features (or lack thereof) contributing to a given volume. Figure 9 provides an example of such an ambiguity.

In the case of shadowing, some volumes can be produced from the intersection process which could be either empty or solid. In general, both types of ambiguity are reduced when more views are used.

3.2 The Algorithm

Volumes are generated using the following procedure: First, an image is smoothed with a Gaussian and processed with a suitable edgel detector (e.g. Canny[20] see figure 10). Then, the edgels are converted to 1-chains (sets of edges) by selecting junctions and points of high curvature (when the edgels are viewed as curves) to serve as vertices. Edgels which are between two such vertices are converted into edges whose endpoints are those vertices. Gaps in the edgels are connected to form cycles using a zero-crossing following algorithm[21] (figure 11). The resulting complex of edges is pruned so that only 1-cycles remain. Each 1-cycle is then extruded to form a 2-cycle. Each edge of the 1-cycle is used to form a back-projected face, as in the range data case. The front and rear faces of the 2-cycle are simply planes set at arbitrary forward and rear distances from the camera. The camera focal length is used to determine the shape of the 2-cycle by inverting the perspective function. The perspective function maps points in E^3 (x,y,z) to points on the viewplane $(\hat{x},\hat{y})$ as follows:

$$\hat{x} = \frac{f\,x}{f-z}$$

$$\hat{y} = \frac{f\,y}{f-z}$$

where f is the camera focal length. Differentiating x and y with respect to z yields:

$$\frac{dx}{dz} = -\frac{\hat{x}}{f}$$

$$\frac{dy}{dz} = -\frac{\hat{y}}{f}$$

This, along with the location of the point in the viewplane, completely specifies the ray on which the actual point in E^3 corresponding to a viewplane point must lie. Two values of z are chosen (z_0, z_1) so that they fully enclose the scene volume in the z direction. These z values fix the positions of the front and rear faces of the 2-cycles. The 2-cycles are finally transformed into a common coordinate system and intersected with each other to get bounding surfaces for objects found in the scene.

Figures 12-16 illustrate this process for a set of blocks on a turntable. In figure 12 the only information is that the object causing the cycle occurred somewhere in the viewing prism. Figure 13 shows two views intersected. The objects are much more tightly located. Using 5 views, the final model of figure 16 is obtained. Comparing this to figure 17, we see that it is a fairly good representation of the actual scene. In this case, no disambiguation was required. Five views were enough to completely specify the approximate shape and position of the objects. A rendered model of the jeep in figure 3, obtained from two intensity views, can be seen in figure 18.

4. Conclusion

The purpose of this paper is to demonstrate the practical use of certain geometric and topological properties of image data to construct three-dimensional models of the environment. The essential information consists both of local, geometric information obtained from step and curvature edges found in the image, and the more global, topological information inherent in the data (the cycles which define surface boundaries). Preserving this information at every step of the model creation process, whether for intensity or range data, results in geometrically and topologically sound volumetric models.

References

1. O. D. Faugeras, "New Steps Toward a Flexible 3-D Vision System for Robotics," *Proc. 7th International Conference on Pattern Recognition*, p. 796, Montreal, Que., Canada, August 1984.

2. D. Lowe, "Perceptual Organization and Visual Recognition," in , Kluwer Academic Publishers, Boston, MA, USA, 1985.

3. C. I. Connolly, J. L. Mundy, J. R. Stenstrom, and D. W. Thompson, "Matching from 3-D Range Models into 2-D Intensity Scenes," *Proc. 1st International Conference on Computer Vision*, London, UK, June 1987.

4. J. E. Mayhew et al, "Keynote address," *IEEE Conf. on Computer Vision and Pattern Recognition*, 1986..

5. H. Baker, "Three Dimensional Modeling," *Proc. 5th IJCAI*, pp. 649-655, Cambridge, MA, USA, 1977.

6. M. Herman, "Generating Detailed Scene Descriptions from Range Images," *Proc. 1985 IEEE Conference on Robotics and Automation*, p. 426, St. Louis, MO, USA, 1985.

7. J. R. Stenstrom and C. I. Connolly, "Building Wire Frames from Multiple Views," *Proc. 1986 IEEE Conference on Robotics and Automation*, p. 615, San Francisco, CA, USA, April 1986.

8. R. J. Popplestone, C. M. Brown, A. P. Ambler, and G. F. Crawford, "Forming Models of Plane-and-Cylinder Faceted Bodies from Light Stripes," *Proc. 4th International Joint Conference on Artificial Intelligence*, pp. 664-668, Tbilisi, Georgia, USSR, September 1975.

9. M. Brady, J. Ponce, A. Yuille, and H. Asada, "Describing Surfaces," *Computer Vision, Graphics and Image Processing*, vol. 32, p. 1, 1985.

10. W. Snyder and G. Bilbro, "Segmentation of Three Dimensional Images," *Proc. 1985 IEEE Conference on Robotics and Automation*, p. 396, St. Louis, MO, USA, April 1985.

11. C. I. Connolly and J. R. Stenstrom, "Construction of Polyhedral Models from Multiple Range Views," *Proc. 8th International Conference on Pattern Recognition*, vol. 1, Paris, France, October 1986.

12. H. W. Guggenheimer, *Differential Geometry*, Dover, New York, NY, USA, 1977.

13. H. G. Barrow and J. M. Tenenbaum, "Recovering Intrinsic Scene Characteristics from Images," in *Computer Vision Systems*, ed. A. Hanson, pp. 3-26, Academic Press, New York, 1978.

14. G. Markowsky and M. A. Wesley, "Fleshing Out Wire Frames," *IBM Journal of Research and Development*, vol. 24, pp. 582-597, September 1980.

15. H. Asada and M. Brady, "The Curvature Primal Sketch," AIM-758, MIT.

16. T. Noyes, M. S. Thesis, Rensselaer Polytechnic Institute, 1986.

17. C. H. Chien and J. K. Aggarwal, "A Volume/Surface Octree Representation," *Proc 7th International Conference on Pattern Recognition*, vol. 2, pp. 817-820, Montreal, Que., CAN, August 1984.

18. M. A. Wesley and G. Markowsky, "Fleshing Out Projections," *IBM Journal of Research and Development*, vol. 25, pp. 934-954, November 1981.

19. W. E. L. Grimson, *From Images to Surfaces*, MIT Press, Cambridge, MA, USA, 1981.

20. J. F. Canny, "Finding Edges and Lines in Images," TR 720, MIT, 1984.

21. C. I. Connolly, "Obtaining Closed Curves from Directional Edge Detectors," *GE IU Project Technical Memo 900587*, June 1987.

A TRIANGLE BASED DATA STRUCTURE FOR MULTIRESOLUTION SURFACE REPRESENTATION

Leila De Floriani

Istituto per la Matematica Applicata del C.N.R.
Via L.B.Alberti, 4
16132 Genova (Italy)

ABSTRACT

A hierarchical model for approximating 2-1/2 dimensional surfaces is described. This model, called Delaunay pyramid, is a method for compression of spatial data and representation of a surface at successively finer levels of detail. The Delaunay pyramid is based on a sequence of Delaunay triangulations of suitably defined subsets of the set of data points.

INTRODUCTION

Surface representation plays an important role in computer graphics, CAD, computer vision and geographic data processing. The surface reconstruction problem in 2-1/2 dimensions is considered here for a specific application to digital surface modeling. This problem can be mathematically stated as the interpolation of a bivariate function defined either at the vertices of a uniformly spaced grid or at a set of irregularly distributed points in the x-y plane. In applications like computer vision or geographic data processing, when large amounts of data are available, a surface model capable also of compressing spatial data according to an accuracy-based criterion is required.

Hierarchical structures have been used as models of point data, planar regions, surfaces and three-dimensional objects. Such models allow the manipulation of an entity at increasingly higher levels of resolution as well as the application of efficient algorithms based on divide-and-conquer. Hierarchical models of 2D or 3D objects can be classified into domain-dependent representations, like the quadtree [14] or the octree [12], which are based on the decomposition of the space occupied by the object, and object-dependent models, like the prismtree [8], the hierarchical triangulations [4,10] or the Delaunay tree [2], which provide object descriptions in an object-centered coordinate frame. Hierarchical surface models belong to this latter group and provide representations of a surface at different levels of resolution, thus allowing also a reduction in the number of points needed to describe the shape of the surface.

Existing hierarchical surface models, however, either are only well-suited either to solve the surface reconstruction problem when the data points are regularly sampled [6] or produce a triangle-based surface approximation which is inaccurate in the x-y coordinates [4]. Being based on a fixed splitting rule, all these models introduce artificial edges on the resulting surface, which are strictly dependent on the subdivision and do not disappear even by increasing the level of accuracy. On the other hand, Delaunay triangulation has been extensively applied to build surface approximations at fixed accuracy because of its local definition and of the equiangularity property. Here, we define a hierarchical Delaunay-pyramid as an attempt to combine the useful features of Delaunay triangulation with the advantages of a hierarchical surface description.

The properties of the proposed model are studied and compared with a hierarchical search structure defined by Kirkpatrick [11]. An algorithm is presented for the construction of a Delaunay pyramid, which has a time complexity comparable with that of any incremental Delaunay method or of any algorithm which builds a hierarchical triangulation.

THE HIERARCHICAL SURFACE APPROXIMATION PROBLEM

Hierarchical models for surface description have been recently proposed in the literature [3,4,6,8,10]. Such models provide data compression mechanisms to reduce the number of points needed to describe a surface: fewer points, for instance, will be used to represent large surface areas of constant slope. Other benefits include the possibility of a fast surface rendering by an efficient application of ray tracing algorithms and the capability of producing local refinements of the surface at any level of resolution. On the other hand, these models have some drawbacks which make them not completely adequate as a terrain model in a geographic information system. A ternary triangulation [4] is capable of handling arbitrarily distributed points, but it does not satisfy the equiangularity property. A quadtree-like triangulation [10] is composed of equiangular triangles only in the case of regular data sampling, and also may produce discontinuous approximations. Quadtree-based rectangular models [3,6] are capable of dealing only with uniformly sampled data points. Even when the surface is sampled at regular intervals in both of the coordinates, it is highly desirable to include in the model surface specific points, like peaks, pits or passes, which characterize the surface independently of the data sampling and thus are usually irregularly distributed [5,9]. Hence, a triangle-based model provides the most appropriate and flexible description of a topographic surface. Also, such a model must be based on a triangulation of the surface domain in which the triangles are as equiangular as possible so as to avoid thin and elongated triangular facets. The Delaunay triangulation is optimal with respect to the previous requirement and thus has been extensively used as a basis of triangular surface models. The two-dimensional Delaunay triangulation of a set V of points in the plane is defined as the dual graph of the Voronoi diagram of V, which is a subdivision of the plane into polygonal regions, each of which is associated with a point P of V and is defined as the region which is closer to P than to any other point of V [13].

A model of a functional surface should be capable of dealing with arbitrary distributed points, it should be based on a Delaunay tesselation of the domain so as to satisfy the equiangularity property and, finally, it should provide a hierarchical description of a surface based on an error criterion. A strictly hierarchical structure based on some fixed triangle-splitting rule cannot be imposed on a Delaunay triangulation, since the modification of a triangle t in a Delaunay

278

tesselation, caused by the insertion of one or more data points, may produce a triangulation which does not satisfy the Delaunay criterion. Hence, we propose a hierarchical triangle-based surface representation obtained by superimposing a pyramidal structure on a suitable collection of planar surface approximations based on a Delaunay triangulation of subsets of the given data set.

A hierarchical structure based on an arbitrary two-dimensional triangulation has been proposed by Kirkpatrick as a search structure for the planar point location problem [11]. Given a set V of points in the plane, this hierarchical structure consists of a sequence T1,T2,...,Tk of triangulations of subsets of V, such that T1 is an arbitrary triangulation of V, Tk is a triangle and Ti is obtained from Ti-1 by eliminating a maximal set of independent vertices from Ti-1 and their adjacent triangles, and retriangulating the resulting polygons. It has been proven that k=O(logn) and, hence, the point location problem can be solved in O(logn) steps on this structure. For surface approximation problems, the selection of the points to be eliminated should be done according to an error directed strategy, which could produce a hierarchy with O(n) levels in the worst case. Furthermore, a Delaunay triangulation should be reproduced at each step in the bottom-up reduction process. According to the Delaunay criterion, the retriangulation of a polygon could have an effect which extends beyond the boundary of the polygon itself, thus resulting in an O(n) algorithm. Thus, the hierarchical structure proposed by Kirkpatrick is essentially a search structure and cannot be used as a multiresolution surface model.

THE MULTIRESOLUTION SURFACE MODEL

Let S be the set of the data points, each of which is expressed as a triple $Pi=(xi,yi,zi)$, where $zi=F(xi,yi)$ and $z=F(x,y)$ is the function to be approximated. Let V denote the set of the data point projections on the x-y plane. Let T denote the Delaunay triangulation of V and Ti the Delaunay triangulation of a subset Vi of V. $F(x,y)$ is approximated on Ti by a network of planar triangles in the 3D space, each defined by the plane passing through the three points of S whose projections are vertices of a triangle of Ti. If $z=fi(x,y)$ denotes the resulting piecewise linear approximation built on Ti, then we define the error associated with Ti as
e(Ti) =max {abs(fi(x',y')-z') such that P=(x',y',z') and (x',y')
          belongs to V-Vi}

Given the set V of data point projections, let n and b denote the number of points and of boundary points of V respectively. The multiresolution Delaunay-based surface model built on V, called Delaunay sequence of V, is a sequence DS of Delaunay triangulations of subsets of V, denoted DS=[T0,T1,...,Tm] where m<=n-b+1 such that
(i) Ti is a Delaunay triangulation of a subset Vi of V, i=1,2,...,m
(ii) Vi-1 is a subset of Vi, i=1,2,...,m
(iii) Vm=V
(iv) e(Ti)<=e(Ti-1), i=1,2,...,m

Hence, T0 is the coarsest surface approximation, which will contain at least the points defining the boundary of the domain, while Tm will be the Delaunay triangulation of V. Alternatively, a tolerance value Eps could be defined and condition (iii) could be replaced with e(Tm)<=Eps. In this case, Vm could be a proper subset of V (when Eps>0), and thus DS will provide a sequence of approximations defined by a tolerance value Eps.

The relationships between two consecutive triangulations in DS are defined as follows. Let T' be a Delaunay triangulation of a set V' and let T" be the Delaunay triangulation obtained from T' by insertion of a new point P. Then, we call polygon of influence of P, denoted PI(P), the polygon formed by the edges of T' (and T") which bound the region of the plane affected by the insertion of P into T'. We call region of influence of P in T', denoted RI(P,T'), the collection of all triangles in T' which do not belong to T". Similarly, we call refined region defined by P in T", denoted RR(P,T"), the set of triangles of T" which do not belong to T'. Like RI(P,T'), RR(P,T") is a Delaunay triangulation covering polygon PI(P). If R denotes the region of the plane inside the boundary polygon of T', then each triangle in R-RI(P,T') will belong to both T' and T". RI(P,T') is a star-shaped region containing P in its kernel. Figure 1 shows two Delaunay triangulations T' and T": T" has been obtained from T' by insertion of a point P. The polygon of influence PI(P) is indicated by bold line segments in both T' and T".

In a Delaunay sequence DS a triangulation Ti is obtained from Ti-1 by insertion of the points belonging to Vi - Vi-1. Hence, we call difference set between Ti-1 and Ti, denoted Di-1, the union of the regions RI(P,Ti-1) of the points P belonging to Vi - Vi-1, i.e., Di-1 = {RI(P,Ti-1) such that P in Vi-Vi-1}. Di-1 is the collection of the triangles of Ti-1 which do not belong to its successor Ti. Such triangles define a region in the plane not necessarily connected. Similarly, we define the difference set between Ti and Ti-1, denoted Di', as the union of the refined regions RR(P,Ti) defined in Ti by the points P belonging to Vi - Vi-1. According to the above definitions we can associate two regions, Dj and Dj', with any element Tj of a Delaunay sequence DS, where Dj is the difference set between Tj and Tj+1 and Dj' the difference set between Tj and Tj-1. Note that both Dj and Dj' are subsets of Tj, and D0'=$\emptyset$ and Dm=$\emptyset$. Figure 2 gives an example of a Delaunay sequence.

The Delaunay sequence DS=[T0,T1,...,Tm] can be represented as a pyramid with m+1 levels, called Delaunay pyramid and denoted DP, such that
(i) T0 is the root of DP
(ii) The level i in DP describes Ti
(iii) For each pair of consecutive elements Ti-1 and Ti in DS, there exists a conceptual link from each triangle t of Ti-1 not belonging to Di-1 to the triangle t' of Ti having the same vertices as t, and there exists a conceptual link in DP from each triangle t in Di-1 to all the triangles t' of Ti, which belong to Di', such that t∩t'≠$\emptyset$.

Figure 3 shows the links of the Delaunay pyramid defined by the sequence of Delaunay triangulations depicted in figure 2.

AN ALGORITHM FOR BUILDING A DELAUNAY PYRAMID

The method we propose is based on a top-down construction strategy, which performs a stepwise refinement of an initial Delaunay triangulation T0. In general, T0 will be a constrained Delaunay triangulation, since it will include also surface specific points (peaks, pits and passes) and lines (ridges and valleys) [5]. This is the main motivation for using a top-down building strategy instead of the bottom-up method based on point elimination which has been employed by Kirkpatrick for defining arbitrary triangulations [11].

A triangulation Ti is obtained from its predecessor Ti-1 by sequentially inserting ni data points: each time the farthest point from the surface is selected and inserted into the triangulation which is

280

updated as a consequence. The basic problem is how to pick a sequence of
triangulations. In principle, we could have an 'exhaustive' pyramid by
considering all the triangulations obtained from T0 by adding one point
at a time. A more efficient way of picking a subset of all possible
triangulations produced by a stepwise point insertion is to define a
sequence of threshold values E0,E1,...,Em with Ei<Ei-1, i=1,2,...,m
such that

(i) e(Ti)<=Ei   i=0,1,...,m
(ii) e(Ti-1)>Ei   i=1,2,...,m
(iii) if there exists a Delaunay triangulation Ti', different from Ti,
which satisfies condition (i) and is obtained from Ti-1 by the
error-directed insertion strategy described above, then the number of
vertices of Ti' must be greater or equal to that of Ti. Sometimes E0 is
not defined a priori, but a subset of the data points and a set of edges,
which must be included into T0, are specified. In this case, we assume
that E0=e(T0). Another possibility for selecting a Delaunay sequence
consists of fixing the number of vertices in each Ti. In this case,
however, the approximation error is not guaranteed to decrease
monotonically.

In the following, we describe a construction algorithm based on a
sequence of threshold values. In the algorithm description, a
triangulation Ti is viewed at an abstract level as a collection of
triangles. The Delaunay pyramid is defined as a collection of 5-tuples
Lj=(Tj,Ej,Sj,Dj,Dj'), where Tj is the Delaunay triangulation
corresponding to level j in the pyramid, Ej the threshold value defining
Tj, Sj is the subset of the data points whose projections are vertices of
Tj, Dj is the difference set between Tj and Tj+1 and Dj' the difference
set between Tj and Tj-1. Note that D0' and Dm are undefined.

The kernel of the Delaunay pyramid construction algorithm is
procedure BUILD_SEQUENCE described below, which builds Tj from Tj-1
according to the threshold value Ej defining Tj. The complete 5-tuple Lj
is defined from Lj-1 as follows. Sj is obtained from Sj-1 by adding
those points of S whose projections are vertices of Tj. The difference
Dj' between Tj and Tj-1 is computed as the union of the refined regions
defined by the points of S inserted into Tj-1. Dj is undefined until
Tj+1 is constructed and is computed as the union of the regions of
influence in Tj of the points belonging to Sj+1 - Sj.

Procedure BUILD_SEQUENCE (S,Lj-1,Ej,Lj);
(* initialization of Lj and Dj-1 *)
    Sj:=Sj-1; Tj:=Tj-1;   Dj':=∅;
    Dj-1:=∅;   e(Tj):=e(Tj-1);
    while e(Tj)>Ej and S-Sj<>∅ do
        let P be the point in S-Sj with the maximum error;
        (* e(P)=e(Tj) *)
        let t be the triangle of Tj in which P falls;
        (* update Tj by insertion of P and Dj' and Dj
            as a consequence *)
        insert P into Tj by joining it to each of the
        vertices of t;
        optimize the resulting triangulation to produce
        a Delaunay tesselation Tj;
        Sj:=Sj U {P};
        compute the region of influence RI(P,Tj-1) of
        of point P in Tj-1;

```
      merge RI(P,Tj-1) into Dj-1;
      compute the refined region RR(P,Tj);
      merge RR(P,Tj) into Dj';
      (* compute the error associated with the new triangulation
         Tj: the error at any point outside RR(P,Tj) remains
         unchanged *)
      for every point Q in S-Sj which belongs to a
                    triangle t in RR(P,Tj) do
         compute e(Q) as the absolute value of the difference
         between its given z value and the interpolated value
         at Q defined by t;
         if e(Q)>e(Tj) then e(Tj):=e(Q)
      end for
   end while;
   Lj:=(Tj,Ej,Sj,Dj,Dj')
end BUILD_SEQUENCE.
```

In the above algorithm RR(P,Tj) is initialized with the three triangles of Tj resulting from the splitting of the triangle t of Tj-1 to which P belongs, and is then updated at each step of the optimization process. Similarly, RI(P,Tj-1) is initialized with the triangle t of Tj-1 to which P belongs, and is updated each time an edge swapping occurs in the optimization process.

Based on procedure BUILD_SEQUENCE, we define the Delaunay pyramid construction algorithm BUILD_PYRAMID as described below.

```
Algorithm BUILD_PYRAMID (S,T0,S0,E0,E1,...,Em,DP)
   D0':=∅;   L0':=(T0,E0,T0,D0,D0');
   DP:={L0};   Dm:=∅;
   for i:=1 to m do
      BUILD_SEQUENCE(S,Li-1,Ei,Li);
      DP:=DP U {Li}
   end for
end BUILD_PYRAMID.
```

For simplicity, critical cases are not considered in algorithm BUILD_PYRAMID. It could happen that all the data points must be added to the structure to produce a triangulation Tr such that e(Tr)<=Er. In this case, a pyramid of height r+1 (with r<m) will be produced. The worst case time complexity of the whole algorithm is $O(n^{**}2)$, which is the same as the time complexity of any incremental Delaunay triangulation algorithm applied to the given data set S [5]. Incremental algorithms are the most widely used in the applications: they are easier to code than divide-and-conquer ones and many of them do not require that all of the data points be known at the beginning of the computation.

CONCLUDING REMARKS

A hierarchical model of a 2-1/2D surface based on Delaunay triangulation has been presented. This model, called Delaunay pyramid, consists of a sequence of Delaunay triangulations of a given set of points, which are linked together to form a pyramid. The Delaunay pyramid is a method for the compression of spatial data according to a criterion based on the accuracy required for a given application. It provides a representation of a surface at increasingly finer levels of detail. Further advantages of such a model include the possibility of a fast rendering of a surface by applying ray tracing algorithms and of an efficient detection of surface intersections [7]. The Delaunay pyramid is invariant through arbitrary translations, rotations or other geometric transformations. Compared with hierarchical models based on a quadtree approach, the Delaunay pyramid has the advantage of dealing with arbitrarily distributed data points, thus allowing the inclusion of surface specific points and lines in the model. Unlike the ternary triangulation, which is defined by a fixed triangle splitting rule, our model satisfies the equiangularity property required for numerical interpolation, thus providing a more accurate surface approximation. Generally speaking, an advantage of the Delaunay pyramid over any hierarchical surface model described by a segmentation tree is that the pyramidal representation does not irreversibly insert artificial edges in the surface model.

The algorithm for building a Delaunay pyramid proposed exhibits a quadratic worst case time complexity in the number of data points. This is the same as the complexity of algorithms, which build hierarchical models on irregularly distributed data, or of incremental algorithms, which construct a 'flat' Delaunay triangulation.

The Delaunay pyramid concept could be extended to construct geometric models of three dimensional objects for applications to computer vision and graphics. A pyramidal triangle-oriented representation can be used as a multiresolution description of the boundary of a solid object or, combined with a three-dimensional Delaunay triangulation [1], as a hierarchical volumetric object representation [2].

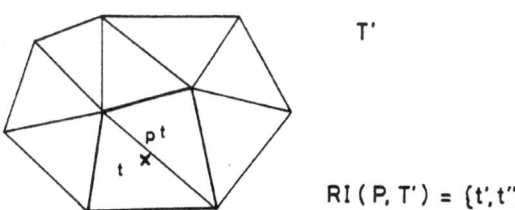

$$RI(P, T') = \{t', t''\}$$

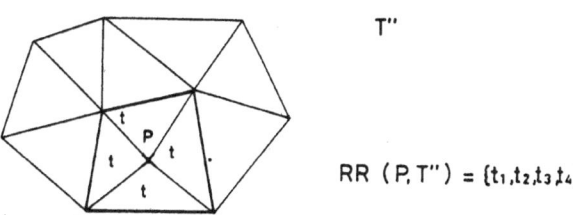

$$RR(P, T'') = \{t_1, t_2, t_3, t_4\}$$

Figure 1
Polygon of Influence PI, Region of Influence RI, Refined Region RR in a pair of Delaunay triangulations T' and T"

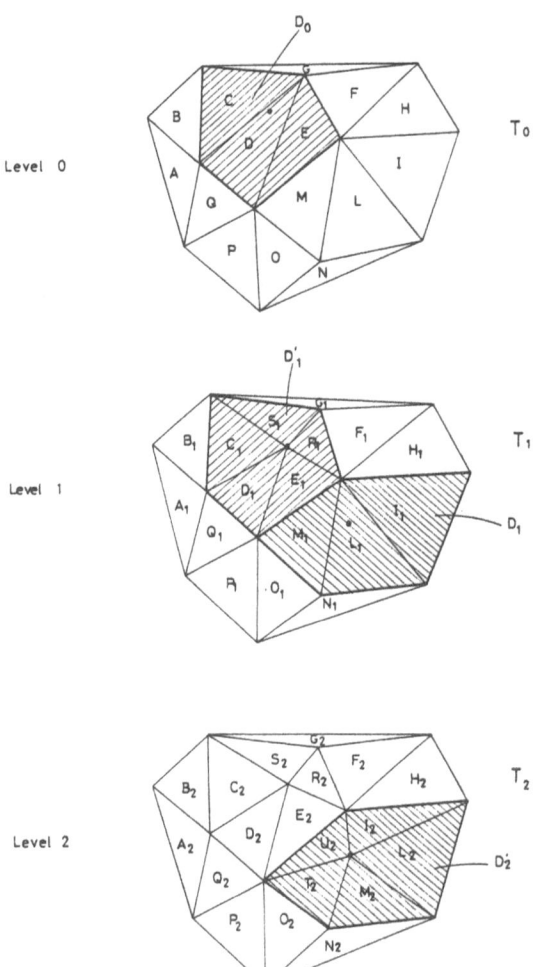

Figure 2
An example of Delaunay sequence

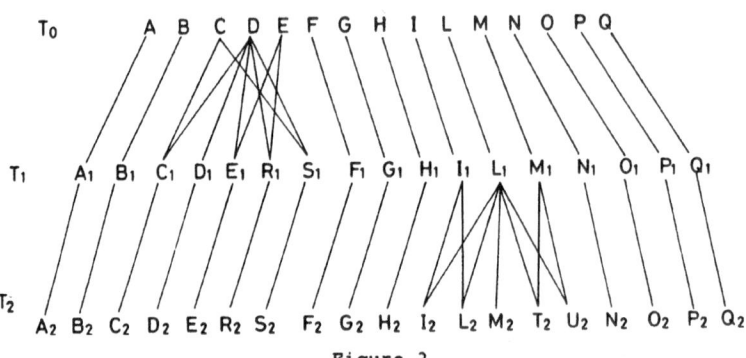

Figure 3
The Delaunay pyramid defined by the Delaunay sequence of figure 2

REFERENCES

1. Boissonat,J.D., Geometric Structures for Three Dimensional Shape Representation, A.C.M. Trans. on Graphics, 3, 4, 1984, pp.266-286.
2. Boissonat,J.D., Tellaud,M., A Hierarchical Representation of Objects: the Delaunay tree, Proceedings Second A.C.M. Symposium on Computational Geometry, Yorktown Heights, June 1986, pp.260-268.
3. Chen,Z.T., Tobler,W.R., Quadtree Representation of Digital Terrain, Proceedings AUTOCARTO 86, London, 1986, pp.475-484.
4. De Floriani,L., Falcidieno,B., Nagy,G., Pienovi,C., A Hierarchical Structure for Surface Approximation, Computer and Graphics, 8, 2, 1984, pp.183-193.
5. De Floriani,L., Falcidieno,B., Pienovi,C., A Delaunay Based Representation of Surfaces Defined over Arbitrarily Shaped Domains, Computer Vision, Graphics and Image Processing, 35, 1985, pp.127-140.
6. De Floriani,L., Mantero,D., Quadtree Based Surface Models, Tech. Rep. I.M.A., n.219, Genova, 1986 (in Italian).
7. Dobkin,D.P., Kirkpatrick,D.G., Fast Detection of Polyhedral Intersections, Theoretical Computer Science, 27, 1983, pp.241-253.
8. Faugeras,O.D., Ponce,J., Prism Trees: A Hierarchical Representation for 3D Objects, Proceedings Eight Int. Conference on Artificial Intelligence, Karsruhe, 1983, pp.982-988.
9. Fowler,R.F., Little,J.J., Automatic Extraction of Irregular Digital Terrain Models, Computer Graphics, 13, 1979, pp.199-207.
10. Gomez,D., Guzman, Digital Model for Three Dimensional Surface Representation, Geo-Processing, 1, 1979, pp.53-70.
11. Kirkpatrick,D.G., Optimal Search in Planar Subdivisions, SIAM J. Computing, 12, 1, 1983, pp.28-35.
12. Meagher,D., Geometric Modeling Using Octree Encoding, Computer Graphics and Image Processing, 19, 2, 1982, pp.129-147.
13. Preparata,F.P., Shamos,M.I., Computational Geometry: An Introduction, Springer Verlag, 1985.
14. Samet,H., The Quadtree and Related Hierarchical Data Structures, A.C.M. Computing Surveys, 16, 2, 1984, pp.187-260.

ALGORITHMIC INFORMATION OF IMAGES (*)

O. Martinoli, F. Masulli, and M. Riani

Department of Physics
University of Genova
Genova (Italy)

ABSTRACT

Men and animals instantly detect "regularity" and "constancy" in visual patterns (and, in general, their structural aspects) with high efficiency. Our approach to evaluating the structural content of a pattern starts from the definition of algorithmic information or complexity, given by Kolmogorov and Chaitin, in which we distinguish two parts containing the metric and structural aspects of the pattern. SIT theory, developed by Leeuwenberg et al. in the area of visual perception, allows one to evaluate efficiently the structural complexity of a linguistic pattern code. We analyse the formal properties of SIT in the context of the theory of reduction calculi.

INTRODUCTION

One of the most peculiar properties of the process of vision in men and animals is their ability of selecting the "regularity" and "constancy" features underlying the perceptual bounding of discrete and recognizable objects from an unstructured "ground" associated with the intensity changes in the images falling on the retinas.

An example of this ability is given by the "disparity" detectors in the visual system of men and monkeys: they permit a structured pattern to be perceived in depth, before or behind a "ground", by the binocular fusion of a proper couple of random dots "stereograms", each of which has no information about the pattern itself. In this case, the perceptual system disregards the large amount of information about the random distribution of dots of different intensities, and focuses itself on the "regularity" that the disparity detectors can extract from a comparison of the two fused images; this allows pattern perception.

From this example, it appears evident that only the detection of the "regularities" present in a visual input is able to elicit a "useful" perception of the environment, that is, to lay the basis for the process of producing responses to particular stimuli encountered by an animal or a human organism in the environment.

(*) This work was partially supported by ESPRIT Grant No.P940

A mathematical framework, in which this problem can be suitably analyzed, is the theory of algorithmic information developed in the sixties by Chaitin [5], Kolmogorov [9] and Solomonov [12]. For instance, the problem of detecting an object with a regular form in an image can be considered as complementary to that of discriminating the random or "patternless" nature of a finite binary sequence of 0's and 1's. Chaitin and Kolmogorov suggested that random sequences are the ones that require the longest programs in order to be generated by a computer.

More generally, Chaitin defined the complexity of a binary string as the length of the shortest program for a Turing Machine able to generate the string itself. Independently, Kolmogorov gave an analogue definition of string complexity based on the Goedel and Kleene formalism of partial recursive functions. Then, the complexity of a string is the length of the minimal program to generate it in a formal system equivalent to the Turing Machine, in the sense of computability theory.

These approaches avoid the need for a probabilistic scheme of the source of strings, and focus our attention on the active process in the receiver.

METRIC AND STRUCTURAL ASPECTS OF IMAGE INFORMATION

Following Kolmogorov and Chaitin, one can define the complexity of an image as the length of the shortest binary computer program that will cause the printout of the image to a given accuracy.

For example, if we consider the image of a circle drawn on a screen with m*m pixels of g levels of gray, we can measure its complexity as follows:

$$C \leq K + L$$

where $K = 3*2*\log_2 m + \log_2 g$ is the amount of information necessary to assign three points identifying the circle through their gray level, and L is the length of the minimal program that uses the three points to generate the circle. We see that K grows with the screen dimension in pixels (m*m), while L is independent of the screen grain and is mostly related to the structural properties of the pattern.

In this way, in evaluating the complexity of an image (or in general, of a string) we can separate the structural and metric aspects of the computational information, which are respectively dependent mostly on the pattern form and on its metric aspects related to the screen resolution.

This distinction is useful to the perception and recognition of patterns either by natural or artificial systems. In fact, to perceive a "square" on a ground, the structural aspects of the figure are the salient ones, while the length of the sides or other metric aspects are less important. Moreover, in human vision, the physical dimensions of an object moving in depth are usually "seen" as constant, although their measures change considerably in the retinal images.

In a more general context, a similar distinction between different aspects of information was made by Mc Kay [11] in the fifties. He called these different aspects "logon content" (or dimensional aspect) and "metron content" (metric aspect).

STRUCTURAL INFORMATION AND LINGUISTIC CODES OF OBJECTS

The analogy between the structure of patterns and the syntax of formal languages led many researchers to follow a linguistic approach to the problem of complex patterns recognition.

In the so-called "structural approach", Fu [6] described a complex pattern in terms of a hierarchical (tree-like) composition of simpler subpatterns, each of which can in turn be described in terms of even simpler subpatterns, etc.; in this way, one can analyse the structural organization of the pattern.

This linguistic approach to pattern recognition permits us to describe a large set of patterns by using small sets of primitives and of grammatical rules.

In this work, we try to associate syntactical approaches to pattern analysis with the complexity theory of Kolmogorov and Chaitin, following the approach based on the "Coding Theory" or "Structural Information Theory" (SIT) developed by Leeuwenberg and Buffart [3,10] for the study of human visual perception.

The latter authors focused their attention on the measurement of the structural information of patterns. They chose to use, as primitives, the lines and angles of the picture outline, and to assign a unitary information value to each of them.

We note that this kind of choice is arbitrary but it is consistent with the fact that our perceptual system contains both efficient detectors of lines in different directions, and angle detectors.

Apart from this aspect, the syntactical structure of the theory and its application to the evaluation of pattern complexity are more general, even if the theory shows traces of the phenomenological approach on which it is based.

In SIT, the visual pattern is coded in a sequence of lines and angles, and a string of symbols of the linguistic alphabet ("primitive code") is associated with it; in addition,a set of "coding operators" is applied recursively to the pattern. In most cases, the "coding operators" extract regularities, and each step of this process reduces the code length. After a finite number of steps, the code is no more compressible with the SIT operators. This code is the "end-code" for that derivation.

Pattern complexity is the number of variables in the shortest end-code. Note that, in general, the SIT operators don't have any information load.

The original operators chosen by Leeuwenberg and Buffart were much more numerous, but recently Van der Helm and Leeuwenberg [13] showed that three operators are sufficient to obtain .an accurate evaluation of pattern complexity. The operators are Iteration, Symmetry and Alternation (see Table 1), and constitute the ISA-form of SIT.

TABLE 1 : THE ISA-FORM OPERATORS

OPERATOR	PRIMITIVE CODE	END-CODE
Iteration	ab ab ab	3*(ab)
Symmetry	a bc d bc a	S[(a)(bc),(d)]
	a bc bc a	S'[(a)(bc)]
Alternation	ab acd ae	<(a)>/<(b)(cd)(e)>
	aa ca dea	<(a)(c)(de)>/<(a)>

As we can see, SIT uses an empirical set of operators, but it gives a quantitative evaluation of the pattern complexity "L" for different intepretations of a pattern[8], consistently with human perceptual behaviour. In many cases[4], it permits one to guess the perceptual

decisions of the observers, i.e., their preferred interpretations of a
line drawing. For example, in Fig.1 the preferred interpretation of the
pattern is (a), that is the simplest. An implementation of this minimum
principle in computer vision was made by Adorni et al.[1].

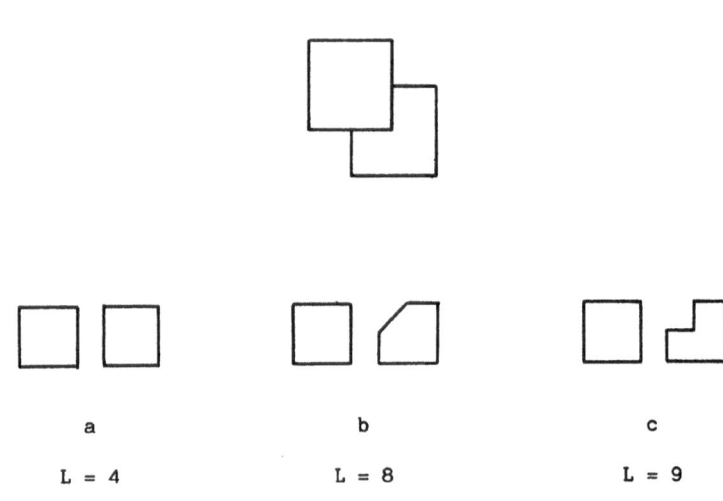

a

b

c

L = 4

L = 8

L = 9

FIGURE 1
Complexities of different interpretations of a pattern

FORMAL PROPERTIES OF THE SIT SYNTAX AS A REDUCTION CALCULUS

    In this paragraph, we try to consider the SIT syntax within the
framework of reduction calculi [2,7], in order to obtain a well founded
theory whose properties may easily be expressed in accordance with that
framework.   This kind of analysis is of major importance for·a possible
computer implementation of an effective automatic system for evaluating
the complexity of patterns characterized by a large number of elements.
    A reduction calculus R is defined by a set of well-formed formulas
(R-wffs) and by a binary relation "immediately reducible to" (R-imr)
acting on the R-wffs. Examples of reduction calculi are Lambda-calculus
and SKI-calculus.
    A reduction calculus R can exibit the following properties:
a) termination - if, for each R-wff, the R-reduction ends in a finite
   number of steps, that is, an R-wff has been reached that is no more
   R-reducible; this well-formed formula is called an R-wff in "normal
   form";
b) confluence or Church-Rosser property - if two different R-wffs, X and
   Y, can be derived from the same R-wff U, then there exists the R-wff Z
   to which both X and Y can be reduced;
c) local confluence - if the R-wffs X and Y are R-imr from the same R-wff
   U, then both X and Y are reducible to Z. (In connection with this
   property, we recall a theorem that states: if R is characterized by

both the local confluence and termination properties, it is also confluent);

d) R is deterministic iff (if and only if) R-imr is a partial function.(Note that, if R is deterministic, then R is Church-Rosser).

The syntax of SIT can be regarded as a formal system in which:

a) the alphabet S is the union of five sets of symbols: the set A of angular measures ranging from -180 to 180 degrees, the set B of linear measures belonging to positive real numbers, the set M of multiplicity coefficients belonging to natural numbers, the set R of production rules, and the set P of punctuation symbols such as parentheses, commas, and so on;

b) the axioms are primitive codes;

c) the set of inference rules is made up of "coding operators", which can be limited to the three operators in the ISA-form of SIT;

d) the set of well-formed formulas (wffs) is made up of axioms and reduced formulas, obtained by using the production rules.

We define a transitive and non-reflexive partial ordering on the wffs such that the statement "X R-imr Y" implies that Y contains less primitives (i.e., symbols of the sets A and B) than X. This order gives SIT the property of termination (i.e., starting from a primitive code, the related derivation ends in a finite number of steps).

We note that the end-codes are the normal forms or the theorems of the system.

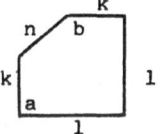

primitive code
a k a l a l a k b n

derivation No.1 :

central code $<a>/<k,1,1,k>$ b n

end-code $<a>/<S'[(k)(1)]>$ b n

I=6

derivation No. 2 :

central code   a k 2*(a 1) a k b n

end-code $<a k>/<2*(a 1), b n>$

I=7

derivation No. 3 :

end-code   a S[(k)(a)(1),(a)] b n

I=8

FIGURE 2
Syntactical ambiguity

The syntax of SIT does not have the confluence property: in actual fact, by means of different derivations, we can reduce the same primitive code to different end-codes ("syntactical ambiguity"). A typical example is shown in Fig. 2. The example also demonstrates that the SIT syntax is not even locally confluent.

THEORETICAL PROBLEMS OF SIT

Avoiding "synctatical ambiguity" in using SIT is of great importance for the implementation of an efficient program for complexity evaluation; to attain this goal, local confluence (and then the Church-Rosser property) must be obtained in the reduction calculus.

From a formal point of view, some restrictions on the theory should be imposed.

The efforts made by Van der Helm and Leeuwenberg [13] to reduce the set of operators to the ISA-form and to characterize their perceptual properties ("holographic regularity" and "transparent hierarchy"), resulted in a restriction on SIT, but the context of the theory remained heuristic.

A drastic but effective way to accomplish the above goal lies in imposing the determinism property on the theory. For example, a restriction should consist in the application of the inference rules to different and disjointed parts of the wffs, following a fixed order from the left to the right.

But to avoid "syntactical ambiguity", in using SIT, this approach is inappropriate: in particular, the most critical operator of the ISA-form to follow this method is Alternation, which is instead suited to detecting symbols placed with regularity among the others in the same formula.

A possible way to reach a solution might be to give up the SIT syntax (in particular, its operators) and to use a formal system characterized by the Church-Rosser property (e.g., SKI-calculus).

REFERENCES

1. G. Adorni , M. Di Manzo, and L. Massone, Coding Theory and Computational Vision, Applied Informatics, 1:164 (1983).
2. H.P. Barendregt, "The Lambda Calculus", North-Holland (1985).
3. H. Buffart, E. Leeuwenberg, Structural Information Theory, in "Modern Issues in Perception," H. Geissler, H. Buffart, E. Leeuwenberg, and V. Sarris, eds., North-Holland, Amsterdam (1983).
4. H. Buffart, E. Leeuwenberg, and F. Restle, Analysis of Ambiguity in Visual Pattern Completion, J. Exp. Psychol.[Hum. Percept.], 9:980 (1983).
5. G.J. Chaitin, On the Length of Programs for Computing Finite Binary Sequences, J.A.C.M., 13:547 (1966).
6. K.S. Fu, Syntactic (Linguistic) Pattern Recognition, in "Digital Pattern Recognition," K.S. Fu, ed., Springer-Verlag, Berlin (1976).
7. K.J. Greene, "A Fully Lazy Higher Order Purely Functional Programming Language with Reduction Semantic," Case Center Technichal Report No. 8503, Syracuse University (1985).
8. G. Hatfield, W. Epstein, The Status of the Minimum Principle in the Theoretical Analysis of Visual Perception, Psychol. Bull., 97:155 (1985).

9. A.N. Kolmogorov, Three Approaches to the Quantitative Definition of Information, <u>Prob. Info. Transmission</u>, 1:1 (1965).
10. E. Leeuwenberg, A Perceptual Coding Language for Visual and Auditory Patterns, <u>Am. J. Psychol.</u>, 84:307 (1971).
11. D.M. Mac Kay, Quantal Aspects of Scientific Information, <u>Philosoph. Mag.</u>, 41:289 (1950).
12. R.J. Solomonof, A formal theory of inductive inference, <u>Information and Control</u>, 7:1 (1964).
13. P.A. Van der Helm, E. Leeuwenberg, Avoiding Explosive Search in Automatic Selection of Simplest Pattern Codes, <u>Pattern Recogn.</u>, 19:181 (1986).

EFFECTS OF HETEROGENEITY OF VARIANCE ON THE PROBABILITY OF CORRECTLY

IDENTIFYING THE BEST NORMAL POPULATION

Adel M. Zaher and Zaki A. Azmi

Faculty of Economica
Cairo University
Cairo, Egypt

ABSTRACT

This study examines the effects of heterogeneity of variance on the probability of making the correct selection when using the means procedure for selecting the population with the largest mean from a set of independent normal populations. The study is conducted by using Monte Carlo simulation techniques for 3, 4, and 5 normal populations as an application of pattern recognition and classification. The population means and standard deviations are assumed to be equally-spaced. Two types of heterogeneity of variane are considered: (1) associating larger variances with larger means, and (2) associating smaller variances with larger means.

1. Introduction of the importance of the application of pattern recognition

    In many practical situations, the problem of selecting one or more populations from a finite number of populations is addressed as an application of pattern recognition. Statistically, selection may be defined in terms of a parameter of the population such as the mean, the variance or some quantile. Populations with large (small) values of the parameters are defined to be best.

    Tests of homogeneity such as the F-test and Barlett's test cannot handle the selection problem since they only tell us whether the populations are equivalent. In fact, what is needed is a procedure which will tell the investigator which population(s) to choose and which will tell the probability of his making the correct selection. The selection procedures were developed to serve this purpose.

    Selecting the best from among K normal populations is one of the most important selection problems and has wide applications. Therefore, selecting procedures for normal populations were developed first and many publications in the statistical literature have been devoted to various aspects of ranking and selecting normal populations. According to Gibbons, Olkin, and Sobel (1977), the application of selection methods to normal distributions was considered first not because the normal distribution occures more often in applications, but because the central limit theorem brings about an important relation between the normal distribution and almost every other distribution.

We have $k(k \geq 2)$ independent normal populations π_i, where the distribution function of each observation from π_i is $\phi(x, \underline{\theta}_i)$, $i = 1, 2, \ldots, k$ and $\underline{\theta}_i = (\mu_i, \delta_i^2)$. It is assumed that all parameters are unknown. Let $\mu_{[1]} \leq \mu_{[2]} \leq \cdots \leq \mu_{[k]}$ and $\delta_{[1]}^2 \leq \delta_{[2]}^2 \leq \cdots \leq \delta_{[k]}^2$ be the ordered μ_i's and δ_i^2's respectively. The best population is the population associated with $\mu_{[k]}$.

A random sample of size n is taken from each of the k populations so that independence holds among and within samples. In general, we determine an appropriate statistic T of the observations. Let T_i be the value of T for the sample taken from π_i, $1 = 1, 2, \ldots. k$. If $T_{[1]} \leq T_{[2]}$ $_{[3]k} \cdots \leq T_{[k]}$ denote the ordered T_i's then the selection procedure based on T selects the population that fives rise to $T_{[k]}$ as the best population. Since the best population does not always produce $T_{[k]}$, the probability of making a wrong selection is always present. The probability of making the correct selection is an indicator of the performance of a selection procedure and provides the most important basis for compairing a set of competing selection procedures.

2. The Problem and Procedures

For selection the population with the largest mean from a set of k independent normal populations, the means procedure which selects the population from which the largest sample mean is obtained is quite reasonable under the assumption of homogeneity of variance ($\delta_1^2 = \delta_2^2 = \cdots = \delta_k^2$). In fact, Hall (1959) showed that under this assumption, the means procedure is optimal in the sense that there does not exist any other procedure that can achieve the same probability of a correct selection with a smaller sample size.

The problem of indentifying the normal population with the largest mean becomes more difficult when the variances are not equal. The reason for this is that under heterogeneity of variance, the differences in sample means may be due to differences in population variances as well as to differences in population means, and it is difficult to distinguish between the contributions of these two types of population differences (Gibbons, Olkin, & Sobel, 1977). The main purpose of this study is to examine the effects of heterogeneity of variance on the probability of making the correct selection when using the means procedure to identify the best normal population.

Let $\bar{X}_{(i)}$ be the sample mean for the population with mean $\mu_{[i]}$, $i = 1, 2, \ldots, k$. The probability of making the correct selection can be written as :

$$P(CS) = P \left[\max(\bar{X}_{(1)}, \bar{X}_{(2)}, \ldots, \bar{X}_{(k-1)}) < \bar{X}_{(k)} \right] \qquad (2.1)$$

Unfortunately, the multiple integral which gives this probability cannot be evaluated in finite terms and numerical approximations have to be used. These approximations however, are unattractive and unreliable (Carral and Cupta, 1977). In this study, no approximations are used for the evaluation of the probability P(CS). Instead, the Monte Carlo simulation technique is used to generate samples from normal populations under different conditions and estimates of the probabilities of correctly identifying the best populations are obtained.

The Minitab Statistical Computing System is used for generating the empirical data and performing the necessary calculations. Therefore the was no need to write a computer program to be specifically used for this study. The Polar method is used for generating the required normal variables. This method has the advantage of having perfect accuracy (Knuth, 1969).

The following two types of heterogeneity of variances are considered:

1. $\delta^2_{[i]}$ is associated with $\mu_{[i]}$, i.e., the largest variance is associated with the largest mean, the second largest variance is associated with the second largest mean, and so on. In this study, this type of heterogeneity of variance is referred to as heterogeneity of Type I.

2. $\delta^2_{[i]}$ is associated with $\mu_{[k-i+1]}$, the largest variance is associated with the smallest mean, and so on. This type of heterogeneity of variance is referred to as heterogenity of Type II.

The means and standard deviations are assumed to be equally spaced, i.e., $\mu_{[i]} = \mu_{[1]} + (i-1)\alpha$, and $\delta_{[i]} = \delta_{[1]} + (i-1)\beta$, $i = 1,2,\ldots,k$, where $\alpha > 0$ and $\beta > 0$ are known constants.

Without loss of generality, $\mu_{[1]}$ and $\delta_{[1]}$ are set equal to 0 and 1 respectively. The assumption of equally spaced parameters is not new with this study; it was also made by Kleijnen (1975) and Gupta and Singh (1980). Here, theis assumption is not a serious limitation on the study since there always exists a linear transformation that transforms any set of k normal variables into a set of normal with equally-spaced means and standard deviations. Since this transformation preserves the order of the means and variances, the performance of a selection procedure under such a restricted parameter space will reflect its performance in general.

The study is conducted for four factors. The following lists the factors and their levels:

1. The number fo populations : k = 3,4, and 5.
2. The size of the sample drawn fro, each population: n = 5,10,25, and 50. These values area chosen to represent very small, small, moderate, and large samples respectively.
3. The difference between two successive means: $\alpha = 0.5\,\delta_{[1]}$, $\delta_{[1]}$, and $1.5\,\delta_{[1]}$.
4. The difference between two successive standard deviations: $\beta = 0$, $0.5\,\delta_{[1]}$, $\delta_{[1]}$, and $2\,\delta_{[1]}$. The value $\beta = 0$ corresponds to homogenous variances. The other three values are chosen to represent small, large, and very large deviations from homogenous variances.

Table 1
Observed Probabilities Correct Selection for the
Means Procedure when n=5

Heterogeneous Variances

k		Homogeneous Variances (β = 0.0)	Heterogeneity of Type I			Heterogeneity of Type II		
			β=0.5	1.0	2.0	β=0.5	1.0	2.0
3	0.5	.7625	.6175	.5775	.5125	.6025	.5675	.4400
	1.0	.9675	.8175	.7100	.6475	.8500	.8025	.6500
	1.5	.9900	.8900	.8400	.7250	.9725	.9150	.7675
4	0.5	.7575	.5850	.5225	.4700	.5900	.4975	.3250
	1.0	.9600	.7150	.6325	.5575	.8375	.7825	.5320
	1.5	.9875	.8675	.7275	.6175	.9675	.9100	.7175
5	0.5	.7550	.5475	.4950	.4100	.5675	.4075	.2175
	1.0	.9500	.6825	.5675	.5375	.8250	.7400	.4275
	1.5	.9775	.7975	.6650	.5700	.9500	.9000	.6625

Table 2.
Observed Probabilities Correct Selection for the
Means Procedure when n=10

k		Homogeneous Variances (β = 0.0)	Heterogeneity of Type I			Heterogeneity of Type II		
			β=0.5	1.0	2.0	β=0.5	1.0	2.0
3	0.5	.8650	.6850	.6575	.5900	.7425	.6250	.4900
	1.0	.9800	.9050	.8325	.7175	.9475	.9050	.7525
	1.5	1.0000	.9675	.8975	.7775	.9950	.9850	.8725
4	0.5	.8600	.6600	.6125	.5050	.7325	.6050	.3950
	1.0	.9750	.8600	.6725	.5875	.9300	.8900	.6775
	1.5	1.0000	.9500	.8150	.6775	.9925	.9800	.8525
5	0.5	.8500	.6350	.5450	.4550	.7200	.5500	.2975
	1.0	.9700	.7600	.6525	.5575	.9200	.8800	.6450
	1.5	1.0000	.8500	.7700	.6000	.9900	.9800	.8425

Table 3
Observed Probabilities Correct Selection for the
Means Procedure when n=25.

Heterogeneous Variances

k		Homogeneous Variances (β = 0.0)	Heterogeneity of Type I			Heterogeneity of Type II		
			β=0.5	1.0	2.0	β=0.5	1.0	2.0
3	0.5	.9525	.8450	.7375	.6300	.8950	.7975	.6075
	1.0	1.0000	.9775	.9075	.7725	.9975	.9875	.9200
	1.5	1.0000	1.0000	.9850	.9150	1.0000	1.0000	.9850
4	0.5	.9525	.7675	.6675	.5475	.8800	.7750	.5375
	1.0	1.0000	.9275	.8125	.7350	.9950	.9775	.9125
	1.5	1.0000	1.0000	.9375	.7600	1.0000	1.0000	.9800
5	0.5	.9500	.7225	.5850	.5150	.8700	.7625	.4675
	1.0	1.0000	.8675	.7950	.6000	.9950	.9675	.9025
	1.5	1.0000	.9875	.8700	.7500	1.0000	1.0000	.9800

Table 4.
Observed Probabilities of Correct Selection for the
Means Procedure when n=50.

k			Heterogeneity of Type I			Heterogeneity of Type II		
			β=0.5	1.0	2.0	β=0.5	1.0	2.0
3	0.5	.9950	.8950	.8350	.7150	.9750	.9400	.7850
	1.0	1.0000	1.0000	.9825	.8825	1.0000	1.0000	.9750
	1.5	1.0000	1.0000	1.0000	.9625	1.0000	1.0000	1.0000
4	0.5	.9950	.8600	.7400	.6750	.9700	.9275	.7800
	1.0	1.0000	.9925	.9325	.7975	1.0000	1.0000	.9725
	1.5	1.0000	1.0000	.98-0	.8750	1.0000	1.0000	1.0000
5	0.5	.9900	.8100	.7050	.5500	.9600	.9200	.7200
	1.0	1.0000	.9675	.8700	.6950	1.0000	1.0000	.9600
	1.5	1.0000	1.0000	.9425	.8375	1.0000	1.0000	1.0000

For each type of heterogeneity of variance and each factorlevel combination (k, n, α, β), 400 simulations of dara are computer generated; this number of simulations was determined so the estimated probability of a correct selection will be within 0.05 of the true probability of a correct selection when using 95% confidence limits. In each simulation, a random sample of size n is generated from each of the k populations, and the mean of each sample is calculated. An estimate of the probability where m is the number of simulations in which the population with the largest mean produced the largest sample mean.

3.   Findings

The resulting empirical data in terms of the observed probabilities of a correct selection for various k, α, β, and r, are displayed in Tables 1 through 4. Each observed probability in the tables is an estimate of the actual probability; the accuracy of each estimate is $\pm$ 0.05 with 95% confidence. P_1 and P_2 are used to denote the observed probabilities of correct selection under homogeneity of variance and heterogeneity of variance respectively. An important feature of the tables is that P_1 exceeds or equals P_2 for all combinations (k,n,α). This indicates that the performance of the means procedure is better under homogeneity of variance (β = 0) than under both types of heterogeneity of variance ($\beta \neq 0$). In other words the ability of the means procedure to correctly identify the normal population with the largest mean is generally descreased when the assumption of homogeneous variances is volted.

Upon examination of the difference P_1 - P_2 for various values of k, one would note that for giving n,α, and β, this difference is generally increasing with k. This demonstates that as the number of populations increases, the lack of homogeneity of variance has larger effect on the probability of a correct selection.

REFERENCES

1.   Bechhofer, R.E. A single sample multiple decision procedure for randking means of normal populations with known variances. The Annals of Mathematical Statistics, 1954 25, 16-39.

2.   Carroll, R.J. & Cupta, S.S. On the probabilities of rankings of k populations with applications. Journal of Statistical Computation and Simulation 1977. 5. 145-157.

3.   Cibbons, J P. Olkin, I. & Sobel, M. Selecting and ordering Populations: A new Statistical Methodology. New York Wiley, 1977.

4.   Gupta, S.S. Selection and ranking procedures: A brief introduction. Communications in Statistics Theory and Methods. 1977, A6. 933-1001.

5.   Gupta, S.S. & Singh, A.K. On rules based on sample medians for selection of the largest location parameter. Communications in Statistics - Theory and Methods. 1980 A9. 1277-1298.

6.   Hall, W.J. The most economical character of some Bechhofer and
            Sobel decision rules. The Annals of Mathematical
            Statistics. 1959, 30. 964-969.

7.   Kleijnen, J.P.C. Statistical techniques in simulation. New York
            Marcel Dekker, 1975.

8.   Knuth, D.E. The art of computer programming, Vol.2. Seminumerical
            Algorithms. Reading PA: Addison-Wesley, 1969.

MULTIFEATURES AND SYSTEM BASED SOLUTIONS

INTEGRATING DISPARITY MEASUREMENTS OVER SPACE

AND SPATIAL-FREQUENCY

Michael R. M. Jenkin, Allan D. Jepson[†] and John K. Tsotsos[†]

Department of Computer Science
University of Toronto
Toronto, Ontario, Canada

Rather than build a stereopsis model based upon determining correspondences between sophisticated monocular features such as zero-crossings or peaks in band-pass versions of the monocular input, we propose that disparity detectors should be constructed that act directly upon band-pass versions of the monocular inputs. Building upon the results of these disparity detectors, we show that a simple surface model based on object cohesiveness and local planarity across a range of spatial-frequency tuned channels can be used to reduce false matches. The resulting local planar surface support could be used to segment the image into planar regions in depth. Due to the independent nature of both the disparity detection and local planar support mechanism, this method is capable of dealing with both opaque and transparent stimuli.

1. Introduction

Many current theories of stereopsis involve three distinct stages (Marr, 1982; Marr and Poggio, 1976, 1977; Mayhew and Frisby 80; Grimson 1981). First, the two images of a stereo pair are processed separately to extract monocular features. One common choice of monocular features is the presence of a zero-crossing in a bandpassed version of the image. Second, the monocular features in one image are matched with corresponding features found in the other image. In practice this second stage cannot be expected to produce only the correct matches, and a third stage must be considered in order to remove the incorrect matches ("false targets"). There are therefore three main issues in the design of such a traditional algorithm for stereopsis, namely *i)* the choice of image features; *ii)* the choice of matching criteria; and *iii)* the way false targets are avoided or eliminated.

In a companion paper (Jenkin and Jepson, 1987) it was shown that symbolic features did not need to be extracted from the monocular images in the first

---

[†]Also, Canadian Institute for Advanced Research.

stages of processing. Rather a technique was presented that measured the local phase difference between the two images. This phase difference could be used to recover the relative disparity between regions in the images.

These disparity detectors are independent detectors of structure of a given spatial frequency at a specific point in (x, y, d)-space (where d is disparity). These detectors combine the first two stages of the traditional approach to stereopsis algorithms. A third stage is still necessary to remove false targets (which now arise as errors in the relative-phase measurements by roughly an integer multiple of 2π.)

This problem of reducing false matches has been considered by many researchers in the classical model for the algorithm. As was noted by Marr and Poggio (76) physical objects exhibit "cohesiveness". This cohesiveness property has been used in a number of different ways as the basis for the reduction of false targets. Marr and Poggio (76) proposed that cohesiveness could be implemented as local surface continuity, while Mayhew and Frisby (80) proposed a "microedge" conjecture. Prazdny (85) argues that local surface continuity cannot be supported by the "cohesiveness" property. Local surface continuity is not valid near object boundaries, nor for non-opaque surfaces. The "microedge" conjecture on the other hand only holds at zero-crossings (which are implicitly assumed to be edges, hence the "microedge" conjecture), and thus also fails to correctly implement the "cohesiveness" property.

If a stereo algorithm is to deal with non-opaque surfaces then non-continuity in the disparity field cannot be used in an inhibitory fashion to disambiguate false targets. With non-opaque surfaces, dissimilar disparities cannot be used to inhibit each other as dissimilar disparity measurements in close physical proximity may correspond to different surfaces. Note that even in the case of opaque surfaces, dissimilar disparities cannot be used to inhibit each other near object boundaries. As Prazdny (85, pp. 95) notes

"Two disparities are either similar, in which case they facilitate each other because they possibly contain information about the same surface, or dissimilar in which case they are informationally orthogonal, and should not interact at all because they potentially carry information about different surfaces."

Finally, we note that all of the techniques for resolving disparities presented here can only disambiguate "local" false targets. Regularly repeated surfaces (such as sine-wave gratings) will give rise to false targets which will have "cohesiveness". The determination of the correct interpretation (or interpretations) from the set of "cohesive" surfaces must be dealt with by later processes.

2. Measurement of Binocular Disparity

In this section we sketch the basic computational approach used to measure local disparity information. For an in-depth treatment see Jenkin and Jepson (87).

The general form of the phase measurement scheme is as follows: The raw images, I_l^0 and I_r^0, are first sent through bandpass filters B_l and B_r, respectively, to produce

$$I_\nu(x) = \int_{-\infty}^{\infty} B_\nu(x-z)\, I_\nu^0(z)\, dz, \text{ for } \nu = l, r. \tag{2.1}$$

The left and right bandpass image are shifted by s, and the phase detector,

$$P(x, s) = \int_{-\infty}^{\infty} L(x-z)\left(I_l(z-s/2)\,I_r(z+s/2)\right) dz,\qquad(2.2)$$

is applied. Here L is a low-pass filter. Finally, zero-crossings of $P(x, s)$ (as s is varied) are detected for each sample point x. For appropriately chosen filters B_l, B_r, and L, the positions of the zero-crossings of P provide local disparity information. We refer to this type of mechanism as a *phase-sensitive demon*. We imagine the initial stereo images decomposed into a multiple spatial scale representation, perhaps through the use of a Burt pyramid (Burt, 1981), or the DOLP transform (Crowley and Stern, 1984). Also there is a similar hierarchical scheme that can be used to construct Gabor-like receptive fields (Fleet and Jepson, 1986). The filters B_l and B_r are taken to be members of such a family, and we assume that their output is resampled appropriately. For each spatial scale there is a demon applied at each sample point x and, like all good demons, they are taken to be independent of other demons at different x's or different scales. In particular, we emphasize that the demons are simply collecting *local* disparity information. A globally consistent interpretation of the computed local disparities is the job of either a subsequent level of processing or a network connecting demons to their neighbours.

As with zero-crossing and peak directed methods, false targets can easily be generated by the above scheme. Ringing from the initial filtering, or the coincidental positioning of objects in three-space can easily result in false matching.

3. Support for Locally Planar Surfaces

The need to reduce false targets from the output of the disparity detectors can best be shown by an example. Consider the application of the disparity detectors (at a given spatial frequency) to a random-dot display. If the spatial-frequency tuning is chosen so that the "dots" in the display are within the detectors spatial-frequency tuning, then a positive response will be obtained for each and every disparity that corresponds to a left-right combination of dots. In terms of a traditional stereopsis algorithm, the measurement of disparity performs the task of the first and second stages of stereopsis; monocular feature extraction and local stereopsis. In order to reduce this large set of measurements (which may contain a large number of "false targets"), a second level of processing is required.

Figure 1 shows the application of a set of disparity detectors to a three-dimensional simulated scene. The scene was rendered with Gouraud shading using a point light source. The stereo-pair was viewed with a non-convergent vision system using a perspective camera model. Disparities were measured with respect to the left eye's view of the scene. Disparity detectors were applied over a wide range, however not all disparities were covered, hence the regular gaps in the disparity dimension. The disparities that were detected are displayed in three orthographic projections; upper-left (x, d), upper-right (d, y), and lower-right (x, y). The d dimension has been expanded for clarity.

Unlike Prazdny (85) the initial disparity detection we have proposed have certain analytic properties which can be exploited in the design of a "coherence" based false target reduction algorithm. Consider an ideal step edge in intensity at position x_0 and at disparity d_0. Disparity detectors will respond positively in the x dimension over the range $(x_0 - \frac{\lambda}{2}, x_0 + \frac{\lambda}{2})$, where λ is the primary wavelength passed by the initial band-pass filtering. In addition, in the d dimension no other ideal step edge can be detected with disparity in the range $(d_0 - \frac{\lambda}{2}, d_0 + \frac{\lambda}{2})$. Thus for any point in (x, y, d) at which (2.6) is satisfied, if that point corresponds to an ideal step edge in intensity, then there must be a range of other positive responses that contains the local neighbourhood $(x_0 - \frac{\lambda}{2}, x_0 + \frac{\lambda}{2})$. Conversely, when looking for support for a point (x, y, d), support can only come from this rectangular region in x and d. Of course if the initial band-pass filtering had included some vertical component, then support could also come from the y dimension. The extension of these algorithms to 2-D filtering is the subject of current research.

A coherence based algorithm could be constructed by building Hough-like voters (Ballard 81). In particular, a positive response at (x, y, d) would vote for $(x+\Delta, y, d)$ where $\Delta \in (-\frac{\lambda}{2}, \frac{\lambda}{2})$, and only cells receiving at least $\lambda * \rho$ votes would be retained. (Where ρ is the density of disparity demons.) Unfortunately such a scheme would strongly favour fronto-parallel surfaces. In order to support planes with varying disparity, it is necessary to consider support from both the x and d (and possibly y) dimensions. Prazdny (85) considered such a scheme, weighting votes by a Gaussian support function. However, by weighting support from similar disparities more strongly than support from dissimilar disparities, his scheme also gave extra support to fronto-parallel surfaces.

A simpler approach, using the above mentioned properties of our disparity detector, can be constructed as follows. For each (x, y, d) at which (2.6) is satisfied, count the number of positive responses within the region $x' \in (x - \lambda/2, x + \lambda/2)$ and $d' \in (d - \lambda/2, d + \lambda/2)$. If there are fewer than $\rho * \lambda$ votes then (x, y, d) can be discarded because the "coherence" principle is not satisfied by the detected structure. However, (x, y, d) does not necessarily belong to any planar structure.

A necessary (but not sufficient) condition that (x, y, d) lies on a planar structure is that d is the mean value of d' over the region $(x - \lambda/2, x + \lambda/2)$. Note that this does not hold near the edges of a planar region. We can test this "planarity" condition by considering the average value of d' over the values of (x', y, d') that gave support to (x, y, d), and reject the point (x, y, d) as not lying on a planar surface if the average value of d' is not nearly d.

This technique can be expanded to integrate information across disparity detectors tuned to different spatial-frequencies quite easily. Let λ_i be the optimal spatial frequency of the i'th set of disparity detectors, and let there be a total of D different spatial-frequency channels used; then for each point (x, y, d) for which (2.6) is satisfied (by at least one of the spatial-frequency tuned operators), count the number of (x', y, d') for which (2.6) (with optimal wavelength λ_i) is satisfied within the region $x' \in (x - \lambda_i/2, x + \lambda_i/2)$, and $d' \in (d - \lambda_i/2, d + \lambda_i/2)$, and weight this vote by $1/(\rho_i * \lambda_i)$. If the point (x, y, d) receives fewer than D votes then (x, y, d) can be discarded. Note that if the image is resampled at different spatial frequencies, the weighting of each vote must be adjusted.

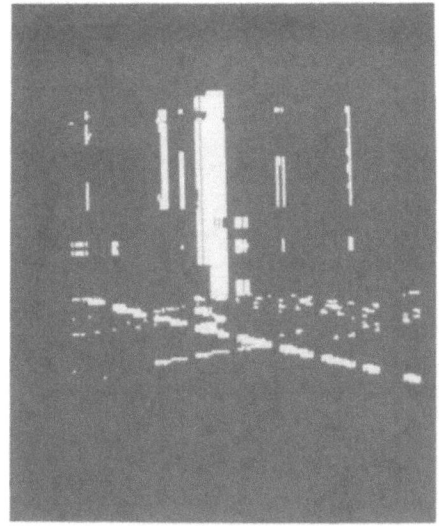

1a - (x, d) Orthographic Projection

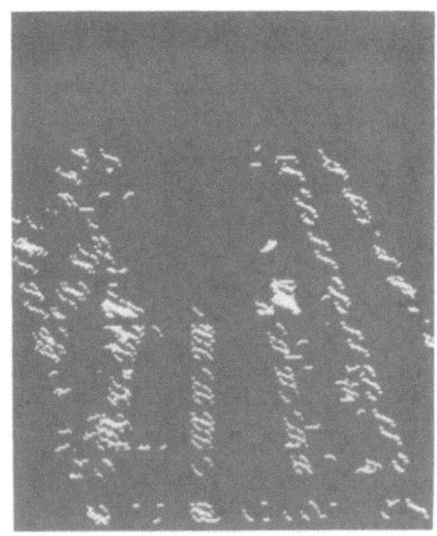

1b - (d, y) Orthographic Projection

1c - (x, y) Orthographic Projection

1d - Left Eye's View

Figure 1. Simulated Scene: Detected Disparities.

In a similar way, the planarity constraint for a given spatial-frequency tuned channel can be generalized to a set of spatial-frequency tuned disparity detectors.

Figure 2 shows the application of this "object coherence" model to the disparity measurements from the simulated scene in Figure 1. Disparity measurements which did not support a coherent planar surface structure have been removed. Not all false targets can be removed in this fashion.

 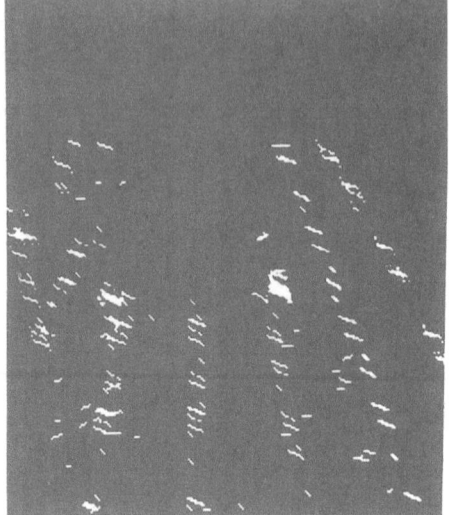

2a - (x, d) Orthographic Projection 2b - (d, y) Orthographic Projection

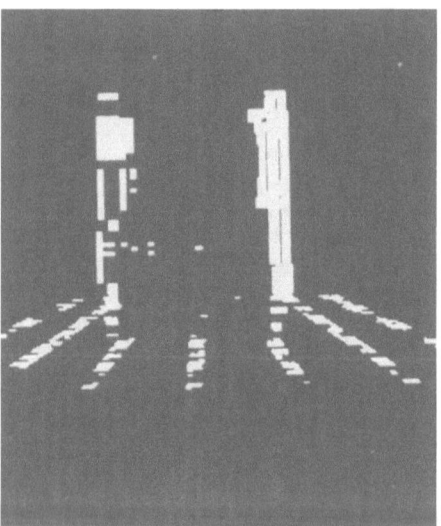

2c - (x, y) Orthographic Projection

Figure 2. Simulated Scene: Coherent Disparities.

As a second example we applied to disparity detection and surface coherence algorithm to a stereo pair of the U.S. Pentagon. The image was processed with one spatial-frequency tuned channel ($\lambda = 13$ pixels, on a 256x256 image). Figure 3 shows the output of the disparity detection algorithm. The general form of the Pentagon has emerged, and in addition a number of spurious responses (false targets) were recorded. A number of these false targets can be eliminated using the object coherence algorithm, and this is shown in Figure 4. Once again, not all false targets are removed. However, the remaining detected features support a

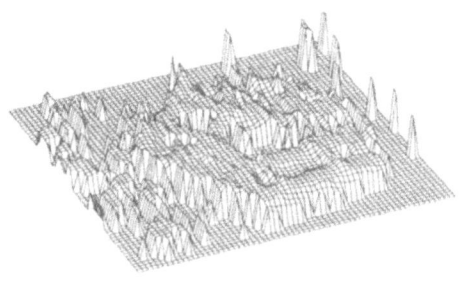

Figure 3. Pentagon: Detected Disparities.

coherent planar surface, and thus would be ideally suited to an object segmentation algorithm whose primitives are based on planar surfaces in depth.

4. Discussion

We have shown that without resorting to the sophisticated monocular token detection and matching of classic computational models of stereopsis, it is possible to extract disparity information from binocular images. In addition, we have presented an algorithm based on a notion of "object coherence" to reduce the number of false targets produced by the disparity detectors. The object coherence algorithm does not rely on either co-operative computation nor on inhibitory connections between disparity detectors to reduce false targets.

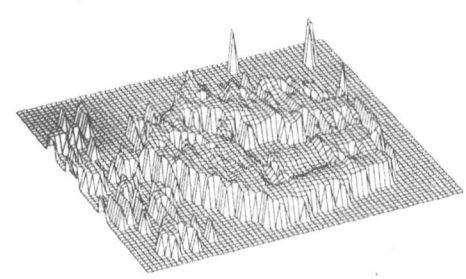

Figure 4. Pentagon: Coherent Disparities.

Both disparity detection, and "object coherence" reduction operate locally. There is no restriction on surfaces constraining them to be opaque. Thus this algorithm can process both transparent and opaque surfaces equally well. The resulting disparity detection shows support for coherent planar surfaces at particular (x, y, d) positions in three space. The resulting votes could be used by a segmentation algorithm to group these votes into surface descriptions.

5. Acknowledgements

We would like to thank the graphics group at the University of Toronto, and in particular John Amanatides for the use of their graphics rendering package "3d".

6. Bibliography

Ballard, D. H., (1981), Generalizing the Hough transform to detecting arbitrary shapes, *Pattern Recognition* 13, 2, 111-122.

Barlow, H. B., Blakemore, C., and Pettigrew, J. D. (1967), The Neural Mechanism of Binocular Depth Discrimination, *J. of Physiology,* 193, 327-342.

Crowley, J. L., and Stern, R. M. (1984), "Fast Computation of the Difference of Low-Pass Transform", *IEEE Trans. on P.A.M.I.,* March.

Fleet, D. J., and Jepson A. D. (1986), "On the hierarchical construction of orientation and velocity selective filters", submitted to *IEEE Trans. Pattern Analysis and Machine Intel.*

Grimson, W. E. L., (1981), *From Images to Surfaces,* MIT Press, Cambridge.

Jenkin, M. R. M., and Jepson, A. D. (1987), The Measurement of Binocular Disparity, this proceedings.

Marr, D. (1982), *Vision,* Freeman Press.

Marr, D., and Poggio, T. (1976), Cooperative computation of Stereo Disparity, *Science* 194.

Marr, D., and Poggio, T. (1977), A Theory of Human Stereo Vision, *AI Memo No. 451* MIT.

Mayhew, J. E. W., and Frisby, J. P. (1980), The Computation of binocular edges, *Perception* 9, 69-86.

Prazdny, K., (1985), Detection of Binocular Disparities, *Biol. Cybern.* 52, 93-99.

IMPROVING BOUNDARY CONTOUR MATCHING USING VIEWING TRANSFORMS

J. Ross Stenstrom

Aule-Tek Incorporated
General Electric Company
Research and Development Center
Schenectady, New York 12345 USA

ABSTRACT

Boundary contour matching typically involves classifying a sequence of curves and deciding what class of object that sequence represents. The geometry of the shapes is ordinarily a factor in the curve classification. Once the curves are classified the shape information is usually ignored. The variation of curve shapes in the object contour boundary should arise from a single consistent viewing transform. In this paper, techniques are developed to insure that boundary curve sequences reflect a consistent viewing transformation.

1. Introduction

A two to three dimensional hybrid boundary contour matching plan is developed and demonstrated. Such system represents an effective compromise between a 2D matcher and a more general but computationally expensive 3D matcher. As noted by Roberts [1], object views may be catalogued by considering visible features visible from different viewing angles. Chakravarty [2] defines characteristic views for objects in stable positions. While an object may occasionally be viewed from any angle, typically only a small number or even a single characteristic view is of interest. An airplane on a runway, for instance, may be seen from the air at a variety of angles but will be well modeled by one set of visible features. Consideration is given to how to improve a purely two-dimensional boundary contour matcher to insure that the boundary curve represents a three-dimensional view of some boundary curve.

The plan begins with the syntactic pattern recognition system first reported in [3]. In [4] it was established that the boundary curve segmentation is perspective invariant for non-trivial views of planar cross-sections. The boundary curve segmentation requires no knowledge of the particular shape or any global characteristics to subdivide a boundary contour. This fact permits automatic model generation and allows efficient and effective model matching. In [5] it is shown that the boundary curve categorization employed varies gently with changes in viewing perspective. A curve segment will have usefully similar curve attribute values under significantly different viewing transforms. Boundary curve fragments are here liberally classified as being any curve category to which they are adequately close in curve attribute space. Other curve categorizations could be used provided that they remain constant over large changes in view angle. The curve sequence or string around the boundary will serve to identify potential matches. The viewing transforms will be employed to verify the integrity of the spatial relationships indicated by the boundary string match.

313

2. Background

Boundary matching has been a central technique in structural pattern recognition. Boundary matching includes many syntactic techniques and some non-syntactic ones as well.

Robert Ledley's work, [6-8] for example, marks the beginning of syntactic pattern recognition. Ledley begins by identification of the chromosome boundary. Each boundary point is considered as the center of a short curve. Through a simple approximation, the angle of curvature is estimated at each edge point. Peaks in curvature as the boundary contour is traversed become the boundary-segment curve centers. The curvature as well as the curve length, a tangent length, and the distance between adjacent curve segment centers permit the categorization of these boundary curves as one of a few basic types. These boundary-segments are combined according to a formal grammar to identify sub-median and telocentric chromosomes.

Another important syntactic classification procedure is from Pavlidis and Ali [9,10]. The idea applies when rotation and scale are known. Each segment of a boundary, as within a particular section of the image, is described by a few parameters. These parameters are made into a short string within the window. The outline of the shape is the concatenation of these shorter strings. There are a large number of individual classifications made and the resulting string is usually long. A grammar necessary to accomodate variations in shape becomes large. However, such a grammar defines a regular language (see [11,12]). A regular language can always be put into a form where it is very easy to determine membership. Also, the automatic generation of a regular grammar is straightforward, even from a very large number of classified examples.

Different applications have seemed to require different choices in boundary curve primitives. This human involvement effectively blocked any hopes of a generally applicable system. You and Fu [13-15] devised a general description for any planar curve based upon four purely numerical attributes. These four attributes are similar to those underlying the curve classification of Ledley. You and Fu provide a normalization that makes these curve attributes invariant to rotation, translation, and scaling. The usage of these attributes for curve classification depended upon being able to determine closed object boundaries. Their error-correcting matching techniques also effectively precluded consideration of complex networks of classified curves.

As an example of a non-syntactic boundary matcher consider a paper by Wallace, Mitchell, and Fukunaga [16]. Boundary contours are segmented at curvature extrema. The boundary is represented by the connecting angles and the boundary curves. The curves themselves are represented by the total distance between curvature extremum or connecting angles. The paper states that the "structural (syntactic) methods" have "failed to solve the three-dimensional problem". It must be noted that neither the connecting angle nor the curve length is invariant under perspective. Actually, Wallace, Mitchell, and Fukunaga characterize the connecting angle using the same feature as You and Fu. The curve characterization is reduced to the You and Fu's single length attribute without normalization.

In fact there is evidence of ability to solve problems that are not purely two-dimensional in the work of Ledley [6-8], You and Fu [13-15], Wallace, Mitchell, and Fukunaga [16], and Stenstrom [3-5]. However, in *none* of these cases was a viewing transformation considered *explicitly*.

In this report a method is developed where each shape primitive is classified with a single label at model building time. When input patterns are considered, each primitive is allowed to have all plausible labels. Possible matches are identified and these possible matches are verified for a consistent viewing transformation.

3. The Use of View Transformations in Matching Contours

Boundary contour matching is usually reduced to the problem of string matching. When a consistent starting point can be defined the problem becomes rotation invariant and translation invariant. You and Fu [13-15] add a normalization to also make the problem scale invariant.

Begin by considering a check for consistent rotation and scale. A method such as that in [3,5] makes use of a scale conformity check in a scaled finite state machine. The size of component curves are checked for an approximate ratio to the total expected perimeter based upon models in the system. The size is not checked as an exact value, however, for such a check would prohibit the implementation as a quick finite-state process. In the end it is desired to check the size more carefully. Also, the string believed to represent the boundary contour cannot be explicitly checked for closure. The method relies upon the shape of the curves matched and the connecting angles. Again speed is the issue. Once a candidate match is encountered there is an opportunity to take some extra time and carefully verify the match.

The use of a curve c_i in an input image as matching model curve c_m implies a simple viewing transform if the object is not tilted with respect to the viewer. The transformation from model coordinates to scene coordinates is simply

$$
S \begin{pmatrix} \cos\theta & \sin\theta \\ -\sin\theta & \cos\theta \end{pmatrix} \begin{pmatrix} x \\ y \end{pmatrix} = \begin{pmatrix} x' - x_0' \\ y' - y_0' \end{pmatrix}
$$

where the model curve sequence begins at (0,0), the input sequence begins at (x_0', y_0'), (x,y) is some model coordinate, (x',y') is some scene coordinate. The angle θ represents the rotation angle between the model and the scene coordinates and can be established from a single matched curve. The usage of a curve in a match sequence distinguishes the two ends of the curve. Consider a vector from the starting endpoint of a curve to the ending endpoint. The angle θ is simply the angle from the vector of the model curve to the vector of the input curve. The scaling constant S can be seen to be either the ratio of the length of the input curve to the corresponding model curve or the ratio of the distances between their endpoints. If the two curves matched exactly except for rotation and scaling, the ratios would be the same. It should be emphasized that the knowledge of a single matched curve is sufficient to establish this transformation

The next coordinate transformation of interest is the projection of an orthogonal transformation. Consider the aircraft shape in figure 1. It is rotated by 2 radians in figure 2. The shape is tilted in figure 3. Finally, the tilted shape is rotated to a particular angle in figure 4. With this simple sequence of transfor-

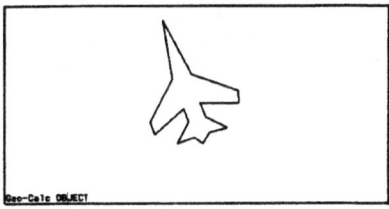

Figure 1. Original shape.

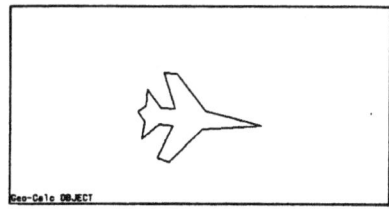

Figure 2. Shape after rotation.

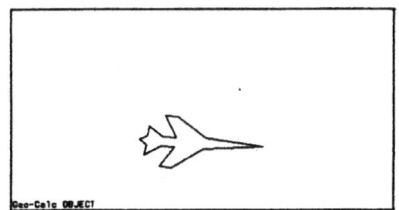

Figure 3. Shape after rotation and tilt.

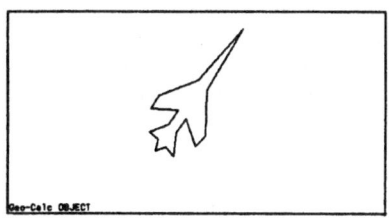

Figure 4. Shape after rotation, tilt, and rotation.

315

mations, boundary contours may be brought into approximate agreement with perspective views. The agreement is improved as the imaging distance is increased. The error becomes totally negligible when the objects are imaged from a distance large in comparison with the scale of the objects.

This basic "tilting" transformation may be seen to be:

$$\begin{pmatrix} \cos\gamma & \sin\gamma \\ -\sin\gamma & \cos\gamma \end{pmatrix} \quad \begin{pmatrix} 1 & 0 \\ 0 & \cos\psi \end{pmatrix} \quad \begin{pmatrix} \cos\theta & \sin\theta \\ -\sin\theta & \cos\theta \end{pmatrix}$$

where the right array is a rotation by θ to align the tilt axis to x=0. The middle transformation is the tilt by angle ψ from being orthogonal to the view direction. In a tilt along x=0, the y coordinate is related to the cosine and the x coordinate is left unchanged. Finally, the leftmost transformation rotates the image by γ in the view plane. As with the simpler case using rotation and scale, the coordinates are offset to an origin before the transformation and a scale factor applies. As derived, the combined system represents a non-linear system of equations. Fortunately, the explicit solution for the angles is unnecessary. The transformation can be captured in the simple linear transformation

$$\begin{pmatrix} a & b \\ c & d \end{pmatrix}$$

directly. As before assume that the model and the input image are shifted so that the first point of each is (0,0). Then the solution of the block diagonal linear system

$$\begin{pmatrix} x_1 & y_1 & 0 & 0 \\ x_2 & y_2 & 0 & 0 \\ 0 & 0 & x_1 & y_1 \\ 0 & 0 & x_2 & y_2 \end{pmatrix} \begin{pmatrix} a \\ b \\ c \\ d \end{pmatrix} = \begin{pmatrix} x_1' \\ x_2' \\ y_1' \\ y_2' \end{pmatrix}$$

for model points (x_i, y_i) and input points (x_i', y_i') produces the a,b,c, and d directly.

The direct solution for the perspective transformation as developed in [1] has 12 parameters. There are at most two data items per point, assuming independence. Thus there would be at least six model and image point correspondences necessary to establish the transformation. The boundary segmentation of choice is presented in [4]. It has the advantage of producing the same segmentation of a planar contour regardless of the imaging perspective (unless the transformation reduces the entire contour to a straight line). Since most objects segment into fewer than 10 curves, and many into fewer than six, virtually the entire match is necessary to fix the transformation. Because of the large number of point correspondences necessary, the perspective transformation has not been implemented at this time.

4. Experience

Boundary contours are segmented at breakable angles as developed in [4]. Using breakable angles to segment a boundary assures the curvature of a contour does not change sign along each component curve. The places where the curvature changes sign are called breakable angles. Figure 5 shows the breakable angle decomposition for several shapes.

The point correspondences for the transformations are between the breakable angles in the model and the input image.

Curves in the model and input images are categorized based upon the attributes of [12-14] as explained in [3-5]. As noted earlier, these curve attributes are not extremely affected by changes in viewing angle.

The algorithm of [3,5] proceeds to identify locations where a classified boundary string matches a model boundary string. The algorithm works for a curve

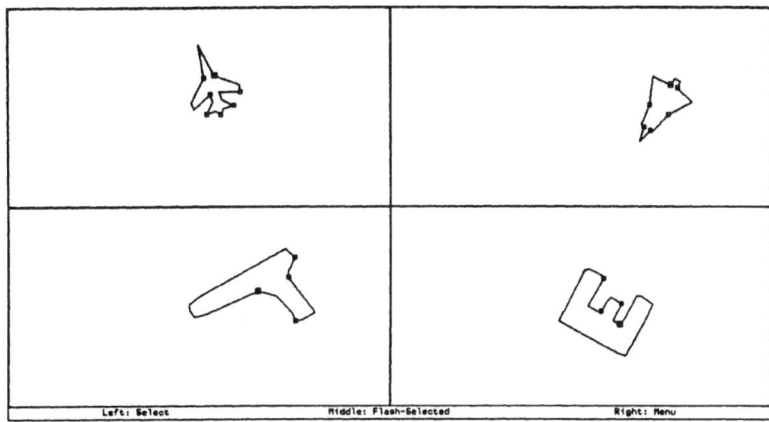

Figure 5. Example models after boundary curve segmentation.
Breakable angles/line segment junctions are
marked with squares.

network and does not require isolation of a closed boundary contour. The
essence of the algorithm is to maintain scale consistent finite-state machines at
breakable curve junctions in an input image. The path to an accepting state is
guaranteed if and only if there is a scale consistent path in the network. The
algorithm terminates knowing only the point at which it discovered the match
and one curve from the input and the corresponding curve in the model. This is
precisely sufficient for the simple rotation and scaling viewing transform. The
original motivation of this research was to provide a way to locate and display
the entire match once the algorithm identifies the existence of a match.

Figure 6 illustrates an airplane model and Figure 7 shows the match of a
simple airplane shape to the model. Figure 8 shows the model correspondence as
a simple matter of rotation and scaling. Since the curve matcher employed produced matches that are rotation and scale invariant, the principle advantage of

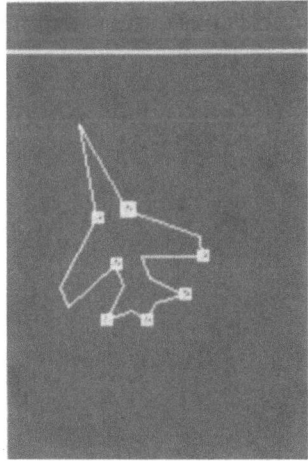

Figure 6. Airplane model.

Figure 7. Match of model to airplane shape.

Figure 8. Comparison of rotated and scaled model to image
boundary contour.

using the rotation and scaling view transform are three-fold. First, it allows the particular string in the image matching the model to be quickly discovered by a geometrically guided search using the view transform.

Another advantage is that the scale was enforced approximately by the scaled finite state machines in the original algorithm. Using the view transform, the scale may be forced to be arbitrarily consistent.

Probably the most significant effect of the post-match verification process is the ability to disambiguate curve strings. With any curve classification procedure using a fixed number of categories to model object shapes, several differing curves will be classified as in the same category. Depending upon the specificity of the categorization, the curve string matched may look little like the original shape. Thus it is ordinarily necessary to use some care in deciding how many different classes of curves to use. Using more categories always leads to fewer incorrect identifications. But using extra categories also leads inevitably to more failures to properly recognize any object shape at all. This becomes a simple variant of the type I and type II errors of elementary statistics. The addition of the post match verification makes the tradeoff much better conditioned. Curve categorization may be made into fewer categories without fear -- only an infinitesimal number incorrect strings will have proper geometric correspondence. Thus the adaption of the method to a class of images is much quicker and less error prone while at the same time the accuracy is improved.

The more complex tilt perspective is also demonstrated. Figure 9 shows the same shape model overlaid on an image with the tilt transform discovered in matching the model. As is illustrated in this example, the tilt transformation closely approximates the general perspective transformation for a large class of images. As the viewing distance becomes large, the transformation becomes a more exact. Figure 10 illustrates a model taken from an aerial image on the left and an aerial image matched to it on the right.

Figure 9. Tilt-rotated model match superimposded on an image.

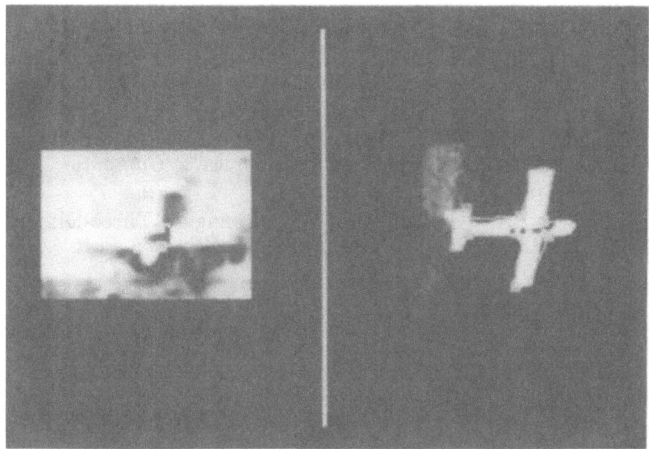

Figure 10. Model taken from aerial image (left) and match to model.

5. References

[1] Roberts, L.G. "Machine Perception of 3-Dimensional Solids", *Optical and Electro–Optical Information Processing*, J. Tippett, D. Berkowitz, L. Clapp, C. Koester, A. Vanderbergh Eds, M.I.T. Press, Cambridge, pp 159-197, 1965.

[2] Chakravarty, I., Ph.D. Thesis, Rensselaer Polytechnic Institute, Troy, New York, 1982.

[3] Stenstrom, J. R., "Syntactic Pattern Recognition for Robot Vision", *IEEE Int. Conference on Robotics*, Atlanta, 1984.

[4] Stenstrom, J. R., "An Improved Segmentation for a Syntactic Curve Network Parser", *Eighth Int. Conf. on Patt. Recog.*, Paris, 1986.

[5] Stenstrom, J. R., "Training and Model Generation for a Syntactic Curve Network Parser", *Proceedings NATO Advanced Workshop on Structural and Syntactic Pattern Recognition*, Sitges Spain, October 1986.

[6] Ledley, R. S., "Automatic Pattern Recognition for Clinical Medicine", *Proc. IEEE*, vol. 57, no. 11, 1969.

[7] Ledley, R. S., "High-Speed Automatic Analysis of Biomedical Pictures", *Science* 146., no. 3641, pp. 216-223, October 9, 1964.

[8] Ledley, R. S. "Practical Problems in the Use of Computers in Medical Diagnosis", *Proc IEEE*, vol 57, no. 11, 1969.

[9] Pavlidis, T. and F. Ali, "Syntactic recognition of handwritten numerals", *IEEE Transactions on Systems, Man, and Cybernetics*, Vol. SMC-7, pp. 537-541, 1977.

[10] Pavlidis, T. and F. Ali, "A Hierarchical Syntactic Shape Analyzer", *IEEE Tr. Patt. An. and Mach. Intell*, vol PAMI-1#1, 1979.

[11] Hopcroft, J. E. and J. D. Ullman, *Formal Languages and Their Relation to Automata*, Reading, Massachusetts: Addison-Wesley, 1969.

[12] McNaughton, R., *Elementary Computability, Formal Languages, and Automata*, Englewood Cliffs, New Jersey: Prentice-Hall, 1982.

[13] You, K.C. and K.S. Fu, "Syntactic Shape Recognition Using Attributed Grammars", TR-EE 78-38, Purdue University, September 1978.

[14] You, K.C. and K.S. Fu, "A Syntactic Approach to Shape Recognition Using Attributed Grammars", *IEEE Transactions on Systems, Man, and Cybernetics*, Vol. SMC-9, No. 6, 1979.

[15] You, K.C. and K.S. Fu, "Distorted Shape Recognition Using Attributed Grammars and Error-Correcting Techniques", *Computer Graphics and Image Processing*, Vol 13, 1-16 1980.

[16] Wallace, T.P., O.R. Mitchell, and K. Fukunaga, "Three-Dimensional Shape Analysis Using Local Shape Descriptors", *IEEE Tr. Patt. An. and Mach. Intell*, vol PAMI-3#3, 1981.

3-D RANGE ESTIMATION

FROM THE FOCUS SHARPNESS OF EDGES

G.Garibotto and P.Storace

Central Research
ELSAG S.p.A.
Genova

ABSTRACT

In the proposed method the focus sharpness of edge points in a recorded image is used to estimate their actual position in the 3-D space. As a matter of fact, the amount of blurring in a picture depends on the distance of the object details from the plane which is "in focus" in the observed scene. Henceforth, from such measures it is possible to roughly provide depth estimates on strong edges of a recorded image. Moreover, sharpness estimates can be effectively used as additional features of the detected contours in other problems of image registration (stereo and motion). This approach can be used to roughly provide depth estimates on strong edges of a recorded image; some examples are enclosed to show its performance on a set of real scenes.

INTRODUCTION

3-D object reconstruction is mainly performed through active techniques (such as ultrasounds, lasers, etc.) or passive stereo vision [1]. In such models it is quite often ignored a fundamental adaptive process performed by the human optical system, in the selection of the optimal focal conditions for vision; in fact, when moving visual attention to a certain position, all details which are included within that region are correctly perceived in focus on the retinal fovea, whilst the surrounding objects are progressively blurred: this is due both to the space-variant distribution of rods and cones on the retina, (from the denser foveal area) and to the relative distance of objects from the viewer. The same phenomenon takes place also for any imaging camera using an optical system with small depth of focus where only the object points which are placed in the focused plane are sharply recorded; all the other details are blurred both in the background and in the foreground. Henceforth, it seems quite promising to take advantage of this feature for depth range estimation in computer vision applications. In the following relative depth position of the objects in the 3-D scene is obtained from focus sharpness estimate. A mathematical model is discussed in section 2 to achieve a better understanding of the physical process and suggests an efficient processing technique, under noise-free assumptions. Some experiments have been carried out on synthetic data where the accuracy of the estimation is tested using finite arithmetic processing tools and

321

noise influence has been evaluated under different conditions of signal to
noise ratios.

1. MATHEMATICAL MODEL OF FOCUS SHARPNESS

A 3-D object in the scene can be ideally sectioned by plane surfaces,
parallel to the imaging plane: one of such sections, the "focused plane",
corresponds to the correct "in focus" position, and its intersection with
the object determines the maximum sharpness in the recorded image; of
course this plane is an approximation of the locus of the exactly focused
points which should be a spherical surface in the 3-D space.
The purpose of this study is to estimate the edge sharpness and, finally,
the position of the corresponding features in the scene. The basic
assumption in the proposed method is a parametric gaussian model of the
blurring function of the optical system, so that standard deviation is
sufficient to fully describe such defocussing process.   Experimental MTF
analysis on real optical systems has proved reasonable reliability of such
an assumption. According to the previously mentioned gaussian model the
recorded picture is supposed to be obtained from an ideal reference image
I(x,y), which represents the whole 3-D scene ideally "in focus", as the
result of a convolution with a space-variant gaussian blurring function
whose standard deviation at point (x,y) in the image is proportional to
the distance of the corresponding point in the scene from the "focus
plane". A gaussian model of the optical recording process has been used
also in [2], by performing a local analysis, within a suitable
neighbourhood, of the Laplacian filtered image. Another approach, based
on local processing of the directional gradient function, has been
proposed in [3]. In this paper a variable resolution model is proposed,
to estimate the rate of change of the edge slope, as a function of the
different scales of the gaussian operator.

1.1 Noiseless step-edge model

Let us consider an ideal 1-D step-edge of contrast δ:

$$I(x) = \delta \qquad \text{for } x => 0 ;$$
$$I(x) = 0 \qquad \text{for } x < 0 \tag{1}$$

According to our model the recorded image F(x) will be obtained by the
convolution of I(x) with a Gaussian blurring function $G(x, s_I)$,

$$F(x) = I(x) * G(x, s_I) \tag{2}$$

where the standard deviation s represents the amount of blurring in the
recording optical system, and is proportional to the distance of the
corresponding edge in the scene from the focus plane. The purpose of the
following analysis is to perform an estimate of such a parameter s_I, along
the most relevant edges of the scene, since they are mainly affected by
this blurring process.
By computing the partial derivative we obtain the 1-D gradient function:

$$F'(x) = d/dx \, I(x) * G(x, s_I) \tag{3}$$
$$= \delta \cdot G(x, s_I) = \frac{\delta}{\sqrt{2\pi} \, s_I} \, e^{-\frac{x^2}{2 s_I^2}}$$

since the derivative of the step edge I(x) is the ideal impulse function
at the edge position x=0. Henceforth, in this ideal situation, the
gradient function computed at x=0 would be inversely proportional to the
parameter s_I,

$$F'(x=0) = \frac{\delta}{\sqrt{2\pi}\, s_I} \tag{4}$$

Unfortunately, actually there are several sources of errors which prevent an accurate estimation from such value. In fact the image function and its partial derivatives are always affected by noise (sampling, quantization, random fluctuations, etc.).
Besides, relation (4) is context dependent, being proportional to the local contrast of the image, and is valid only at the edge position ($x=0$ in our model). As a consequence, any approximation in the edge localization would represent additional source of errors in the evaluation of s_r. Due to these reasons the proposed technique consists in convolving the recorded image $F(x)$ by two gaussian operators having standard deviations s_{1*} and s_{2*}, with $s_{1*} < s_{2*}$ (the choose of s_{1*} and s_{2*} will be justified later) to obtain:

$$F_1(x) = F(x) * G(x, s_{1*}) = I(x) * G(x, s_1)$$
$$\tag{5}$$
$$F_2(x) = F(x) * G(x, s_{2*}) = I(x) * G(x, s_2)$$

where

$$s_1 = \sqrt{s_I^2 + s_{1*}^2}$$
$$\tag{6}$$
$$s_2 = \sqrt{s_I^2 + s_{2*}^2}$$

The rate of change of the edge slope at different scales is obtained as the ratio of the directional gradients of the two functions $F_1(x)$, $F_2(x)$. In our model such ratio, when computed at the edge position, would be approximated by

$$D = \frac{\frac{\partial}{\partial x} I(x) * G(x, s_1)}{\frac{\partial}{\partial x} I(x) * G(x, s_2)} = \frac{s_2}{s_1} \tag{7}$$

and the blurring parameter s_I, which is inversely proportional to the edge sharpness, is estimated from (6) and (7) as:

$$s_I = \sqrt{\frac{s_{2*}^2 - D^2 s_{1*}^2}{D^2 - 1}} \tag{8}$$

This relation will be used in section 3 to implement a processing scheme for the range estimation of edge points using a single image registration.

1.2 Noise analysis

In practical situations noise interference is always present at the different levels of the processing scheme and the optimal selection of the parameters is conditioned by error propagation.
From equation (8) it turns out that:

$$s_I = \sqrt{\frac{F_2'^2 s_{2*}^2 - F_1'^2 s_{1*}^2}{F_1'^2 - F_2'^2}} \tag{9}$$

The amount of uncertainty ds_I can be regarded as the propagation of noise from the first step (image acquisition) to the final step (computation of the amount of blur) of the algorithm.
We can consider the error on s_I as a function of the uncertainties on F_1' and F_2', so:

$$ds_I = \left| \frac{\partial S_I}{\partial F_1'} \right| \Delta F_1' + \left| \frac{\partial S_I}{\partial F_2'} \right| \Delta F_2' \qquad (10)$$

Let be $A = \dfrac{F_2'^2 S_2^2{}_* - F_1'^2 S_1^2{}_*}{F_1'^2 - F_2'^2}$

then $ds_I = \dfrac{1}{2 S_I} \, dA$

where $dA = \left| \dfrac{\partial A}{\partial F_1'} \right| \Delta F_1' + \left| \dfrac{\partial A}{\partial F_2'} \right| \Delta F_2'$

We are going to analyze the dependence of the two factors $\Delta F_1'$ and $\Delta F_2'$ on the initial noise.
Let's consider $\hat{F}$, (the recorded image) as the sum of an ideal image F plus a noise component:

$$\hat{F} = F + dF,$$

where dF is assumed to be a white noise with standard deviation σ_m.
Then:

$$\hat{F}_1 = F_1 + dF_1$$

where dF_1 is still a white noise with variance $\sigma_{m_1}^2$ such that [6]:

$$\sigma_{m_1}^2 = \frac{\sigma_m^2}{2\sqrt{\pi}\, S_{1*}}$$

In the same way:

$$\hat{F}_2 = F_2 + dF_2$$

where the variance of dF_2 is such that:

$$\sigma_{m_2}^2 = \frac{\sigma_m^2}{2\sqrt{\pi}\, S_{2*}}$$

Since dF_1 and dF_2 are supposed to be white noises, denoting by E the energy of a signal, it holds:

$$E[dF_1] = \int_{-\pi}^{\pi} \left| \underset{\omega}{\mathcal{Y}}(dF_1) \right|^2 d\omega = \sigma_{n_1}^2 \qquad (11)$$

It follows that :

$$E[dF_1'] = \frac{\pi}{3}\, \sigma_{m_1}^2$$

and obviously:

$$E[dF_2'] = \frac{\pi}{3}\, \sigma_{m_2}^2$$

Then, it is possible to estimate upper bound values for the error terms:

$$\Delta F_1' = \frac{\pi\, \sigma_m}{\sqrt{6\, S_{1*}}\, \sqrt{\pi}}$$

and $\qquad\qquad\qquad\qquad\qquad\qquad\qquad\qquad\qquad (12)$

$$\Delta F_2' = \frac{\pi\, \sigma_m}{\sqrt{6\, S_{2*}}\, \sqrt{\pi}}$$

Moreover, it's easy to see that

$$\frac{\partial A}{\partial F_1'} = \frac{2\sqrt{2\pi}\ s_1^3\ s_2^2}{\delta \cdot (s_{1*}^2 - s_{2*}^2)}$$

and   (13)

$$\frac{\partial A}{\partial F_2'} = \frac{2\sqrt{2\pi}\ s_1^2\ s_2^3}{\delta \cdot (s_{2*}^2 - s_{1*}^2)}$$

so that the final uncertainty ds_I is estimated as:

$$ds_I = \frac{6_m\ s_1^2\ s_2^2 \cdot (\sqrt{s_{1*}}\cdot s_2 - \sqrt{s_{2*}}\cdot s_1)}{2\,s_I\,\delta \cdot (s_{2*}^2 - s_{1*}^2)\cdot \sqrt{s_{1*}\ s_{2*}}}$$ (14)

where $k = 2\pi\sqrt[4]{\pi}$.

As easily predictable best results (with minimum uncertainty) would be obtained for higher values of the local image contrast δ. The difference between the two selected standard deviations s_{2*} and s_{1*} should be large as much as possible to minimize uncertainty ds_I. On the other hand the individual values s_{1*} and s_{2*} should be small enough, since they are proportional, to the power 1.5, to the error propagation factor.

2. RANGE ESTIMATION FROM A SINGLE IMAGE REGISTRATION

From the discussion of section 1 it is clear that significant blurring estimates are obtained only from sharp image discontinuities, where the gaussian blurring model can be satisfactorily applied. In this section two different algorithms for the same purpose are described.

2.1 First algorithm.

The first algorithm which was developed strictly follows the mathematical model described in section I: the blurring parameter $s_I(x,y)$ is estimated using equation (8). Its main steps are here described:

Edge detection. The edge detection is realized according to "non maxima suppression" [4] on the absolute value of the gradient of the image, which is carried on by means of a convolution between the image itself and the first derivative of a Gaussian.

Chaining. The extracted points are then connected into lists of contiguous samples belonging to the same spatial contours: this connection is realized using a hysteresis tecnique. As a result a limited subset of samples is retained for the next local analysis.

Obtaining the smoothed images. Then the registered image is smoothed in order to obtain F_1 and F_2, by means of a convolution with two gaussian operators of different sizes, and directional derivatives F_1' and F_2' are computed. The sizes of the two gaussian operators, namely s_{1*} and s_{2*}, should be chosen according to the conclusive suggestions of noise analysis (section 1.2), taking into account the constraints of the hardware convolver which is available [5]: the values of the possible standard deviation for a gaussian convolution vary from about 0.5 to 3.73 pixels, corresponding to a mask width from 5 to 31 pixels (just odd values). In the developed experiments s_{1*} and s_{2*} were chosen as 1. and 2.24 respectively.

Computing the sharpness of edge points. The slope s_I is computed for each point according to equation (8). A first test evaluated the edge sharpness of a simple black shape on a white background; in ideal conditions homogeneus results in all directions would be expected, but it was found that in practical situations there is a constant factor between sharpness of horizontal and vertical edges. The acquired image was processed by separating the horizontal edges from the vertical ones. For each subset the sharpness values $s_I(x,y)$ were computed, and the average terms have been obtained as the horizontal and vertical components s_{Ih} and s_{Iv}, with corresponding uncertainties:

$$s_{Ih} = 1.09 \qquad\qquad \Delta s_{Ih} = 0.11$$

$$s_{Iv} = 1.57 \qquad\qquad \Delta s_{Iv} = 0.17$$

Subsequent experiments confirmed a constant ratio (about 1.35) between vertical and horizontal components. This asymmetry is due to the physical geometry of the TV-camera and A/D converter; the imaging camera was a GE4TE80 with the following characteristics: the solid state sensor is 6.6 cm. x 8.8 cm. which is partitioned into 577 X 374 elements, with individual size 11.5μm x 23μm. According to this remark, the sharpness $s_I(x,y)$ of equation (8) is always corrected by a factor depending on the orientation of the edge at point (x,y). At last, noise effects are minimized by local weighted averaging along the connected edge contours.

2.2 Second algorithm

Another numerical algorithm for the same problem has been further investigated to prove the correctness of our approach. Significant advantages are obtained in terms of processing time and minimum data storage. In the following the whole process is described step by step.

Edge detection. The edge-extraction process in the recorded image is not different from the previous one. The gradient of the image is obtained through the convolution with the first derivative of a Gaussian; the gradient matrix and partial derivatives are kept in memory for the last step.

Chaining. The edge points are organized into lists of contiguous points, by means of a hysteresis technique as before.

Computing the spread of the gradient. The spread of the absolute value of the gradient is computed in the direction perpendicular to the edge point. This value is contrast-independent, but depends on the Gaussian used to obtain the gradient of the image. The spread of the gradient, computed point by point, is the same as the sharpness term s_I evaluated by the previous algorithm, plus a known bias equal to the standard deviation of the Gaussian used for detecting edges. The computed values are locally averaged along the connected edge contours the same way as before.

2.3 Performance of the two algorithms

Both algorithms have been applied on a set of real images representing some black shapes laying on different plane surfaces, parallel to the imaging sensor, in the 3-D scene. The estimated values for sharpness and spread are comparable (provided the subtraction of the bias due to convolution from the spread values of the second algorithm): from these results it's possible, without further processing, to establish a spatial arrangement of edges in the scene with respect to the TV-camera. The range distance R(x,y) from the lens position to the object can be obtained from $s_I(x,y)$ through a non-linear function determined by optical lens properties, as described in [2]

$$R(x,y) = F\,v\phi \; / \; (\; v\phi \; - \; F \; - \; s_I(x,y)\cdot f) \qquad\qquad (15)$$

where F is the focal length of the lens, f is the f-number of the lens and
v0 is the distance between the lens and the image plane. Anyway, some
additional corrections are required in the practical application of this
relation. In fact, let's suppose to change the spatial position of the
focused plane so that a certain blurred edge point E(x,y) is moved to its
focus position (maximum sharpness) and then it is blurred again. The
estimated parameter $s_I(x,y)$ decreases up to a minimum non-zero value
(intrinsic width as called in [3]), and then increases again. That
minimum value, which can be obtained by calibration of the optical system,
should be subtracted from the current estimate to be correctly used in
relation (15).

3. APPLICATION EXAMPLES

An example of the different steps of the two described algorithms is
given in this last section.

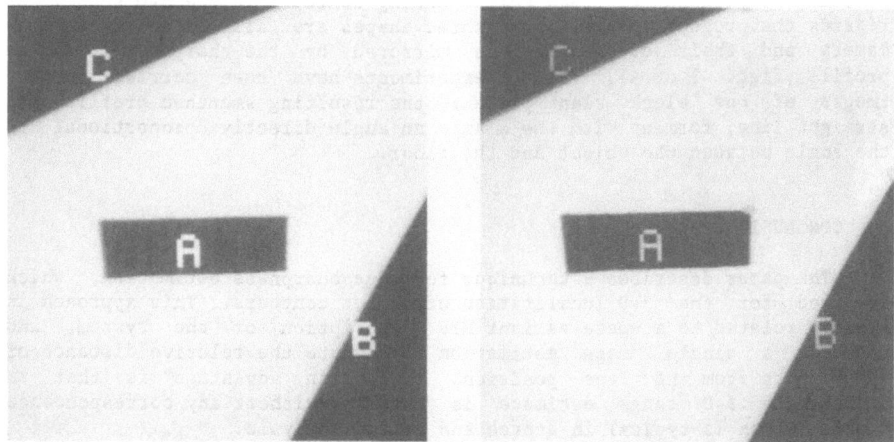

Fig. 1 original image   Fig. 2 edge detection
                                                 and chaining

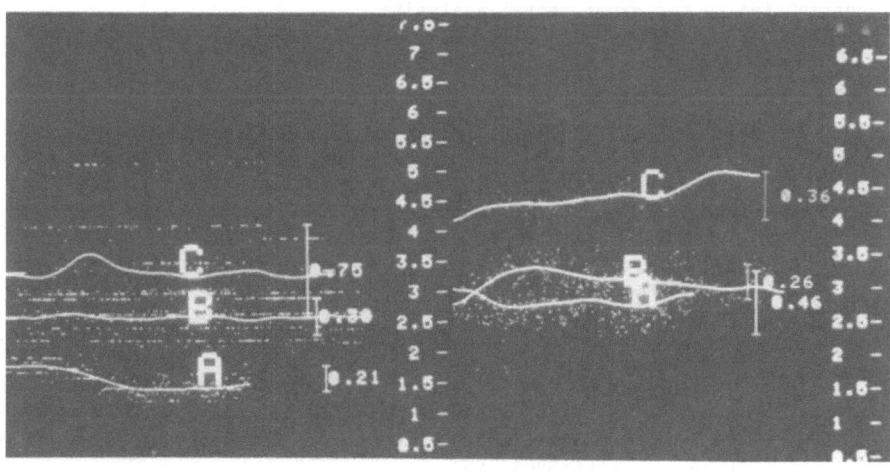

Fig. 3 results of 1st algorithm   Fig. 4 results of 2nd algorithm

The original image (fig.1) represents three black shapes laying on three different planes in respect to the TV-camera. Shape 'a' lays on the floor, which approximately corresponds to the focus plane, whose position is known by the setting of the optical system (focus shift, diaphragm aperture); shapes 'b' and 'c' lay between the floor and the TV-camera, with 'c' over 'b'. Fig.2 represents edge detection and chaining, which are common to both algorithms. In fig.3 and fig.4 the results of the first and the second algorithm are presented: the x axis is regarded as the curvilinear abscissa along each contour, while on the y axis the relative sharpness (in fig.3) or spread value (in fig.4) is plotted for each point of the contour. In this way, the set of estimated values for each contour becomes a discrete function; in order to understand the global behaviour of sharpness (or spread) along a contour and minimize the effects of noise, that discrete function was smoothed using a recursive gaussian filtering, that carries out a weighted average on the data. Through the analysis of the smoothed functions, (white coloured in figg. 3 and 4) on different edges in the image it's possible to distinguish relative positions of the corresponding contours in the 3-D scene. As regards the proposed example, the three shapes are all parallel to the camera and their orientation is mirrored by the sharpness or spread profile (figg. 3 or 4). Various experiments have been carried out on images of one black slant plane : the resulting smoothed profile is a straight line, forming with the x axis an angle directly proportional to the angle between the object and the floor.

4. CONCLUSIONS

The paper describes a technique for edge sharpness estimation, which is used for the 3-D localization of object contours. This approach is closely related to a space-variant MTF description of the system, and requires a single image estimation to compute the relative distance of edge points from the lens position. Its main advantage is that an approximate 3-D range estimate is obtained without any correspondence problem which is typical in stereo and motion analysis.

ACKNOWLEGMENTS

The authors would like to thank their collegue Stefano Masciangelo for his contribution to the experimental analysis.

REFERENCES:

[1] D. Ferrari, A. Pardo, "A 3-D Contour Description using range data", Proc. of the 2nd Int. Workshop on Time-varying Image Processing and Moving Object Recognition, Firenze, 8-9 Sep. 1986.
[2] A. Pentland, "A new Sense for Depth of Field", IJCAI, vol.2 pp.988-994, 1985.
[3] P. Grossmann, "Depth from focus", to be published on Pattern Recognition Letters, 1987.
[4] J. F. Canny, "Finding Edges and Lines in Images", MIT Rep. June 1983.
[5] L. Borghesi, E. Giuliano, G. Musso, F. Cabiati, P. Ottonello, "Programmable Modified Systolic Array for Fast One- and Two-dimensional Convolutions", J. Opt. Soc. Am., Vol. 3, No. 9, September 1986.
[6] X.Zhuang, T.S.Huang, "Multi-scale Edge Detection with Gaussian Filters", ICASSP 1986, TOKIO, pp.2047-2050.

OBJECT RECOGNITION AND LOCATION BY A BOTTOM-UP APPROACH

V. Cantoni (*), L. Carrioli (**), M. Diani (*),
M. Ferretti (*), L. Lombardi (*),and M. Savini (*).

(*) Dipartimento di Informatica e Sistemistica,
    Universita' di Pavia, Italy
(**) Istituto di Analisi Numerica del CNR, Pavia,
    Italy

ABSTRACT

The advantages deriving from a hierarchical implementation of the "Labeled Hough Transform" are pointed out. It offers flexibility, confidence and a reduced computation time. Moreover it furnishes more information, useful when we face the partial occlusion problem. After a description of the technique and its main characteristics, we present some encouraging experimental results obtained with artificial images.

1. Introduction

In this work we present a generalization of the labeled Hough methodology for object recognition by shape recognition. Our goal is to distinguish a known object from others without regard to position and orientation.

The labeled Hough transform is based on the following approach. Not all points of the edge have the same importance: some points, called "critical" points [1], are situated in particular positions and are unaffected by change of scale and orientation of the object [2, 3]. If we are able to classify edge points we can build a table for every class.

If the edge point labeling is performed in an inexpensive way, the computation time, which is proportional to the reference table cardinality, is reduced because the reference table is divided into smaller tables formed by points of the same class. Only one sub-table is used by each voting point, so that, if N is the number of points of the original reference table and K is the number of classes into which contour points can be divided, the average length of the sub-tables is N/K. Supposing that all sub-tables are used the same number of times, the voting process is reduced by a factor K. Moreover, it is to be noted that the votes accumulated on the reference center are in the same number, as the whole reference table is used by each voting point, as in the traditional generalized Hough transform [4, 5], so that the signal-to-noise ratio is increased.

This approach may be used in a different manner: we can describe an object in term of sub-patterns (that is features that are characteristic in the contest we are studying), we may then do the same for the sub-patterns, too. Only the last level is described in term of contour points. The recognition process, then, matches the description of the pattern and proceeds in data driven mode from bottom to top. It can be represented as an oriented graph in which each node is a "parameter space", relative to one kind of feature, containing information about the location of the instances. The edges entering a node mean that the feature relative to the node is described in terms of the sub-features which the edges come from. We can note that the features of the first node are the edge points and the last node's one is the shape we are seeking.

In this work, after a detailed description of the hierarchical Hough methodology, we present same experimental results in which we try to recognize an airplane by finding its characteristic sub-patterns.

2. Hierarchical Labeled Hough Transform
    ------------ ------- ----- ---------

The hierarchical implementation of the labeled Hough transform is a useful extension of the technique illustrated in [6]. The main advantages offered are relative to the confidence of the recognition process and to usage flexibility. Moreover it allows a reduction of the computation time in comparison with the previous one.
Fig. 1 shows a hierarchical decomposition of an object's model. A value giving the number of sub-patterns of a specified class is associated to each node of the graph, while there is a value for each edge between two nodes giving the number of sub-patterns of the lower level (first node) composing the pattern represented by the second node. The example of the figures illustrates a 3-level decomposition, in which the first level features are angles of different width, the second ones are couples of angles, the third ones are significant parts of the object (airplane).

The recognition process aims at finding position and orientation of all instances of the pattern described by the terminal node of the graph. It is accomplished by means of a sequence of labeled Hough vote processes that allow to visit the nodes of the graph along one or more paths connecting the first level (critical contour points) to the terminal node.
Reaching a node means to measure position and orientation of all sub-patterns of the class described by the node. Moreover the modelization in the graph form is suitable of a

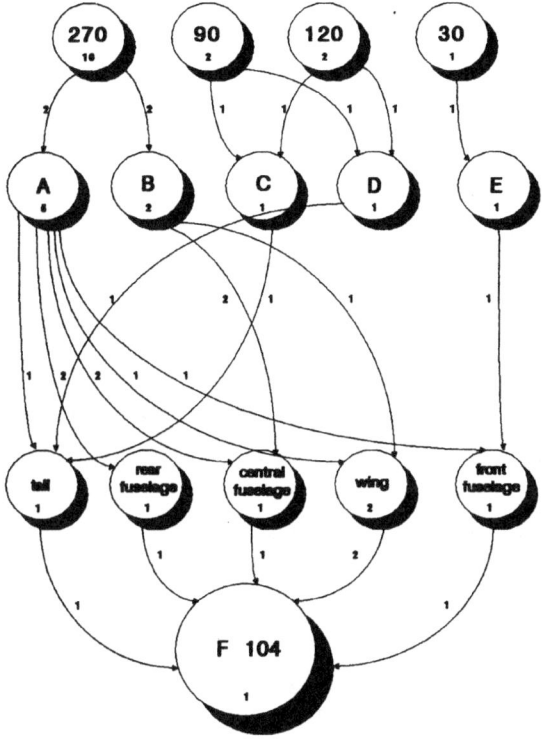

Fig. 1 - Hierarchical decomposition of a model.

working interpretation: each oriented edge represents a reference-table relating the sub-patterns of the lower level node with the reference center of the pattern associated to the higher node. The cardinality of such a table is less or equal (depending on the symmetries) than the value associated to the edge.

The advantages of the hierarchical approach are related to the following aspects. First of all usage flexibility is increased because the hierarchical description of the models can be specialized depending on the context. Moreover, when a part of the object is occluded, it is easy to detect what the missing part is. This is accomplished by using the information coming from the intermediate levels. In fact the information extracted from the whole hierarchical process does not concern only the pattern (considered as a unique entity), but it is structured in more levels.

In the recognition process two kinds of noise appear: the first one is related to the input signal degradation, which causes alteration of the objects contours with respect to their models; the second one is intrinsic to the process itself because it is due to the votes out of target in the parameter space. They can be generated either by the object we are looking for or by other objects in the scene. The hierarchical methodology allows the minimization of the intrinsic noise effects because the vote processes are performed by using reduced reference-tables so that the parameter spaces are cleaner.

Another advantage obtained by the hierarchical technique concerns the computation time. In fact it is possible to prove that the total number of votes expressed in the whole hierarchical process is less than or equal to the number of votes furnished by the direct labeled process if the multiplicity of the intermediate levels sub-patterns is greater or equal than 2. The conditions needed for the proof are:

- all sub-patterns of each level must be utilized for building next level sub-patterns;
- sub-patterns must not overlap each other and must be formed by at least 2 sub-patterns of the lower levels.

3. Optimal Recognition Strategy
    ------- ----------- --------

In a hierarchical recognition process the strategies are decisive for what concerns precision, speed and reliability of the method.
First of all it is necessary to build the trees that represent the hierarchical models of the objects that we want to recognize.

There is not a unique technique to generate the models, but it is always necessary to build models with regard to the context in which we are working; in this context, formed by completely known objects, we build the models putting in evidence the characteristic features of each object.
This flexibility allows an optimized recognition, minimizing the time and increasing the confidence of the measures. In fact if we are going up through the tree and we identify one or more salient features that uniquely identify the object we are searching for, it will not be necessary to proceed further.

As previously mentioned, one of the advantages in the hierarchical approach is a lower dependence of the degree of uncertainity in the recognition process from possible partial occlusions. In order to exploit this potential advantage it is necessary to adopt some tricks in the hierarchical decomposition of the model object.
One of these tricks that helps exploiting the power of the hierarchical approach can be summarized in the choice of a decomposition that uses spatially limited object sub-patterns, such that the probability of having one of them completely visible be maximized. In addition, the sub-patterns can be partially overlapping; in this way we can further reduce the problems due to occlusions.
As to the off-line phase of the Hough process, we should make another observation on the choice of the reference points [7]: since we start from labeled contours the number of votes is relatively low, so we should try not to spread

them over a wide area. This can be achieved by choosing the
reference points on some axis or center of symmetry of the
considered feature. In order to obtain also the spatial
orientation of the feature we need either an analysis of the
texture around the peaks or at least two reference points
[8]: the direction of the straight line passing through the
points gives the sub-feature orientation. This is uniquely
determined if and only if the two points are distinguishable
after the voting process, thus only one of the reference
points may be put in a center of symmetry.

Actually the critical issue is the choice of the
hierarchical models because it affects every aspect of the
recognition process.
Until now the decomposition of an object into sub-parts has
been based on subjective judgements of the operator. It is
possible to achieve a rationalization of this procedure
through the description of a decomposed object by a tree in
which the root represents the entire object and the nodes
hierarchically represent the sub-parts down to the
elementary features. By building a tree for each candidate
decomposition of each object and by defining a measure of
the distance between two trees that keeps into account the
discriminating aspects and features it is possible to find
the optimal decomposition for each object in a limited and
completely known set. We can also think of a complete
automatization of the entire process by building an expert
system that uses the heuristic knowledge of an experienced
operator to define the appropriate measure of distance.

4. Experimental Results
    ------------ -------

We used the method described on some airplane patterns.
We can see (fig. 2) that the contour of these patterns is
piecewise. So it is very helpful to describe the edge using
width and orientation of the angles. We performed the angle
classification using a program simulating heat diffusion.
We put a heat impulse on the pattern border and then we
observe its diffusion inside. The border temperature at time
T is an index of the width of the angles. Figure 3 shows
the hierarchical decomposition procedure of one of the
airplane patterns. In figure 3a we can see features of the
first level: we have four different classes of angles (fig.
1): angles of 270, 90, 120, 30 degrees. In figure 3b we see
features of the second level: couple of angles (except
feature E). It is worth noting that we use all the features
of the first level to form the second level ones and we
allow partial overlapping in features of the second level;
so we have a better description of the feature of the next
level and a better robustness with regard to partial
occlusions. Third level (fig. 3c) is made up by five
features (tail, rear fuselage, central fuselage, wing, front
fuselage). In figure 3d we see the airplane pattern and the
angles we found. Figure 4a shows the voting procedure for a
picture including arbitrary placed airplane patterns.
Figure 4b and 4c show the results from one voting procedure
of intermediate level. Figure 4d shows the result for the
final level. It must be noted that even if we have partial
overlapping features and some of them lying in the picture

Fig. 2 - Models.

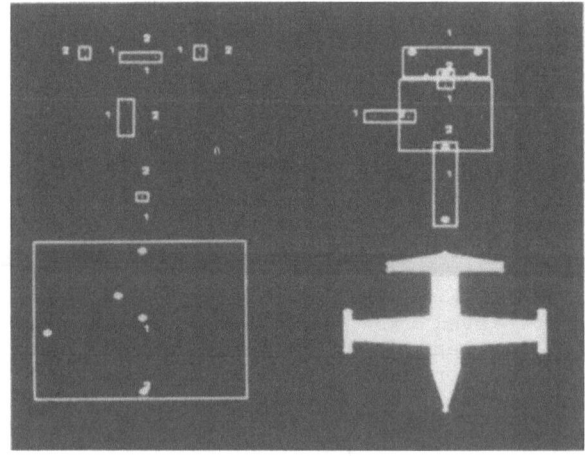

Fig. 3 - Hierarchical decomposition of a model.

only with one pattern, the number of votes of the global
procedure is less than using labeled Hough procedure.
Really:

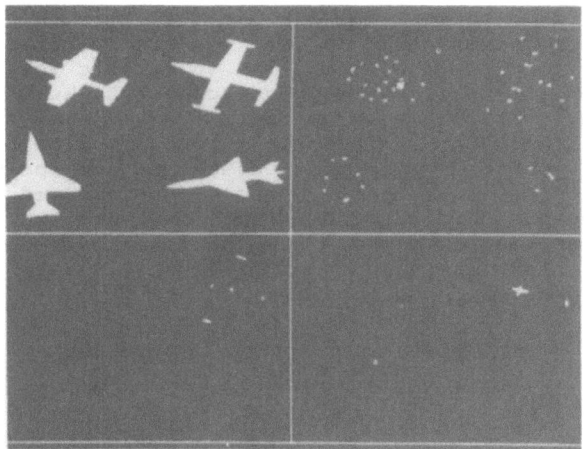

Fig. 4 - Example of Hough vote processes.

Labeled Hough:

```
 2    2    2    1
10  +2   +2   +1  =105
```

Hierarchical Hough:

```
10*(2+2)+2*(1+1)+2*(1+1)+1*(1)=49 first level votes
5*(1+2+2+1+1)+2*(2+1)+1*(1)+1*(1)+1*(1)=44 second level
votes
1*(1)+1*(1)+1*(1)+2*(2)+1*(1)=8 third level votes
101 global votes
```

If we see the vote of the last level we find only a peak in
addition to the one pointing at the position of the airplane
itself, moreover their ratio is very good: the "position"
peak is 12, the expected value, the other peak is 1. So we
have a large margin of safety for dealing partial occlusion
and noise in the images we have to study. As far as the
calculus of pattern and sub-pattern   orientations   are
concerned we use a very simple method: we perform two
different Hough processes for each feature using   two
different reference points. The angle between a reference
axis and the straight line connecting the two points
furnishes the orientation of the feature [9].

5.   Conclusions

      In this work a new approach of hierarchical data-driven
recognition,   based   on   the Hough technique, has been
presented. The solution obtained for artificial images seems
to be very promising and suggests further investigations
either for what concerns the preliminary learning off-line

phase, to determine the hierarchical strategy (on the base of the knowledge about object to be recognized and the context), or for the capability to exploit this approach on the new multiprocessor image processing systems in order to obtain near-real-time solutions.
In this connection is is worth to quote the matching between the hierarchical processes described and the hierarchical architectures (compact and distributed pyramids, heterogeneous hierarchical machines, etc.); in fact some operations (as for example the labelling process [10], and the feature detection and metrication [6]) have already been specialized for these machines.

References
----

1. H. Freeman, "Shape description via the use of critical points", Pattern Recognition, 10, (3), 1978, 159-166.

2. J.L. Turney, T.N. Mudge, R.A. Volz, M.D. Diamond, "Experiments in Occluded Parts Recognition Using the Generalized Hough Transform", Proc of the Conference on A. I., Oakland University, Rochester, 1983.

3. J. L. Turney, T. N. Mudge, R. A. Volz, "Recognizing partially occluded parts", Center for robotics and integrated manufactoring, University of Michigan, RSD-TR-20-83, 1983.

4. D. H. Ballard, "Generalizing the Hough Transform to Detect Arbitrary Shapes", Pattern Recognition, vol. 13 N. 2, 1981, 111-122.

5. K. R. Sloan Jr, D. H. Ballard, "Experience with the generalized Hough transform", Proc. V Int. Conf. of Pattern Recognition, Miami (FL), 1980, 174-179.

6. V. Cantoni, L.Carrioli, "Structural Shape Recognition in a Multiresolution Environment", Signal Processing April 1987.

7. V. Cantoni, G. Musso, "Shape Recognition Using the Hough Transform", Cybernetic systems: recognition, learning, self organization E. R. Caianiello and G. Musso, eds. Wiley & Sons, 1981, pp. 121-128.

8. A. Arbuschi, V. Cantoni, G. Musso, "Recognition and location of mechanical parts using the Hough technique" Digital Image Analysis, S. Levialdi (ed), Pitman, 1984, 373-379.

9. O. Skliar, M.H. Loew, "A new method for characterization of shape", Depart. of Electrical Engineering and Computer Science, George Washington University.

10. V. Cantoni, S.Levialdi, "Contour Labelling by Pyramidal Processing", in Intermediate-level Image Processing, M. B. Duff ed., pp. 179-188, Ac. Press, 1986.

THE DYNAMIC PYRAMID

A MODEL FOR THE MOTION ANALYSIS WITH CONTROLLED CONTINUITY

J. Dengler and M. Schmidt

Department of Medical and Biological Informatics

German Cancer Research Center, D-6900 Heidelberg

Abstract

The Dynamic Pyramid is a model to solve the correspondence problem of image sequences. A robust estimation of local displacements is combined with controlled continuity constraints. At the heart of the model is the functional of an elastic membrane whose elastic constants are subject to variation. The continuity control function is derived from the tension in the displacement vector field at grayvalue edges. The displacement term of the functional is based on robust local binary correlations derived from the signs of the bandpass filtered images. The basic representation of the model is the pyramid: The original images are converted into Laplacian pyramids, the signs of which are the features to determine the local displacements as well as the continuity control function. The vector field is built up as a pyramid from coarse to fine, giving the final displacement vector field at the finest level.

Introduction

This paper presents an approach to solve the correspondence problem in moving realworld scenes as well as in sequences of the biological and medical field, where there can be considerable differences of lighting or recording conditions between two frames. The determination of the displacement vector field is local so that independently moving objects can be dealt with. The approach is completely bottom up, and no scene specific apriori knowledge is used.

The described approach is related to other methods based on optical flow (Horn and Schunck 1981, Hildreth 1984, Nagel and Enkelmann 1986). The basic ideas of this approach have been reported elsewhere (Dengler 1886), but here it is put into a consistent theoretical framework, and the method to determine the continuity control function has been newly developed.

The method makes use of some very general physical constraints of realworld scenes, like rigidity of objects, grayvalue changes as potential object edges and therefore discontinuities. It exploits only locally available information , making maximum use of information from edges and lines as well as from corners, saddle points, minima, or maxima. These aspects are all incorporated into the concept of the Dynamic Pyramid, at the core of which there is the method of elastic matching based on the physical model of the elastic membrane. The membrane model was first introduced at the University of Pennsylvania for matching CT-slices (Bajcsy et al. 1983). It has been modified here completely to make it appropriate for motion stereo.

The Functional of the 2D-Membrane Model

The basic idea of this model is to consider one of the 2 frames to be fixed, and the other attached to an elastic membrane, representing the displacement vector field. An important part is to allow the membrane to tear, where there are occluding edges in the images. In the framework of controlled continuity stabilizers for the regularization of inverse visual problems (Terzopoulos 1986) this model is a 2D-vector field application of a controlled continuity functional enforcing C_0 continuity. The functional describes the balance between the deformation energy $D(u_x, u_y, v_x, v_y)$ and a measure of similarity between the images $V(u,v)^T$, where $U=(u,v)^T$ is a displacement vector field:

$$E(u,v) = \int\int \{ V(u,v) + D(u_x, u_y, v_x, v_y) \} \, dxdy$$

The solution $U(X)$ with $X=(x,y)^T$ minimizes this functional.

Determination of the displacements

The part of the functional $V(U)$, that describes the displacements in flow or motion problems, is usually directly derived from the grayvalue difference of the images . As Schunck (1984) points out, however, using the grayvalues directly is very sensitive to noise and sampling effects in space and time. Also changes of lighting conditions between the frames can cause arbitrary large errors. Taking into account these problems, a correlation based approach is preferred. Nishihara (1984) discussed the properties of various types of correlation regarding noise sensitivity, controllable range, and sharpness of the correlation peak, with the conclusion that the binary correlation has rather optimal properties. The calculation of a binary correlation can also be done a lot faster than a grayvalue- or feature-based correlation (Kass 1983, Nishihara 1984).

The Binary Cross-Correlation Signal as Similarity Measure

The cross-correlation signal CC between the sign matrices S_1 and S_2 is

$$CC_{12}(U) = CC(X - X_{ij}) = \left(\frac{1}{|G|} \right) \sum_{X' \in G} S_1(X+X') \cdot S_2(X_{ij} + X')$$

A natural choice for the signals, the sign of which are S_1, S_2 are the Laplacian filtered images. The Laplacian gives a bandlimited signal which is a necessary precondition to control the searching range (Marr and Poggio 1979). Further, given sufficient bandpass channels, based on gaussian filter masks, the zero-crossings of them are a rather complete representation of the image content. The question of exact localization of contours is of minor importance w.r.t the correspondence problem, as both images are treated exactly the same way. The sign of the bandpass images contains the same information as the zerocrossing contours, they are less sensitive to noise, however, because they are essentially 2D signals (Nishihara 1984). The correlation signal is approximated with a bivariate quadratic polynomial. The best fit is achieved with the model

$$V(U) = (CC(U) - 1)^2 = V_0 + V_1^T \times U + +U^T \cdot V_2 \times U$$

where $V_1 = (V_x, V_y)^T$ and V_2 the Hessian Matrix of the local curvatures

338

V_{xx}, V_{xy}, V_{yy}, all determined in a 3×3 neighbourhood. The curvature parameters V_2 are best determined from an autocorrelation signal, where ideally $V_1 = 0$.

It is assumed that any consecutive frames are similar enough so that the curvatures of the cross-correlation signals are well represented by those of the auto-correlation signals. For reasons of symmetry, the signals for the binary autocorrelation are the signs of the sum of both bandpass images.

From the cross-correlation signal the gradient components V_1 are determined. In order to compensate for asymmetries of the correlation signals, the gradient of the auto-correlation signal – which should ideally be 0 – is subtracted.

Now F(U), the displacement term of the Euler-Lagrange equations, can be determined:

$$F(U) = - \nabla_U V(U) = - V_1 - V_2 \cdot U$$

For reasons of consistency of the resulting equations the vectors V_1 are projected onto the images of the curvature matrices V_2. This procedure is based on the Moore-Penrose-Inverses of V_2 (Dengler 1986), so that the components of V_1 pointing to directions associated with vanishing eigenvalues of V_2 are set to zero:

$$V_1 \leftarrow (V_2 \times MPI\ V_2) \times V_1$$

This means that a point contributes actively only to that component of the displacement where its correlation signal has significant negative curvature.

The Deformation Part of the Functional

The deformation part $D(u_x, u_y, v_x, v_y)$ of the original mambrane functional consists of two contributions, one describing isotropic stretch, and the other one essentially transverse contraction, which is in fact shear.

$$D(u_x, u_y, v_x, v_y) = C_1(u_x^2 + u_y^2 + v_x^2 + v_y^2) + C_2(u_x + v_y)^2$$

The shear term, which is present in real membranes should not be considered in the determination of displacement vector fields. There should be no interaction per se between the two components. Any coupling between the 2 components must have external origin. With $C_1 = C_1(U(X))$ and $C_2 = 0$ the remaining elastic constant becomes a continuity control function. The resulting Euler-Lagrange equations for $U = (u, v)^T$ are

$$C(U)\Delta U + (\nabla C(U))^T \times \nabla U + F(U) = 0$$

As there are wellknown difficulties to solve nonlinear systems of equations like this (Terzopoulos 1986), it is simplified. In order to achieve linearity of the system of equations, it is solved for U(X) with a given C(U(X)).

The system of linear equations is solved by an iterative multigrid algorithm on the basis of the conjugate gradient method (Hestenes and Stiefel 1952) which is known to be the most reliable and the one to converge best, besides being conceptually parallel.

Determination of the Continuity Control Function C(X)

C is considered to be constant between possible lines of discontinuity, which are supposed to be closed. At these lines there are jump transitions to a function G(X) ($0 \leq G(X) \leq C$) with the consequence that the gradient of C is not defined there. Therefore the limits of the partial derivatives of U orthogonal to the line of discontinuity must vanish, resp. should be small in the discrete case. In order to find the lines of discontinuity, all possible grayvalue edges are considered. As it can be assumed, that real image edges at one scale are essentially reflected by zerocrossings of the Laplacian at that scale, (Torre and Poggio 1986), the reductions of continuity is restricted to those zerocrossing lines.

If a more precise localization of the lines of discontinuity is needed, they could be detected with other operators, e.g. with the second directional derivative. This, however, would disturb the elegance of the whole scheme.

The continuity control function C(X) is now taken as

$$C(X) = \begin{cases} G(X) & \text{if X is zero-crossing} \\ 1 & \text{otherwise} \end{cases}$$

Without loss of generality, any scaling factor can be included in F(U), so we assume here C=1.

The two smoothness terms of the simplified Euler-Lagrange equations can be integrated into one discrete operator, the weighted regionoriented Laplacian. Laplace filtering is partially suppressed at the possible lines of discontinuity, depending on G(X), and at these region edges only onesided gradient terms are active. Given a labeling of separated regions, the discrete realization of the weighted regionoriented Laplacian is

$$U(X) = \sum_{X_i \in R} G(X, X_i)^* \langle U(X_i) - U(X) \rangle$$

where R is the local 3×3 neighbourhood, centered at X. $G(X, X_i)$ is a symmetric function:

$$G(X, X_i) = \begin{cases} 1 & \text{if X and } X_i \text{ share the same label} \\ \text{MEAN}\langle G(X_i), G(X) \rangle & \text{otherwise} \end{cases}$$

In the model described here the region labels are just the signs of the Laplacian-filtered image.

In the elastic membrane model the function G(X) should indicate places of high tensions, where the membrane is likely to tear. Therefore the solution $U_c(X)$ of the equation with C(X)=1 is analysed for its local variance. The local discontinuities of the underlying physical process are assumed to occur along fairly straight smooth lines. They cause strong gradients in the vectorfield $U_c(X)$. So we have to distinguish between variance that is caused by a linear change of $U_c(X)$ and variance caused e.g. by single point outliers. This is done by a local polynomial expansion of U_c:

$$U_c(X) = U_0 + U_1^*(X-X_0) + U_R(X-X_0)$$

at X_0, where $U_R(0)=0$, U_R being the residual of the linear fit.

Notated discretly we define

$$G^2(X_0) = \frac{\sum |U_R(X-X_0)|^2}{\sum |U_1 \cdot (X-X_0) + U_R(X-X_0)|^2}$$

where the sums are taken over $X \in W(X_0)$ which is a small neighborhood of X_0 and $|.|$ is the euclidean norm. $G^2(X_0)$ represents the proportion of the residual square sum of a linear fit to the local variance. It follows immediatly that $0 \leq G(X_0) \leq 1$. G is also consistently defined to be 1 at points where the denominator vanishes.

So at real edges, where $U_c(X)$ has a significant gradient, discontinuities are possible. On the other hand zerocrossings of the Laplacian represent also artefacts due to noise. Here the sign correlation is robust against noise (Nishihara 1984) and we get flat solutions $U_c(X)$ in noisy areas. G(X) is significantly larger in noisy regions and only real image edges at the given scale remain as potential discontinuities of the displacement vectorfield to be estimated.

The Dynamic Pyramid, a consistent multigrid approach

Various reasons motivate a multigrid approach. It is well known, that the convergence of iterative algorithms can be accelerated enormeously by applying a multigrid technique (Terzopoulos 1983). The other point is, that the original image data are also well represented in a pair of Laplacian pyramids to separate coarse from fine features (Burt 1984). Finally the controlled search range remains constant in a pyramidal representation. If at each level the registration is correct within $\pm^1/_2$ pixels, the theoretical search range is one pixel at the next finer level. This can be covered reliably with a correlation neigbourhood of 5×5 pixels. The images are first transformed into a Laplacian pyramid (Burt 1984), the sign of which is the basis for the local correlations as well as the segmentation to determine the continuity control function C. The matching process starts at a sufficiently coarse level, where the controlled continuity membrane equation is solved on the basis of the sign images of the correspondig level of the Laplacian Pyramid. The initial vector field is assumed to be zero. As the images are small, convergence is reached very fast. The resulting vector field of this level is transformed to the next level by a bicubical interpolation scheme. Here it is taken as an initial field, on the basis of which the local correlations are determined and the controlled continuity membrane equation is solved. This process is continued up to the finest level.

A sketch of the Dynamic Pyramid is shown in Fig. 1. The circles stand for the membrane equation and the vertical arrows for the interpolation scheme.

Results and Discussion

The algorithm has been applied to the 'Taxi-sequence' from Hamburg, in order to be able to compare the results with those of other algorithms. Fig. 2 shows a contrast-enhanced pair of Laplacian pyramids with 3 levels of the two frames under investigation. At the coarsest level the displacements are within one pixel, and Uc was determined with the membrane equation after 20 iterations. With C being computed from U_c final convergence of U was reached after another 10 iterations. Fig. 3 shows U at this coarse level. The calculations of the other levels took about the same number of iterations. Fig. 4 shows U at the intermediate level and Fig. 5 shows the final displacement vector field at the finest level. At Fig. 4-5 only every second vector in both dimensions is shown.

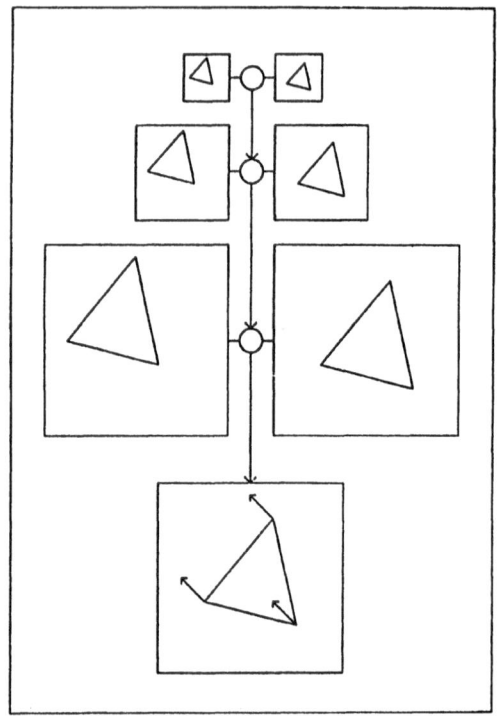

Figure 1. The Dynamic Pyramid

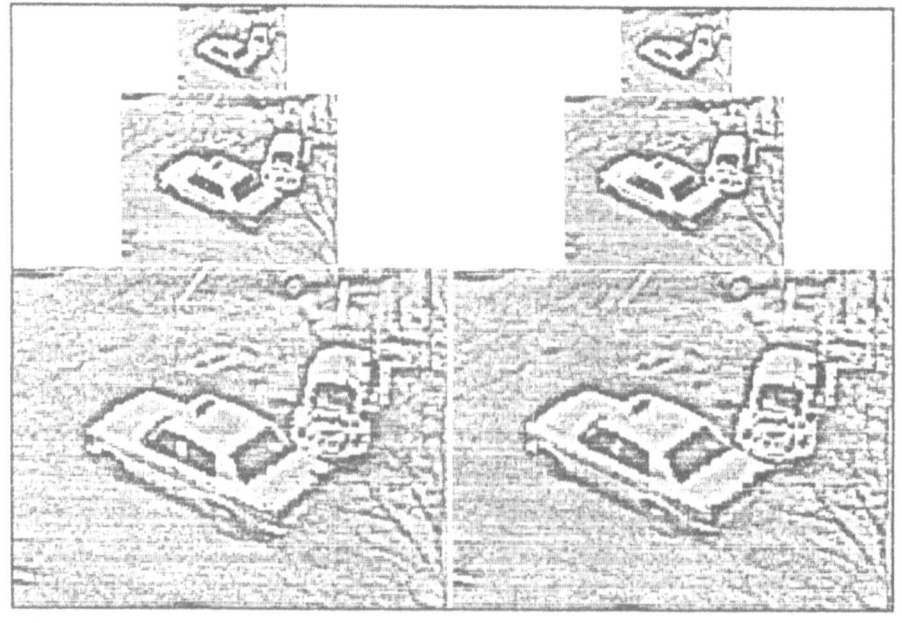

Figure 2. Pair of contrast enhanced Laplacian pyramid

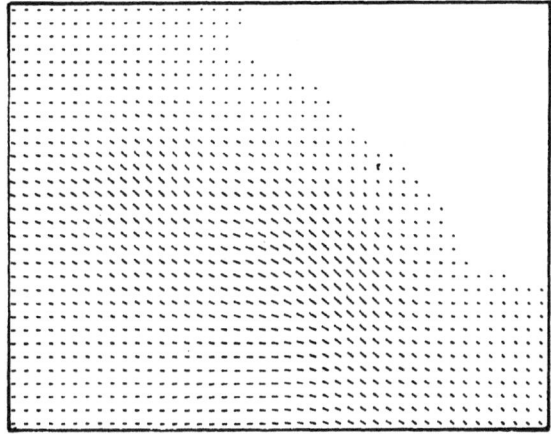

Figure 3. Displacement vector field at coarse level

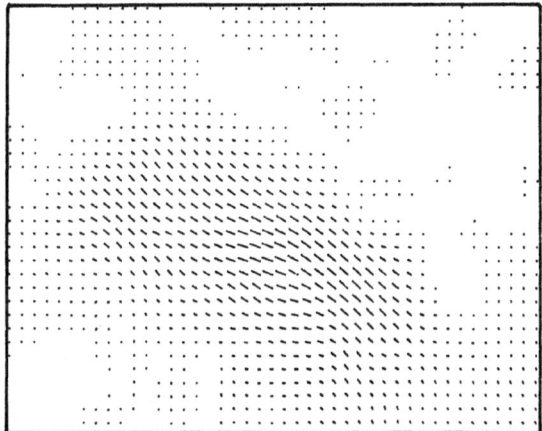

Figure 4. Displacement vector field at intermediate level

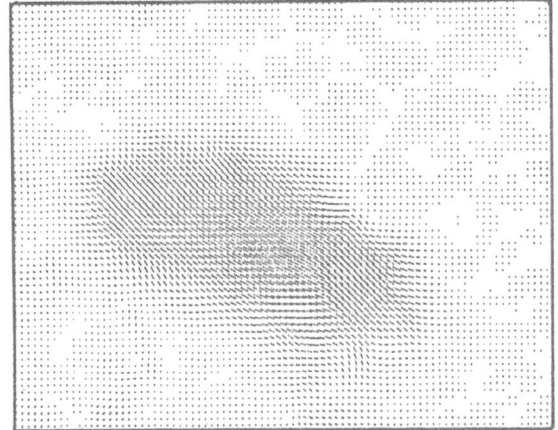

Figure 5. Displacement vector field at finest level

Acknowlegdements

We thank the Deutsche Forschungsgemeinschaft (German Research Council) for their partial support of this work and Prof. Radig as well as Dr. Dreschler-Fischer from the Hamburg Institute of Informatics for providing the tape with the image sequence.

References

Bajcsy,R.; Lieberson,R.; Reivich,M. (1983),"A Computerized System for the Elastic Matching of Deformed Radiographic Images to Idealized Atlas Images", Journ. of Comp. Ass. Tomography 7(4),pp. 618-625

Burt,P.J. (1984),"The Pyramid as a Structure for Efficient Computation " in Rosenfeld,A.(ed.),Multiresolution Image Processing and Analysis, Springer,Berlin

Dengler,J. (1986)," Local Motion Estimation with the Dynamic Pyramid", Proceedings 8th ICPR, pp.1289-1292

Hestenes,M; Stiefel, E. (1952),"Method of Conjugate Gradients for Solving Linear Systems", J. Res. Nat. Bur. Standards 49,pp. 409-436

Hildreth,E.C. (1984),"The Computation of the Velocity Field", Proceedings R. Soc. Lond. B 221, pp. 189-220

Horn,B.K.P.; Schunck,B.G. (1981),"Determining Optical Flow", Artificial Intelligence Vol. 17, pp. 185-204

Kass,M. (1983),"A Computational Framework for the Visual Correspondence Problem", Proc. IJCAI 83,pp. 1043-1045

Marr,D.; Poggio,T. (1979)," A Computational Theory of Human Stereo Vision" Proc. R. Soc. London B 211,pp. 301-328

Nagel,H.H.; Enkelmann,W. (1986), "An Investigation of Smoothness Constraints for the Estimation of Displacement Vector Fields from Image Sequences", IEEE Trans. PAMI 8, No 5, pp.565-593

Nishihara,H.K. (1984),"PRISM: A Practical Real-Time Imaging Stereo Matcher", MIT A.I. Memo 780, Cambridge, USA

Schunck,B.G. (1984),"The Motion Constraint Equation for Optical Flow", Proc. 7th ICPR, pp.20-22

Terzopoulos,D. (1983),"Multilevel Computational Processes for Visual Surface Reconstruction", CVGIP Vol. 24, pp. 52-96

Terzopoulos,D. (1986),"Regularization of Inverse Visual Problems Involving Discontinuities", IEEE Trans. PAMI 8,No 4, pp.413-424

Torre,V.; Poggio,T.A. (1986),"On Edge Detection", IEEE Trans. PAMI 8,No 2, pp. 147-163

COMBINING LAPLACIAN-PYRAMID ZERO-CROSSINGS:

FROM THEORY TO APPLICATIONS TO IMAGE SEGMENTATION

G.Gerig

Institute for Communications Technology, ETH-Zentrum
CH-8092 Zürich, Switzerland

Introduction

Image segmentation is a transformation of the original pixel array into a much more compact description, whose primitive elements should both represent complete information and capture significant properties of the physical world. A basic problem with image segmentation is that on one side we want to gain global properties of structures but on the other side don't want to loose information about local details.

A hierarchy of computational processing provides a structure by which information at coarser levels can guide more detailed processing at finer levels. Thus the use of global information allows one to add constraints to local operations. What is needed is a **multiresolution representation** of the image as a structure where the information is successively condensed into layers of decreasing resolution.

Physical edges are most important properties of objects in a scene and may cause intensity changes in the image. The zero-crossings (ZC) of the Laplacian pyramid represent information about intensity changes at various resolution levels, which favours a multiscale expansion of images for segmentation purposes.

It is a result of the scaling theorem that zero-crossings can be tracked from coarse to fine. One interesting application for example is the refinement of edge positions. In this paper we want to use the possibilities of accessing local and global descriptions in terms of physically evident units for segmentation applications.

Laplacian Pyramid

There are many reasons why for segmentation purposes it may be useful to transform an original image into a multiresolution Laplacian-Gaussian pyramid:

$$(\nabla^2 G_i \otimes \ I \ \rightarrow \ B_i \ )_{i=1..n}$$
$$G_i \ : \ \text{Gaussian of width } \sigma_i$$
$$B_i \ : \ \text{Bandpass channel}$$

Important events are intensity changes of the image surface, which can be detected by differentiation. Numerical differentiation of original image data is an ill-posed mathematical problem [1] and needs regularization. The Gaussian filter has strong regularization properties, such that the combination of a Gaussian and Laplacian yields a simple, but powerful operator. When applied to a multiresolution scheme, the Gaussian filter plays two roles: It acts as a smoothing operation **for generating a pyramid** of various resolutions and is at the same time a **regularization filter** for transforming the subsequent differentiation into a well-posed problem:

The Logan's theorem shows for the one-dimensional case that the set of one-octave bandpass channels contains complete information about the filtered signal. Yuille and Poggio [2] and Babaud et.al. [3] extended this analysis on bandlimited functions into two dimensions, suggesting that the set of zero-crossing (or level-crossing) contours across scales is a nearly **unique** and **complete representation** of the image. This result may be useful for a reconstruction of the image, but is also very important for a segmentation scheme.

If the original image has been filtered with a Laplacian-Gaussian filter the resulting ZC-contours are **always closed curves** or curves that terminate at the boundary of the image [1]. Each bandpass image will be divided into a set of regions bounded by zero-crossing contours, a property which is most likely to be desired for image segmentation.

There are some recently published mathematical proofs about the **nice scaling behavior** of the Gaussian: The Gaussian is the only filter that does not create generic zero-crossings as the scale increases.

Two further qualities favour the use of a Laplacian and a Gaussian with respect to implementation: Using the Laplacian gives the advantage of applying **only one** orientation independent differential operator, whereas the Gaussian is **separable** allowing fast computation.

Combining ZC of multiple resolution channels

Objects are not always bounded by strong physical edges detectable at one scale, so that only a combination of different resolution channels may give correct results.

Summary of previous research

Because zero-crossings of the second derivative can be generated by physically evident intensity changes in images Marr and Hildreth [4] suggested that ZC-segments from different channels are not independent. In their spatial coincidence assumption they proposed that intensity changes of single physical phenomena have about the same position in each channel. Witkin [5] proved for the one-dimensional case the well-behavedness of the Gaussian filter. Babaud et.al. [3] and Yuille and Poggio [2] extended the scaling theorem to the two-dimensional case. They again obtained the result that using a Gaussian filter ZC's are never created as the scale increases, which allows a coarse to fine tracking of ZC-contours in scale space. The knowledge about the scaling behaviour of ZC-contours can guide a linking strategy, but there is one important difficulty when looking at the two-dimensional case: Although ZC-contours do not vanish, they are free to split and merge at increasingly fine scale. In this sense they cannot be described by a simple tree structure as in the one-dimensional case.

Additional results concerning the properties of Laplacian-Gaussian filters may be important for the development of a strategy for linking zero-crossings of subsequent resolution channels: The ZC of the Laplacian are displaced from the true edges by less than σ [2]. By analyzing a noise model Lunscher and Beddoes [6] predicted a noise free interval of 2σ around detected edge positions. They studied the response of the Laplacian-Gaussian to stair-case and sqare-wave signals in detail and could give results for resolution measures limiting the minimum distance between neighboring ZC's.

Linking regions bounded by ZC

In Torre und Poggio [1] it is pointed out that edges derived through rotational operators are generally smooth, closed curves. In our linking scheme we use **regions bounded by zero-crossing contours** instead of the contours themselves. These regions represent a set of primitive elements containing complete information about the image. ZC-regions may often be related to objects or part of objects of a real scene and therefore can represent meaningful primitives for segmentation.

Using regions bounded by ZC's instead of ZC-contours offers many advantages in developing linking strategies. The most important is that linking overlapping regions between

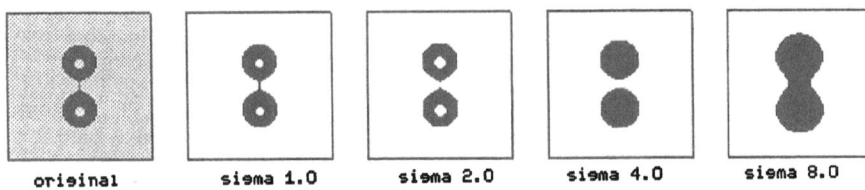

Figure 1: Regions bounded by zero-crossings in the Laplacian-Pyramid

subsequent slices in scale-space is much simpler than linking curves. This is because parameters like degree of overlap, containment and sign are easy to derive. Even when regions are splitted up or merged, they are always bounded by closed curves, which is a most desired quality in segmentation algorithms.

Generation of the multiresolution region-structure

The transformation of an image into a set of bandpass channels can be done independent of the image content. We use bandpass channels in steps of half octaves which creates a certain overlap in the frequency domain and allows the linking of zero-crossings in subsequent channels.

Zero-crossing contours are defined as imaginary boundaries between positive and negative Laplacian outputs. In our region approach we divide the output value range into the three classes positive (P), negative (N) and zero (Z), thus obtaining regions of one of the three types. A subsequent connected component labelling assigns to each region a unique label. Each closed region has then as attributes a *type*-label and a *level*-number indicating the slice in scale-space. Additional attributes are the connected-component labels (*label*) and various shape features (size, form etc.):

REGION (type / level / label / attribute)

Finding initial regions of interest

Images are transformed into a pyramid of labelled regions independent of their content. We now introduce world-knowledge for finding regions of interest depending on the recognition task. This world-knowledge has to be reformulated in terms of entities of the region-label description, e.g. looking for objects of a certain size that are darker than the surrounding background means searching for compact P-type regions of about that area in scale-space (pyramid) and storing them in a list. The resulting regions fulfill some global properties of the desired structures and therefore are candidates for more detailed processing.

Linking subsequent bandpass levels

It is often desired to extract more detailed information about objects represented as rough global structures at low levels, such as a more accurate boundary description (considerable deviation of ZC-contours from the true boundaries can occur with large smoothing). We use our labelled regions for linking corresponding elements in next finer resolution channels. A simple strategy is to link regions of the same type (P,N,Z) (or arbitrary type if completely enclosed by the region above) showing a high degree of overlap in subsequent levels. A link-list containing the connected component labels is constructed (see fig. 2). The label set of the finest resolution level that is obtained after coarse to fine tracking represents the fine detailed structure of a selected low-level region.

Problems may arise when many unwanted merges occur by tracking from coarse to fine (regions are free to **split and merge** as the scale decreases). Merged regions keep a common label at a certain resolution level, but are linked to separate regions in the level above. This observation would lead to split-rules, which are not implemented yet and an object of further investigation.

Introducing maximum deviation constraints

ZC-contours of strong and important edges survive over a relatively large range of scales with only a slight change of positions. One of the problems of the link algorithm proposed

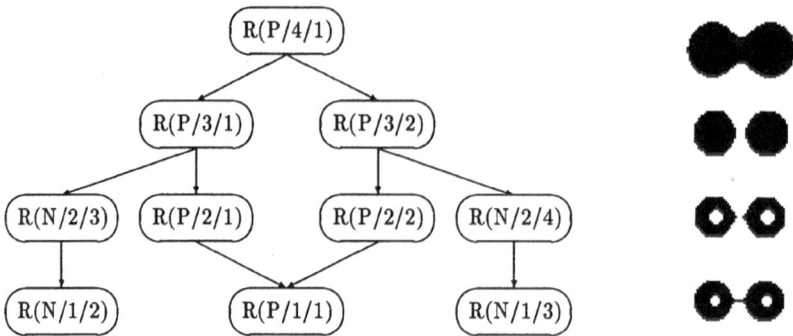

Figure 2: Label-linking strategy

above is, that regions may be **merged** that correspond to different physical units. Using the label-link strategy the merged regions can be tracked down to the next level, because they share the same component-label independent of their spatial extension. One possibility to control linking even with respect to spatial extension is to constrain the maximum deviation of ZC-regions between successive bandpass channels. Applying such constraints to ZC-contours would cause broken lines and missing boundary parts of tracked contours. An important advantage of our approach of using ZC-regions is that we have the possibility to always preserve closed boundaries. We use an erosion/dilation algorithm (Serra [7]) and logical operations between channels for controlling maximum deviation. The complete procedure is similar to the simple one described in the previous section, but with additional steps to control spatial behavior:

The erosion and dilation procedures may be described as expanding resp. shrinking operations on regions in a nearly euclidian sense. We now use the result of Lunscher and Beddoes [6] concerning the minimum distance between two neighboring zero-crossings, which is 1.15σ for the square-wave model, and the result, that zero-crossings deviate by a maximum amount of σ from the true edge positions. These results determine the search radius for our linking process, which is chosen to be σ_{i-1}, the filter width of the next finer level. The following procedure describes the link of overlapping regions of subsequent channels by allowing a certain amount of deviation of their boundaries:

- Dilation of regions on level l_i by a radial amount of σ_{i-1}.

- Binary .AND.-operation between dilated regions of level l_i and regions of level l_{i-1}

- Erode (original) regions of level l_i by σ_{i-1}

- Binary .OR.-operation between eroded regions of level l_i and resulting regions of previous .AND.-operation

Linking level-crossings instead of zero-crossings

Marr and Hildreth [4] proposed to use the slope of zero-crossing as additional information about intensity variations, because slope can be directly related to contrast and width of intensity changes. But the slope measure turned out to be sensitive to noise.

Yuille and Poggio [2] have shown that not only zero-, but also level-crossings of the Laplacian images have nice-scaling behaviour across scales. The level-crossings may therefore be used exactly the same way as zero-crossings, with the additional advantage of suppressing weak edges and details that originate from noise.

We have the possibility to suppress noise and weak intensity variations by only a small change in our linking scheme: When dividing the output of bandpasses into the three types P, N, and Z, we now declare values within certain thresholds around zero $(+/- th)$ as Z-type regions (fig. 3).

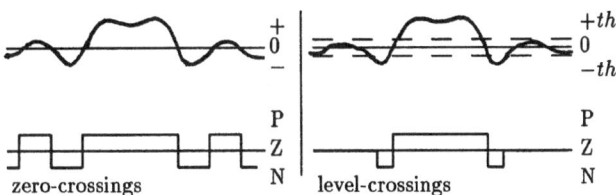

Figure 3: Illustration of level-crossings

Implementation

The DOG (difference of Gaussians) is a good approximation to the $\nabla^2 G$-filter ($\frac{\partial}{\partial \sigma} G \propto \nabla^2 G$). We choose a width ration of $1/1.1$ for the positive and the negative Gaussian. This represents a compromise between a $\nabla^2 G$-operator approximation on one side and the need for sufficient filter volume on the other side. The DOG is preferred because its implementation yields an efficient procedure due to the separability of the Gaussians.

A step width in frequency domain of half an octave will generate eight bandpass channels between sigma (1.0) and sigma (11.0), a range which will be sufficient for most applications.

The value range of the DOG-output is classified into the three categories (P)ositive, (N)egative and (Z)ero, resulting in a subdivision of each level-plane into regions of one of the three types. A connected-component labelling is applied to these regions of each bandpass channel. We obtain a structured pyramid where each level consists of regions which are uniquely described by a level, a type information (P,N,Z) and a connected component label.

Applications

The first image is an industrial scene containing several isolated objects (fig. 4). A human observer may easily detect the positions, the number and the coarse shapes of the objects by shortely looking at the image. More detailed information can be obtained by focussing to certain regions of interest. We try to simulate such a two-step recognition of objects using the Laplacian pyramid. After having generated the bandpass channels we introduce a priori knowledge about the approximate size of the objects and then extract compact regions in a certain low-level bandpass channel. This first step gives information about the number and position of the objects in a very efficient way, because detailed processing is not necessary (fig. 4). The regions are marked and tracked down to the finest level using the proposed link-procedure. Depending on the recognition task we have applied two versions of the link strategy: Following exactly the given procedure will result in closed object boundaries of high positional accuracy, where important details are retained (fig. 4 c,e). Important may be the position of the nose of the two knobs, or the thickened part of one of the T-objects. If we are interested not in the object as a whole, but in the set of structural elements describing the specific objects on the finest level, we modify the algorithm by omitting the erosions and logical-.OR. operations (fig. 4 d). Using level crossings instead of zero-crossings significantly eliminates noise without destroying the important zero-crossing information (fig. 4 f).

The second category are medical X-ray CT data (fig. 5). This image class contains nearly all the difficulties which impede successful segmentation by standard segmentation techniques: Low signal to noise ratio, blurred background, slow varying intensity function of the global objects, overlapping structures of finer details, no large homogeneous regions, no sharp edges and a typical problem in medical images: Organs of different human beings may significantly differ in appearance. Models for calculating the lung-volumes using manually extracted boundaries from a front- and a side-view CT image are successfully tested. The basic goal is the automatic extraction of the contours of the left and right lung as a small but important step towards building an expert system that combines knowledge acquired from different sources for diagnostic purposes. Again, a global/local technique is important

a) original image
b) Laplacian pyramid, σ-parameters:
   top: 1.0/1.4/2.0/2.8, bottom: 4.0/5.6/8.0/11.0
   frame lower right contains recognized objects
c) tracking regions from (11.0) to (2.0)
d) tracking similar to c), without
   erosions and .OR.-operations
e) overlay of contours of c) and original image
f) 'level-crossings' (σ = 1.0 - 2.8)

a

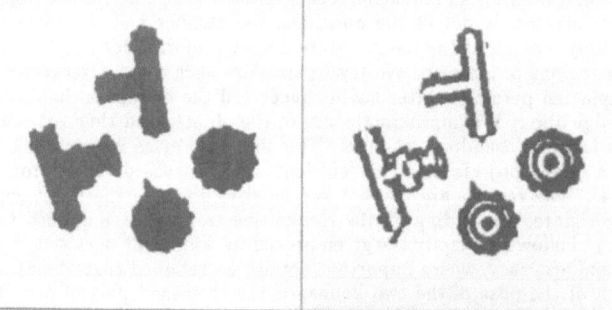

b

c

d

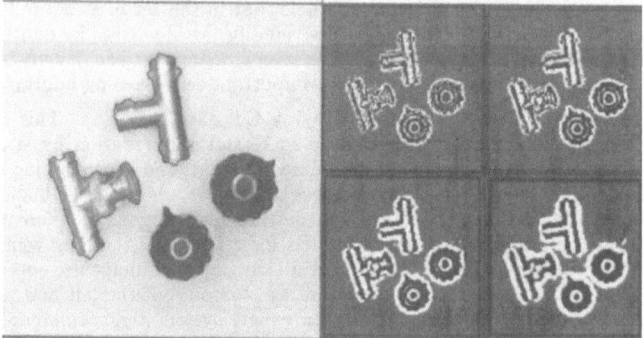

e

f

Figure 4: Detection of industrial objects

a) original image (CT)
b) Laplacian pyramid, σ-parameters:
   top: 1.0/1.4/2.0/2.8, bottom: 4.0/5.6/8.0/11.0
c) recognized left and right object
   extracted from level (8.0)
d) tracking regions from (8.0) to (1.4)
e) 'erosion/dilation'-procedures applied to d)
f) overlay of post-processed contours
   with original image

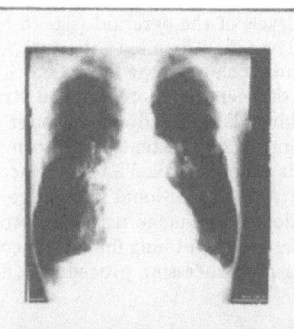

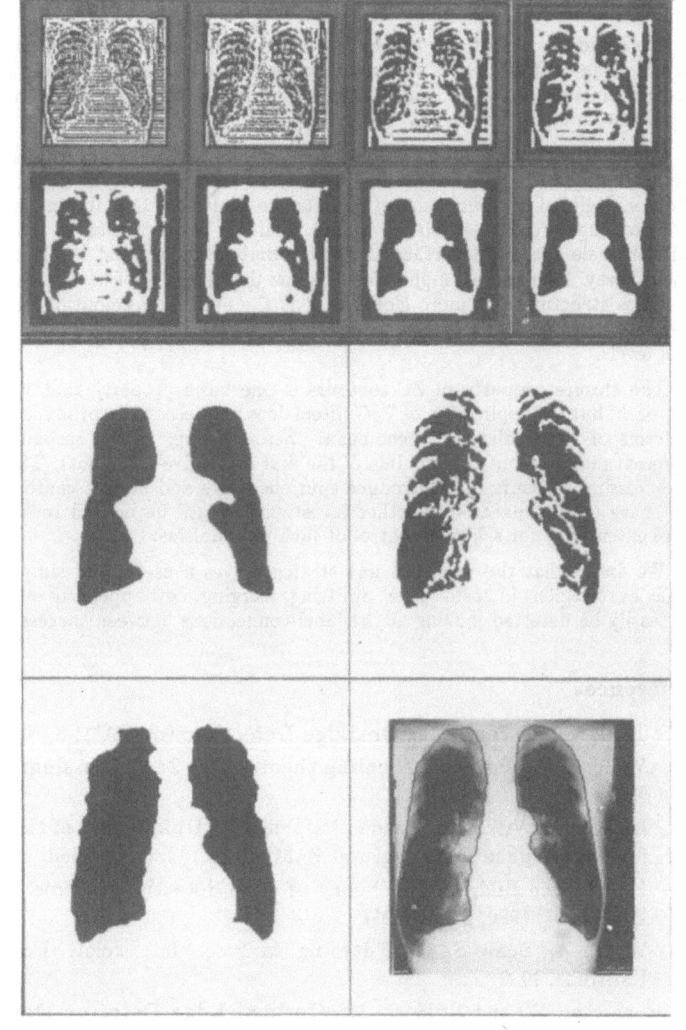

Figure 5: Segmentation of left and right lun

to guide the segmentation. The coarse global shape of the two objects can be detected in low levels of the pyramid (fig. 5 b), as the information at this level is not disturbed by the overlayed fine detailed structure which makes a recognition quite difficult. But the rough boundary shows large deviations from the correct boundary due to both the strong blurring and the merge of more detailed structures which do not belong the the lungs. Coarse to fine tracking allows to adjust the outer contour. First a priori knowledge is used to automatically recognize large, extended and compact left and right objects (fig. 5 c). The linking process yields a more detailed and more accurate description of the main objects (fig. 5 d) and allows to introduce additional knowledge about the desired objects (anatomic shape). Erosion and dilation of the image results in two compact objects (fig. 5 e). The outer boundaries of the left and the right lung must form convex curves, which leads to a refinement of the boundaries using postprocessing procedures (fig. 5 f).

Conclusions

We could show that our concept of using regions bounded by zero-crossings instead of zero-crossing contours offers several advantages in developing a linking scheme. We use the closure principle of zero-crossings as one important property, which is most likely desired in segmentation algorithms.

A coarse to fine tracking of zero-crossings to obtain more accurate edge positions is one interesting field of application. Linking regions always preserves closed boundaries, so that additional global grouping algorithms can be omitted. Our linking procedure applied to level-crossings of the Laplacian-Pyramid results in a description of the behavior of evident units in scale space. It further allows to combine global and local information in a very efficient way. The two examples clearly show the necessity of having global and local access to image structures: A more global task is the efficient recognition of the rough shape of desired structures, whereas detailed information is obtained by tracking boundaries down to finer levels.

The closure property of ZC-contours is one basic property that we use in our linking approach, but the application of $\nabla^2 G$ filters does not necessarily produce correct segmentation in terms of visual distinct phenomena. Zero-crossings of the second derivative not only represent maxima, but also minima of the first derivative (gradient). The closure property of zero-crossing contours may introduce spurious edges and merges contours of nearby regions that have to be separated. Further investigations will be needed to study the split/merge more extensively for a better control of such mechanisms.

We found that the proposed link-strategy gives a useful and simple description of the behavior of regions in scale-space. Splitting, merging, new appearing or disappearing regions will easily be detected looking at the label-connections between successive levels.

References

1. Torre, V. and Poggio, T., **On Edge Detection**, vol. PAMI-8, No.2, March 1986
2. Yuille, A. and Poggio, T., **Scaling theorem for Zero Crossings**, vol. PAMI-8, No.1, January 1986
3. Babaud, J., Witkin, A., Baudin, M., Duda, R., **Uniqueness of the Gaussian Kernel for Scale-Space Filtering**, vol. PAMI-8, No.1, January 1986
4. Marr D. and Hildreth E., **Theory of edge detection**, in Proc. Royal Soc. London, B, vol. 207, 1980, pp. 187-217
5. Witkin A., **Scale-Space Filtering**, in Proc. Int. Joint. Conf. Artificial Intell. Karlsruhe, 1983
6. Lunscher, W. and Beddoes, M., **Optimal Edge Detector Design I: Parameter Selection and Noise Effects / II: Coefficient Quantization**, vol. PAMI-8, No.2, March 1986
7. Serra J., **Image Analysis in Mathematical Morphology**, Academic Press, 1982

A PARALLEL PYRAMIDAL ALGORITHM

TO DETERMINE CURVE ORIENTATION

Ph.Clermont [1]   A.Belaid[2] and A.Merigot [3]

[1]ETCA/LITP 16bis Av. Prieur de la Côte D'Or F94114 Arcueil
[2]CRIN Campus Scientifique F54106 Vandoeuvre les Nancy
[3]IEF Université Paris XI F91405 Orsay

ABSTRACT

We present in this paper a parallel algorithm to orientate curves, which executes efficiently on a pyramidal machine. Independently of its direct interest for image processing, it gives example of the methodology needed to conceive new algorithms for massively parallel architectures. Some deviations are presented, which extend the method to a class of algorithms.

1 INTRODUCTION

As parallel machines are becoming more and more widespread, the problem of conception of parallel algorithm becomes crucial. Among all parallel machines projects, pyramidal systems are of special interest in the area of image processing and computer vision, and algorithmic research for these machines is now developing. The first massively parallel machines developped mainly for image processing were CLIP4 (Duff, 1976), MPP (Batcher, 1980), DAP (Flanders, 1977). These were cellular arrays allowing efficient pointwise or local operations, thanks to their mesh organization. But these structures however lack of mechanisms for global data processing.

This has led to the addition of another dimension of processor network resulting in a new structure: the introduction of hierarchical tridimensional connection. We refer to this structure as pyramidal. The architecture of pyramidal systems is based on a treelike network of processing elements, connected as layered stacked arrays of decreasing size. Each processor is connected to four or eight neighbours in the same layer, one father processor and two or four children.(Tanimoto, 1983; Cantoni, 1985; Scheafer, 1986).

This work is related to the software aspect of a pyramidal machine project named SPHINX, under joint development at the Institut d'Electronique Fondamentale, university of Paris Sud and ETCA defence research laboratory (Merigot, 1983; Merigot, 1984). The software studies include programming methodology for pyramidal machines and adaptation of a high level image processing language called LPSI (Belaid, 1985), developed by the Centre de Recherche en Informatique de Nancy. Much work has been done in the field of finding algorithms to solve classical problems on pyramidal machines. In fact, a large class of quad-tree algorithms (Tanimoto, 1975; Shapiro, 1979) can be adapted and extended to a more general multi-resolution class, well adapted for use on pyramidal machines.(Mult, 1984; Tanimoto, 1986). Hong and al. (Hong, 1983) present a pyramidal algorithm to detect curves with respect to geometric constraints.

2 USING PYRAMIDS TO DETERMINE GLOBAL ORIENTATION OF CURVES

We consider in this paper the problem of orienting curves, because of its intrinsic sequential nature and of its interest in image processing. For example, it provides the basis for Freeman curves representation without need to follow the curve. The algorithm works upon binary images containing pixel wide curves, stored in the bottom layer. The algorithm can handle multiple closed curves, opened curves as well as intersecting ones.

The model of abstract machine considered here is the following: processors have four neighbours in local layer, one father cell and four sons. On layer level, the machine works in SIMD mode, while interlayer mode is MIMD. For such a control, all the processors of a given layer behave in similar manner. In the following, we number the layers starting from bottom layer as zero. The original binary image is stored one pixel in each processor on the bottom layer. The orientation algorithm is recursive: each processor of a pyramid receives from its four sons lists of chains which it tries to merge. Then, it tries to merge the chains of some of its four layer neighbours.

In section 3 we define a total ordering on the processors. It is this ordering that will govern the recursive processes. This relation defines the common orientation of the plane for all the processors. Section 4 presents a first version of the merging algorithm - without considering plane neighbours - and discussions. Section 5 presents the final version - considering as neighboorhood four sons and four plane neighbours-. In these two last sections, we suppose temporarily for clarity that the original image contains only one curve without junction points. Section 6 explains the orienting process, following the merging pass. Section 7 analyses the algorithm on some caracteristic curves.Section 8 explains how to modify the algorithm to deal with multiple curves and with intersecting ones. Section 9 is conclusion and discusses various aspects of algorithm.

3 DEFINITION OF PROCESSOR NUMBERING

We want to number the processors in such a way that, for two adjacent processors (plane neighbours, father or sons neighbours), it is possible to determine their closest common father. At times, we use the term name or coordinate as synonyms for processor number.

3.1 Definition

Def: Given two processors p1 and p2, we define the smallest recovering pyramid (denoted SRP(p1,p2)) as the smallest pyramid whose root is a common ancestor of p1 and p2. In the following, we will confuse a pyramid and its root. See examples in fig. 1.
The numbering is defined recursively, in a pseudo-CSP notation (Hoare, 1978) as:

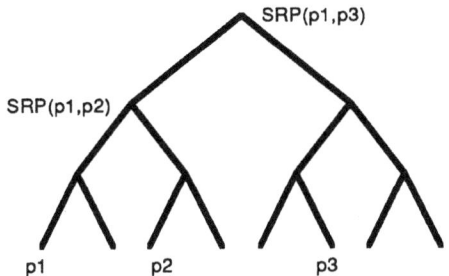

Fig. 1. Examples of SRP function.

- for the top processor of whole pyramid:

```
        number := e;   /* null word */
        son_NW ! 0;
        son_SW ! 1;
        son_SE ! 2;
        son_NE ! 3;
- for others:
        father ? number;
        son_NW ! concat(number,0); /* word number and 0 concenated */
        son_SW ! concat(number,1);
        son_SE ! concat(number,2);
        son_NE ! concat(number,3);
```

This procedure defines a variable length naming for all processors.

3.2 Properties

1- Given two processors p1 and p2 in the pyramid, the name of SRP(p1,p2) is the longest common left part of the names of p1 and p2.

2- The first differing letter after this common part identifies the different son pyramids of SRP to which processors p1 and p2 belong. This defines the following total order relation (<=) on the set of processors:

given 2 processors p1 and p2,

p1 <= p2 iff   - p1 = p2
           or - p1 <> p2 and o(p1) < o(p2)
                where o(p1) and o(p2) represent the two first
                differing letters as defined above.

3- The relation between two neighbours can be locally calculated, depending only on names of neighbours.

3.3 Definition of successors of a processor

We derive here of this order definition a relation between processors. The disadvantage of the order relation is that it cuts the plane at the limit of a half-line: thus, with notations of preceding section, if o(p1) = 3 and o(p2) = 0, we assert p1 < p2 and contradict the intuitive notion of direct rotation.

Given two processors p1 and p2 , we say that p2 is a successor of p1 (denoted p1 << p2) iff:

                - either o(p1) = 3 and o(p2) = 0
                - or p1 < p2.

This formalizes in some way the intuitive idea that one rotates in the positive direction to join p1 to p2. Fig. 2 gives examples.

3.4 Adaptive local orientation

In the former section, we supposed that each processor was ordering its four sons by turning in a conventionnaly positive direction, the same for all processors on the pyramid. We can suppose also that the choice of

0	1	4	5
3	2	7	6
12	13	8	9
15	14	11	10

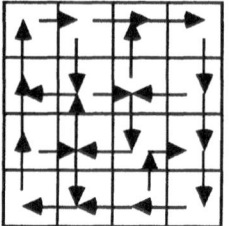

Fig 2. Example of numbering   Restriction of successor relation to plane 4-neighboorhood.

this orientation depends of the processor. Thus, we can define in each processor a boolean variable meaning that the local orientation is positive or negative. If this variable is the result of a calculus on the image, we say we have calculated an adaptive local orientation.This is the basis of the algorithm described here: we construct for each chain in the bottom image a local orientation in each processor of the pyramid.

4 A FIRST VERSION OF MERGING ALGORITHM.

In the two following sections, we suppose to simplify description that the original binary image contains only one curve (closed or opened) without intersections.

4.1 Principle of merging

In the following, each working processor contains lists of subchains. To each processor p corresponds a subimage in the bottom layer, composed of all descendants of p in this layer. The subchains contained in a processor represent the chains obtained by intersecting the original image and the corresponding subimage.

Subchains are represented by coordinates of their extremities in the bottom layer. We have advantage to take coordinates defined by numbering: we have then to keep only the last differing digits of numbers, the first digits being common between these processors. The order of extremities is significant: the first point is the origin of the curve when following it in a positive sense. Thus, merging two subchains can lead to reorientate one of them. (fig 3). The merging process is only concerned with subchains intersecting with some frontiers of the bottom subimage: thus the subchains to be considered for particular merging is a subset of subchain list.

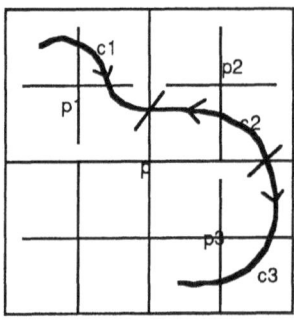

Fig 3. Subchains oriented as determined by processors p1, p2, p3. c2 must be reoriented by p.

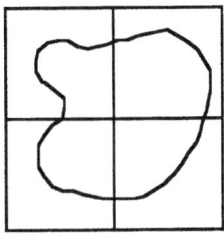
4.a. Smooth curve

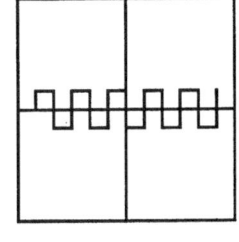
4.b. Snake curve.

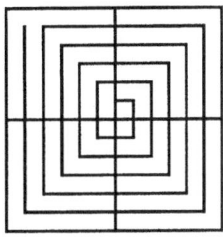
4.c. Spiral curve

Fig. 4. Typical evaluation curves

4.2 Algorithm

The algorithm is recursive: each processor receives from its sons a list of subchains, and merges them. When terminated, it transmits to the sons which subchains were reoriented. We omit in the following the algorithm executed by the bottom layer (send to father trivial chains composed of one point).

```
for layer = bottom + 1 to top do
    for each pe on layer , in synchronous mode, do
        acquire chain lists from sons.
        merge local chains on the basis of positive orientation of
        four sons.
        send to sons the reorientation information.
        send new chains to father, which intersect the subimage
        frontier of pe.
        (* ie chains not wholly contained in pe subimage *)
```

4.3 Remarks

The principle of algorithm implies that a processor may hold many subchains related to the same chain in the bottom layer. Chains wholly contained in a particular subimage are kept in the corresponding processor, which does not transmit them to its father. This algorithm is little efficient on some problems: the spiral, which exhibits intrinsic global properties, and snake-like curve (fig. 4.b) at middle of image.

In this last case, the cost is specially high: we can merge the whole curve only at top layer, and we have a lot of chains (app. N) to merge. In the following section, we propose an improved version of algorithm which solves this kind of problem by adding consultation of layer neighbours for each pe.

5 FINAL VERSION OF MERGING ALGORITHM

5.1 Algorithm

We keep here the same general idea of algorithm. But consultation of plane neighbours allows to do more merging in bottom layers, where the number of processors is more important. Thus, we decrease the number of subchains transmitted to upper layers. When merging subchains between two processors on the same layer, we have to decide which processor will own the new subchain. This is made on the basis of successor relation.

```
for layer = bottom to top do
    for each pe on layer , in synchronous mode, do
```

```
acquire chain lists from sons.
merge local chains.
send to sons reorientation information
(* resolve plane neighbours *)
enable( domaine1) (* see 5-2. *)
for each neighbour pe1 in a 4-neighboorhood do
      if ( pe << pe1)
            modify local list in pe by merging with local list in
            pe1
            remove  from  pe1  merged  chains  and  set  in  pe1
            reorientation information
enable( domaine2)
for each neighbour pe1 in a 4-neighboorhood do
      if ( pe << pe1)
            modify local list in pe by merging with local list in
            pe1.
            remove  from  pe1   merged  chains  and  set  in  pe1
            reorientation information
send new chains to father, which intersect subimage frontier of
pe.
```

The merging pass of lists received from sons is not trivial: we
consider only in a plane a 4-neighboorhood. Thus, diagonal neighbours are
treated in upper layers through this pass.

5.2 Mutual exclusion on local chain lists

Without caution, a processor and its neighbour working in parallel
can modify simulteanously the processor local chain list. This can lead
to the following incoherent result of figure 5. To solve this problem, we
partition the processors of a layer in two classes:
 domain1 and domain2, wich correspond respectively to a chessboard and to
its negative. We allow first the processor belonging to domain1 to work
in parallel, then domain2. Thanks to the 4-neighbourhood considered, we
ensure thus mutual exclusion.

6 GLOBAL COHERENT ORIENTATION OF CHAINS

This step constructs the final data structure representing
orientation, and is bound to the choice of one particular curve, which
determines the local orientation of 2X2 sided subimages contained in
bottom level (corresponding to the sons of each processor in level 0).

This algorithm implies a top-bottom layer data flow.The top
processor sends 1 (positive orientation) to its sons. Each processor in
the underlying layers receives from its father a binary information
representing the orientation as known by the father (ie the relative
order of sons of father cell). Upon consideration of the curve variable
indicating reorientation during merging step, its sends to every son the
same value, upon this decision criterion:

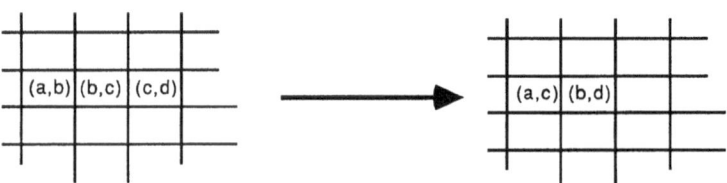

Fig 5. Overlapping subchains incoherent result.

- if it receives from father negative orientation, send to sons the opposite of merging-reorientation variable. This is due to the fact that merging pass was reorienting chains thinking a priori to a positive orientation of sons.
- else send merging reorientation variable.

7 PERFORMANCES STUDY

It is difficult to give general results, because performances are extremely dependant on curves and position of curves in the plane. Then, we will just give general results and analyse on some typical (smooth curve) and worst configurations (spiral, snake). The merging process between two subchains lists can be assimilated to a sort. After a sort, a linear algorithm can be applied to determine merging. The algorithm on a layer then works in $O(n \log n)$, where n is a bound for the number of subchains for the lists in each processor of layer. (depends of the layer) .

We give examples of n calculus on some typical cases:
- Smooth curve (fig 4.a): each processor contains at most 1 subchain. Merging is trivial. The algorithm time is proportionnal to the number of layers. Performance: $O(\log N)$ N= image side. The algorithm is more efficient than curve following.
- Snake curve (fig 4.b): thanks to the plane neighboorhood consultation, merging happens in low level layers, leading to processors containing at most one chain. Result is identical. With algorithm of §4, merging would have happen only at top layer, leading to a number of around N/2 subchains in top processor and a performance at best of $O(N \log N)$.
- Spiral curve (fig 4.c): the number of chains in each processor is of the order of the side of the corresponding subimage: in layer i, $n = 2^i$, merging on layer i requires $O(i\ 2^i)$, and the global time is $O(N \log N)$. A propagation algorithm would be in $O(N^2)$. The consultation of plane neighbours brings here no significant improvement.

8 EXTENSIONS TO MULTIPLE INTERSECTING CURVES

Each processor in this algorithm has to keep a bind between a merged chain and the two corresponding subchains, residing either in itself or in plane or son neighbour, so as to transmit reorientation information. Thus, the same algorithm can deal as well without changes with multiple curves.

We deal with intersecting curves by adding a special attribute to point definition. Junction points are defined as being center of some neighboorhood configurations in the bottom image (for example, center of a cross). This point attribute can be locally calculated in parallel in the bottom layer. We add to each extremity of a subchain a boolean information meaning that this point is a junction point. The merging process is modified so that we don't continue a subchain from an extremity which is a junction point. Thus a junction point is terminal, and an intersecting curve is decomposed in many non intersecting subchains.

9 CONCLUSION

This algorithmic example demonstrates the use of multiple types of data flow over the pyramid, and proves that the SIMD by layer control mode let programmers deal whith relatively asynchronous algorithms, as far as programs on the same layer are not too different. In addition, it gives examples of data structures used when programming cellular machines, and of new distributed representations of classical objects.

It is of interest in image processing, as it allows to give parallel version of Freeman chains construction. The classical sequential algorithm lacks of global information related to curve orientation. The algorithm shown here adds contextual information to the points, and allows thus to determine locally (with access to layer neighbours and to father) which point is next point with respect to curve orientation.

Using appropriate attributes for subchains, we can derive many efficient interesting algorithms: length of subchains, properties of extremities (eg slope)... The algorithm can also be used to differentiate curves in bottom layers: in the final top-bottom step, instead of orienting information, we send numbers, different for each curve. This lead to more efficient algorithms than plane signal propagation techniques.

Acknowledgements
This work was partially supported by French Ministry of Defence DRET/ETCA research laboratory.

REFERENCES

Batcher, K. E., 1980, Design of a Massively Parallel Processor IEEE Transactions on Computers, Vol. C-29, No 9.
A.Belaid, A, Z.Boufriche and R.Mohr, 1985 LPSI: Un langage de programmation structurée en traitement d'images. Congrès AFCET-INRIA-ADI (RFIA) Grenoble Nov. 1985.
V. Cantoni,M. Ferreti , S. Levialdi and F. Maloberti, 1985 A pyramid project using integrated technology, in S. Levialdi (ed.) Integrated technology for parallel image   processing, Academic Press, 1985
Duff, M. L.B. , 1976, CLIP4, a large scale integrated circuit array processor Proc. 3rd ICPR, 1976.
Flanders, P. M., Hunt , D. J., Reddaway , S. F., Parkinson D., 1977 Efficient high speed computing with the Distributed Array Processor, in High speed computer and algorithm organization Kuck D.J., Lawrie H.L.,Sameh A.H. ed., Academic press, 1977.
C.A.R. Hoare, C. A. R., 1978, Communicating sequential processes Comm.ACM 21, 8 (1978).
T. Hong-hong, M.O.Schneier, R.L.Hartley and A.Rosenfeld, 1983 Using pyramids to detect good continuation IEEE transactions on systems, man and cybernetics, vol SMC-13,4,1983.
A. Merigot, 1983, Une architecture pyramidale d'un multiprocesseur cellulaire pour le   traitement des images. These de 3eme cycle.Universite Paris Sud-Orsay. France.
A.Merigot, B.Zavidovique and F.Devos, 1984, Pyramidal algorithms for image processing, 7th ICPR, Montreal, Jul. 84.
Multiresolution image processing and analysis, Ed. A.Rosenfeld, Springer Verlag, 1984.
Linda G. Shapiro, 1979, Data structures for image processing: a survey Computer Graphics and Image processing, Vol. 11,1979.
Scheafer D., May 1986, The GAM system,Communication in workshop held in Maratea, Italy.
S.Tanimoto and T.Pavlidis, 1975, A hierarchical data structure for picture processing Computer Graphics and Image processing,4,1975.
Tanimoto S.L., 1983, A pyramidal approach to parallel processing 10th annual Internatinal Symposium on Computer Architecture 83.
Tanimoto S.L., May 1986, Lecture notes on pyramid algorithm design Communication in workshop held in Maratea, Italy.

PARALLEL IMAGE PROCESSING IN A CSP-ENVIRONMENT:

PERFORMANCE EVALUATION*

A. Chianese, L.P. Cordella, M. De Santo,
A. Marcelli,and M. Vento

Dipartimento di Informatica e Sistemistica
Universita' di Napoli
Via Claudio,21 80125 Napoli

INTRODUCTION

The field of Computer Vision is characterized by the need of processing very large amounts of data in a time that, for many applications, is extremely short. Moreover, many of the algorithms proposed for solving the so called low level vision tasks exhibit a high degree of parallelism while, as far as one ascends to higher perception levels, it becomes evident the need to resort to complex reasoning schemes, where information derived from a variety of sophisticated computations, involving sequential processing, has to be combined with a knowledge base. In any case, all has to be performed fast enough to interact with the real world changes. Hence the demand for computer arrays whose structure reflects the problem's structure, and for powerful tools that allow an optimal mapping of logical to physical architectures.

A number of specialized architectures for processing images have been proposed in the recent past [1], but, because of their specialization, they are not equally efficient in implementing the different parts of a complete vision task [2,3]. The execution of a complex vision task requires performances that could be achieved by subdividing the task into subtasks, and by committing any of these to a suitable architecture. Instead of having different machines in order to attempt a best matching between algorithms and architectures, it seems convenient, for many reasons, to have a system fit for multiprocessing, but reconfigurable [4-6].

To allow both reconfigurability and the possibility of facing complex tasks, it seems necessary that the elementary processors constituting the array have adequate processing and memory capacity. Moreover, it is necessary to define a software and hardware environment that allows communication among processors and among processes, allocation of processes on processors and system reconfiguration. Furthermore, criteria have to be defined that allow to decide about distribution of processes on processors, in order to optimize system performance in any different configuration.

In this framework the Transputer processing unit, by Inmos, and the CSP-like environment Occam, developed for it, seem very interesting [7,8].

_____

* This work was supported by Ministero Pubblica Istruzione

Our approach to the problem starts by taking into account some computational structures defined in terms of process organization and relations among processes, and intends to assess the possibility of efficiently mapping such structures on a Transputer array, in a CSP-like environment. More specifically, we are interested in :
i) assessing the optimal dimension of every physical structure supporting a given computational structure;
ii) evaluating the cost of reconfiguring the array in order to match different computational structures.

Therefore we started a study to ascertain the actual possibility of profitably using the quoted processors and their communication environment, in order to realize a reconfigurable multiprocessor system, especially suitable either for low and high level processing typically required in the field of image understanding.

Our first aim has been to evaluate the performance of a system of processors as a function of the type and number of processes allocated on every processor, and of the different possible communication requests among processes. We illustrate here a model that allows to evaluate the performance of a two dimensional array of mesh connected Transputers, and we refer about the results of a set of measures devised for validating it. We also discuss the practical implications of the model.

ARRAY ARCHITECTURE

The today's configuration of our system is made by an acquisition and monitoring device, the Transputer Array Processor (TAP) [9] and a Vax 750 host, supporting an Occam based design and development environment.

TAP consists of a two dimensional adjustable size array of processors, based on the Transputer. The Transputer has been designed with a language first approach, using the Occam language. The Occam provides a computational model substantially similar to the CSP [10], which allows to implement an algorithm as a set of concurrent processes communicating through channels [11]. One of the most interesting feature of the Occam/Transputer environment is the possibility it provides to develop the project of a parallel algorithm by three successive steps. As a matter of fact, it is possible to enter a first step, whose aim is to realize the logical structure of the program by decomposing the algorithm into a set of processes and by defining its topological structure. The second step is devoted to test the correctness of the logical organization of the processes, through the simulation tools provided by the Occam environment. During both these steps, the programmer has to take care only of the logical structure of his project. Eventually, when the third step is entered, the programmer has to face the non trivial problem of mapping the logical structure on the physical one. An efficient mapping requires both the allocation of processes on processors and of logical communication channels on physical links. This mapping becomes more and more difficult as the complexity of the logical structure increases. In order to free the programmer from this problem, and to take into account that an optimal mapping really affects both efficiency and flexibility of the communication mechanism, we have realized an extension of Occam, implemented by primitives. We describe here two of them, developed under the following hypotheses: i) the logical environment is made by a 2D array of eight- connected processes ; ii) any process can communicate with any other process within a window of variable size; iii) any process can require information about the state of all the processes of the array; iv) the physical environment is realized by a 2D adjustable size array of four-connected Transputers.

The mentioned extension allows to support an important class of computer vision algorithms that mainly use local operations. These

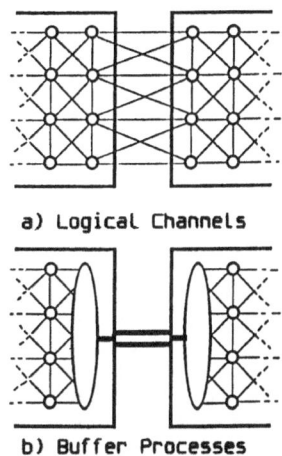

a) Logical Channels

b) Buffer Processes

FIG. 1 - Logical/Phisical
          Correspondence

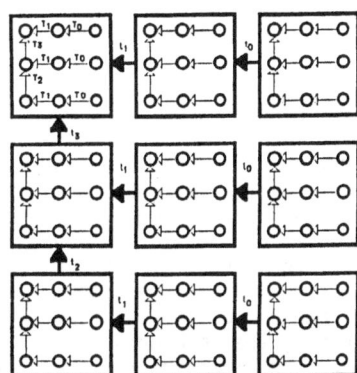

FIG. 2 - The Convergence
          Mechanism

operations allow to modify the value of any pixel of the image array, depending on its current value and the value of its neighbours (often limited to the pixels belonging to a 3x3 window centered on the pixel itself). Moreover, they allow to verify any global condition (i.e. conditions relative to the state of the whole array or to the state of a window larger than 3x3) which can control processing. To implement the algorithms belonging to this class, the system has to be able to perform the following basic operations: i) neighbours value acquisition; ii) pixel value processing; iii) global convergence verification.

The implementation of the primitives is based on a model in which every pixel of the image array is associated to a process.

Primitive for Neighbours Acquisition

Neighbours acquisition involves communication between the process P_i operating the acquisition and the set of processes $\{Q_j\}$ which provide neighbouring pixel values. The communication is implemented in different ways, depending on the allocation of the processes P_i and Q_j on the processors. More precisely, if P_i and all the Q_j processes are allocated on the same processor, the communications are implemented using logical channels directly managed by Occam. On the contrary, if at least one of the Q_j processes is allocated on a different processor with respect to P, implementing communication requires the use of physical links. Depending on the number of processes allocated on a processor, more than one channel has to be mapped on the same physical link. If this is the case, the execution of the primitive activates a process which handles a buffer to multiplex the channels on the link (see fig. 1).

Primitive for Global Synchronization

Often, in parallel processing, algorithms need periodic evaluation of a global convergence function (GCF) in order to determine if the computation must be terminated or not. In many cases the GCF is a logical expression whose terms are variables associated to single processes. The implementation of such a global convergence mechanism on TAP requires the acquisition of the value of the variable associated to each process, the evaluation of GCF, and the passing of the calculated value to all the processes, which are in this way enabled to decide about their own

363

termination. Our primitive first evaluates the convergence function relative to the processes which are on a single Transputer and then the GCF. Figure 2 shows the way the evaluation of GCF is realized. Note that, for practical reasons, in its first implementation, the GCF has not been optimized; however this is not meaningful for our scopes.

THE MODEL FOR PERFORMANCE EVALUATION

One of the basic problems when dealing with non conventional architectures, is that of defining suitable tools for performance evaluation. Evaluating the performance of a system implies the selection of a set of parameters adequately characterizing the system under evaluation with respect to a given type of performance, and the definition of the functional relationship linking the quality of the performance to those parameters. In this way an analytical model describing the performance of the machine can be obtained. The model can be used to dimension the whole machine, since it allows to determine the minimum number of processors needed to maximize the performance, for a given structure of processes.

With reference to our system, we have assumed as significant parameters the degree of parallelism among processes and the number of processors available. The former parameter is represented by the linear dimension D_p of the array of processes, while the latter is represented by the linear dimension D_t of the array of processors.

Our goal is to determine the functional relationship between the response time and the chosen parameters, once an algorithm expressed in terms of a set of communicating sequential processes is given. To this purpose we need to characterize any process in terms of the basic operations it requires.

In the following, we will indicate by:
-T^*_{tot}, the response time of a D_t^2 mesh connected transputer array running an algorithm decomposed in a two dimensional array of D_p^2 processes;
-T^*_{cpu}, that part of T^*_{tot} due to the execution of cpu operations involved by the algorithm;
-T^*_{na}, that part of T^*_{tot} due to the execution of neighbours acquisition operations involved by the algorithm;
-T^*_{gc}, that part of T^*_{tot} due to the execution of the global convergence operations involved by the algorithm.
We assume that:

$$T^*_{tot}(D_p,D_t) = T^*_{cpu}(D_p,D_t) + T^*_{na}(D_p,D_t) + T^*_{gc}(D_p,D_t) \tag{1}$$

Note that the characteristics of our environment allow the decomposition of an algorithm of the considered class into a set of identical concurrent processes. Moreover, the correspondence between the transputer mesh structure and the process logical organization, allows a uniform distribution of the computational demand among processors. In these hypotheses, it is possible to disregard the overhead of synchronization among processes. Then, after the first communication among processes has taken place, the response time is the same for each process, so that the time to process the same number of equal processes is the same for each processor. In this way the response time of the transputer mesh is equal to the response time of any of the component processors.

If we assume that T_{tot}, T_{cpu}, T_{na}, and T_{gc} are the parameters respectively corresponding to T^*_{tot}, T^*_{cpu}, T^*_{na}, and T^*_{gc}, but referred to a single processor of the mesh, we have:

$$T^*_{tot} \equiv T_{tot} = T_{cpu}(D_p, D_t) + T_{na}(D_p, D_t) + T_{gc}(D_p, D_t) \tag{2}$$

The dependence of any of the terms of (2) from D_p and D_t can be expressed in the following way.

If T_{seq} is the cpu time required for the execution on a single processor of the whole set of processes constituting the algorithm, we have:

$$T_{cpu} = T_{seq} / D_t^2 \tag{3}$$

Let N_{gc} be the number of global convergence operations executed by a process and T'_{gc} the time needed for the execution of any of these operations by the set of processes allocated on a same processor. The time spent by a processor to execute the requested global convergence operations is given by:

$$T_{gc} = N_{gc} * T'_{gc} \tag{4}$$

With reference to the convergence mechanism illustrated in fig. 2, the global convergence function (GCF) requires the collection of data belonging to processes allocated on the same processor and, in a second time, the collection from different processors.

Let be $r = D_p/D_t$.

For each processor, we have r rows and $(r - 1)$ data exchanges for each row; furthermore we have to consider $(r - 1)$ data exchanges corresponding to the collection along the first column. Therefore the first part of the collection mechanism, that works in parallel on each processor, requires $(r(r - 1) + r - 1)$ data exchanges. If k_{1gc} is the time needed to execute a single exchange (in our hypotheses this time can be regarded as constant) we can write the following expression:

$$T'_{1gc} = k_{1gc} [r (r - 1) + r - 1] = k_{1gc} (r^2 - 1) \tag{5}$$

The second part of the collection mechanism introduces a term corresponding to the time for exchanging data among processors belonging to the same row, and another term taking into account the collection on the first column. Data exchanges along different rows are executed in parallel.

Let now K_{2gc} be the time needed for the execution of a single data exchange between two processors. We have:

$$T'_{2gc} = 2 k_{2gc} (D_t - 1) \tag{6}$$

which yields:

$$T'_{gc} = T'_{1gc} + T'_{2gc} = k_{1gc} (r^2 - 1) + 2 k_{2gc} (D_t - 1) \tag{7}$$

By substituting (7) into (4), we obtain the wanted explicit expression of T_{gc}.

In order to evaluate the expression of T_{na} the following considerations hold. If we denote by N_{na} the number of neighbours acquisition operations executed within a process and by T'_{na} the time needed for the execution of any of these operations by the set of processes allocated on a same processor, we can write:

$$T_{na} = N_{na} * T'_{na} \tag{8}$$

The neighbours acquisition time is composed by the time needed for communicating among processes allocated on the same processor, and by the time relative to processes allocated on different processors.

As regards the first term, consider that all the processes allocated on a single processor run sequentially and that $(r - 2)^2$ is the total number of processes communicating only with processes allocated on the same

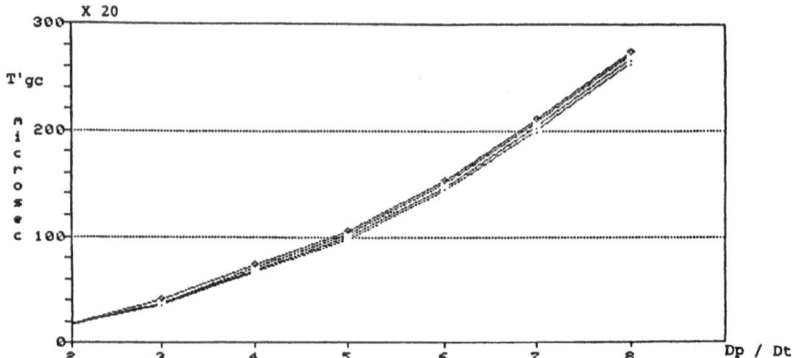

FIG. 3 - Plot of T'_{gc} as a function of r, for values of D_t ranging from 2 to 5.

processor.

If k_{1na} is the time corresponding to a single communication between two processes on the same processor, we have:

$$T'_{1na} = k_{1na} \ (r - 2)^2 \tag{9}$$

To derive the expression of the second term, let k_{2na} be the time corresponding to a single communication between two processes allocated on different processors. The number of processes operating this kind of communication is $4 \ (r - 1)$, and therefore we have:

$$T'_{2na} = 4 \ k_{2na} \ (r - 1) \tag{10}$$

which yields :

$$T'_{na} = T'_{1na} \ + T'_{2na} \ = k_{1na} \ (r - 2)^2 \ + 4 \ k_{2na} \ (r - 1) \tag{11}$$

EXPERIMENTAL RESULTS

In order to validate the proposed model, we realized a set of tests that allows to determine the value of the terms in equation (2), and to verify the two fundamental hypotheses of our model: i) the independence of the response time from synchronization time among processes and ii) the dependence of every time component characterizing the system, only on the parameters r and D_t.

A first test, whose aim was to verify the adequacy of the model as far as T'_{gc} is concerned, has been realized by implementing a suitable algorithm only constituted by global convergence operations. Figure 3 shows the measured times for different values of D_t and r. The figure points out the quite perfect overlap among the curves, confirming the independence of T'_{gc} from D_t, as we assumed. Furthermore, each curve shows a squared trend indicating that the term T'_{2gc} is negligible with respect to T'_{1gc} , within the examined range of variation of r.

A second test devoted to ascertain the adequacy of the model as regards T'_{na}, has been implemented by running an algorithm exclusively made by neighbours acquisition operations. Figure 4 shows the trend of T'_{na} with respect to the variation of the r ratio, assuming D_t as a parameter. In this case, the perfect overlap of the four curves, in the considered range, highlights that the parameter D_t weakly affects T'_{na}. Furthermore, although it is not evident in figure 4, T'_{na} is more influenced by the linear term of the equation (11) if r is less than 6, while the dependence is more similar to a squared one if the value of r

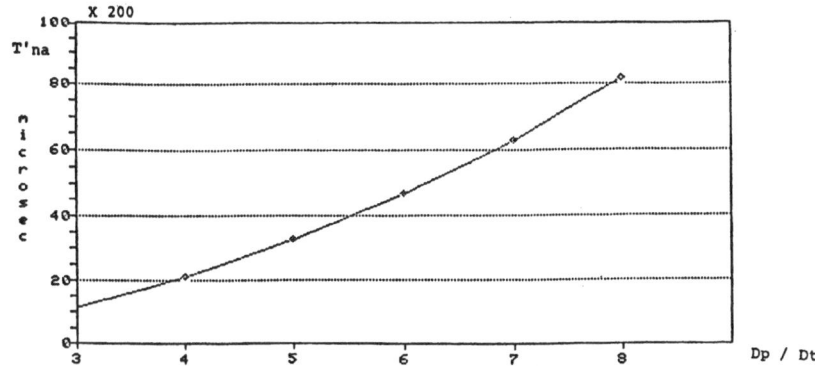

FIG. 4 - Plot of T'$_{na}$ as a function of r, for values of D_t ranging from 2 to 5. Note that the four curves perfectly overlap

is greater than 6. As a consequence, in the equation (11), neither the first term nor the second one are negligible, in the examined range of parameter values.

The figures fully confirm the proposed model which estimates a linear dependence of T_{gc} and Γ_{na} on D_t, and a squared one on r.

DISCUSSION AND CONCLUSION

As it has been anticipated in the introduction, one of the main purposes of the proposed model is to allow the sizing of the devised distributed machine, in order to get a given performance.

The response time of the mesh for executing a certain set of processes, can be expressed as:

$$T_{tot} = (T_{seq} / D_t^2) + f(D_p, D_t) \tag{12}$$

obtained by substituting T_{cpu}, T_{na} and T_{gc} in equation (2) and grouping into $f(D_p, D_t)$ all the terms related to the communication times.

In this equation T_{seq}, due to its definition, can be considered invariant for our problem, whilst the second term of (12), represents the overhead introduced by the communication mechanism.

We want to note explicitly that the communication mechanism is used in different moments and in different ways, depending on the processes logical organization, as well as on the mapping between logical and physical structures and on the type of operations the algorithm requires. In ideal conditions, when the overhead due to the communication mechanism is zero, the response time could be reduced by increasing the number of processors, as suggested by the first term of (12). In the real case, such overhead is not zero and therefore it leads to decrease of the system performance. Obviously, the influence of such overhead upon the response time increases as much as the ratio between number of communication operations and number of cpu operations increases. The analytical expression of the function $f(D_p, D_t)$, can be evaluated through the model, once the parameters N_{na}, N_{gc}, D_p and all the k's have been set. In figure 5 the two terms of equation (12) are qualitatively plotted with respect to D_t: the former exhibits a decreasing trend, while the latter, at least since a certain point, is monotonically increasing, as it can be derived by their analytical expressions.

The model can be used for dimensioning the array. In fact, if we denote by T the desired response time, and by D_p^2 the number of processes of a given algorithm, figure 5 shows how, when increasing the number of

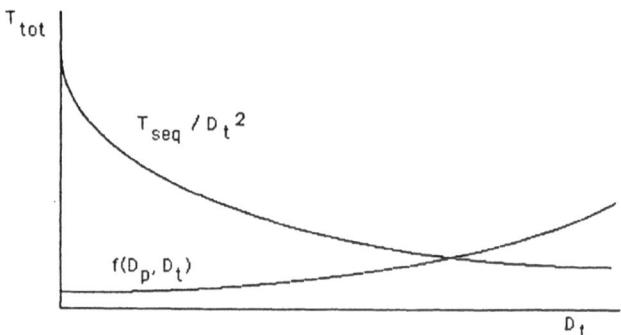

FIG. 5 - The dependence of T_{tot} on D_t can be obtained by
linearly combining curves of the illustrated type.

processors, the response time decreases according to the first term of
(12), but at the same time, it increases, due to the increasing overhead
in communication management. Then, by simple derivative operations, it is
possible to determine the optimal value of D_t , that is the value over
which the ratio between the percentage increase of D_t and the percentage
decrease of T_{tot} is over a given threshold. This threshold is obtained by
characterizing the algorithm in terms of D_p, T_{seq}, N_{ns}, and N_{sc}, and by
using the model to calculate the value of $f(D_p, D_t)$.

ACKNOWLEDGMENTS

The IRSIP of CNR is gratefully acknowledged for providing the Transputers
used in the experiments.

REFERENCES

[1] L. Uhr, "Algorithm-Structured Computer arrays and networks",
     Academic Press 1984
[2] L. P. Cordella, M.B. Duff and S. Levialdi "Thresholding: a Challenge
     for Parallel Processing ", C.G.I.P., Vol.6, n. 3, June 1977, pp.
     213-220
[3] M.B. Duff, "Pyramids - Expected Performance", Proc. of the NATO ARV,
     Maratea, Italy, May 1986
[4] H.J. Siegel   and others, "PASM: A Partitionable SIMD/MIMD System
     for Image Processing and Pattern Recognition", IEEE Trans. on
     Comp., Vol. C-30, no.12, 1981, pp. 934-946
[5] S. Yalamanchili and J.K. Aggarwal, "Reconfiguration Strategies for
     Parallel Architectures", Computer n.12, December 1985, pp.44-61
[6] A. Nikitas, "Adaptable Software and Hardware: Problems and
     Solutions", IEEE Trans. on Computers, n. 2, Feb. 1986, pp. 29-39
[7] INMOS "Occam Programming Manual", Prentice Hall, 1984
[8] INMOS "Transputer Reference Manual", Sept. 1985
[9] A. Chianese, M. De Santo, and M. Vento, "TAP: Transputer Array
     Processor for Computer Vision", Proc. of IASTED Conf. on Computer
     and Their Applications for Development, Taormina, Italy, Sept. 3-5,
     1986, pp. 55-59
[10] C.A.R. Hoare, "Communicating Sequential Processes", Comm. of ACM,
     Vol. 21, no. 8, Aug. 1978, pp. 666-677
[11] D. May, "Communicating Sequential Processes Transputer and
     Occam", Esprit Summer School on Future Parallel Computer, Apr. 1986

IMAGE RESTORATION BY FAST LOCAL CONVOLUTION

Erkki Oja and Jouko Lampinen

Kuopio University
Department of Computer Science and Mathematics
POB 6, 70211 Kuopio, Finland

ABSTRACT
A local spatial convolution filter for the restoration of one- and two-dimensional signals is suggested whose design is based on approximating any global linear restoration filter, which might be the Wiener, pseudoinverse, constrained least squares, or projection filter. The local filter provides a restoration that is as close as possible to the global restoration. It is shown by an example using a blurred standard image that the restorations are satisfactory even when the filter size is quite small. Quantitative properties of the suggested localization filter are discussed.

1. INTRODUCTION

A number of global restoration filters have been proposed for restoring blurred images contaminated by noise, either linear like the Wiener, pseudoinverse, and constrained least squares filters /1/ or nonlinear /9/. For a recent review, see /6/. Recently, the Projection Filter (PF) and the Parametric Projection Filter (PPF) were introduced /4,6/. The latter one minimizes a combined criterion in which one part is the error in the noiseless restoration compared to the original image and the other part is the energy of additive noise passed through the restoration.

A problem in the actual implementation of even the linear global type of restoration filters, like the Wiener and the PPF filters, is that at best the computations are carried out in frequency domain with FFT techniques if circulant approximations are made throughout to the shift invariant degradation operator and the additive noise autocovariance. Usually this involves complete 2-D FFTs of very large matrices. On the other hand, most practical convolution filters for e.g. edge detection or noise cleaning are of the local neighbourhood type, applied directly in spatial domain and allowing very fast implementation with special signal processors. Such a property would be highly desirable for image restoration tasks, too.

Various methods have been suggested earlier for local spatial domain restoration. A least squares fit of the restored object can be made with the original object /3/;

the central core of the composite point-spread function can be made small /8/; or its largest sidelobe can be minimized /2/. A straightforward method is to use impulse response truncation, possibly with a smoothing window /7/.

In /5/, the present authors proposed a local approximation to the global linear restoration filters that takes advantage of the design principles of those filters. The localized filter suggested here provides a restoration that is as close as possible to the global restoration in the topology of the usual l^2 norm. This ensures closeness to the original image, too.

In addition to reviewing in Section 2 the basic localization solution already given in /5/, the present paper emphasizes the procedure as valid for *any* linear global filter, gives in Section 3 a shortcut method to solve the filter coefficients which reduces computational complexity in filter design, and gives in Section 4 new results on the filter performance.

2. THE LOCAL FILTER AS A LEAST SQUARES APPROXIMATION

We denote the original $N^2 \times 1$ stacked image vector by f and the observed image, blurred by the linear $N^2 \times N^2$ shift invariant degradation operator A and additive noise n, by g:

$$g = Af + n \tag{1}$$

The restored image f_1 is obtained from g by a linear restoration operator B_1 :

$$f_1 = B_1 g \tag{2}$$

where f_1 is also a stacked image vector and B_1 an $N^2 \times N^2$ matrix.

Let now $f_2 = B_2 g$ be the restoration achieved with a shift invariant local spatial filter B_2, and denote by $v = f_1 - f_2$ the difference between the local and global PPF restorations. The problem

$$\|v\| = \|f_1 - f_2\| = \|B_1 g - B_2 g\| = minimum \tag{3}$$

has a straightforward solution in frequency domain /5/. Assuming circulant approximations throughout, both B_1 and B_2 are diagonalized by the block Fourier matrix W /1/ and we obtain by multiplying $B_1 g = B_2 g + v$ by W:

$$D_1 G = D_2 G + V \tag{4}$$

with $D_1 = W B_1 W^*$, $D_2 = W B_2 W^*$, $G = Wg$, and $V = Wv$. Matrices D_1 and D_2 are diagonal. Equivalently, $D_1 G$ can be written as

$$D_1 G = D_g L_1 \tag{5}$$

where D_g is the diagonal matrix whose diagonal elements are the elements of G and L_1 is an $N^2 \times 1$ vector made up from the diagonal of D_1 . When L_2 denotes the vector on the diagonal of D_2 , (4) yields

$$D_g L_1 = D_g L_2 + V. \tag{6}$$

Since B_2 is a local filter,its DFT only contains part of the possible frequencies.

Hence, L_2 is of the form $L_2 = W_t h$ with W_t a submatrix of W and h the stacked vector of spatial filter coefficients. We have

$$D_g L_1 = D_g W_t h + V. \tag{7}$$

Now the least squares problem (3) is equivalent to

$$\|V\|^2 = \|D_g L_1 - D_g W_t h\|^2 = minimum \tag{8}$$

which corresponds to the normal equations for the unknown h:

$$(W_t^* D_g^* D_g W_t) h = W_t^* D_g^* D_g L_1. \tag{9}$$

This gives the point coefficients of the local spatial PPF. On the other hand, if the original image is regarded as a sample from some random image f, then we can take averages with respect to both f and the additive noise n and minimize $E_{n,f} \|f_1 - f_2\|^2$ instead of (3). It follows from above that the solution will be in this case (provided that the coefficient matrix is nonsingular)

$$h = (W_t^* K W_t)^{-1} W_t^* K L_1 \tag{10}$$

with $K = E_{n,f}(D_g^* D_g)$. On the diagonal of matrix K there is the power spectrum of the degraded image g, averaged over the noise and image fields n and f.

3. COMPUTATION OF THE LOCAL FILTER COEFFICIENTS

The actual computation involves the following steps: constructing the truncated block Fourier matrix W_t ;either computing $D_g^* D_g$ or estimating the power spectrum matrix K using a suitable statistical model; computing the global filter in Fourier domain with proper values for the pertinent parameters (e.g. in case of the PPF);and solving h from Eq.(9) or (10). For simplicity, results are presented only for the case that the point spread function of the degradation A is symmetric, e.g. Gaussian blur or rectangular symmetric motion blur.

1-D Localization

In some special cases restoration can be carried out in one dimension, for example if the degradation is a linear motion in row or column direction. Since the degradation PSF was assumed symmetric, only the non-redundant $N/2 + 1$ first elements are needed in the normal eguations (9). Thus L_1 is an $(N/2 + 1) \times 1$ vector containing the frequency response of the basic filter and h is an $(L + 1) \times 1$ vector containing the $L + 1$ nonredundant filter coefficients of the local filter of length $2L + 1$. The truncated Fourier matrix W_t is an $(N/2 + 1) \times (L + l)$ matrix whose elements are

$$W_t(i, j) = s(i) * cos(2\pi i j / N), i = 0, \ldots, N/2; j = 0, \ldots, L, \tag{11}$$

where $s(i)$ is the scaling function

$$s(i) = \begin{cases} 1, & \text{if } i = 0; \\ 2, & \text{if } i > 0. \end{cases} \tag{12}$$

The term $D_g^* D_g$ is an $(N/2 + 1) \times (N/2 + 1)$ diagonal matrix containing the nonredundant part of the power spectrum of the degraded signal or image on the diagonal. It can be computed directly with FFT or estimated using a suitable statistical model for f. We used a 1st order Markov field approximation for f, yielding the power spectrum of f analytically and hence also the power spectrum of g because A is known. This saves one FFT computation. In the 1-D case the saving is in fact not important but in the 2-D case, to be discussed below, it is considerable.

2-D localization

In the two dimensional case images are $N \times N$ matrices. Defining images as columnwise stacked vectors yields the vector form

$$F = Wf \tag{13}$$

where F and f are $N^2 \times 1$ vectors and the $N^2 \times N^2$ transform matrix W is the Kronecker product of columnwise and rowwise transform matrices $W1$ and $W2$. In the case of the Fourier transform of a symmetric 2-D sequence, the truncated transform matrix is

$$W_T(I, J, i, j) = W_t(I, J) * W_t(i, j) = s(I)cos(2\pi IJ/N) * s(i)cos(2\pi ij/N)$$

$$I, i = 0, \ldots, N/2; J, j = 0, \ldots, L, \tag{14}$$

where I and J are the indices of the block, i and j are the indices of the element in the block, and W_t is a 1-D Fourier cosine transform matrix, according to Eq. (11).

In the 2-D case, it is now the truncated block matrix W_T given in Eq. (14) that must be used in the normal equations (9) or (10) in place of W_t. Let us denote the matrix $W_t^* D_g^* D_g W_t$ in the normal equation by T and the right side $W_t^* D_g^* D_g L_1$ by b. Now T is an $(L+1)^2 \times (L+1)^2$ block structured matrix, with $(L+1) \times (L+1)$ size blocks. Writing out the matrix products in (9) yields an equation for the elements of T and b:

$$T(I, J, i, j) = \sum_{k,l=0}^{N/2} W_t(k, I)W_t(k, J)x(k, l)W_t(l, i)W_t(l, j), \tag{15}$$

$$b(I, i) = \sum_{k,l=0}^{N/2} W_t(k, I)W_t(l, i)x(k, l)L_1(k, l), \tag{16}$$

where I, J, i, j are the indices of the block and the element in the block, respectively. x denotes now the nonredundant part of the power spectrum of the degraded image and L_1 is likewise part of the frequency response of the basic global filter.

Computation of the double summation in eq. (15) takes a large amount of CPU time. Since the computation is carried out off-line only when designing the filter, the time may not be critical. Still, it is convenient that with some manipulation of eq. (15) the computation time can be greatly reduced. This reduction will now be outlined.

We denote by X the Fourier transform of the power spectrum x of the degraded image. Since x is symmetric, the Fourier transform and the inverse transform are formed identically, and thus X is the autocorrelation of the degraded image. Also we denote the 1-D Fourier transform of the j-th column of x by $X_i(k, j)$, where subscript i defines the direction of the transform, j is the index of the column and k is the discrete spatial frequency. Similarly, we denote the Fourier transform of the i-th row of x by $X_j(i, l)$, with subscript j the direction, index i the row, and index l the frequency.

By definition we have

$$\sqrt{1/N}X_i(i,j) = x(0,j) + (-1)^i x(N/2,j) + 2\sum_{k=0}^{N/2-1} x(k,j)cos(2\pi ki/N), \qquad (17)$$

$$\sqrt{1/N}X_j(i,j) = x(i,0) + (-1)^j x(i,N/2) + 2\sum_{k=0}^{N/2-1} x(i,k)cos(2\pi kj/N), \qquad (18)$$

and

$$W_t(i,j) = s(j)cos(2\pi ij/N).$$

After some manipulation we can write eq. (15) in the form

$T(I,J,i,j) =$

$f(I)*f(J)*f(i)*f(j)*(N*(X(I-J,i-j)+X(I-J,i+j)+X(I+J,i-j)+X(I+J,i+j))$

$+2\sqrt{N}*(X_j(0,i-j)+X_j(0,i+j)+(-1)^{I+J}X_j(N/2,i-j)+(-1)^{I-J}X_j(N/2,i+j)+$

$X_i(I+J,0)+X_i(I-J,0)+(-1)^{i+j}(X_i(I-J,N/2)+X_i(I+J,N/2)))$

$+4*[x(0,0)+(-1)^{I+J}x(N/2,0)+(-1)^{i+j}(x(0,N/2)+(-1)^{I+J}x(N/2,N/2))]]). \qquad (19)$

This equation includes the autocorrelation of g, the Fourier transforms of the 0-th and the $(N/2)$-th rows and columns of x and the corners of the nonredundant power spectrum $x(i,j)$, $i,j = 0$ or $N/2$.

If a statistical model for image autocorrelation is used, it is possible to compute X, X_i and X_j analytically. However, since X is the autocorrelaton of the degraded image, the degradation PSF must be convolved with the image autocorrelation to yield X. A more practical method is to construct a model for the image power spectrum, multiply it by the power spectrum of the degradation PSF and Fourier transform it with FFT.

The double summation of eq. (15) was replaced by FFT computa- tion in eq. (19). Thus the computational saving is of the same order as direct 2-D DFT as compared with FFT, which is quite considerable for practical size images.

Summary of the Algorithm

The main steps of the localization procedure were:

1) Construction of the basic global filter (and storing the DFT of the PSF for step 2)

2) Computing the power spectrum of the degraded image g by multiplying the power spectrum of the degradation PSF and that of the original image (using e.g. a Markov random field model for the original image).

3) Assembling the normal equations from eq. (16) and (19).

4) Solving the normal equations, possibly using pseudoinversion techiques.

Computational Complexity of Local vs. Global Filtering

The computational complexity of local convolution filtering is now compared to that of Fourier domain filtering using FFT-techiques, to show the advantage gained with the localization. The filter coefficients are assumed to have been computed beforehand.

Usual degradation PSFs are symmetric like defocusing blur, or can be made symmetric, like motion blur, with rotation and shift of the origin. For such PSF, localization of the standard filters will produce a symmetric convolution filter, having only half a quadrant of independent coefficient values. Computation of the convolution requires then $(L^2 + L)/2$ multiplications per pixel, where L is the size of the mask, including the zero element. For a 5×5 filter, for example, $L = 3$ and 6 multiplications are required.

Implementing the filter in Fourier domain will require three 2-D Fast Fourier Transforms. Since the image is real and the degradation PSF is real and symmetric, the number of multiplications in FFT can be reduced resulting in the total number of about $8 \log N$ per pixel.

The computational effort for a 256 by 256 image on an ordinary serial computer is plotted in Fig.1 as a function of L. The horizontal line in the figure corresponds to the Fourier domain filter. The figure shows that the practical range of L, where the local filter is considerably faster than the FFT-techniques, is roughly $L < 7$ in this case. For larger images, the range of L will be larger. With special convolution processors or parallel processor arrays the limit of L is somewhat higher, but it is largely dependent on the hardware and algorithms used.

4. RESTORATION RESULTS

The 256×256×8 bit digitized "Mandrill face" was blurred by Gaussian defocusing blur. The point spread function decreased to about 0.35 of the peak value at a distance of 1.1 pixels. The only noise was quantization noise. Part of the blurred image is shown in Fig. 2a. The images were zoomed to the eye area to make individual pixels visible. Global restoration using the Parametric Projection Filter /6/ with a nearly optimal value for parameter γ is shown in Fig. 2b. For such a modest blur and low noise level, the result is quite good as expected; results on global PPF restoration

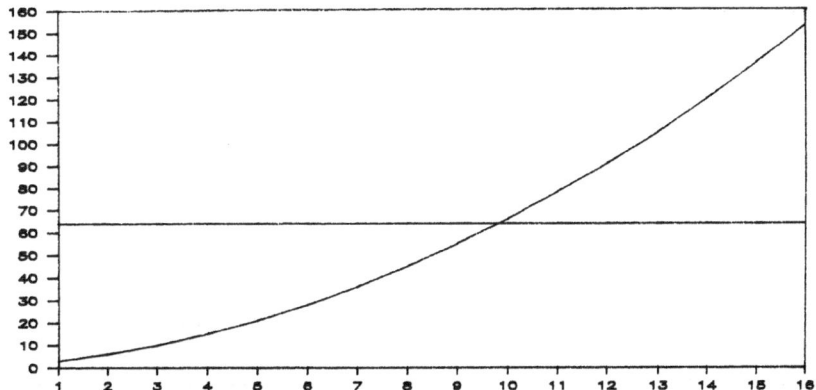

Fig. 1. Computational complexities of local filtering (curved line) and global filtering (horizontal line) as functions of filter length L. Vertical axis: number of multiplications per pixel. Horizontal axis: filter length L.

Fig. 2. Image restoration. a) Zoomed part of the blurred image. b) Global restoration with the PPF filter. c) Local restoration with the corresponding 5×5 spatial convolution filter.

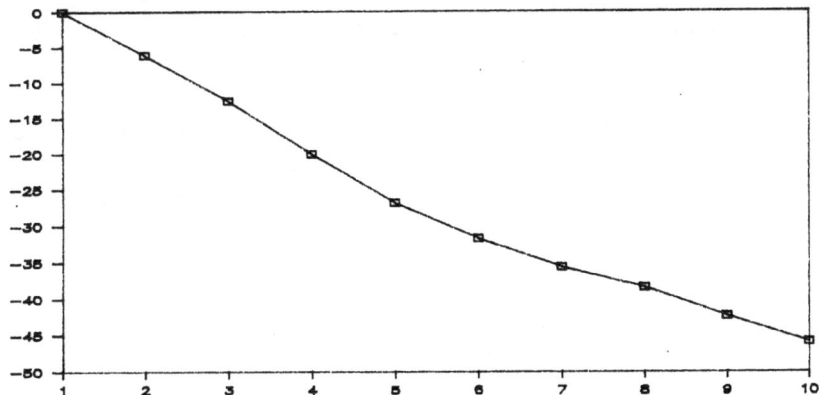

Fig. 3. Localization error as function of filter length. Vertical axis: error in log scale, 0 dB level corresponding to no filtering. Horizontal axis: filter length L.

375

of more drastic degradation were given in /6/. Fig. 2c shows the restoration with a symmetric local PPF of total size 5×5, whose coefficients were computed from Eq. (10). The power spectrum (or matrix K) was approximated from a Markov random field model.

Since the aim of restoration is, usually, to produce a suitable image for human observation, there are no simple mathematical criteria to measure the quality of restoration quantitatively. We computed the localixation error as a norm of the difference of the restoration results with the global Fourier domain filter and the corresponding local filter. The measure that was used is the ordinary Euclidean norm, since it is the mini- mization criterion in the filter design procedure. The experimental results for various filter sizes are plotted in Fig. 3. The length $L = 1$ corresponds to the degraded image with no restoration. The vertical axis shows a logarithm of the localization error, with the degraded image as the reference point 0 dB. The results indicate that the error decreases nearly exponentially at practical filter sizes, $L < 8$. In this case it turns out that even the filter with $L = 3$ (i.e. 5x5 convolution mask) will produce a visually excellent restoration, as can be seen from Fig. 2c. As a thumb rule, 3 or 4 are sufficient values for L in the case of moderate degradation.

REFERENCES

/1/ H.C.Andrews and B.R.Hunt,Digital Image Restoration. Prentice- Hall, Englewood Cliffs, NJ, 1977.

/2/ R. Frieden, *Image restoration by discrete convolution of minimal length*, J. Opt. Soc. Am. **64**, pp. 682-686, May 1974.

/3/ M.J.Lahart,*Local image restoration by a least-squares method*, J. Opt. Soc. Am. **69**, pp. 1333-1339, Oct. 1979.

/4/ H. Ogawa and N. Nakamura, *Projection Filter restoration of degraded images*, Proc. 7th ICPR, Montreal, July 30 - Aug.2, 1984, pp. 601-603.

/5/ E. Oja and J. Lampinen, *A fast local PPF restoration filter*, Proc. **1986** ICASSP, Tokyo, April 7-11, 1986, pp. 1497-1500.

/6/ E.Oja and H.Ogawa, *Parametric Projection Filter for image and signal restoration*, IEEE Trans. ASSP, vol. ASSP-34, pp. 1643-1653, Dec. 1986.

/7/ L. R. Rabiner and B. Gold, Theory and Application of Digi- tal Signal Processing. Prentice-Hall, Englewood Cliffs, NJ, 1975.

/8/ B.E.A.Saleh, *Trade-off between resolution and noise in restoration by super-position of images*, Appl.Opt.17, pp. 2186-2190, 1978.

/9/ H.J.Trussell and M.R.Civanlar, *The feasible solution in signal restoration*, IEEE Trans. ASSP, vol. ASSP-32, pp. 201-212, April 1984.

A COST-EFFECTIVE ARCHITECTURE FOR VISION

A. Alcolea, A. Roy, A. Martínez, P. Laguna, J. Navarro, T. Pollán, and S.J. Vicente

Department of Electrical Engineering and Computer Science
University of Zaragoza E.T.S.I.I.
María Zambrano, 50   50015 - Zaragoza, (Spain)

Abstract

Parallel processing appears as the only solution to the performance bottleneck of current industrial vision systems, mostly based on Von-Neumann processors. In this paper two of the most common approaches to date are analyzed from a point of view of cost-effectiveness; SIMD arrays are more flexible, simpler to design, and have lower latency than pipeline processors. FAMA, a new SIMD array based architecture is introduced whose main features are cost-effectiveness at the processor element level, and applicability to different levels of processing.

INTRODUCTION

Most of current industrial computer-vision systems consist of a general-purpose Von-Neumann processor, in many cases assisted by a simple special-purpose preprocessor performing some low-level processing steps. Software is specifically developed for every new application, resulting in a substantial portion of the total system cost.

Nevertheless, many potential applications remain beyond the capabilities of those systems. Applications requiring grey-level processing, three-dimensional and partially occluded object recognition, and perhaps colour and motion analysis render Von-Neumann architectures useless when real-time processing is a necessity.

Parallel processing appears as the only solution to the performance bottleneck of Von-Neumann architectures. Until now, most efforts in the computer-vision field have been focused on early or low-level processing, but as complexity of applications increases, and the world of visual objects and relations expands, high-level processing, consisting basically of pattern-recognition and symbolic processing, gives rise to new bottlenecks where the possibilities of parallelism should be exploited if real-time must be achieved.

Complex applications have requirements close to human vision capabilities in many respects, and consequently computational power at all levels far higher than that currently available in industrial systems is needed. Software and hardware in advanced systems should implement the models based on existing theories on human vision (1.), though much theoretical work remains to be done, particularly in the high level vision modelling.

This work has been supported in part by the CONAI of the Diputación General de Aragón

Flexibility and robustness should feature these systems, allowing easy adaptation to classical and new applications.

In industrial applications, cost-performance ratio is one of the most important parameters to be optimized. It is possible to find a function aproximately relating hardware cost with variables such as area, volume, complexity of interconnection, power dissipation... Therefore, optimization consists in finding the number and complexity of processor elements, the complexity of interconnection, and the technology of integration that maximize performance for a given cost. Many trade-offs are possible between values for those variables, and choosing the best one is the challenge for every new architectural design effort. Software development costs will not be considered in this discussion.

Two main parallel architectures for image processing have been proposed: pipelined processors and SIMD array processors (operator parallel processors and image parallel processors in another terminology (2.)). A brief comparison between them from a cost-performance viewpoint follows.

In a pipelined processor, each processing stage can be optimized for the specific subtask it has assigned; on the other hand, processing elements in SIMD arrays must perform diverse subtasks and, consequently, efficiency decreases. The pipelined processors developed to date (3,4) make use of one or two different processor stage designs, and their high efficiency in some applications has been shown (5); but in advanced systems, it is not clear that all the low level tasks and, of course, the high level ones, could be efficiently partitioned in such a simple way. Therefore, the design effort, considering that in general each stage should be individually optimized, is higher than in SIMD arrays, where a single processor element design is repeated throughout the array. Besides, in pipelined processors, the memory requirements can be, in some cases, higher.

I/O inefficiency has been considered the main drawback to SIMD arrays, but adequate interface circuitry introducing data through all the processor elements on one of the sides of the array, and independent paths within the array, allowing simultaneous flow of input data and data interchanged between processor elements, permit an overlapping (pipelining) of I/O operations with processing without any performance degradation. This approach is common in recent designs.

When latency, instead of throughput, is the performance measure, pipeline architectures can be disadvantageous. Its formula:

$$t_l = n \cdot \alpha \cdot t_I + n \cdot \beta \cdot t_p \text{ where}$$

$n =$ number of stages

$t_I =$ input time for data from the preceding stage

$t_p =$ processing time

$\alpha =$ fraction of input time before the start of processing $(0 \leq \alpha \leq 1)$

$\beta =$ fraction of processing time before the start of data output $(0 \leq \beta \leq 1)$ (for simplicity all the parameters are supposed to be equal for all the stages)

shows that for long pipelines (large n), latency can be remarkably higher than processing time. In an SIMD array, assuming, for simplicity, equal hardware efficiency than in a pipeline processor (considerations made above should be noticed in this respect), the same processing time is obtained (assuming I/O overlap) with the same cost; but in this case, latency is equal to tp increased by tI and consequently, lower values can be obtained.

Hierarchical or pyramidal architectures (6,7) share features with both pipeline and SIMD array processors. The bottom-up flow naturally concentrates information at the high levels, as many computer vision algorithms demand. Besides, matching the processor element structure to the subtasks assigned at each level allows further optimization. On the other hand, the latency issue must be considered, and the interconnection complexity increases cost, compared to SIMD arrays, keeping constant the number of processor elements. Three-dimensional structures being searched (8) can help in this respect.

In the pipeline versus SIMD array issue, underlies the problem of matching the architecture to the task (9,10,11). The computer-vision algorithms selected will ultimately give an advantage to one of the alternative architectures, depending on which kind of partition, horizontal or vertical (image or operator parallelism) is favoured by these algorithms.

THE FAMA ARCHITECTURE

The advantage of SIMD array architectures in respects such as flexibility of size, programmability, simplicity of design and custom integrability, besides cost-effectiveness were decisive in the selection of an initial approach to task-oriented architectures for vision. The FAMA (Fine granularity Advanced Multiprocessor Architecture), currently being developed in the Department of Electrical Engineering and Computer Science of the University of Zaragoza, is based on a custom IC containing several processor elements. SIMD arrays of different sizes can be obtained through direct bidimensional interconnection of the IC.

Optimization at the processor element level is one of the keys to cost-performance efficiency in an SIMD array architecture; a careful design of the processor element in FAMA results in very high performance with moderate complexity. Pure SIMD architectures suffer from inefficiency when implementing algorithms where conditional data-dependent operations are frequent; FAMA has been provided with mechanisms that largely overcome this problem. I/O operations efficiency and reconfigurability to fit the requirements at different levels of processing have also been considered in FAMA.

Several design rules for hardware efficiency have been taken into account: circuitry and interconnections must be doing useful work for a fraction of the time as large as possible; as a consequence, redundancies arising when different circuits are performing the same task should be eliminated. Pipelining, SIMD array structure, and 1-bit processor elements are direct consequences of these rules. Pipelining allows, with the segmentation of the data path, efficient operation of the circuitry in the different segments. SIMD structure permits the elimination of most control circuitry within the processor element by making use of only one common controller that broadcasts instructions to the whole array. 1-bit processor elements allow full use of hardware resources whichever the size of operands; in operations with carry (borrow) propagation, they behave, in a sense, as a segment of a pipeline, further improving efficiency. In addition, memory bandwith in every processor element is high enough to permit the load of operands and the store of results in the same clock cycle, making possible a performance of one useful ALU operation per cycle. A brief description of the structure (fig. 1) and main characteristics of FAMA follows.

REGISTER FILE

It has some similarities with CLIP (12) and GRID (13) architectures; two read and two write parts provide enough memory bandwith to allow one useful ALU operation per clock cycle in both local and neighbourhood operations. In local operations, two operands can be read and one result can be written in the same cycle; in addition, input data can be written simultaneously through the second write port. In neighbourhood operations, one operand can be read, the second operand can be received from a neighbour, and one result can be written in the same cycle; in addition, data sent to neighbours can be read from the second read port and data received from a neighbour can be given intermediate store through the second write port simultaneously. The data in intermediate store can be sent to neighbours in later processing steps, allowing optimum efficiency even if neighbourhoods are larger than 3 x 3.

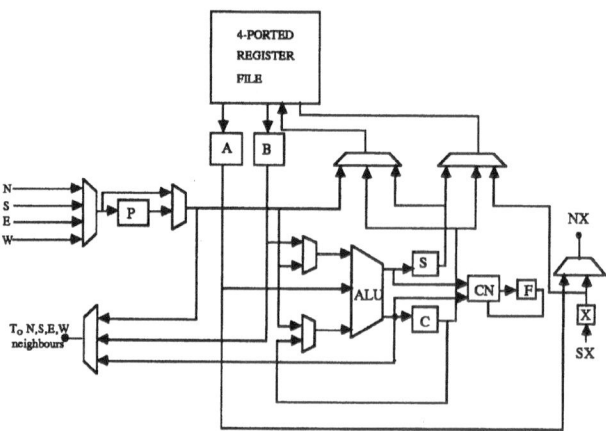

Fig. 1 Simplified block diagram of a processor element of FAMA. Control signals and control circuitry. have been omitted. P,A,B,S,C,X,F are synchronous registers. CN is a combinational network N,S,E,W are connections with the neighbours. NX,SX are connections with neighbours through the second S-N path.

Two alternatives exist for the implementation of the 4-port register file; the first one, full physical implementation of the four ports, is very expensive in hardware resources but allows optimum speed; the second one, time multiplexing of read and write ports, could slow down the processing if one write and one read operation are not possible through the same physical port in one clock cycle, whose length is established by the propagation delay of the segment of the pipeline that contains the ALU. This second alternative allows significant saving of hardware resources and its convenience will depend on each particular implementation of the architecture.

The number of address lines is high, one set for each of the four ports; if the most significant bits of the addresses are stored in registers loadable by specific instructions, and only the least significant bits must be specified in every instruction, a significant reduction of interconnections complexity can be achieved with little loss of performance.

For efficient processing, the register file should have enough capacity to store the portion of input image data corresponding to one processor element and its intermediate and final results. This allows the overlapping of input/output operations with processing, and the avoidance of the need to partition the input image in windows for processing, with its associated problems of I/O bottlenecks and inefficiency at the borders of the windows. A configuration with 4096 processors would require a memory in the 256-512 bits-per-processor range for images of 256 x 256 8-bit pixels. The amount of memory needed in the processor element for a constant image size varies with the number of processors; if internal memory is used exclusively , it is not possible to fit the memory needs for configurations varying in the number of processor elements. In the initial implementation, the register file will be in external memory; this allows an increase in the number of processor elements per IC and greater flexibility for adaptation to different image and array sizes. The external memory is implemented with two very fast or four moderately fast 1-bit wide memories interfacing with the processor element through two 1-bit I/O ports; in a clock cycle, during the first semicycle, data are written in memory, and during the second one, data are read from memory.

INTERCONNECTIONS

Every processor element is connected to four neighbours. One of the four inputs can be selected as an input to the ALU or written in the register file, showing very efficient neighbourhood operations. The register P can introduce an additional segment in the pipeline if needed to match the delay of the different segments; besides, in tasks which require extensive communication with far neighbours, this can be overlapped with processing by using P as intermediate store, avoiding the need of access to the register file.

In addition, a by-pass function is provided that allows direct communication between distant processor elements; when the flag register F is in a predetermined state in a processor element, this is by-passed: the output to other neighbours follows the input from the selected neighbour. It allows row or column broadcasting from a selected neighbour, and efficient accumulation along rows or columns (fig. 2); the speedup in this operation is not 0 (n/logn) because delays introduced by by-passed processor elements must be taken into account; but a performance increase by a factor in the 2-3 range is easily achievable.

A second South-North path is included that allows full overlapping of I/O operations with processing. Input data enter the array through the southern side, and output data leave the array through the northern side.

PIPELINES

For local operations, the data path has 3 segments, the first one is the register file seen from its read ports; it ends at the registers A or B; the second one includes the ALU and its input multiplexers; the ALU can perform the usual arithmetic and logic operations; it ends at the register S; the third one is the register file seen from its write ports. The frequency of operation will be limited by the slowest segment when set-up and delay times for registers are taken into account. In neighbourhood operations there are two possibilities; if the register P is by-passed, the pipeline containing the data received from the neighbour increases its second segment by three multiplexers, consequently the delay increases. If the P register is not by-passed a new segment appears; in this case the second segment includes the output multiplexer of the neighbour from which data are received and the input multiplexer of the processor element considered; the third segment includes the multiplexer at the output of P, and the ALU and its input multiplexer. Delay times should be balanced in all the segments whichever the operation performed if optimum efficiency is to be achieved.

The flow of data in the pipelines is shown in fig. 3.

CONDITIONAL OPERATIONS

Conditional operations are executed only by processor elements whose flag register F is in the state specified in the conditional instruction. The flag register can remain unaltered, be loaded from the ALU output, or be loaded with the OR function of its contents and the ALU output. The last possibility allows efficient comparisons between words of any length; the two words are substracted or XORed, if the flag register has been initialized to Ø, it will remain in that state only if the two words are identical.

If external memory is used for the implementation of the register file, the alteration of the register file cannot be performed exclusively in enabled processor elements, because all the control signals are common for the whole array; to overcome this problem, the masking of ALU inputs as a function of the contents of the flag register is possible in this implementation. In most cases, there is no loss of efficiency.

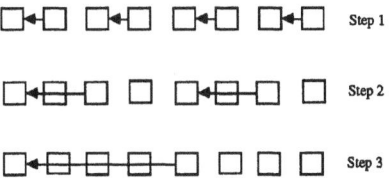

Fig. 2 Accumulation in a file using the by-pass function of FAMA. It allows a binary-tree accumulation, reducing the number of operations to $\log_2 n$

Cycle n	Cycle n+1	Cycle n+2	
OPA n	OPA n+1	OPA n+2	Segment 1 : input to register A
OPB n	OPB n+1	OPB n+2	Segment 1 : input to register B
—	RES n	RES n+1	Segment 2 : input to register S
—	—	RES n	Segment 3 : input to write port of register file

(a)

Cycle n	Cycle n+1	Cycle n+2	Cycle n+3	
OPN n	OPN n+1	OPN n+2	OPN n+3	Segment 1 of the pipeline coming from the neighbour: input to neighbour's register B
—	OPN n	OPN n+1	OPN n+2	Segment 2 of the same pipeline: Input to local register P
—	OPA n	OPA n+1	OPA n+2	Segment 1 of local pipeline: input to local register A
—	—	RES n	RES n+1	Segment 2 of local pipeline: input to local register S
—	—	—	RES n	Segment 3 of local pipeline: input to local write port of register file

(b)

Fig. 3 Data flow in the pipelines for local operations (a) and neighbourhood operations (b)

The possibility of conditional operations at the controller level is provided by the output of the OR function of the contents of all the flag registers in the array.

RECONFIGURABILITY

If an SIMD array must be applicable to different levels of processing it should be reconfigurable to match the requirements of processing at each level. At high levels, levels, fewer, more powerful processors are needed than at low levels. FAMA allows the reconfiguration of a n 1-bit processor array into a n/m m-bit processor array (fig. 4); the link pattern is almost arbitrarily established by three 1-bit registers (not shown in fig. 1) loadable from the register file by specific instructions, allowing on the fly reconfigurability. Addition and substraction can be performed in two modes: flow-through mode, in which the carry output of one processor element is directly connected to the carry input of the next one, and pipeline mode, in which the register P separates the ALUs. In the first mode, carry propagation delay results in a substantial performance degradation. In the second mode, performance does not suffer, but the different bits of a word must enter the multibit adder at different clock cycles as shown in fig. 5.

The reconfiguration capability of FAMA allows the implementation of a virtual pyramid, with the advantage over physical pyramids of reduced interconnection complexity and, probably, better cost-effectiveness.

PERFORMANCE

In the following expressions the number of clock cycles required in FAMA to perform some usual tasks is shown. The image has $\sqrt{N}$ x $\sqrt{N}$ pixels of b bits; the array has $\sqrt{n}$ x $\sqrt{n}$ processor elements; $\sqrt{N}$ is a multiple of $\sqrt{n}$.

I/O operation $\dfrac{N}{\sqrt{n}} \cdot b$

Local operation $\dfrac{N}{n} \cdot b$

Neighbourhood operation
(neighbourhood size m) $m \cdot \dfrac{N}{n} \cdot b$

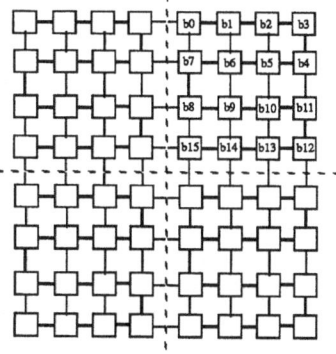

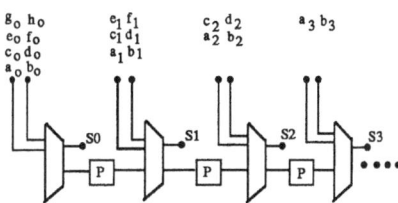

Fig. 4 Reconfiguration of a 1-bit 64-processor array into a 16-bit 4-processor array

Fig. 5 Data flow when additions are performed in a multibit processor in pipelined mode

Histogram

(2b is a multiple of $\sqrt{n}$) $\dfrac{2b}{\sqrt{n}}(2 + \dfrac{N}{\sqrt{n}}(b\text{-}1) + \log\dfrac{N}{\sqrt{n}}(1 + \dfrac{N}{\sqrt{n}} + \dfrac{1}{2}\log n)) + 3/4 \log n$

As can be seen, in local and neighbourhood operations optimum performance is achieved, and the speedup efficiency is 1. On the other hand, histogram is very inefficient, as in most architectures of this kind; the possibility of access to register file with local data-dependent addresses could improve performance substantially, but it is doubtful that the increase of performance for the whole vision task could compensate the increase of hardware complexity.

IMPLEMENTATION

The initial implementation is based on a custom 3mm CMOS IC containing four processor elements. The package is a 48-pin DIP. As mentioned above, the register file is implemented with fast external memories. The operation frequency expected for the IC is 10 MHz. The array of processor elements is complemented with I/O interface circuitry and a sophisticated controller capable of exploiting the processor elements to their ultimate limit of performance.

A 1024 processor element system would fit in a 19" double eurocard height rack. Hosted by a VME based CPU or PC, it would have a performance of 10,24 1-bit Gops.

For 256 x 256 8-bit pixel images the processing times would be:

I/O operation : 1,6 msecs.
Local operation : 51 msecs.
(one 8-bit logic or arithmetic operation per pixel)
Neighbourhood operation : 346 msecs.
(Prewitt's operator 3x3 neighbourhood)
Histogram : 30 msecs.

CONCLUSION

FAMA provides the power and cost-effectiveness required in advanced industrial vision systems. Its efficiency at the processor element level, and its flexibility in interconnection functions result in an excellent cost-performance ratio; its reconfigurability allows the implementation of virtual pyramids with reduced interconnection complexity. Its applicability to high level processing will contribute to the elimination of bottlenecks in advanced systems.

The initial implementation, though limited in the number of processor elements per IC, will allow extended experimentation with the set of computer- vision algorithms selected for an advanced industrial vision system, and the fine tuning of the the architecture for future implementations with improved cost-effectiveness.

REFERENCES

1. Marr, D. (1982). Vision. Freeman, San Francisco.
2. Danielsson, P.E. & Levialdi, S. (1981). Computer Architectures for Pictorial Information Systems, IEEE Computer, Vol. 14, Nº 2, pp. 53-67.
3. Lougheed, R.M., McCubbrey, D.L. (1980). The Cytocomputer: a Practical Pipelined Image Processor, Proc. 7th Annual International Symposium on Computer Architecture.
4. Kent, E., Schneier, M., Lumia, R. PIPE (Pipelined Image Processing Engine), Report of the National Bureau of Standards, Industrial Systems Division.
5. Stenberg, S.R. (1985). An Overview of Image Algebra and Related Architectures, In Integrated Technology for Parallel Image Processing (S. Levialdi ed.). Academic Press, London, pp. 79-100.
6. Tanimoto, S.L. & Klinger, A. (eds.). (1980). Structural Computer Vision: Machine Perception through Hierarchical Computation Structures. Academic Press, New York.
7. Cantoni, V., Ferretti, M., Levialdi, S. & Maloberti, F. (1985). A Pyramid Using Integrated Technology, In Integrated Technology for Parallel Image Processing (S. Levialdi ed.). Academic Press, London, pp. 121-132.
8. Nudd, G.R., Grinberg, J., Etchells, R.D. & Little, M. (1985). The application of Three-dimensional Microelectronics to Image Analysis, In Integrated Technology for Parallel Image Processing (S. Levialdi ed.) Academic Press. London pp. 167-185.
9. Cantoni, V. & Levialdi, S., (1982). Matching the Task to an Image Processing Architecture, Proc. 6th International Conference on Pattern Recognition, pp. 254-257.
10. Cantoni, V., Guerra, C., Levialdi, S. (1983). Towards on Evaluation of an Image processing System, In Computer Structures for Image Processing (M.J.B. Duff ed.). Academic Press, London, pp. 43-55.
11. Offen, R.J. & Scharbach, P.N. (1985). From Algorithms to Architectures, In VLSI Image Processing (R.J. Offen ed.). Collins, London, pp. 128-187.
12. Duff, M.J.B. (1978). Review of the CLIP Image Processing System, Proc. National Computer Conference, pp. 1055-1060.
13. Parker, I.N. (1985). VLSI Architecture, In VLSI Image Processing (R.J. Offen Ed.). Collins, London, pp. 99-127.

HIERARCHIAL SPECIFICATION IMAGE DATA TYPES AND LEVEL LANGUAGE

A. Belaid and Z. Boufriche

CRIN - B.P. 239
54506 Vandoeuvre-Les Nancy, Cedex, France

1. INTRODUCTION

Several reasons have led to the structuring of image processing and the formulation of this structuring by programming languages. The most important of these reasons is the need for providing clearer specification, and better control of associated image data structures, and for greater clarity in program structures, in order to increase the readability and make debugging easier. Another important reason is the desire to exploit new possibilities offered by recent advances in hardware technology. Many dedicated processors can now cooperate in parallel execution to provide efficient parallel implementation for some of the most commonly used image processing algorithms, and this for a manageable level of investment.

The current literature on computer graphics[8,14], on pattern recognition[5,15,12] and on image processing[13,11], provides a wide range of data structures and newly an emerging programming style. New languages[7,10] with specific concepts, new constructs and adapted control structures are proposed. Many of the constructs appearing in these new languages are not always neatly specified. Moreover, they are often designed for expressing high-level image tasks. Some are system dependent or ad hoc solutions in particular algorithms.

In this paper, we propose a rough summary specification of the kernel of some of the special concepts commonly used in image processing languages. The proposed specification is machine and system independent. To avoid redundancy in the definition of data types, concepts are grouped together according to their underlying data structures. For each group of concepts, the list of operations applicable to the underlying data structure is specified. This list is then completed by the specification of concept specific operations to particularize the underlying structure for each of the concepts embraced. Section two defines the basic concepts and presents formal schemes for representing them. Associated operations are then defined in section three, while section four discusses problems occuring at implementation time. These ideas are successfully implemented in Pascal. An ADA version taking advantage of the packaging and the genericity capabilities of that language is under development.

2. IMAGE CONCEPTS

Concepts in use in recent image processing languages are closely related to the characteristics of the underlying data types. Thus an image represents a scene, a map represents objects in a scene, an edge represents object boundaries, etc. This close association between concept and data type present two major advantages: first, it makes it easy to understand the program; secondly, a program can be made aware of the data type operations it is handling[1].

We believe that only formal specification can ensure a coherent definition of these concepts and reduce redundancies to a minimum. Formal specification imposes a careful consideration of each data type and of the choice for its representation. For this, the properties of the data type, those of the operations defined on it and the mode of access to the information it models must all be taken into account. Each concept is seen as a set of definitions (transparent to the user) and a collection of operations for creating, accessing, interrogating and modifying the data type. The formal specification establishes the relationships between data type operations.

We use the algebraic specification approach[4]. In this approach it is important to distinguish the basic constructors from the other types of operations. In our case, these constructors are grid, list, and table. More specifically, we have[2] adopted N. LEVY's definitions to specify our image concepts[6]. These form a kernel for our abstract data types for image processing.

The formal specification led us to concluding that we could do better by structuring not only the data types, but also the concepts themselves. This approach has the advantage of setting at higher levels basic operations that can simply be instanced at lower levels. As one moves down the hierarchy, more specific operations are added to the upper level operations. The main purpose of this paper is to show the importance of the hierarchical structure and to illustrate its implementation in any applied (basic) language.

3. DEFINITION OF CONCEPTS

We define the following concepts: **IMAGE MAP MASK EDGE FTAB.** They form a complete kernel for the most commonly used objects in image processing algorithms. They belong to two fundamental structures : the **planar structures** used to represent two-dimensional pictorial information for human interpretation and analysis like IMAGE, MAP, and MASK ; and the **linear structures** used in almost all the algorithms, to store data that must be created dynamically and processed in a linear order. This class involves the EDGE concept.

An obvious hierarchy of concepts emerges (see figure 1). The underlying structures - GRID - LIST - TABLE are placed at the top. They are abstractions of the objects (IMAGE - MAP - MASK), (EDGE) and (FTAB), which form the second level. We start by defining the concept of POINT with associated list of elementary operations applicable to it. We then built the other concepts therefrom (cf. fig. 1).

The GRID concept is parametrized at the first level of the hierarchy by WINDOW and INFO. WINDOW is a cartesian product of two intervals, *interx, intery*, which give the x and the y index intervals. INFO provides numeric values on which an order relation as well as boolean or arithmetic operations are defined. INFO can be instanced by a simple or structured data type.

On the next level, a monomorphism is defined for deriving instances of GRID on IMAGE, MAP and MASK. For example in IMAGE, INFO is instanced by INFO-IM, which groups all information specific to the image, e.g. grey-level (GL$\in\{0,1,...,2^n-1\}$), color (COL: red, green, blue $\in$ GL), binary (BIN $\in \{0,1\}$).

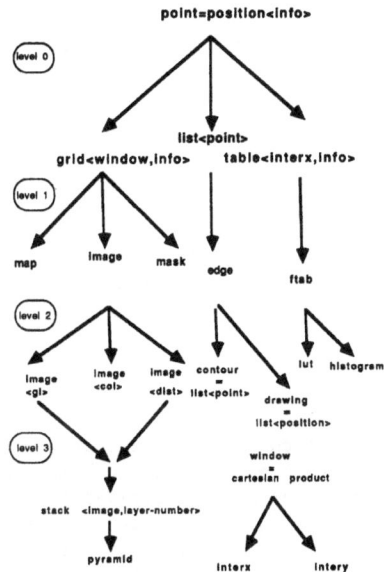

Figure1: **Hierarchical Structures of Image Concepts**

An IMAGE is defined as a GRID of points whose values (grey-level, color, distance, etc.) depend on the nature of the image and hence, on the information being manipulated. Each point (pixel) is given by its position in the grid and by its value. An IMAGE is thus defined as a function (a grid) which to each position in the grid assigns information of a given type.

The concept of IMAGE may be viewed from two angles: IMAGE as a GRID with the usual associated operations (consultation, and modification), and IMAGE as an instance of a grid on which global characteristic operations are defined. To this point, the concept of IMAGE remains quite general and can support only general operations. For more specific types of images, such as grey-level images, binary images, etc. it will be necessary to define more specific operations on the instanced information.

387

This hierarchical process can be repeated for MAP and MASK. Indeed the three concepts have a common basic structure (= GRID). However they differ from IMAGE by the nature of the instances they support. We define a MAP as a grid of object indicators in an image. The MAP plays a double role: it serves as means for direct access to objects in an associated image, and also contains a list of information on the nature of the objects present in an image, e.g. position, surface, homogeneity, boundary, etc. The MAP essentially manipulates integer values. MASK, like IMAGE, is a GRID with the same index interval, except that INFO is instanced as INFO-MS. In many situations, a MASK can represent parts of an image such as pixel neighbourhoods in local operations. It holds real or integer values.

The LIST structure parametrized by POINT supports linear objects such as EDGE. An EDGE is considered as a succession of points representing a region boundary in image (contours) or drawings of a sketch.

The TABLE structure inludes the FTAB concept which is devoted to define particular objects such as histograms, look-up tables etc., whose representation needs a monodimensional table.

We regroup into a special class objects which do not have a particular basic structure. WINDOW is not only a parameter for a GRID but also provides a tool for selecting rectangular zones in images.

The FILTER as a selection tool is more general than WINDOW. The filter concept, often used in a restricted sense in several specialized languages, is generalized here to give a wide range of ways for accessing points in a grid.

4. DEFINITION OF OPERATIONS

The importance of hierarchy is two-fold: it groups operations level by level and it allows a coherent and complete parametrization of these operations for correct instancing of all the derived types. These operations are classified into two principal groups: on the one hand we have consultation which allows access to the basic structure of the concept and its manipulation, and modification operations on the other.

4.1 POINT OPERATIONS

A POINT is a POSITION parametrized by INFO. The elementary operations on INFO are :

consultation : Consultation regroups test and arithmetic operations which are realized respectively with logical and arithmetic operators.

    info-test: INFO x INFO → BOOLEAN
    info-comp: INFO x INFO → INFO

modification : Modification operation replaces an INFO with another one.

    info-modif: INFO x INFO →INFO

More POINT specific operations are :

consultation : **info-access: POINT** → **INFO**, gives the value of a point. In this case, no constraint is imposed on the point.

pcond-pos-access: POINT → **POSITION**, here the POINT is compelled to verify some conditions imposed on INFO (P(V =...)) and thus access to the POINT will be possible if such conditions are true; if not, the empty position is returned.

pcond-info-access: POINT → **INFO**, access to INFO is possible if a given criterion on POSITION is satisfied (P((X >=...) and (Y =...))). If the point is not selected, an empty information is returned. We shall see later that this criterion could be a general condition called FILTER.

p-neighbour: POINT x POINT → **BOOLEAN**, p-neighbour returns a boolean information about the neighbourhood of two points. This concerns points in a grid structure and is valid both for the 4 or 8-neighbourhood.

test : **first-test/last-test: POINT** → **BOOLEAN**, this states whether a position is the first or the last one in a LIST.

succ-test/pred-test: POINT x POINT → **BOOLEAN**, this states whether a position succeeds or precedes another. In an edge, besides the list order, succeeding positions are connected.

succ-access/pred-access: POINT → **POINT**, gives the position successor or predecessor of another point, if it exists. It returns an empty value otherwise.

modification : **info-modif: POINT x INFO** → **POINT**, modifies the value of a point. Info-modif checks the compatibility of information values involved in this operation.

Position modification operations consist of geometric transformations principally used in graphics. Only translations and rotations are considered here.

p-rotation: POINT x POSITION x INTEGER → **POINT**, moves a point of a given angle from an origin; the sign of the angle gives the rotation direction.

p-translation: POINT x INTEGER x INTEGER → **POINT**, shifts a point along interx and intery by a given number of positions. The sign of this number determines the direction of the translation.

4.2 GRID OPERATIONS

These operations include array functions and the operations common to the derived types. They cover access and modification operations. Access operations include direct, sequential and conditional accesses. The conditional access is parametrized by a filter (function: point → boolean, true if point can be accessed). The possibility of selecting subimages increases the power of these global filters: transfer of regions may for instance be written as a simple assignment and selection. In fact, a binary image is a special kind of filter. Selection may be formulated, syntactically, in a variety of ways:

- by WINDOW operations: WINDOW selects a new image as being a full subimage of the image on which it is applied. A possible syntax could be: I(W<p>). This selects the subimage from I whose size is constrained by W, with origin at the point p by imposing constraints on the positions or by imposing constraints on pixel values.

- Imposing constraints on pixels may be achieved through boolean expressions applied to position and value. If i\$, j\$ and v\$ are the generic names for line column and pixel value, we could for instance write: I(i\$-j\$<0 and v\$>9). This selects the subimage of I situated above the main diagonal with pixel values are greater than 9. Through the application of the window W, we can select any desired set of pixels in an image (see fig. 2): I(i\$+j\$ ≥ 4 and W<p> and v\$ ≥10). We introduced a new object: FILTER. Like all other objects, it must be declared before it can be manipulated.

Regions are created by region growing procedures or by mapping a filter on an image. Consider the map M. The following assignment creates the region i and modifies the map: M[i] := I (S), where S is a filter.

In most cases, each grid operation makes use of a point operation on a reduced scale. The first step consists of selecting a point in the grid. The appropriate point operation is then applied to the point selected within the intervals demarcated by the grid.

Here are the basic operations on GRID:

consultation : **wind-access: GRID[INFO]** $\rightarrow$ **WINDOW**, returns the frame of the grid (i.e. its breadth and its length).

seq-access: GRID[INFO] x POINT x (info-access(p)) $\rightarrow$ **INFO**, we first select the point p from the grid (GRID][INFO]] $\times$ POINT $\rightarrow$ POINT), and then apply info-access to obtain its value (POINT $\rightarrow$ INFO).

neigh-access: GRID[INFO] x POSITION x WINDOW $\rightarrow$ **MASK**, selects a position neighbourhood.

gcond-info-access: GRID[INFO] x POINT x (pcond-info-access(p)) $\rightarrow$**INFO**

gcond-pos-access : GRID[INFO] x POINT x (pcond-pos-access(p)) $\rightarrow$ **POINT** pcond-access is applied locally on the point resulting from the first cartesian product.

test : **grid-test: GRID[INFO] x INFO x info-test(infol,info2)** $\rightarrow$ **BOOLEAN**, compares the value of every selected point pl in the grid to a given value info2.

double-grid-test: GRID[INFO] x GRID[INFO] x (info-test(infol,info2)) $\rightarrow$ **BOOLEAN**, compares every selected point pl in the first grid to its correspondant p2 in the second.

modification : **simple-modif: GRID[INFO] x INFO x (info-modif(p, info))** $\rightarrow$ **GRID[INFO]**, takes a punctual modification function (info-modif) as parameter and replaces the value of every point in the grid with a given value.

double-modif: GRID[INFO] x GRID[INFO] x (info-modif(pl,info2)) $\rightarrow$ **GRID[INFO]** ; this is a more general operation. Operations involve corresponding points pl and p2 in two different grids.

The following operations modify intervals in which the GRID is defined.

rotation: GRID[INFO] x POSITION x INTEGER x (p-rotation(p1,p2,a)) → **GRID[INFO]**, all the grid points are rotated about the origin (p2) through the angle (a).

translation: GRID[INFO] x INTEGER x INTEGER x (p-translation(p,x,y)) → **GRID[INFO]**, all the grid points (p) are shifted along interx and intery by an integer distance.

4.3 IMAGE OPERATIONS

IMAGE is an instance of GRID: IMAGE = GRID[INFO-IM, INFO-IM] could be GL, COL, DIST or BIN. GL defines an interval of grey-levels (minint..maxint) characterized by the highest and the lowest levels. In all modification operations, controls must be provided to check the values that go out of bound. COL represents colored images. Pratically, it is a cartesian product of 3 values (G: GL, B: GL, R: GL). DIST is used for images in a 3-D system. The information manipulated is a distance represented by a real. BIN describes particular images with very specific objects: characters, blobs, medical and industrial data (cells or tools), etc. In these cases, we have to indicate the presence or absence of infor-mation, thus a binary value (0 or 1, true or false) is sufficient for representing BIN.

IMAGE operations are obtained from the set of operations defined on the GRID. This set is completed by IMAGE specific operations.

<u>consultation</u> : **dynamic: GRID[INFO-IM]** → **INFO-IM**, calculates the difference between extreme pixel values:maxint-minint.

region-growing: IMAGE[INFO-IM] x (fgrow(m,p)) → **MAP[INTEGER]**, where fgrow profil is **fgrow : POINT x neigh-access(p,w)** → **INTEGER**. Region-growing uses the local function fgrow to assign to each pixel (p) a region indicator. "fgrow" scans the pixel neighbourhood obtained using the neigh-access.

<u>modification</u> : we now specify two kinds of modification operations :

1) instanciation of grid modification operations: simple-modif becomes thresholding:

thresholding: GRID[INFO-IM] x INFO-IM x (info-modif(p1,info2) → **GRID[INFO-IM]**. The parenthesized function is a local thresholding function that sets each pixel whose value (p) is less than a given threshold value to zero.

2) newly defined operations on IMAGE. These operations act on a pixel neighbourhood. Zoom operations modify the IMAGE size and its values.

red-zoom: GRID[INFO-IM] x WINDOW x (p-red (m,info)) → **GRID[INFO-IM]**, where **p-red: neigh-access(p,w) x INFO-IM -> INFO-IM**. A reduced zoom is parametrized by a function which replaces an area with a point in the image. The area (m) is characterized by its center (p) and its size (w). The value of the result point is the mean of area values. The function is repeated for all the grid points.

ext-zoom: GRID[INFO-IM] x WINDOW x (eigh-access(p,w)) → GRID[INFO-IM], An
extended zoom is the reciprocal of the reduced one. Here the function
parameter will transform a point into an area by successive duplications,
for example.

filtering: GRID[INFO-IM] x MASK x (neigh-operation(m1,m2)) → GRID[INFO-IM], where **neigh-operation: MASK x MASK-> INFO-IM.** The principal operation
here is neigh-operation. It applies a mask (m1) on a selected neighbourhood
(m2) in the image and gives a value. If neigh-operation is a convolution,
the result will be an edge information, while in the case of matching, the
result will be a boolean.

Third level operations are not explicitly specified here, partly
because they are few in number and partly because most of them are
instances of inherited operations from higher levels (statistical operations as variance, mean etc.).

5. IMPLEMENTATION

In section 2, we presented a hierarchical specification of image data
types. The hierarchical structuring was based on collections of concepts
grouped together on the basis of their underlying data structures. This
approach led to a non-redundant definition of data types. But the most crucial outcome, in our view, is the ease with which the specified concepts
can be implemented in any language endowed with a typing mechanism. In an
earlier work, we design an experimental language called LPSI[3]. The use of
PASCAL as a host language was to satisfy the conditions dictated by the
ICOTECH machine and to reach a wide public. Though LPSI could handle high
level structures with great ease, it is less suited for the implementation
of the hierarchies. With LPSI, we were however able to resolve some major
implementation problems, such as type conversion, passage of parameters
between procedures, etc.

We are currently implementing the hierarchy of concepts using ADA and
CLU as host languages. From the first results, it appears that these
languages are more convenient for handling hierarchical structures, mainly
owing to their packaging and genericity capabilities.

5.1 SYNTACTIC DEFINITIONS

LPSI structures are based on the foregoing specification. It is an
extension of Pascal. The image object declarations are done as in PASCAL,
through the use of types and variable types. Every image object type
declaration specifies the size of the object, the type of the component
value or simple-type, and the attributes.

Examples of LPSI declarations :

```
TYPE TI = IMAGE[64,64] OF GL(8);
VAR
    P : POSITION; W : WINDOW; I : TI RESIDENT; J: TI IN support;
    R : MAP[64,64] RESIDENT; D1,D2 : EDGE;
    H : FTAB[256] OF INTEGER; PR : PYRAMID[TI,8]; (* where 8 is
            the layers number *)
```

where GL(8) denotes gray-level on 8 bits, 64 and 256 determine the higher

boundary of the interval domain (interx or intery) of the object; the lower
boundary is equal to 1. RESIDENT and IN are attributes of both images I and
J. There are two kinds of attributes:

a) environmental attributes which provide information about the
environment of the object. They provide a flexible mean for communicating
with external devices: RESIDENT underlines that the object has to reside in
memory, under the control of the user. RESIDENT attribute is a system
default attribute. IN support indicates that "support" is the name of the
file medium for the object to be created by the program with the SUPPORT
directive and which carries the following indications: the nature of the
medium, the input/output mode and the image element type. In one program
session, objects created are bound together by some relations (histogram
of.., edges of ..); ASSOCIATE provides a convenient way of stocking these
relations.

Example :

ASSOCIATE H : histogram(I), declares H as an histogram of I and ensures
that the attribute remains related to the object throughout its execution.
Whenever the object is used, its values correspond to its definition; this
may call for implicit recomputation if needed.

b) Implicit attributes include frame, type of information, number
of points, coordinates (in the case of positions), etc. They are directly
associated with an object at its declaration and can be referenced in the
following way: <name-of-object>.<name-of-attribute>.

5.2 INSTRUCTIONS

LPSI allows three types of instructions:

1- assignement with arithmetic and logical operations,
2- flow control statement,
3- input-output statement.

1- LPSI includes several arithmetic and logical operations applied on
sets of points^{10}. It consists of global operators able to manipulate
objects of unequal size. As a rule, we operate on the smallest frame.

2- For control of flow, LPSI allows the following two kinds of control
statements: global conditional statement and sequential and parallel iteration. The conditional statement is said to be global because it affects
sets of points. A global test yields a true result if its elementary con-
stituent tests are all true. For example, in the statment :

IFG I1 > I2(F) **THENG** module1 **ELSEG** module2

module1 will be executed only if each point of I is greater than the
corresponding point of I2 filtered by F, otherwise module2 will be executed.

The iteration processes a set of points in an object. It can be tackled using either parallel or sequential processing. A large number of

applications treat points of an image object independently. This is especially true for punctual and local operations like filtering, thresholding etc. It is possible to compute all points in parallel. This necessitates a special architecture where each processor takes care of a point. In LPSI, it is expressed by the following simplified syntax:

FOR ALL p **IN** I(v$>200) **DO** I[p] := 256;

This forces all values of I exceeding 200 to 256. But other applications impose a sequential treatment; the calculation of a point depends on the results of the computations on its neighbours (regions growing for instance), and thus a single control structure is insufficient to describe them. Such applications follow the general scheme:

FORSEQ {p, L$, c$} **IN** object **UNTIL** condition **DO** module

where position indicates that object will be scanned position by position; L$ and C$ impose respectively a row by row or a column by column scanning.

3- LPSI allows two types of I/O statements which are READG and WRITEG.

READG or **WRITEG**(lsupport)lvariables;

where "lsupport" is a list of variables defined with the SUPPORT directive. The directive must have given the physical characteristics of the media on which the I/O operations are to be carried out. It identifies the file from which images are read or written.

6. CONCLUSION

We presented a method for structuring common image processing data types based on the notion of hierarchy of concepts. Concepts were grouped together according to their underly of data structures. Starting from the formal specification of image concepts, we derived a complete, standardized and a non-redundant kernel of image types. Each type was represented by an appropriate data structure with the associated list of operations. The resulting kernel is both machine and application independant. Control structures allowing both parallel and sequential processing were also explored. A translation scheme was successfully implemented using an extended PASCAL. Codes whose generation require specialized hardware were identified. These include calls to such functions as convolution, linear image combination, etc. The extended Pascal, LPSI, which provided the earliest software environment for testing the ideas presented, was implemented on a sequential machine, SM90 with SMX (UNIX-like) system. An image processing equipment, connected to the SM90, was simply used to output the results of programs. It is obvious that for others machines, the appropriate programs for the translation scheme will have to be written.

Acknowledgments : The authors wish to express their thanks to Prof. Roger Mohr from CRIN and the ETCA group for their helpful advice, P. Doh and K. Tombre for their constructive comments on this paper.

7. REFERENCES

1. A. BELAID, "An Image Processing Language Based on Abstract Types",
   Technology and Science of Informatics, vol. 4, no. 3, pp 309-323, 1985.
2. A. BELAID and N. LEVY, "Spécification de types de données et leur
   intégration dans des langages spécialisés", Bigre + Globule no. 45,
   pp 141-155, Evry, October 28-29, 1985.
3. A. BELAID, Z. BOUFRICHE and R. MOHR, "LPSI: un langage de Programmation
   Structurée en traitement d'Images", Proceedings 5th AFCET-INRIA conference
   on Artificial Intelligence and Pattern Recognition, vol. II, pp 649-667,
   Grenoble, 1985.
4. J.V. GUTTAG, "The algebraic specification of abstract data types",
   Acta Informatica vol. 10, no. 1, pp 27-52, 1978.
5. A. KLINGER, "Data Structures and Pattern Recognition", Proceedings of First
   International Joint Conference on Pattern Recognition, Washington, 1973.
6. Nicole LEVY, "Outils d'aide à la construction et transformation de types
   abstraits algébriques", Thèse de 3ème cycle, NancyI, 1984.
7. S. LEVIALDI, A. MAGGIOLI-SCHETTINI, M. NAPOLI, G. TORTORA and G. UCCELLA,
   "On the implementation of PIXAL, a language for Image Processing",
   Languages and Architectures for Image Processing, ed. by M. J. B. Duff
   and S. Levialdi, Academic Press, 1981.
8. W. R. MALLGREN, "Formal Specification of Graphic Data Types", ACM Trans.
   on Programming Languages and Systems, Vol. 4, no. 4, pp 687-710, 1982.
9. A. MERIGOT, B. ZAVIDOVIQUE, F. DEVOS, "Pyramidal Algorithms for Image
   Processing", Proceedings of 7th International Conference on Pattern
   Recognition, 1984.
10. T. RADHAKRISHNAN, R. BARRERA, A. GUZMAN and A. JINICH, "Design of a high
    level language (L) for Image Processing", In Languages and Architectures
    for Image Processing, ed. by M. J. B. Duff and S. Levialdi, Academic
    Press, 1981.
11. A. ROSENFELD, "Quad trees and Pyramids for Pattern Recognition and Image
    Processing", Proceedings of 5th International Conference on Pattern
    Recognition, pp 802-811, 1980.
12. L. G. SHAPIRO, "ESP: A high-level graphics language", Proceedings of
    second Annual Conference on Computer Graphics and Interactive Techniques,
    1975.
13. L. G. SHAPIRO, "Data Structures for Picture Processing: A Survey",
    Computer Graphics and Image Processing, vol. 11, pp 162-184, 1979.
14. C. J. VAN WYK, "A High-level Language for Specifying Pictures",
    ACM Transactions on Graphics, Vol. 1, no. 2, pp 163-182, 1982.
15. C. T. ZAHN Jr, "Data structures for pattern recognition algorithms A
    case study", Proceedings of conference on Computer Graphics , Pattern
    Recognition, and Data Structures, pp. 191-195, 1975.

GEOMETRIC TRANSFORMATIONS OF RASTER IMAGES

ON SIMD PROCESSORS

Wolfgang Wilhelmi

Academy of Sciences of the GDR
Central Institute for Cybernetics
and Information Processes
Kurstrasse 33, GDR Berlin 1086

INTRODUCTION

The geometric rectification of images is an important preprocessing
task. Dewarping is necessary especially in multispectral and multitemporal
image analysis to achieve registration with subpixel accuracy. Lines,
edges, and textures are essential features for segmentation and object
recognition and should not be destroyed by coarse signal interpolation
techniques. This paper shows that the Single Instruction Stream Multiple
Data Stream (SIMD) processing scheme is well suited for the geometric
transformation task.

The most general model of a SIMD machine (Evans 1982) is a control
unit (CU) attached to a number of identical processing elements (PE) all
of which work in parallel. The CU will broadcast identical instructions
or start addresses for internal stored identical program sections. Then all
PE's execute the instructions in a step-lock fashion. The PE's are connected
to one another by a communication network. Limit cases of such a network
are the global memory (GM) and the local memory (LM). The main problem of
every multiprocessor architecture is the management of possible access
conflicts caused by the communication network. The most effective approach
is the application of the SIMD processing principle. The next simplification
of the problem is given by a sparse communication structure, e. g., processing is done fully in the LM, data exchange occurs only between two
neighboring PE's during one time slice, or there is a mode in which data are
broadcasted along busses to many PE's.

Two implementation schemes are of special interest for the processing
of raster images:

2D Array Processors. Different image pixels are assigned to different
PE's and stored there in the LM. The neighborhood of pixels or tiles is
mapped onto the neighborhood structure of the array. Usually the image has
much more pixels than the array has PE's. In this case the pixels will be
stacked to ensure fragmentwise operation. A well known representant of
this kind of machine is the Massively Parallel Processor (MPP) by Goodyear
Aerospace (Batcher 1980). Coletti (1982) describes a procedure for geometric
correction. As in all other proposals it is assumed that the integer portion
of the displacement vector was loaded together with the pixel value. In
every PE a mask is generated where the only set bit is on a location which

is defined by the displacement vector. The pixels are routed in the 4 permissible directions in parallel. With each routing step the mask is shifted through every PE's write inhibit flipflop. If it is enabled then the pixel value is saved because it has arrived at its destination. If more neighbors have to be saved for resampling purposes then this write enable bit is replaced by the corresponding number of set bits. The weak point seems to be the definition of the routing strategy so that travelling pixels do not collide in their intermediate locations and the number of routing steps is smaller than the number of different displacement vectors. Wilhelmi (1982) has proposed a scheme where data routing is reduced to a space-variant window operator.

Arraylogical Processors. This notion corresponds to the well known of Cellular vs. Cellularlogical systems introduced 1979 by Preston et al. The pixels are read out from a multidimensional memory and fed into a programmable network which alters its function after comparatively long time and at predefined checkpoints. The relations of this kind of processors to Array Processors is explained by the following considerations. The PE's are activated in a scanning mode corresponding to the sequence of reading the arraylogical memory. Because only one PE is busy at one time there is the opportunity to shift only one PE across the memory saving the other. However this equivalence is limited to the case where the internal state of the PE are not of interest and any updating of neighboring cells does not occur. Examples of such arraylogical processors are display or video processors which have found wide application in interactive image processing and computer graphics. Kempe et al. (1982) have described a procedure for geometric correction on the video processor K2027. Here again a shift vector is assumed to be stored together with the pixel values. For every shift vector a corresponding address offset is programmed. All pixels with identical shift vector are transferred to the destination memory in parallel retaining already arrived pixels. However the degree of parallelism is low if the number of different vectors is very high like it happens for a centered rotation.

ALGORITHM

Mapping

Let x,y - the coordinates of the input or the source image and u,v - the coordinates of the output or destination image. It is assumed that $0 < x,y,u,v < m$. The transformation is given by

$$u=f(x,y); \qquad v=g(x,y) \qquad (1)$$

The general topological (rubber sheet) transform is characterized by continuous functions $f(.,.)$ and $g(.,.)$. Usually additional smoothness constraints are desired. In practice f and g are defined by their values on a set of reference points:

$$f(x_i,y_i),g(x_i,y_i) \quad i=1,\ldots,I \qquad (2)$$

So we have an interpolation problem which in the following is considered only for f without loss of generality. The best approach for SIMD computing is given by the interpolation scheme

$$f(x,y)= \sum_{i=1}^{I} f(x_i,y_i)h_i(x-x_i,y-y_i). \qquad (3)$$

Equation (3) can be interpreted as a weighted sum of interpolator functions h_i. This implementation scheme ensures that for every pixel x,y the computation of f does not depend on any of the others pixels. Thus the parallelization on processor arrays and arraylogical machines is easy to achieve. The speedup is linear if the number of processors is less than the number of pixels. If the support points are fixed during different appications then the interpolator functions $h_i(x-x_i,y-y_i)$ should be precomputed. I multiplications and (I−1) additions are required per pixel.

A good interpolator was proposed by Shepard (1965).

$$h_i(x-x_i,y-y_i)=\begin{cases}\dfrac{\left[(x-x_i)^2+(y-y_i)^2\right]^{-\mu}}{\displaystyle\sum_{j=1}^{I}\left[(x-x_i)^2+(y-y_j)^2\right]^{-\mu}} & \text{for } x{\neq}x_j \text{ and } y{\neq}y_j\\[4mm] 1.0 & \text{else}\end{cases}\qquad(4)$$

Functions of C_∞ are generated with $\mu=2$. The interpolation is an isotropic one. A separable version could give some simplifications. A remarkable drawback of (4) is the high requirement for accuracy due to the division of two great numbers near x_i,y_i. The computation of the h_i (i=1,...I) can again implemented with full linear speedup. Further it is possible to reduce the amount of computation by truncation of the nominator sum by terms where current x,y are far away from x_j,y_j.

Our experience (Eichhorn et al. 1986) has yield that an equidistant sampling of the distortion field with an independently on i defined interpolator ensures a good compromise between accuracy and efficiency. Obviously (4) is not suited for this approach because the nominator sum had to be expanded to infinity. The following interpolators should be preferred.

(a) The ideal resampler:

$$h(x,y)=\text{sinc}(\pi x/\triangle x)\cdot\text{sinc}(\pi y/\triangle y)\qquad(5)$$

The generated functions are Fourier finite. Regarding symmetry m^2 function values must be computed.

(b) The Fourier resampler:

$$h(x,y)=\frac{\sin(\pi x/\triangle x)}{(m/\triangle x)\sin(\pi x/m)}\cdot\frac{\sin(\pi y/\triangle y)}{(m/\triangle y)\sin(\pi y/m)}\qquad(6)$$

The generated functions are Fourier finite and periodic. This type should be applied in arrays with toroidal structures. Due to symmetry and periodicity only $m^2/4$ function values are required.

(c) The bilinear interpolator:

$$h(x,y)=\begin{cases}(1-|x|/\triangle x)\cdot(1-|y|/\triangle y) & \text{for } |x|<\triangle x \text{ and } |y|<\triangle y\\[3mm] 0.0 & \text{else}\end{cases}\qquad(7)$$

This interpolator gives a good control over normal geometric distortions by avoiding oscillations. It has a sparse structure and can be described by $\triangle x\triangle y/2$ function values. For any pixel the computation of (3) requires only 4 multiplications and 3 additions.

The proposed interpolator are separable in x,y thus allowing the implementation of functions by 2 table lookups followed by 1 multi-plication. They can be calculated in parallel. The interpolator values for every pixel results from a regular routing under central control which is complete after I steps.

The estimated destination points generally do not coincide with integer locations in the interior of the image. So they will be applied as support points of a bilinear interpolation of subpixel source locations only. We use the notation of fig. 1 for a clique of 4 adjacent signal samples and a vector representation with $r=(u,v,)^T$ and x as the outer product operator.

$$\xi(r) = S(r)+ \left(1 + \frac{(r11-r01)x(r \ -r10)}{(r01-r10)x(r11-r10)}\right) \cdot \left(1 + \frac{(r00-r01)x(r \ -r10)}{(r01-r10)x(r00-r10)}\right) \tag{8}$$

$$\eta(r) = S(r)+ \left(1 + \frac{(r11-r10)x(r \ -r01)}{(r10-r01)x(r11-r01)}\right) \cdot \left(1 + \frac{(r00-r10)x(r \ -r01)}{(r10-r01)x(r00-r01)}\right) \tag{9}$$

$$S(r) = \left(1 + \frac{(r01-r00)x(r \ -r11)}{(r00-r11)x(r01-r11)}\right) \cdot \left(1 + \frac{(r10-r00)x(r \ -r11)}{(r00-r11)x(r10-r11)}\right) \tag{10}$$

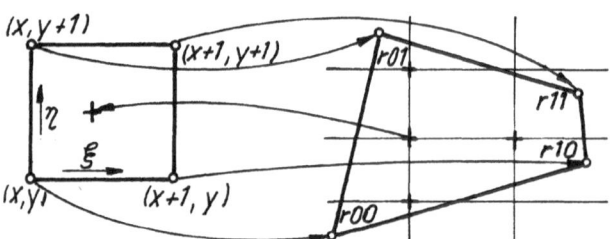

Fig. 1.Mapping of 4 adjacent samples

With these equations 12 vector products, 9 vector subtractions, 3 multiplications, 6 divisions, 8 additions and 3 routing steps are required for the computation of ξ,η from given u.v. All integer lattice points with

max(0,min(u00,u10,u01,u11)) < u < min(m,max(u00,u10,u01,u11))
max(0,min(v00,v10,v01,v11)) < v < min(m,max(v00,v10,v01,v11))

are inserted into the equations (8)-(10). Only results satisfying $0<\xi<1$, $0<\eta<1$ are accepted. It may occur that more than one destina-tion point falls into the quadrangle r00,r01,r10,r11 (local enlargement). Therefore the correponding data are pushed on a stack which will be exhausted during routing. The pixels arel labeled as empty if no integer solution u,v. exists. The ξ,η -values serve as arguments for the following interpolation procedure.

Assuming the signal as a Fourier finite function in x,y the resampling is ruled by

$$A(x,y)= \sum_{i=-\infty}^{\infty} \sum_{j=-\infty}^{\infty} A(i,j)\,\text{sinc}\,\pi(x-i) \cdot \text{sinc}\,\pi(y-j)$$

(11)

with $x=k+\xi$; $y=l+\eta$

This infinite sum will be approximated as follows corresponding to an earlier proposal of Wilhelmi (1982).

(a) Truncate the infinite lattice of support to a finite square window around the current k,l.

$|k| < K$; $|l| < K$

(b) Approximate the sinc terms for $k=1,\ldots K$

$$\text{sinc}\,\pi(x-i)=(-1)^{K-i}\,\sin\pi\xi/\pi(k+\xi) \approx c_K \sin\pi\xi$$

(12)

(c) and for $k=0$

$$\sin\pi\xi/\pi\xi \approx 1-\xi+c_0\sin\pi\xi.$$

(13)

A Chebyshev norm approximation yields for the first 5 coefficients:

$c_0 =\; = 0.14319$; $c_1 =c_{-1}=-0.21974$

$c_2 =c_{-2} =0.12889$; $c_3 =c_{-3}=-0.09151$

$c_4 =c_{-4} =0.07100$

The following equation estimates the signal amplitude for every pixel

$$F(i,j,\xi,\eta):=(1-\xi)(1-\eta)A(i,j)+(1-\xi)\eta A(i+1,j)$$
$$+ \xi(1-\eta)A(i,j+1)+\xi\eta A(i+1,j+1)$$
$$+\sin\pi\xi \sin\pi\eta\, G(i,j)$$

(14)

Equation (14) illustrates that the signal interpolation is reduced to a bilinear approximation corrected by the output of a FIR filter weighted by $\sin\pi\xi \sin\pi\eta$. Fortunately this filter is separabel. The signal interpolator requires that the 4 values $A(i,j)$, $A(i+1,j)$, $A(i,j+1)$, $A(i+1,j+1)$ and the highpass filter output $G(i,j)$ are available. There they meet the subpixel vector allowing the estimation of the signal. Then the subpixel vector ξ,η will be discarded and replaced by F as many times as destination points are in the quadrangle of fig. 1. The filtration process yielding $G(i,j)$ is accomplished by 2 times 2K rounting steps, 2 times 2K+1 multiplications and 2 times 2K additions. This can be done in parallel with linear speedup. Three routing steps are required for collecting the neighbors $A(i+1,j)$, $A(i,j+1)$, $A(i+1,j+1)$; 10 multiplications, 2 table lookups and 4 additions are required for computation of (14).

Routing

Every pixel corresponds to a memory location B where a grey level will be saved when it arrives at its final destination. Further there is a stack for 3-word items interpolated by the before decribed procedures.

F   – the resampled grey value
$X=u-x$ – the horizontal routing distance
$Y=v-y$ – the vertical routing distance

It is obvious that sufficient local memory is a supposition for this data structure. Let us declare the special symbol "e" for X,Y. Pixels with X=e or Y=e are empty and at disposal for storing intermediate values F, X and Y. Usually "e" is identified as the lowest possible value in a 2-complement representation of X or Y.

The following proposal presents a solution well suited for both kinds of SIMD schemes. It is similar to an algorithm developed by Fanya Montalvo in the context of list processing by associative machines (Foster 1976). The reader will observe that there are many similarities to parallel sorting algorithms too. The main steps of the procedure are the following:

(a) All pixels having X=Y=0 send the corresponding grey value into the cell B. Then X and Y are updated from the top of stack. An empty stack results in setting X=Y=e. This action can be done in parallel without communication between PE's. No more than 2 comparisons and 1 stack pointer updating per pixel are necessary. If there are only empty cells then the algorithm finishes.

(b) Shift the pixels along columns in parallel so that as many as possible Y vanish, i.e. a maximal number of pixels arrive at their destination row. The odd-even transposition scheme is applied for this (Baudet et al. 1978). We define disjoint pairs of adjacently arranged PE's and update their data if this contributes to the before mentioned goal. Otherwise nothing is done. Then we select new pairs of PE's not having joined in the step before (Fig. 2). This procedure is repeated until a stable state has been reached. It has been proved that no more than m cycles are needed if the dimension of the array is m. This is true for linear and cyclic arrays as well. The proof is based on the observation that after P cycles the distance of every item from its final destination is bounded by (m-p) supposed that the destinations are different. The fact that the last supposition is not fulfilled can be regarded by introduction of dummy locations where no match is required.

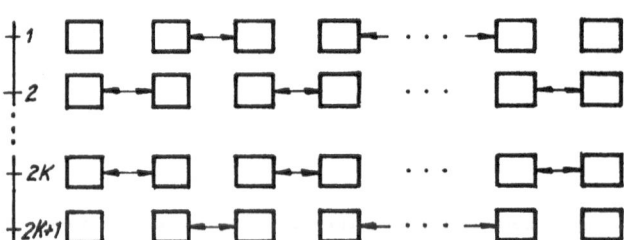

Fig. 2. Odd-even transposition scheme

There are 3 conditions for an interchange:

1. Both cells are not empty and $Y(i).GT.Y(i+1)+1$.
2. The upper cell $(i+1)$ is empty and $Y(i).GT.O$.
3. The lower cell (i) is empty and $Y(i+1).LT.O$.

If a nonempty cell changes its place, the corresponding grey value and the vector X,Y are copied. Y is updated as follows: $Y:=Y-1$ for moving upward, $Y:=Y+1$ for moving down.

The described procedure allows linear speedup because data dependence exists only within disjoint pairs of picture elements. Not more than 2 comparisons, 3 data exchanges and 2 increments/decrements are required per pixel in the worst case.

(b) Proceed as (a).

(c) Now we shift the pixels with their escorting data along the rows. The odd-even transposition scheme is applied again. The following conditions are necessary for an interchange:

1. $Y(i)=Y(i+1)=0$ and $X(i).GT.X(i+1)+1$.
2. $Y(i)=0$ and $X(i).GT.0$; the other cell with $Y(i+1)\neq0$.
3. $Y(i+1)=0$ and $X(i+1).LT.0$; the other cell with $Y(i)\neq0$.
4. For $Y(i)\neq0$ and $Y(i+1)\neq0$
   4.1 Both cells are not empty and $X(i).GT.X(i+1)+1$.
   4.2 The right cell $(i+1)$ is empty and $X(i).GT.0$.
   4.3 The left cell (i) is empty and $X(i+1).LT.0$.

If a nonempty cell changes its place, the corresponding grey value and the vector X,Y are copied. X is updated as follows: $X:=X-1$ for moving to the right, $X:=X+1$ for moving to the left.

Again full parallelism is possble with the worst case time for one cycle resulting from 5 comparisons, 3 data exchanges and 2 increments/decrements.

(d) Go to step (a).

The investigation of this procedure seems quite complicated in general. Lower and upper worst case bounds can be derived from the following observations. We assume a snake-like row-major ordering of the 2D-array yielding a linear one with m elements. Further a one to one mapping of source to destination points should exist (e.g. a rotation by 90°). Then our problem is equivalent to sorting in a linear array. It is well known (Thompson et. al. 1977) that the odd-even transposition scheme needs m^2 comparison/interchange steps in the worst case. The proposed algorithm must have better performance because it ensures additional routings over the distance m from row to row. On the other hand, a lower worst case bound on the number of comparison/interchange steps on a 4-connected processor array is $2(m-1)$.

Numerical simulations with m=64, m=128, m=512 and rotations by 90° have corroborated the conjecture, that the number of comparison/interchange steps q may be estimated by

$$q = m \sqrt{m} . \tag{15}$$

If the distortion is small, then we observe a decreasing q. In the case of a simple translation the algorithm works on its lower bound because collisions of pixels cannot occur.

The maximal time for one comparison/interchange cycle T is given by 5 comparisons, 3 exchanges and 2 increments/decrements. Due to the linear speedup of our algorithm the time for processing the whole image with m^2 PE's is qT. If the number of PE's is smaller than the number of pixels the expected time T decreases significantly by averaging over sequentially in time occuring decision tree traversals. Another reason for better performance than estimated by the right side of (15) is the preordering of shift values in rows and columns due to the continuity of $f(u,v)$ and $g(u,v)$.
As an example may serve the before mentioned translation.

The relations of routing for geometric dewarping and of sorting algorithms on mesh-connected processor arrays is obvious. So we can expect even more effective algorithms which exploit the possibilities of more sophisticated arrangements allowing perfect shuffle, multi-way and bitonic merge (Thompson et al. 1977). The extension to multidimensional arrays may yield further improvements too. On the other hand the simple idea of saving already arrived items and providing empty cells can improve sorting algorithms in processor arrays.

CONCLUSION

The performance of the algorithm should be shown under the following assumptions:

m=1024; I=7*7=49   i.e. $\triangle$ x= $\triangle$ y=146
K=5   i.e.   a 9*9-window
Local magnification less than 10.
Time for add, multiply, compare, exchange, table lookup: 0.4 µs
Time for vector product execution and division 1.2 µs
Time for a parallel routing: 4 µs

Then the following balance of computation is derived.

Generation of interpolator (7)	: 2*0.4 µs	=0.8 µs
Routing the interpolator	:49*0.4 µs	=196 µs
Interpolation corresponding to (3):(49+48)*0.4 µs		=38.8 µs
Computation according to (8)-(10) :		
10*(12*1.2+18*0.4+6*1.2+8*0.4)	µs=320 µs	
Routing the 3 neighbors	:3*4 µs	=12 µs
FIR filtering 9*9 routing	: 64 µs	
FIR filtering 9*9 add/multiply	:(18+16)*0.4 µs	=13.6 µs
Collecting the 3 neighbors	:3*4 µs	=12 µs
Interpolation (14)	:10*(10+2+4)*0.4 µs=64 µs	
Time for a comparison/interchange:T=(5+3+2)*0.4 µs		=4 µs
qT less than	:32*1024*4 µs	=131.1 ms

All but the routing times are given per pixel. For the example we estimate 131.8 ms/pixel including coordinate and signal interpolation. This values are comparatively high but allows full parallelism. Thus the performance of a 64*64 array is 33.7 s per image and of a 128*128 array 8.4 s per image.

REFERENCES

Evans, D:J:(Ed.): Parallel Processing Systems. Cambridge University Press, Cambridge 1982.
Batcher,K.E.: Design of a Massively Parallel Processor. IEEE Trans. on Computers C-29 (1980) pp. 836-840.
Coletti,N.B.: Image Processing on MPP-like Arrays. Rep. No.UIUCDCS-R-83-1132 Univ. of Illinois Urbana 1982.
Wilhelmi,W.: Geometric Transform on a SIMD-System - a Case Study. Int. Report LiTH-ISY-I-0527 Univ. of Linköping 1982.
Preston, K.;Duff,M.I.B.;Levialdi, S.;Norgren,P.;Toriwaki,I.: Basics of Cellular Logic with some Applications in Medical Image Processing. Proc. of the IEEE vol. 67 (1979) No. 5 pp. 826-856.
Kempe,V.;Rebel,B.;Wilhelmi,W.: The Interactive Image Processing Console A6471. Proc. 6th ICPR Muinch Oct. 1982 pp. 607-609.
Shepard,D.: A Twodimensional Interpolation Function for Irregularly Spaced Data. Proc. 23th Nat.Conf.ACM 1965, pp.517-523.
Eichhorn,N.;Kovalevski,V.A.;Saedler,J.;Wilhelmi,W.: Algorithmen auf dem Displayprocessor K2067. Bild und Ton 1986.
Foster, C.C.: Content Adressable Parallel Processors. Van Nostrand Reinhold Comp., MY 1976.
Baudet,G.;Stevenson,D.: Optimal Sorting Algorithms for Parallel Computers. IEEE Trans. on Computers C-27 (1978) No. 1, pp 84-87.
Thompson,C.D.;Kung,H.T.: Sorting on a Mesh-Connected Parallel Computer. Comm. of the ACM vol.20 No. 4 April 1977, pp. 263-271.

A PICTORIAL AND TEXTUAL IR ENVIRONMENT

BASED ON IMAGE DESCRIPTION

Isabella Gagliardi*, Dora Merelli*, Fulvio Naldi*,
Piero Mussio**, Marco Padula*, and Marco Protti**

(*)   Servizio Informatica Area di Milano - CNR
      Via Ampere, 56
      20131 Milano

(**) Dip. di Fisica, Sez. di Fisica Medica
      Universita' degli Studi di Milano
      Via Viotti, 5
      20133 Milano

INTRODUCTION

The availability of more and more on-line mass storage has
made feasable the realisation of new applications in the
field of the management of image archives by means of
Information Retrieval (IR) techniques.   These applications
can be realised by using documents that contain both
formatted and unformatted information that describes the
image itself.   In addition, the documents contain suitable
pointers that allows to have the images back from an external
memory.   It doesn't matter the recording support: microfilm,
magnetic or optical disks, and so on.   Formatted information
and descriptions are processed with IR techniques in order to
retrieve the documents and obtain a reproduction of the
desired images.

At present, the choice of formatted information and of the
descriptions is done manually, with a procedure similar to
that used by traditional systems by means of manual indexing.
In this work we describe a system that allows one to obtain
automatically a description of the image in a semi-natural
language and to use this description as an instrument for the
automatic classification and indexing of the document (fig.1)
associated with the image (fig.2).

In Section 1, we describe the approach on which this work
is based as well as the reasoning behind this approach.   We
also relate this to current literature and compare it with
other possible approaches.

In Section 2, we introduce the formal concept of
description, at least in its external form, that is to say,
how it is seen logically by the user.

```
      MOSTRA IL DOCUMENTO 1
AREA          195
PERIM          56
XB             19
YB             19
**** SOGLIA
032
***** NOME
PAVO
**** TIPO
GALASSIA A SPIRALE
**** DESCRIZIONE
BRACCIO DAL PUNTO (23,13) AL PUNTO (23,18) ED ESTREMO NEL
PUNTO (29,5)
NUCLEO ELLITTICO CON VERTICI NEI PUNTI (9,24) (17,25)
(23,15) (13,14)
BRACCIO DAL PUNTO (9,19) AL PUNTO (9,23) ED ESTREMO NEL
PUNTO (3,25)
****    PRIMAL-D
7 25 15  7  27 136  8  24  6  11  23  48  8  23 143 12  23 24 13 24 12
9 23  3 13  25  62 12  22  6  10  21  15 14  25  24 15  26 60 14 17  6
14 16 15 17  26 120 18  12 135  18  25  48 19  12   3 19  25 120 20
11 135 20 24  48  21 11 195  21  24 120  22 12 129  23 22 240 23
21  96 23 12 195  24 13 129  25  19 240  25 18  96  26 13  3 27 12
135  27 16 240 27  15 96 28  12  35  29  11 102 28  14  240 28 13
96  29 10 47  29  12 242 30 10 242  30   9 103 31   8 246 31  6 239
***** ORIGINE
FILE4-NASI005
****    FINESTRA
38,41-32,30
```

Fig.1

    In Section 3, this definition is justified on the basis of
the uses that we intend to make of it.

    In Section 4, the internal form of the description will be
presented along with the functions that permit it to be used
in the way desired.   At the end we have drawn some
conclusions.

1. THE PROPOSED APPROACH

    From our point of view, the problem of the automatic
indexing of images, even if it is technically more complex,
is not conceptually different from that of the automatic
indexing of texts.

    In practice it consists of extracting from the entire
image the significant structure and producing a uniform
description specially suited to the experiment. According to
our approach the descriptions, organised on several levels of
detail and complexity, are expressed in a form understandable
to the user. In order to do that, we use words and rules
that, although standardised internally to the application,
can be defined and modified by the users themselves.

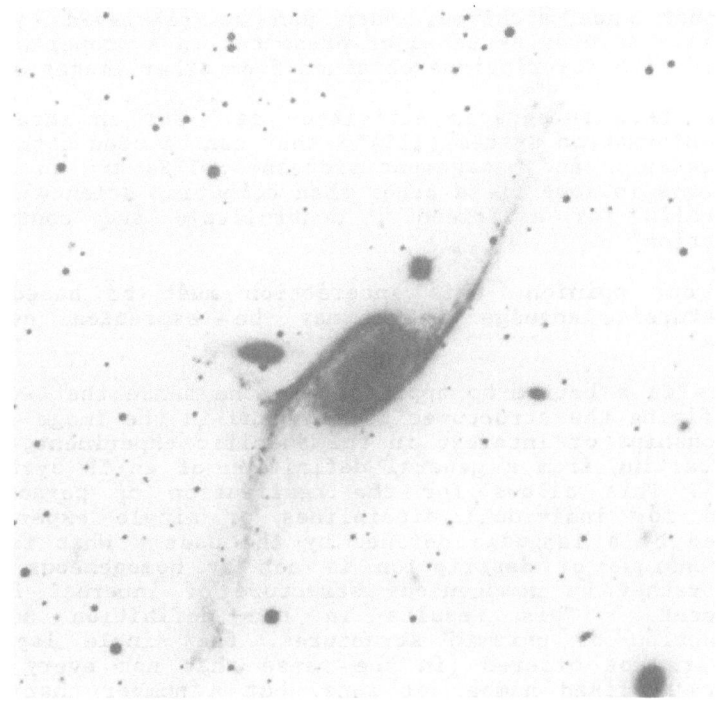

Fig.2

Similarily with what happens for IR Text Management Systems, this process of structure extraction and description allows to talk, also for image descriptions, about dictionaries of terms and a thesaurus of iconic descriptions, that can be used for automatic indexing and retrieval[0]. In addition, it is necessary to be able to automatically synthesise from the description an external representation (sketch) of the images.

The integration of recognition and retrieval tools allows firstly, a proper interpretation of the collected pictorial data and its subsequent meaningful use. A common requirement of these two activities is the need to model and access pictorial data with ease.

It is now generally recognized[1] that the use of pictorial data is not only based on the exploration of the image data and of pattern recognition algorithms, but also on the use of knowledge about the experiments, on the progressive and interlaced processing by both parametric and structural tools[1,2,3], on the comparison and merging of results of data obtained by different methods, and on the concurrent use of ancillary data merged with pictorial ones[1,4].

A description summarises the relevant characteristics of the image. A practical work generally does not end with the interpretation of images: the collected information is the starting point for some other activity (e.g. a Cartography project). Therefore descriptions must be structured in such a way, that once archived, they can be retrieved by their pictorial features sketched or presented in a proper way, and compared with descriptions obtained from other images etc.

All these integrated activities call for an intergrated Image Information System (IIS)[5,6] that can be used both in the interpretation and management processes. IISs are to be used by experts in some field other than computing science[7,8]. This fact calls for a friendly, controllable and controlled[9] interaction[5].

In our opinion, this interaction must be based on a semi-natural language which may be expressed even by symbols[9].

Ours is a bottom-up approach, in the sense that we start off defining the structures to be found in the image and the relationships of interest in the specific experiment, rather than starting from a general definition of an IR system for images[6]. This allows for the realisation of personalised systems for individual disciplines or single experiments, governed by a language defined by the user. What is more, the technique of description is not by homogeneous zones, but[2,3] rather by homogeneous structure of interest for the experiment. This results in the definition and the construction of pyramid structures, the single layers of which are not ordered (in the sense that not every father clusters a fixed number of sons, but a number that varies according to the context). This approach appears to provide a greater capacity for describing complex structures, it is synthetic and appears to be easily adaptable to the treatment of images of different classes, even if it is more difficult to translate for specialised image processing hardware structures.

In this framework we have tackled the problem of the interpretation and use of images. This approach has brought us to the definition of a system scheme that allows us for each separate application to:

- define the rules and the structures to look for in the images and the lexicon to use;
- create the documents from the data (images) on the basis of the rules defined, structured on descriptions at different levels of complexity, synthesis and resolution;
- generate the database with the description and the associated images;
- use the system with graphical restitution of the documents found with the resolution required, on the basis of the description;
- define new structures or relationships;
- update the descriptions starting from those already generated in the database or those associated with new structures extracted from the images.

The system is already partially implemented, with the singular parts under test in different experiments. In this work we dwell particularily on the process of constructing the document.

2. DESCRIPTIONS AS TOOLS OF THE INTERPRETATION PROCESS

In our approach descriptions are derived by the recognition of structures and the computation of their properties. A 'structure' is a set of pixels, which satisfies some relations. A 'feature' of a structure is a characterizing subset of the structure itself. So, for instance, a 'blob'[10] may be viewed as a structure composed of all the pixels connected to a given one. Inlets and outlets are in this case features of the blob. Then an object is a structure composed of blobs, which satisfy some relationships. In this case the related blobs are features of the objects. A structure as well as a feature, is characterized by a set of variables named 'attributes', each of which may assume a value ('property'), in a well defined set. A property may be a multi-dimensional vector.

Attributes and properties are used to build a description. A description is a triplet D=<N,P1,P2> where:

N is the name of the description and denotes the entity being described.
P1 is the set of global descriptors
P2 is the set of the names of descriptions, which are concerned with the features of the entity.

A descriptor is a couple <n,v> where n is the name of an attribute and v is the associated property. The names of the set P2 are in their turn names of descriptions. P2 may be partitioned into subsets, say P2i, i=1,...n, n being specified for each descripion D. P2i collects the names of the features of descriptions taken from a specific point of view. P2 is void at the lowest level of description, that is the level at which the structures are no more decomposed. Also P1 may be empty.

In the astronomical example, a description could be:

N: Pavo galaxy
P1: <area=1000>; <X position=100>; <Y position=100>
P2: <bracciol, nucleol, braccio2>
where bracciol (arm1), braccio2 (arm2), and nucleol (nucleousl) are names of the descriptions of different features.

Four kinds of descriptions are the basis for our approach to the interpretation: intentional, discriminant, measured and 'a posteriori'. First the intentional description is defined. For each class and for each kind of object in the class, an intentional description collects all the descriptors relevant to the experiment in hand. In other words, the intentional description derives from the union of all the descriptions of the known instances of objects in the class, described following the models which are specific to the discipline to which the experiment refers. Remember, that within one experiment, multiple points of view may exist. From each point of view a different intentional description of the same object may be derived. From every intentional description one or more discriminant descriptions are derived. A discriminant description collects a set of properties which guarantees that a structure enjoying them belongs to the class identified by the name of the discriminant description. In the measured description the properties are the results of the measurements performed on structures identified in the input images. Not every attribute present in the intentional description and of interest in the experiment belongs to the discriminant set and/or can be measured before understanding the meaning of the observed structure. The 'a posteriori' description is defined to be the one which collects all the properties of interest in the experiment, comprising those which are measured after the interpretation process has assigned every structure to a meaningful class. The intentional and discriminant descriptions represent the 'a priori' knowledge of the experiment and of the kind of data collected. This knowledge is the basis on which the specific interpretation tool is built-up.

The input images subjected to the measurement activity are mapped into the measured description. This mapping is performed by parallel exploitation of the set of rules derived by the intentional description. These rules specify what is to be measured and how. The interpretation process is performed exploiting a set of rules which drives a description process and multi-valued logical trees (MVLT) which summarise the set of discriminant descriptions previously stated. A multi-valued logical tree maps a measured description into an interpreted class.

As properties in the measured description can be partial, inexact or even missing, the plausibility of interpretation is also evaluated. If the plausibility is judged sufficient, according to some experiment-dependent acceptance rules, the last step is performed. The knowledge about the meaning associated with each structure in the image by the interpretation process is exploited to complete the 'a posteriori' description. That is, properties required by the intentional description, but whose measurement was meaningless or impossible before, are now evaluated as required by the experimenter.

The 'a posteriori' description is, therefore, the one which is archived in the information retrieval system, to be used in the management of the experimental data. It must be possible to retrieve it by its name, by its characterizing features, to update it on the basis of new collected data, and to use it with other kinds of descriptions. Moreover, in some situations, it can be requested that a new interpretation session, on archived pictorial data, is executed to compute newly defined characteristics.

3. DESCRIPTION REQUIREMENTS

In order to be used in the realisation of an archive for the management of images, a description must satisfy certain requirements. These depend substantially on the fact that the archive must be managed efficiently by even a user that is not necessarily an expert in computer science. One must, therefore, remember that the internal description, reserved for the machine and used in computation are constrained by

PAVONE
NUCLEO ELLITTICO CON ESTREMI NEI PUNTI (9,24) (17,25)
(23,15) (13,14)
BRACCIO DA (23,13) A (23,18) CON ESTREMO IN (29,5)
BRACCIO DA (9,19) A (9,23) CON ESTREMO IN (3,25)

Fig.3

the requirements of the external descriptions subject to
human interpretation. On the other hand, the internal
description must also be easily manipulated by the computer
in order to increase the efficiency of the system.

A description must, therefore, be synthetic, that is,
include all the aspects of interest to the structure
described, but occupy the minimum amount of memory. It must
also possess the capacity to evoke, in the observer, a
precise idea of the shape of the object described and of the
parts that it is composed of. Another essential
characteristic is the suggestibility, that is the capacity to
induce the observer to associate the object described with
others already known to him. Naturally, this capacity to
associate and suggest varies with the experience and
occupation of the observer, but in the ambit of an experiment
on a particular class of images, it is possible to produce
descriptions that are suggestive, in general, for a certain
category of experts.

As far as it is concerned the internal description of the
image is represented by a data structure, that is a
collection of data organised following precise rules. This
description must be brief and formal (that is unambiguous and
processable by the computer) whilst not generally suggestive
to the user nor evocative of the structure of the object
described.

Between the forms of the external description we can still
destinguish verbal descriptions (fig.3) and graphical
descriptions (fig.4)[9]. The verbal description consists of a
collection of words and phrases in which one can make use of
the typical terminology of the experimenter. It, therefore,
possesses all the characteristics that are requested by a
description. The graphical form of the description may
reproduce the exact version of the original image or an
approximation (sketch) (fig.3), both possessing a notable
suggestibility for the observer.

4. DATA STRUCTURES

In Section 2, one can see that a description of a
structure is a triplet <N,P1,P2> where P1 is the collection
of the global descriptors (attribute-value) and P2 refers to
a description of sub-parts of the structure.

The data structure adopted reflects this organisation. It
is formed out of N levels (with N variable according to the
experiment). The elements described at the highest level are

411

```
SONO PRESENTI NELLA DESCRIZIONE I SEGUENTI TIPI DI STRUTTURE:
NUCLEO
BRACCIO
VUOI I SIMBOLI SCELTI AUTOMATICAMENTE O MANUALMENTE?
.MANUALMENTE
DAMMI 2 SIMBOLI PER IL DISEGNO
.ox
                         5    10   15   20
                     ---------------------
           0 |                             |
           1 |                          x  |
           2 |                        x    |
           3 |                       xxx   |
           4 |                       xxx   |
           5 |                      xxxx   |
           6 |                 oααααααo     |
           7 |               ooαααooooo     |
           8 |             oooooooooooo     |
           9 |             oooooooooooo     |
          10 |            ooooooooooooo     |
          11 |           oooooooooooooo     |
          12 |           ooooooooooooo      |
          13 |           ooooooooooooo      |
          14 |           ooooooooooooo      |
          15 |           oooooooooooo       |
          16 |           ooooooooooo        |
          17 |           oooooooooo         |
          18 |           oooooooooo         |
          19 |           ααααααααx          |
          20 |           xxααααααα          |
          21 |           xxxxxxx            |
          22 |           xxxxxx             |
          23 |           xxxxx              |
          24 |           xxx                |
          25 |           xx                 |
                     ---------------------

                SIMBOLO   STRUTTURA
                   o      NUCLEO
                   x      BRACCIO

                      Fig.4
```

the structures that one intends to search for (e.g. star,
galaxy). To each element is associated a description,
composed of n global descriptors, and it is placed in
relation (with a key) to the substructures (the names of
which are indicated in P2) present in lower levels (examples:
braccio, nucleo). In their turn, these elements will be
associated to a description formed from n global descriptors
and their substructures, and so on, until arriving at the
last level formed from the structure nolonger sub-divisible.
The data structure also allows for the management of
different descriptions of the same elements (multiple
descriptions).

The structure of the data and the archiving and retrieving
operations made on them are based on the following actions:

- construction of the document to store ('a posteriori' description);
- archive creation and management;
- retrieval and the graphical representation of the documents.

At present we have implemented a prototype of the whole system where the first point is achieved in the way described in this work (ISIID)[11]; the last two points are carried out by means of IRS-G⁰, IR system designed for the management of multimedia documents and specially of pictorial data.

REFERENCES

0. F.Naldi, I.Gagliardi, P.Gallitognotta, Description of an Information Retrieval System with Graphic Capabiblities, in : "Proceedings of the Information Computing Symposium 1985", G.Valle, G.Bucci eds, North Holland, (1985).
1. A.Tailor, A.Cross, D.C. Hogg, D.Mason, Knowledge based interpretation of remotely sensed images, Image and Vision Computing, vol.4, no 2, pp. 67-83, (1986).
2. M.A. Eshera, K.S. Fu, An Image Understanding System using Attributed Symbolic Representation and Inexact Graph-matching, IEEE T-PAMI, vol. SMC-10, no 5, pp. 604-618, (1986).
3. W.H.Tsai, K.S. Fu, Attributed Grammar - A Tool for Combining Syntactic and Statistical Approaches to Pattern Recognition, IEEE Trans. on Systems, Man and Cybernetics, vol SMC-10, no 12, pp. 873-885, (1980).
4. C.F.Hutchinson, Techniques for Combining Landsat and Ancillary Data for Digital Classification Improvement, Photogrammetric Engineering and Remote Sensing, vol.48, no 1, pp. 123-130, (1982).
5. S.K.Chang, Image Information Systems, Proceedings of the IEEE, vol. 73, no 4, (1985).
6. S.K.Chang, S.H.Liu Picture Indexing and Abstraction Techniques for Pictorial Databases, IEEE PAMI, vol. PAMI-6, no. 4, (1984).
7. P.Mussio, R.Rabagliati, Analysis of Water Remote Sensed Data: Requirements for Data Bases and Data Bases Interactions, in : "Data Base Techniques for Pictorial Applications", A. Blaser ed., 1980.
8. D.M. McKeown, Jr, Digital Cartography and Photo Interpretation from a Database Viewpoint, in : "New Applications of Data Bases", G. Gardarin, E. Gelenbe eds., Academic Press, (1984).
9. M.Beretta, P. Mussio, M. Protti, Icons: Interpretation and Use, 1986 IEEE Computer Society Workshop on Visual Languages, IEEE Computer Society Press, (1986).
10. D.Merelli, P.Mussio, M.Padula, An Approach to the Definition, Description and Extraction of structures in Binary Digital Images, Computer Vision Graphics and Image Processing, no. 31, pp. 19-41, (1985).
11. U.Cugini, M.Dell'Oca, D.Merelli and P.Mussio, A Computer Aided System for Interactive Definition of Digital Image Interpretation, in : "Digital Image Analysis", S.Levialdi ed., Pitman, pp. 270-279, (1984).

IMAGE ANALYSIS – APPLICATIONS

A LOW COST 3-D VISION SYSTEM FOR ROBOTIC ASSEMBLY

Maurice Poulenard and Georges Stamon

Laboratoire d'Intelligence Artificielle
Institut Universitaire de Technologie
B.P. 527 90016 Belfort Cedex, France

ABSTRACT

A 3-D vision system designed to enable automatic acquisition of mechanical parts in a bin by means of an assembly robot is presented. Hardware
cost is kept low by using standardised parts: a microcomputer, a frame
grabber, a translating table and one or two video cameras. Yet versatility
is preserved and real time industrial rate is achieved. A very fast recognition method has been designed which is based on local features of the
objects to be identified in the bin and metric information. Early processing
steps avoid lengthy treatments such as systematic edge detection. Rather a
few local features are firstly identified in an image in order to achieve
a rough estimate of the correspondence between an object in the bin and a
stored model of the same object. This correspondence is then confirmed and
refined iteratively. Depth information is obtained by stereoscopy. An experiment involving the acquisition of snubber valves is presented.

INTRODUCTION

One of the most common tasks to be completed for achieving automation of
industrial processes is to move objects from one place to another. Frequently it is necessary to feed a workstation with parts presented in a specified position and orientation. Numerous approaches of this problem have been
used. It may be possible to control the position and orientation of parts
all along the production line by means of specially designed packing and
conveyors. An other method involves mechanical prepositionning devices such
as vibrating bowls. These methods often have drawbacks such as generating an
agressive environment or being costly. Further they lack flexibility. So the
use of a robot equipped with an adequate perception module will be prefered
in many cases. Artificial vision is attractive in this context. Indeed
vision is our most powerful sense. It does not require physical interaction
or excessive proximity and it frequently provides enough information to
enable useful interaction with our surroundings. In an economical context
the large diffusion of video cameras provides us with a relatively inexpensive captor for artificial vision. However human vision is of a great complexity. In order to achieve efficient programs able to run repetitively and
reliably in an industrial environment, the vision task must be precisely
restricted. The system presented here is meant to acquire the location and

orientation of partially visible objects jumbled in a bin. The objects are three dimensional and may have arbitrary orientation. However the algorithms presented here make use of problem dependent knowledge which comprises the presence of local features such a holes and corners in sufficient number in a same plane, as well as metric data. One of the factors inhibiting wide diffusion of vision systems in industrial automation is cost. It is generally admitted that computer hardware cost is only a small part of mechanical investment in robotic systems. Similarly programs should be easely adapted to another similar practical task if flexibility is to be preserved. An architecture and algorithms that tend to meet this needs in the context of robotic assembly has been attempted. In section 1 a "low cost" hardware configuration is presented along with experimental data which is used to examplify the method sketched in sections 2. Timing statistics have been gathered in section 3.

1. HARDWARE CONFIGURATION

Image workstations constructed around a standardized microcomputer exist in many laboratories and extension of their use has been advocated[7]. Many reasons make this concept attractive. Among them the cost effectiveness of microcomputers, their computing power which is approaching the needs of more and more graphics processing and vision tasks, an increase in portability for the realizations. Further the development of compatible frame

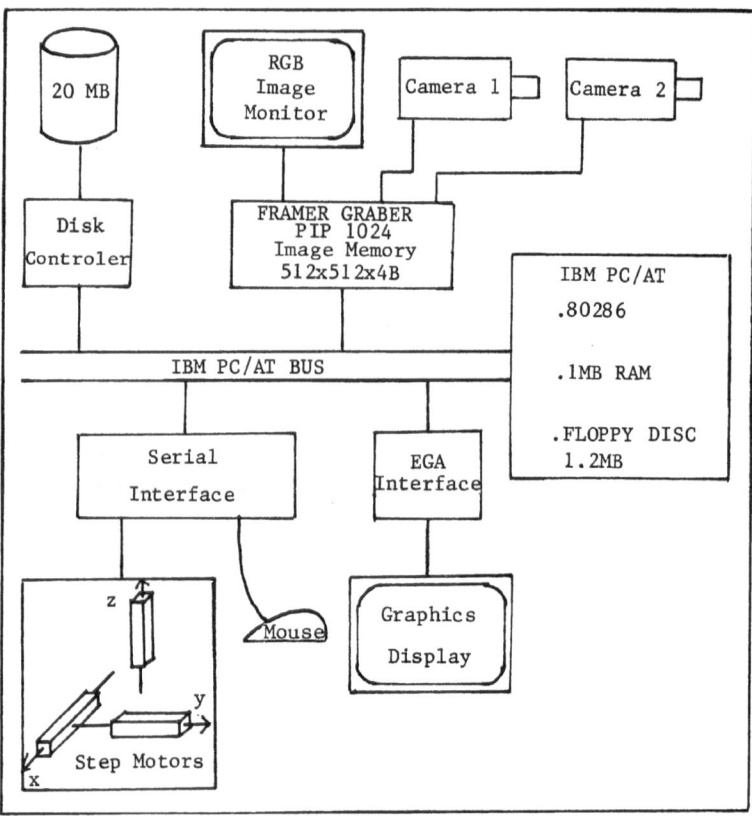

Fig. 1 . Hardware Configuration.

grabbers with enhanced capabilities should be encouraged by the prospect of a wider market.

Video and Computer Hardware

For the binocular stereo experiment presented here we used two CCD video cameras (SYSTEME SUD MICAM HRS) with 500x582 photoelements. Focal length was 16 mm. and baseline was set at 100 mm. Distance between camera and objects was about 400 mm. resulting in a pixel width in the order of 0.4 mm. The images were digitized in 512x512 pixels with a resolution of 8 bits/pixel by means of an image digitizer board (MATROX PIP 1024) plugged in an expansion slot of a microcomputer (IBM AT) with 1024 ko memory and a 20Mo fixed drive.

Motorization

The bin out of which parts had to be picked up was larger (45x29 cm) than image size so that cameras had to be displaced over it. This was achieved by means of a translation table (CHARLYROBOT) equipped with step motors controlled through the computer. This way the bin could be moved in the X and Y horizontal directions by increments of increments of 0.01 mm. and in the vertical direction by increments of 0.005 mm.

Lighting

Since high quality captors were used it was expected that major image defects could be attributed to lighting variations. Nevertheless we delibe-rately made no great effort to improve this because it was thought difficult (or equivalently costly) to maintain a finely tuned lighting in an assembly shop. Two 40 watt neon tubes were placed both sides of the bin with some distance in such way that the bin could be moved or accessed easely and light would hit with about 45° slant. We did not want the images to be invaded by shadow but some local features like holes are most easely recog-nised as small shadow spots and would be less visible with vertical lighting.

The parts in the bin

A snubber valve shown on fig.2 was to be picked from a bin shown on fig.3. This part can coarsely be described as constituted of a circular disk

   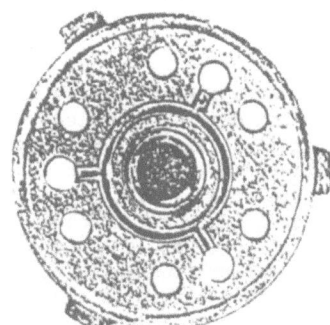

Face B   Face A

Fig. 2. The parts.

with nine small holes bored in it. One face (A) shows a bigger screwed hole while the other (B) has three small protruding wings. The disk diameter is about 40 mm. while the overall heigth with the wings is 20 mm. Numerous less useful details can be seen on fig.3. The part is to be seized in its center by a robot equiped with a three digit gripper which can be inserted in the center hole of face (A) or between the wings of face (B).

Fig. 3. The bin of parts: It can be seen that the parts on the right side of the image cannot be seized because they are too much slanted. When neighbouring parts will be removed they probably will have a more manageable position. If not, a robot move will be needed to displace the parts.

II. OUTLINE OF THE METHOD

The bin of parts problem, or to enable a computer controlled industrial manipulator to acquire a part from a bin or a pile where identical parts are randomly placed has been the subject of increased research in the last decade. Many different methods have been experimented such as using structured light, painted textures or photometric gradients.

Stereo Matching

The system presented here uses the binocular passive stereo paradigm where the six degrees of freedom of the pose of a solid in 3-D space is obtained through analysis of the parallaxe disparity of two images of the same static scene shooted from a slightly different viewpoint. Twin images can be correlated at different levels. In edge-based stereo edge elements found in each image are matched. Combinatorial explosion must be avoided in this process. One way to achieve this is implemented materially within a process of calibration where the scan lines of the cameras are aligned. Matching is then attempted along corresponding "epipolar lines" only. This reduction to a one dimensional process is correct if the distances from the object to the cameras do not differ too much. Other heuristics of continuity and ordering are used. The stereo used here is based on local features which are elaborated from edges and from previous knowledge of the physical properties and arrangement of the objects in the bin. These local features

are much less numerous than edge elements and are structurally linked, so
that matching complexity is alleviated. Once features have been located in
the twin images and matched range information is easely computed. Determination of the position and orientation of the different parts seen in both
images follows using the object knowledge available and usual image segmentation cues[6]. Overall accuracy relies mainly on the precision of edge
location and on the image formation model obtained in an offline calibration
procedure.

Some favorable features of our experiment allow a reduction in computing effort. Firstly observe that the bin contains many identical parts,
but only one is to be picked at any time (new snapshots are taken after each
maneuver because parts may have been jumbled). Experimental setup is arranged in such a way that the probability of finding several parts within image
bounds is great. This enables fast focus of attention after some tests have
been effected (If only a few parts are left in the bin segmentation against
a contrasting background is even more easy). Secondly range variation is
limited. This limitation is material in our setup where the height of the
bin is 5 cm. 1/8 of the distance from object to camera. Range limitation
stems from the picking strategy too, when priority is given for picking
pieces on the top of the pile. This strategy enhances robot accessibility at
the same time. So the parallaxe disparity is limited. If further local
features are far enough away from one another matching combinatorics become
very low. These two observation are rather general. More particular to this
experiment is that local features are coplanar. These features are the
openings of the 4mm. diameter holes bored in the discshaped portion of the
parts. Two models are available for face A and B and discrimination is easy
once the plane has been located.

Hough Transform

Any plane searching method could be used. Hough transform was chosen
for its flexibility. It was originally introduced to detect straight lines
in a noisy environment and later was generalized to general curves and even
empirically defined patterns. The Hough transform is a mapping of feature
points on subsets of a parameter space. For example let M_i (x_i, y_i, z_i)
for i = 1 to n be points supposedly belonging to a plane of equation
$$z = Ax + By + C \qquad (A,B,C) \text{ in a parameter space P}$$
then each M_i is mapped on a subset P_i of P defined as
$$P_i = \{ (A, B, C) \text{ in } P \mid Ax_i + By_i + C = z_i \}$$
An approximation to (A, B, C) will be obtained by means of a voting
procedure. The parameter space is partitionned into a number of cells. In
each cell the number of P_i that intersect it is accumulated. Parameter
(A, B, C) is then supposed to be in the cell with the highest score. The
drawback of Hough transforms is storage and computing time consumptions
therefore a hierarchical implementation has been chosen . This "Fast Hough
Transform[6]" uses an octtree representation of the parameter space and a fast
incremental setup of the distance calculations required for the vote
counting. Strategic choices were made for tree traversal and pruning in
relation with the facts that the resolution is variable and that only one
solution is needed. This procedure is used to achieve both the selection of
feature points and the computation of the face plane equation.

Algorithm

The procedure involves a sequence of steps alternating feature extraction and selection. Extraction proceeds mainly by edge detection. Skimming
stems from the picking strategy, the size information on parts and holes and
the Hough transform. Main points are:

Step 1: First detection. The overlay part of the left image is scanned at low resolution (1 pixel out of 4 in each direction) for detecting large gray level discontinuities. A temporary array of the size of a line is used to save results so that the continuity of edge segments can be maintained on the flight. The process constructs edge segments of nearly vertical orientation which are useful for parallaxe estimation and ignores horizontal edge portions. Two classes of transitions, black to white (a) and white to black (b) are recorded in the same pass along with some information concerning edge segments such as beginning and end points and the label of the nearest element of class (b) on the left of each element of class (a).

Step 2: First selection. Step 1 yields many edge segments which cannot correspond to small hole features. Modeling the holes as black spots it is an easy task to eliminate them using information gathered in step 1. This way a list of potential holes as seen from the left camera is created. To each item of the list is associated a small horizontal window where the same work is effected resulting in potential stereo pairs. Matching and range evaluation is then performed. A cluster of supposed holes with maximal z coordinate is selected to feed

Step 3: Hierarchical Hough transform. This step is mainly seen as a selection process used to eliminate wrong stereo pairs as well as features belonging to other parts than the one selected. The backtracking feature of the FHT from parameter space to physical space is most useful here.

Step 4: New detection. Missed holes are searched and other edge information such as shadows and borders is looked for in selected directions and vicinity. In this step edge detection is conducted with full accuracy but only a few small spots in the image are affected.

III. EXPERIMENTAL RESULTS

An experiment was conducted with 50 scenes of the bin of parts described in section I. The PC AT was clocked at 6 Mhz. and the programs were written in Pascal. Parts were extracted from the bin manually with the help of the image monitor because we did not have a robot at our disposal. Therefore no testing of geometrical accuracy was possible. Two scenes were rejected, one of them in step 2 (no cluster of features satisfying conditions). Average computing time was 5.43 s. with 1.32 s. . Step 1 which was very carefully implemented ran in 1.27 s. average time and 0.12 s. standard deviation. In fact these figures are only indicatory since many parameters of the implementation were arbitrarily chosen. For example the baseline between the cameras was set to 100 mm. and the optical axes kept parallel with the result that only 55% of the images overlaid. It would have been possible to focus the cameras on some point of the scene, but only with an increase in the proportion of occlusion for slanted parts. Clearly a better compromise had to be found between this extremes. Similarly a clustering procedure is called before FHT in step 3. The clusters correspond to overlapping portions of the physical space and are sorted in decreasing order of the likelihood that they contain a solution. After a few unsuccessful trials the scene is rejected because it is decided that a part can be found more surely and quickly in another portion of the bin. The exact number of trials is a matter of choice and affects both the timing statistics ant the reject rate. Therefore we think that an environment enabling and guiding systematic experimentation is necessary in order to foresee the behaviour of the application when several parameters are varied. We are currently working in this direction.

ACKNOWLEDGEMENTS

We are thankful to PSA-CERA and particularly to Monsieur F. Caillot for supporting this project.

CONCLUSION

A low cost 3-D vision system for the bin of parts problem has been presented. Implemented on a microcomputer it can control a robot used in an assembly line for picking parts out of a bin. A fast recognition method is used involving generation and evaluation of hypothesis from a model emphasizing clusters of local features of the object to be located. Stereopsis is used for the evaluation of 3-D pose parameters.

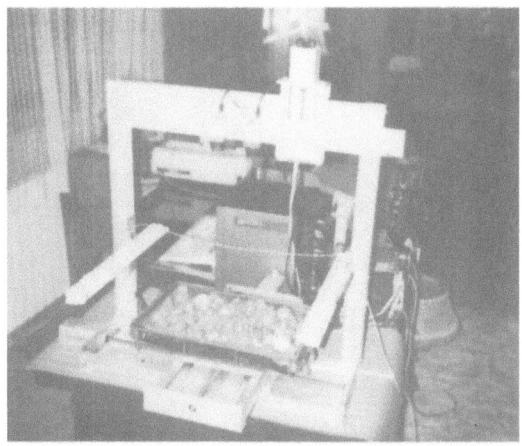

Fig. 4. Image acquisition system.

REFERENCES
1. N. J. Ayache et A. Darmon, Reconnaissance récursive et localisation de formes planes partiellement visibles dans une image, 9° Colloque GRETSI, Nice 1983.
2. C. M. Brown, Inherent bias and noise in the Hough Transform, IEEE Trans. on PAMI-5 n°5, sept. 1983, p. 493-505.
3. S. Clerc, Algorithmes de localisation et de reconnaissance de pièces partiellement cachées, dans un contexte de vrac volumique. Memoire ingenieur CNAM. Belfort 1985.
4. P. V. C. Hough, Method and Means for Recognizing Complex Patterns, U.S. Patent 3069654, 1962.
5. L. Hungwen, M. A. Lavin and R. J. Le Master, Fast Hough Transform: A Hierarchical Approach, CVGIP vol. 36 (1986) pp.139-161.
6. D. Marr, Vision, W. H. Freeman & Co. New York(1982).
7. E. G. Ramo, and J. Salminen, An experimental study of low Cost Image Workstation Architectures.in: Proc. 8th ICPR Paris Oct. 1986 pp. 34-37.
8. K. R. Sloan, Jr, Dynamically quantized pyramid, in Proc. 7th IJCAI, pp. 734-736.

DETERMINATION OF EGOMOTION AND ENVIRONMENTAL LAYOUT FROM

NOISY TIME-VARING IMAGE VELOCITY IN MONOCULAR IMAGE SEQUENCES

John L. Barron, Allan D. Jepson* and John K. Tsotsos*

Department of Computer Science, University of Toronto

Toronto, Ontario, Canada, M5S 1A4

1.1 Introduction

In this paper, we present a algorithm for computing the motion and structure parameters that describe egomotion and environmental layout from image velocity fields generated by a moving monocular observer viewing a stationary environment. Egomotion is defined as the motion of the observer relative to his environment and can be described by 6 parameters; 3 depth-scaled translational parameters, $\vec{u}$, and 3 rotation parameters, ω. Environmental layout refers to the 3-D shape and location of objects in the environment. For monocular image sequences, environmental layout is described by the normalized surface gradient, α, at each image point. To determine these motion and structure parameters we derive nonlinear equations relating image velocity at some image point $\vec{Y}(\vec{P}',t')$ to the underlying motion and structure parameters at $\vec{Y}(\vec{P},t)$. The computation of egomotion and environmental layout from image velocity is sometimes called the **reconstruction** problem; we reconstruct the observer's motion, and the layout of his environment, from (time-varying) image velocity. A lot of research has been devoted to devising reconstruction algorithms. However, a little addressed issue concerns their performance for noisy input: how accurate does the input image velocity have to be to get useful output?

1.2 Previous Work

The most common approach to monocular reconstruction involves solving (generally nonlinear) systems of equations relating image velocity (or image displacement) to a set of motion and structure parameters ([Longuet-Higgins 81], [Tsai and Huang 84], [Prazdny 79], [Roach and Aggarwal 80], [Webb and Aggarwal 81], [Fang and Huang 84a,b], [Buxton et al 84], [Williams 80], [Dreschler and Nagel 82], [Lawton 83]). Some of the issues that arise for these algorithms are the need for good initial guesses of the solutions, the possibility of multiple solutions and the need for accurate input. The latter is by far the most important issue if the reconstruction approach is to be judged a success. As Waxman and Ullman [85] and others have noted, reconstruction techniques that use image velocities of neighbouring image points require accurate differences of these similar velocities. That is, solving systems of equations effectively requires subtraction of very similar quantities: the error in the quantities themselves may be quite small but since the magnitudes of their differences are quite small, the relative error in them can be quite large. Hence, such techniques can be expected to be sensitive to input errors.

A second approach to reconstruction involves solving nonlinear systems of equations relating local image velocity information (one image velocity and its 1st and 2nd order spatial derivatives) to the underlying motion and structure parameters ([Longuet-Higgins and Prazdny 80], [Waxman and Ullman 85]). The rationale is that using local information about one image point means that the problem of similar neighbouring image velocities can be averted. However, this is replaced with the problem of computing these 1st and 2nd order spatial derivatives. Waxman and Wohn [85] propose that these derivatives be found by solving linear systems of equations: where each equation specifies the normal component of image velocity on a moving non-occluding contour in terms of a Taylor series expansion of the x and y components of image velocity. In effect, their reconstruction algorithm divides the computation into two steps: use a normal velocity distribution to compute image velocity and its 1st and 2nd order spatial derivatives at an image point and then use these as input to an algorithm that solves the non-linear equations relating motion and structure to the image velocity and its 1st and 2nd order derivatives.

Only recently, have researchers begun to address the use of temporal information, such as temporal derivatives, in reconstruction ([Subbarao 86], [Bandyopadhyay and Aloimonos 85]). We note that others' use of temporal derivative information and our use of time-varying image velocities are effectively equivalent; image velocity fields (at least locally) can be derived from one image velocity and its 1st and/or 2nd spatial and temporal derivatives and vice-versa. Indeed, image velocity fields are often used in the derivation of spatial and temporal image velocity information.

It is somewhat disappointing that almost none of these reconstruction techniques have been successfully applied to flow fields calculated from realistic scenes. Primarily, the problem is the difficulty in computing accurate flow fields. There has been little or no error analysis in previous monocular reconstruction work, although some researchers, such as [Waxman and Ullman 85], [Buxton et al 84], [Aloimonos and Rigoutsos 86], [Snyder 86] and [Subbarao 86] have begun to consider the inherent sensitivity of their algorithms to random noise in the

---

* Also, Canadian Institute for Advanced Research

input. See [Barron 84,87] for a more detailed survey of reconstruction techniques and their problems.

1.3 Contributions of this Work

Most approaches to reconstruction use an image velocity field at one time instance only; We propose that time-varying image velocities arising from a moving surface be used. Using image velocities distributed over both space and time means the range of image velocity magnitudes may be much larger than the range of image velocity magnitudes distributed over space at one time only. Hence, the problem of similar neighbouring velocities may be reduced. Intuitively, this makes sense, as using image velocities over a time interval is, in effect, simply extending the spatial extent of the surface (and, generally, the larger the spatial extent, the greater the range of image velocity magnitudes). As well, the use of time, may allow the recovery of motion and structure parameters from flow fields that are not unanalyzable at one time only. A couple of other advantages result from using a spatio-temporal distribution of image velocity; namely, sparse flow fields may be used (often accurate image velocity can only be computed at a few points in each image) and occlusion of parts of a surface at different times can be handled as we can use image velocities on the visible parts of the surface at any given time.

In order to relate a spatio-temporal distribution of image velocities to the motion and structure parameters at some image point we need to make some assumptions:

(a) The use of projective transformations requires the assumption of **rigidity**; objects are assumed to be rigid. This ensures that the image velocity of an object's point is due entirely to the point motion with respect to the observer and not due to changes in the object's shape. There is much psychological evidence, for example, [Gibson 57] and [Ullman 79] to support the premise that humans have a tendency to use the rigidity assumption to analyze world scenes.

(b) The 3-D surfaces of objects can be described locally as a plane. The **local planarity** assumption means curved surfaces are treated as collections of adjacent planes.

(c) The observer rotates with a constant angular velocity for some small time interval. Webb and Aggarwal [81] call this the **fixed axis** assumption.

(d) The spatio-temporal distribution of image velocity results from 3-D points on the same planar surface. We call this the **same surface** assumption.

(e) The observer translates with a constant velocity with respect to some frame of reference. This assumption is necessary for relating $\vec{u}$ values on a planar surface over time and will be explained in more detail below.

The use of a spatio-temporal distributions of image velocity means motion and structure is computed using local spatio-temporal data; thus we are not necessarily restricted to stationary environments as we can treat each independently moving surface as stationary relative to the moving observer.

The reader may question of the validity of using these kinds of assumptions, especially when real world imagery is used as the input. As Thompson and Kearney [86, p17] comment: "Unrealistic assumptions are only justifiable when it can be shown that useful answers can be obtained in realistic situations despite violations in the assumptions". Violation of the assumptions listed above is treated as one type of error in the input data. We believe that artificial situations such as the one created by our assumptions need to be fully analyzed before the even harder problem of using real imagery input can be addressed. As well, we expect that the use of separate image velocities measured at separate individual times will increase the robustness of the motion and structure calculation in most cases. At the very least, it will allow the recovery of motion and structure in situations not possible using one flow field only.

The use of the local planarity and fixed axis assumptions means that the point-to point correspondence problem does not have to be solved, i.e. we do not have to use velocities of the same 3-D points at different time intervals, as it is now mathematically possible to relate image velocities distributed in space and time at any point and time on a 3-D planar surface to the motion and structure parameters of any other point on the planar surface at any other time (where these assumptions are reasonably satisfied) [1]. Other researchers, such as [Kanatani 85] and [Aloimonos and Rigoutsos 86], have also advocated a correspondence-less approach. The computation of image velocity may require the solving of correspondence although there is a group of techniques based on the relationship between spatial and temporal grayvalue gradients, for example, [Horn and Schunck 81], for determining image velocity without the need to compute correspondence.

The algorithm presented in this paper involves solving nonlinear system of equations that relate a spatio-temporal distribution of image velocity to a set of motion and structure parameters at some image point at a particular time. Newton's method is used to solve the equations. As a result, an initial guess to the actual solution is needed to start the convergence calculation. This paper investigates two important questions: how good does the initial guess have to be and how accurate must the time-varying image velocity input be?

2 Mathematical Preliminaries

In this section we present a brief description of our algorithm. Complete details are in [Barron 87]. We use notation $\vec{P}(t;\tau)$ to indicate a 3-D point measured at time t with respect to a coordinate system $\vec{X}(\tau)$. Similarly, $X_3(\vec{P},t;\tau)$ is the depth of $\vec{P}(t;\tau)$. $\vec{Y}(\vec{P},t)$ is the image of $\vec{P}(t;t)$. We adopt a right-handed coordinate system as in Longuet-Higgins and Prazdny [80]. $\vec{U}=(U_1,U_2,U_3)$ is the translational velocity of the observer, centered at the origin of the coordinate system $\vec{X}(t)$ and $\vec{\omega}=(\omega_1,\omega_2,\omega_3)$ is the angular velocity of the observer. The image of $\vec{P}$ is located at $\vec{Y}=(y_1,y_2,1)$. The origin of the image plane is $(0,0,1)$, i.e. the focal length, f, is 1. The X_3 axis is the line of sight.

2.1 The General Monocular Image Velocity Equation

We can write an equation relating image velocity at some image point $\vec{Y}(\vec{P}_j,t')$ to the monocular motion and structure parameters at some image point $\vec{Y}(\vec{P}_i,t)$ as

---

(1) Of course, we must still be able to solve surface correspondence, i.e. we must be able to group together all image velocities distributed locally in space and time that belong to the same planar surface. See [Adiv 84] for one approach to this problem.

$$\vec{v}(\vec{Y}(\vec{P}_j,t'),t') = A_1(\vec{Y}(\vec{P}_j,t'))\; h(\vec{Y}(\vec{P}_i,t))\; Q_i(\vec{\omega},t,t')\; \vec{u}(\vec{Y}(\vec{P}_i,t),t;t)\; S_M(\vec{Y}(\vec{P}_i,t),\vec{Y}(\vec{P}_j,t'),t')\; T_M(\vec{Y}(\vec{P}_i,t),t,t')$$

$$+ A_2(\vec{Y}(\vec{P}_j,t'))\; \vec{\omega}(t;t). \qquad (2.1\text{-}1)$$

where $\vec{P}_i$ and $\vec{P}_j$ are 3-D points on the same planar surface and generally $\vec{Y}(\vec{P}_i,t) \neq \vec{Y}(\vec{P}_j,t')$. In the above equation

$$A_1(\vec{Y}(\vec{P},t)) = \begin{bmatrix} -1 & 0 & y_1 \\ 0 & -1 & y_2 \\ 0 & 0 & 0 \end{bmatrix} \quad \text{and} \quad A_2(\vec{Y}(\vec{P},t)) = \begin{bmatrix} y_1 y_2 & -(1+y_1^2) & y_2 \\ (1+y_2^2) & -y_1 y_2 & -y_1 \\ 0 & 0 & 0 \end{bmatrix}, \qquad (2.1\text{-}2,3)$$

$h(\vec{Y}(\vec{P},t))$ is the **perspective correction function** that specifies the ratio between the depth of $\vec{P}(t;t)$, $X_3(\vec{P},t;t)$ and its 3-D distance from the observation point, $||\vec{P}(t,t)||_2 = (\vec{P}(t;t) \cdot \vec{P}(t;t))^{1/2}$, i.e.

$$h(\vec{Y}(\vec{P},t)) = \frac{||\vec{P}(t;t)||_2}{X_3(\vec{P},t;t)} = ||\vec{Y}(\vec{P},t)||_2 \qquad (2.1\text{-}4)$$

and $\vec{u}(\vec{Y}(\vec{P},t),t;t)$ is the depth-scaled translational velocity of the observer,

$$\vec{u}(\vec{Y}(\vec{P},t),t;t) = \frac{\vec{U}(t;t)}{||\vec{P}(t;t)||_2}. \qquad (2.1\text{-}5)$$

One of the advantages of using a single instantaneous image velocity field is that no assumptions about the observer's motion, for example his acceleration, have to be made. However, the use of a spatio-temporal distribution of image velocities requires that we relate the motion and structure parameters at one time to those at another time. Hence, we need to make assumptions about the observer's motion. In this paper, we consider two specific types of motion, although we emphasize that our treatment can be generalized to other motions as well. The two types of motion considered are:

Type 1: (Linear Motion, Rotating Observer): A vehicle is moving with constant translational velocity and has a camera mounted on it that is rotating with constant angular velocity.

Type 2: (Circular Motion, Fixed Observer): A vehicle with a fixed mounted camera is moving with constant translational and angular velocity.

$Q_1(\vec{\omega},t,t') = R^T(\omega,t')R(\vec{\omega},t)$ and $Q_2(\vec{\omega},t,t') = I$ (the identity matrix), for Types 1 and 2 motion respectively. $R(\vec{\omega},t)$ is an orthogonal matrix specifying the rotation $||\omega||_2 t$ of $\vec{X}(t)$ with respect to $\vec{X}(0)$. S_M, the **monocular spatial scaling function**,

$$S_M(\vec{Y}(\vec{P}_i,t),\vec{Y}(\vec{P}_j,t),t) = \frac{\vec{\alpha}(\vec{P}_j,t;t) \cdot \vec{Y}(\vec{P}_j,t)}{\vec{\alpha}(\vec{P}_i,t;t) \cdot \vec{Y}(\vec{P}_i,t)} = \frac{X_3(\vec{P}_i,t;t)}{X_3(\vec{P}_j,t;t)}. \qquad (2.1\text{-}6)$$

specifies the depth ratio of two 3-D points, $\vec{P}_i$ and $\vec{P}_j$ on the same planar surface at the same time. The **monocular temporal scaling function**,

$$T_M(\vec{Y}(\vec{P}_i,t),t,t') = \frac{X_3(\vec{P}_i,t;t)}{X_3(\vec{P}_k,t';t')} = \frac{\vec{\alpha}(\vec{P}_i,t'';t'') \cdot \vec{Y}(\vec{P}_i,t)}{\vec{\alpha}(\vec{P}_k,t'';t'') \cdot R^T(\vec{\omega},t'')R(\vec{\omega},t)\left[\vec{Y}(\vec{P}_i,t) - \vec{\Delta d}(\vec{P}_i,t,t'';t)h(\vec{Y}(\vec{P}_i,t))\right]}, \qquad (2.1\text{-}7)$$

specifies the depth ratio of two 3-D points, $\vec{P}_i$ and $\vec{P}_k$ at times t and t', where $\vec{Y}(\vec{P}_i,t) = \vec{Y}(\vec{P}_k,t'')$

In special cases, equation (2.1-1) reduces to either a purely spatial or a purely temporal image velocity equation when $S_M = 1$ or $T_M = 1$. Given eight distinct components of image velocity distributed over space and time, but on the same 3-D planar surface, we can construct and solve a non-linear system of equations to determine the motion and structure parameters.

2.2 The Non-Uniqueness of the Solutions

Because we are solving non-linear systems of equations we need to be concerned about the uniqueness of our solution. Hay [66] was the first to investigate the inherent non-uniqueness of the visual interpretation of a stationary planar surface. He showed that for any planar surface there are at most two sets of motion and structure parameters that give rise to the same image velocity field for that surface. Hay also showed that given two views of such a surface only one unique set of motion and structure parameters was capable of correctly describing the image velocity field. Waxman and Ullman [85] carried this result one step further by showing the dual nature of the solutions: given one set of motion and structure parameters it is possible to derive a second set in terms of the first analytically. If this second solution is then substituted back into the equations specifying the duality, the first solution is obtained. Given one set of motion and structure parameters, $\vec{u}_1$, α_1 and ω_1 at $\vec{Y}(\vec{P},t)$, we can derive expressions for the dual solution, $\vec{u}_2$, α_2 and ω_2, at $\vec{Y}(\vec{P},t)$ as

$$\vec{u}_2(\vec{Y}(\vec{P},t),t;t) = \vec{\alpha}_1(\vec{P},t;t) \frac{\vec{u}_1(\vec{Y}(\vec{P},t),t;t) \cdot \vec{Y}(\vec{P},t)}{\vec{\alpha}_1(\vec{P},t;t) \cdot \vec{Y}(\vec{P},t)}, \qquad (2.2\text{-}1a)$$

$$\vec{\alpha}_2(\vec{P},t;t) = \frac{\vec{u}_1(\vec{Y}(\vec{P},t),t;t)}{||\vec{u}_1(\vec{Y}(\vec{P},t),t;t)||_2} \qquad (2.2\text{-}1b)$$

and

$$\vec{\omega}_2(t;t) = \vec{\omega}_1(t;t) + \begin{bmatrix} 0 & -\alpha_{13}(\vec{P},t;t) & \alpha_{12}(\vec{P},t;t) \\ \alpha_{13}(\vec{P},t;t) & 0 & -\alpha_{11}(\vec{P},t;t) \\ -\alpha_{12}(\vec{P},t;t) & \alpha_{11}(\vec{P},t;t) & 0 \end{bmatrix} \frac{\vec{u}_1(\vec{Y}(\vec{P},t),t;t)h(\vec{Y}(\vec{P},t))}{\vec{\alpha}_1(\vec{P},t;t) \cdot \vec{Y}(\vec{P},t)}. \qquad (2.2\text{-}1c)$$

427

$\vec{\alpha}_1 = (\alpha_{11}, \alpha_{12}, \alpha_{13})$ in (2.2-1c). Obviously, when $\vec{\alpha}(\vec{P}, t; t) = \dfrac{\vec{u}(\vec{Y}(\vec{P}, t), t; t)}{||\vec{u}(\vec{Y}(\vec{P}, t), t; t)||_2}$, the solution is unique as the dual solution reduces to the first solution. Subbarao and Waxman [85] have also showed the uniqueness of the motion and structure parameters over time as well.

These theoretical results suggest that the possibility of multiple (non-dual in the spatial case) solutions is non-existent. However, they only hold when the whole flow field is considered. It is possible for two distinct image velocity fields to have four common image points at four times with the same the image velocity values. Hence, the analysis of the four image velocities may give rise to any of the sets of motion and structure parameters having those four image velocities in common. An example of such a situation is given below.

2.3 Singularity

If $\vec{u} = (0,0,0)$ then the system of equations is singular. In fact, when $\vec{u} \ll \vec{\omega}$, its condition number becomes very large; very small input error causes instability in the solution technique. Also, Fang and Huang [84a] and others have shown that the solution does not exist if image velocities at three or more collinear 3-D points are used (as the determinant of the Jacobian matrix, J, is 0). We have also observed that the solution cannot be obtained when two pairs of time-varying image velocities resulting from the 3-D motion of two 3-D points on the same planar surface are used. The image motion of the points can be caused by an infinite number of motion and structure combinations. As well, there are particular motion and structure combinations that cannot be recovered at one time. For example, if $\vec{U} = (0,0,a)$, $\vec{\alpha} = (0,0,1)$ and $\vec{\omega} = (0,b,0)$ at time 0, then the values of constants a and b can be arbitrarily set to yield the same image velocity field; hence, it is impossible to distinguish the translational and rotational components of image velocity from each other. Other conditions of singularity are under active investigation.

3 Experimental Technique

In this section we discuss the implementation of our algorithm and present some preliminary results.

3.1 Implementation

Newton's method is used to solve the systems of non-linear equations. Since only two components of $\vec{\alpha}$ are independent, i.e. $||\vec{\alpha}||_2 = 1$, we add an extra row to the Jacobian matrix, J to ensure the computed $\vec{\alpha}$ is normalized; hence J is a rank 9 matrix. The 9th value of $\vec{f}_M$, the measured image velocities, is then set to 1.

When $\vec{\omega}$ is known to be zero, i.e. in the case of pure translation (type 1 and type 2 motions are equivalent here) we solve a 6x6 Jacobian instead of a 9x9 one. We compute a 9x6 Jacobian (the 3 columns corresponding to $\vec{\omega}$ are not computed). We let the LU decomposition of J choose the bests 6 rows of J, with the provision that the normalization row is always one of the chosen rows.

3.2 Error Analysis

We compute an error vector, $\vec{\Delta f}_M$ which, when added to $\vec{f}_M$, yields the perturbed input, $\vec{f}_M{'} = \vec{f}_M + \vec{\Delta f}_M$. For $X\%$ random case error, we compute four random 2-component unit vectors, $\hat{n}_j$, $j=1,...,4$, and then compute each ith component of $\vec{\Delta f}_M$ as

$$\begin{bmatrix} \Delta f_{iM} \\ \Delta f_{i+1M} \end{bmatrix} = \frac{X}{100} n_j ||\vec{V}_j||_2, \quad j=1,...,4, \quad i=j\times2-1. \tag{3.2-1}$$

Thus $X\%$ random error is added to each image velocity. Δf_{9M} is 0, i.e. we do not add error to the normalization constant. Using $\vec{\Delta f}_M$ for random error we compute $\Delta f_{norm} = ||\vec{\Delta f}_M||_2$. We use forward and inverse iteration on J to compute normalized best and worst case error directions, $\vec{e}_B$ and $\vec{e}_W$. We compute $\vec{\Delta f}_M = \hat{e}_B \Delta f_{norm}$ as $X\%$ scaled best case image velocity error and $\vec{\Delta f}_M = \hat{e}_W \Delta f_{norm}$ as $X\%$ scaled worst case image velocity error. In both cases $\vec{\Delta f}_M$ is scaled to be the same size as $X\%$ random image velocity error for comparison purposes. Another way of adding worst case error to the various image velocities is to scale the worst case error direction so that the maximum L_2 norm error added to any image velocity pair is $X\%$. We call this worst case relative image velocity error.

We test violation of the algorithm's underlying assumptions by computing $X\%$ error in $\vec{f}_M$, $\vec{\Delta f}_M$, in one of four ways. We violate local planarity by assuming the surface is hinged. We compute $\vec{f}_M{'}$ using gradient $(\sin\theta, 0, \cos\theta)$ for the points on the right side of the solution point and $(-\sin\theta, 0, \cos\theta)$ for the points on the left side. θ is varied from 0° to 90° in 8 steps. We violate the fixed axis assumption by adding random error to $\vec{\omega}$ at each place and time where image velocity is measured; $\vec{f}_M{'}$ is computed using these perturbed $\vec{\omega}$ values. The constant observer translational velocity assumption (necessary for the two types of motion) and the same surface/rigidity assumptions are violated by computing $\vec{U}$ and X_3 at each place and time where image velocity is measured, perturbing these by random amounts and then using $\vec{u}$, computed using these perturbed $\vec{U}$ and X_3, directly in the computation of $\vec{f}_M{'}$. In all four cases, we compute $X\%$ error for each assumption violation by computing Δf_M using the computed $\vec{f}_M{'}$ and perfect $\vec{f}_M$ via (4.2-1) and scale it (using Δf_{norm}) to be the same size as $X\%$ random case error (again, for comparison purposes).

We compute initial guess error by adding $X\%$ random error individually to $\vec{u}$, $\vec{\alpha}$ and $\vec{\omega}$. In purely spatial cases, we also compute the dual of $\vec{s}$, $\vec{s}_D$, using (2.4-1) and compute the output error as the minimum of error in $\vec{s}$ or $\vec{s}_D$. Since α' and $-\alpha'$ specify the same surface gradient, we always "flip" α' before the output error is calculation if the flipped α' is closer to α than the original α', i.e. $||\alpha + \alpha'||_2 < ||\alpha - \alpha'||_2$.

4 Experimental Results

The motion and structure parameters used in this paper are $\vec{U} = (0,0,1000)$, $\vec{\alpha} = (0,0,1)$ and $\vec{\omega} = (0,0.2,0)$. This corresponds to the motion of an observer translating directly towards a wall as he rotates his head (type 1 motion) or moving towards the wall while turning to the right (type 2 motion). This motion is singular in the spatial case. However it is analyzable using a spatio-temporal distribution of image velocity. We measure image velocity at the four image points, $(70,70)$ at time 0, $(-30,70)$ at time $t/3$, $(-30,-30)$ at time $2t/3$ and $(70,-30)$ at time t, i.e. at the four corners of a square centered at solution point (20,20). The solution is computed for time 0. These image coordinates are assumed to be measured on a 256×256 display device and are scaled by 256 to produce realistic f coordinates. Thus, (20,20) in pixels is scaled to (0.078125,0.078125) in f coordinates. The viewing angle of these points is computed as the maximum diagonal angle subtended by the smallest rectangle containing these points, i.e. 30.55^o. We call this the spatial extent. Temporal extent, written as 0–t, refers to the four times used, i.e. 0, $t/3$, $2t/3$ and t as above.

8×8 tables are used to display the output error for runs made by varying temporal extent again either image velocity error or initial guess error. Temporal extent 0–t is varied by varying t from 0 to 1 or 0.3 to 1 in 8 equal steps; a temporal extent of 0-0 corresponds to the purely spatial case. (As we have already pointed out, when ω=(0,0.2,0), the motion is not analyzable for temporal extent 0-0 but it can be analyzed at other temporal extents. When ω is known to be zero we can solve this motion for temporal extent 0-0 provided we use a 6×6 Jacobian.) Table rows (from left to right) correspond to increasing temporal extent. Image velocity error is varied from 0% to 1.4% in 8 equal steps while initial guess error is varied from 0% to 100% in 8 equal steps. Table columns correspond (from top to bottom) to either increasing image velocity error or increasing initial guess error.

Due to space limitations, we can only report a few of the results we have obtained to date. Tables 4-1, 4-2 and 4-3 show the output error for runs where image velocity error ranges from 0% to 1.4% for best, random and worst case error directions for three motions: (1) $\vec{U}$=(0,0,1000), $\vec{\alpha}$=(0,0,1) and ω is known to be zero (we solve a 6×6 Jacobian) and (2) and (3) where ω is (0,0.2,0) for types 1 and 2 motion respectively. All 100% output error values correspond to unsolved runs; all other values, including those over 100%, correspond to solved runs. As we can see, best case error results are quite good, especially when corresponding runs are compared with random and worst case run output. The overall magnification of input to output error was about 0.1. The output error in the random cases is about ½ the output error in the worst case. We include random results only to show the inadequacy of an error analysis that involves performing a few runs with random noise in the input. Unless a particular run is made n times (n a sufficiently large integer) for random noise in the input we cannot not draw any useful conclusions. (The largest output error of the n runs should approach worst case results, the smallest output error of the n runs should approach best case results while the average output of the n runs would comprise average case results.) As we can see, increasing the temporal extent can reduce output error; time appears to increase robustness for these motions. The results indicate that worst case error of as little as 1.4% can produce unusable output, if we assume only output error that is less than 10%-20% is useful. It seems we need image velocity measurements to be quite accurate.

The second experiment involves using perfect image velocity data and varying initial guess error from 0% to 100%. Again, temporal extent is varied from either 0-0 to 0-1 for the 1st motion and 0-0.3 to 0-1 for the 2nd and 3rd motions. 100% output error again indicates unsolved runs while 0% output error indicates solved runs. For the 1st motion (table 4-4a) most runs solved even when the initial guess error was 100%. Motions 2 and 3 (tables 4-4b and 4-4c respectively) exhibit multiple solutions; all output errors not 0% or 100% represent solved runs where the computed solution differs from the correct solution. For type 2 motion and a temporal extent of 0-0.5, we obtain two multiple solutions. The first, occurs when an initial guess of 71.43% is used. The solution obtained is

$\vec{u}_1$	$\vec{\alpha}_1$	$\vec{\omega}_1$
0.3035594	0.0879059	-0.5507448
-0.5278170	0.1632324	-0.1151465
0.5329872	0.9826636	0.0209885

resulting in 79.32% output error. The second solution, obtained with an initial guess error of 85.71%, is specified as

$\vec{u}_2$	$\vec{\alpha}_2$	$\vec{\omega}_2$
-0.2106061	0.4909884	0.1000418
0.1007392	0.5327612	0.4318877
0.4841026	0.8448401	-0.0389389

for a 57.82% output error. These solutions, plus the correct solution of $\vec{u}$=(0,0,0.4969759), $\vec{\alpha}$=(0,0,1) and $\vec{\omega}$=(0,0.2,0) produce the same four image velocities at the four image points and times, i.e.

y_1	y_2	time	v_1	v_2
70	70	0.0	-0.078235	0.121765
-30	70	0.16667	-0.266381	0.154889
-30	-30	0.33333	-0.272344	-0.072344
70	-30	0.5	-0.028700	-0.073414

The two multiple solutions show no apparent relationship with each other or with the correct solution. They are not duals. As well, the absolute translational observer velocities of the two multiple solutions, $\vec{U}_1$=(610.813104,–1062.057557,1072.460811) and $\vec{U}_2$=(–423.775273,202.704403,974.096744) show no apparent relationship with the correct absolute translational observer velocity, $\vec{U}$=(0,0,1000).

A third experiment that investigates the relationship between image velocity error and initial guess error does not produce any unexpected results; the two are usually independent. In those solved runs where the output error did not depend on image velocity alone, it was impossible to tell how much of the output error was due to image velocity error and how much was due to the existence of a multiple solution caused by the initial guess error.

The fourth experiment investigates the violation of the algorithm's underlying assumptions. Tables 4-5a,b,c,d shows the results when the 2nd motion is used. Output error when $\vec{\alpha}$ is perturbed is quite low; it appears that the planarity assumption can be violated to a large degree. The algorithm is more sensitive when the other assumptions are violated but significantly less than worst case image velocity error. Except for $\vec{\alpha}$ error, the assumption error and random case error look similar.

0.00	0.00	0.00	0.00	0.00	0.00	0.00	0.00
0.03	0.03	0.03	0.03	0.02	0.02	0.02	0.02
0.05	0.04	0.05	0.04	0.05	0.04	0.04	0.05
0.06	0.06	0.08	0.08	0.06	0.06	0.05	0.06
0.09	0.10	0.10	0.10	0.11	0.10	0.09	0.08
0.10	0.13	0.10	0.13	0.12	0.14	0.11	0.11
0.16	0.16	0.17	0.15	0.14	0.11	0.14	0.11
0.14	0.15	0.20	0.15	0.16	0.20	0.16	0.14

Table 4-1a (Best Error)

0.00	0.00	0.00	0.00	0.00	0.00	0.00	0.00
0.03	0.03	0.03	0.03	0.03	0.03	0.03	0.03
0.06	0.06	0.06	0.06	0.06	0.06	0.06	0.05
0.09	0.09	0.09	0.09	0.09	0.09	0.09	0.08
0.13	0.12	0.12	0.12	0.12	0.12	0.11	0.11
0.16	0.16	0.15	0.15	0.15	0.14	0.14	0.14
0.19	0.19	0.18	0.18	0.18	0.17	0.17	0.16
0.22	0.22	0.21	0.21	0.21	0.20	0.20	0.19

Table 4-2a (Best Error)

0.00	0.00	0.00	0.00	0.00	0.00	0.00	0.00
2.29	0.64	1.11	1.39	0.83	0.93	0.43	0.16
2.89	1.23	2.18	1.39	1.67	0.38	0.25	1.93
2.84	2.83	4.08	1.28	1.90	1.43	1.54	2.16
5.51	3.97	3.18	4.54	5.48	3.94	1.77	2.37
7.39	1.84	1.29	5.24	2.80	6.86	2.09	3.08
5.47	6.38	5.20	4.14	1.53	5.23	6.49	0.56
10.56	5.75	3.54	3.78	8.25	2.86	5.67	3.05

Table 4-1b (Random Error)

0.00	0.00	0.00	0.00	0.00	0.00	0.00	0.00
4.61	2.57	5.23	4.43	5.91	7.00	1.01	0.75
5.52	5.45	5.08	5.10	8.03	3.64	7.92	4.76
12.53	12.55	11.03	10.14	10.28	5.87	13.93	11.88
15.08	9.88	12.44	16.80	8.01	9.00	1.88	100.00
15.65	18.14	15.62	12.65	17.81	12.78	14.65	15.68
18.71	12.23	10.56	4.56	15.76	57.14	20.00	21.08
12.28	18.25	12.06	19.62	62.60	6.92	5.75	5.76

Table 4-2b (Random Error)

0.00	0.00	0.00	0.00	0.00	0.00	0.00	0.00
3.76	1.52	1.65	1.43	1.27	1.32	1.16	0.99
6.66	2.57	2.78	2.45	2.68	2.02	2.08	2.53
8.19	3.53	4.47	4.46	3.54	4.15	2.51	3.49
11.88	6.31	5.96	5.67	6.04	5.51	4.87	4.46
12.57	8.34	5.98	7.68	6.43	7.42	6.00	6.05
18.98	10.17	10.00	8.38	8.11	5.93	7.55	6.38
17.66	9.16	12.30	8.61	9.04	11.01	9.10	8.16

Table 4-1c (Worst Error)

0.00	0.00	0.00	0.00	0.00	0.00	0.00	0.00
8.53	8.19	7.91	7.71	7.57	7.52	7.55	7.68
12.81	12.43	12.11	11.86	11.68	11.60	11.62	11.77
100.00	16.57	16.55	16.57	16.64	100.00	100.00	54.36
118.31	95.57	81.92	72.81	66.30	61.45	57.74	100.00
118.00	95.34	81.78	72.76	66.37	61.65	58.14	55.60
100.00	95.14	81.67	72.75	66.47	61.90	100.00	100.00
26.52	100.00	81.58	72.77	100.00	26.06	100.00	100.00

Table 4-2c (Worst Error)

Tables 4-1 and 4-2. Table 4-1a,b,c show best, random and worst case output error results when $\vec{U}=(0,0,1000)$, $\vec{\alpha}=(0,0,1)$ and $\vec{\omega}$ is known to be (0,0,0) (the 1st motion). Tables 4-2a,b,c show best, random and worst case output error when $\vec{U}=(0,0,1000)$, $\vec{\alpha}=(0,0,1)$ and $\vec{\omega}=(0,0.2,0)$ for type 1 motion (the 2nd motion). Initial guess error is 0% for all runs in both tables while Image velocity error is varied from 0% to 1.4% and the temporal extent is varied from 0-0 to 0-1.

0.00	0.00	0.00	0.00	0.00	0.00	0.00	0.00
0.03	0.03	0.03	0.03	0.03	0.03	0.03	0.03
0.06	0.06	0.06	0.06	0.06	0.06	0.06	0.06
0.09	0.09	0.09	0.09	0.09	0.09	0.09	0.09
0.13	0.12	0.12	0.12	0.12	0.12	0.12	0.12
0.16	0.16	0.15	0.15	0.15	0.15	0.15	0.15
0.19	0.19	0.18	0.18	0.18	0.18	0.18	0.18
0.22	0.22	0.22	0.21	0.21	0.21	0.21	0.21

Table 4-3a (Best Error)

0.00	0.00	0.00	0.00	0.00	0.00	0.00	0.00
0.00	0.00	0.00	0.00	0.00	0.00	0.00	0.00
0.00	0.00	0.00	0.00	0.00	0.00	0.00	0.00
0.00	0.00	0.00	100.00	100.00	0.00	0.00	0.00
0.00	0.00	0.00	0.00	0.00	0.00	100.00	0.00
0.00	0.00	0.00	0.00	0.00	100.00	0.00	100.00
0.00	0.00	0.00	100.00	0.00	0.00	0.00	0.00
0.00	100.00	0.00	0.00	0.00	0.00	100.00	100.00

Table 4-4a (Initial Guess Error)

0.00	0.00	0.00	0.00	0.00	0.00	0.00	0.00
1.39	1.94	2.65	2.89	2.26	0.59	1.17	0.28
2.68	1.57	2.38	3.44	1.15	0.70	1.74	3.48
7.26	8.15	4.26	2.54	2.77	5.76	6.73	4.01
14.05	4.72	10.94	9.14	1.15	6.06	2.32	7.56
13.50	19.93	19.66	10.55	9.66	5.93	10.23	9.32
17.66	15.00	6.41	4.70	10.12	8.28	5.13	9.42
19.98	15.08	7.49	21.92	11.82	6.66	5.36	4.04

Table 4-3b (Random Error)

0.00	0.00	0.00	0.00	0.00	0.00	0.00	0.00
4.49	6.00	7.50	8.95	0.00	0.00	12.90	0.00
0.00	0.00	100.00	0.00	0.00	11.65	12.90	100.00
0.00	6.00	100.00	100.00	10.34	11.65	0.00	100.00
0.00	0.00	0.00	0.00	100.00	100.00	100.00	0.00
0.00	6.00	7.50	100.00	100.00	100.00	100.00	74.61
4.49	6.00	7.50	8.95	10.34	11.65	12.90	100.00
100.00	0.00	100.00	0.00	100.00	11.65	100.00	100.00

Table 4-4b (Initial Guess error)

0.00	0.00	0.00	0.00	0.00	0.00	0.00	0.00
5.86	5.24	4.80	4.46	4.18	3.95	3.76	3.59
10.48	9.71	9.08	8.56	8.11	7.72	7.38	7.08
14.41	13.64	12.96	12.35	11.80	11.31	10.87	10.47
17.87	17.19	16.52	15.89	15.29	14.74	14.23	13.76
21.02	20.44	19.83	19.21	18.59	18.01	17.45	16.94
23.91	23.46	22.92	22.33	21.73	21.13	20.55	20.01
26.61	26.27	25.81	25.28	24.70	24.11	23.53	22.97

Table 4-3c (Worst Error)

0.00	0.00	0.00	0.00	0.00	0.00	0.00	0.00
0.00	0.00	0.00	0.00	0.00	0.00	0.00	0.00
0.00	0.00	0.00	0.00	0.00	0.00	0.00	0.00
0.00	100.00	0.00	58.30	100.00	100.00	0.00	0.00
0.00	0.00	0.00	0.00	0.00	100.00	100.00	69.41
0.00	64.07	79.32	0.00	100.00	100.00	0.00	100.00
0.00	100.00	57.82	58.30	100.00	100.00	0.00	100.00
47.74	59.23	0.00	100.00	100.00	100.00	100.00	100.00

Table 4-4c (Initial Guess Error)

Tables 4-3 and 4-4. Table 4-3a,b,c show best, random and worst case output error results when $\vec{U}=(0,0,1000)$, $\vec{\alpha}=(0,0,1)$ and $\vec{\omega}=(0,0.2,0)$ for type 2 motion (the 3rd motion). Initial guess error is 0% for all runs while image velocity error is varied from 0% to 1.4% and the temporal extent is varied from 0-0 to 0-1. Tables 4-4a,b,c show tests that involve varying initial guess error from 0% to 100% and temporal extents from 0-0 to 0-1 from (a) and from 0-0.3 to 0-1 for (b) and (c). The motions tested are: (a) $\vec{U}=(0,0,1000)$, $\vec{\alpha}=(0,0,1)$ and $\vec{\omega}=(0,0,0)$ and (b) and (c) $\vec{U}=(0,0,1000)$, $\vec{\alpha}=(0,0,1)$ and $\vec{\omega}=(0,0.2,0)$ for type 1 and type 2 respectively. All runs use perfect image velocity error.

0.00	0.00	0.00	0.00	0.00	0.00	0.00	0.00
6.13	3.41	4.08	7.69	6.11	8.55	3.24	7.72
5.70	7.09	11.65	8.55	8.14	10.46	7.05	0.87
8.34	4.57	10.95	11.21	14.34	7.69	5.31	5.15
7.64	11.43	13.52	12.68	13.42	8.09	10.41	10.69
6.63	8.56	18.22	20.28	16.66	8.83	13.90	18.29
9.25	16.22	79.99	11.04	19.76	100.00	14.94	11.52
27.00	14.92	17.88	19.22	13.70	100.00	12.69	22.71

Table 4-5a (Perturbing $\vec{U}$)

0.00	0.00	0.00	0.00	0.00	0.00	0.00	0.00
0.57	0.60	0.63	0.66	0.70	0.74	0.78	0.83
1.14	1.22	1.30	1.38	1.47	1.55	1.64	1.73
1.72	1.88	2.03	2.17	2.31	2.46	2.60	2.75
2.38	2.63	2.87	3.10	3.32	3.54	3.76	3.98
3.26	3.65	4.01	4.35	4.69	5.02	5.35	5.70
5.15	5.73	6.24	6.74	7.24	7.76	8.33	8.98
19.03	16.20	7.46	19.32	19.60	17.53	14.96	12.57

Table 4-5b (Perturbing $\vec{\alpha}$)

0.00	0.00	0.00	0.00	0.00	0.00	0.00	0.00
8.55	4.33	5.63	5.39	3.72	6.47	5.06	100.00
100.00	8.17	11.20	10.09	9.42	5.69	11.51	4.34
10.77	5.47	6.95	14.66	13.11	100.00	12.67	9.78
14.62	16.73	15.13	8.69	13.38	15.71	7.66	9.48
11.16	15.92	17.81	15.30	13.96	8.84	16.19	7.95
21.64	8.93	14.05	11.06	66.12	11.37	16.35	15.38
119.71	18.70	9.79	13.46	14.39	26.07	56.62	22.70

Table 4-5c (Perturbing $\vec{\omega}$)

0.00	0.00	0.00	0.00	0.00	0.00	0.00	0.00
5.28	4.95	4.34	0.53	4.84	1.57	1.53	3.26
6.30	0.91	4.11	4.12	8.93	3.68	1.26	4.11
4.79	8.73	9.86	11.79	4.65	8.86	3.92	5.95
7.49	6.35	14.55	6.88	6.22	10.92	2.79	8.36
8.06	11.13	9.19	10.60	2.56	7.83	18.63	13.97
2.39	2.40	18.23	17.31	100.00	21.69	14.90	12.12
12.18	2.98	19.31	3.20	9.46	8.73	12.64	12.48

Table 4-5d (Perturbing X_3)

Table 4-5. Tables 4-5a,c,d involve violating (a) the constant observer translational velocity assumption by perturbing $\vec{U}$, (b) the local planarity assumption by varying the angle between the two hinged planes from 0° to 90° in 8 steps, (c) the fixed axis assumption by perturbing ω and (d) the same surface and rigidity assumptions by perturbing X_3. In all cases, the resulting image velocity error is scaled by Δf_{norm} to be the same size as 0%-1.4% random case image velocity error. The temporal extent is varied from 0-0.3 to 0-1 all for the 4 tests.

100.00	0.00	0.00	0.00	0.00	0.00	0.00	0.00
36.25	8.42	4.69	2.62	1.86	1.59	1.54	1.60
63.38	15.99	8.84	4.95	3.52	2.99	2.87	2.94
80.26	22.80	12.57	7.06	5.02	4.25	4.05	4.12
90.72	28.93	15.98	9.00	6.40	5.40	5.12	5.18
97.53	34.49	19.14	10.80	7.68	6.47	6.12	6.16
102.20	39.56	22.08	12.49	8.89	7.48	7.05	7.07
105.56	44.19	24.86	14.09	10.03	8.43	7.93	7.92

Table 4-6a (Varying Spatial Extent)

105.56	100.00	100.00	100.00	100.00	100.00	100.00	100.00
44.19	45.32	46.76	48.96	51.49	53.30	54.00	53.99
24.86	24.74	24.36	23.98	23.67	23.48	23.44	23.52
14.09	13.83	13.44	12.97	12.48	12.03	11.68	11.45
10.03	9.32	8.75	8.24	7.76	7.30	6.87	6.45
8.43	7.46	6.73	6.14	5.65	5.21	4.81	4.42
7.93	6.79	5.92	5.25	4.70	4.24	3.84	3.48
7.92	6.71	5.75	4.99	4.38	3.88	3.46	3.09

Table 4-6b (Varying Spatio-Temporal Extent)

Table 4-6. (a) show the output error for the 3rd motion when worst case relative error of 0-1.4% is used and the spatial extent is varied from 0° to 70° for a fixed temporal extent of 0-0.3. (b) shows the output error for the 3rd motion when the spatial extent is varied from 0° to 70° and the temporal extent is varied from 0-0.3 to 0-1 for a fixed worst case relative image velocity error of 1.4%.

As we have already seen in experiment 1, increasing temporal extent can reduce output error. (We have also observed the opposite; output error can increase with temporal extent but less often so and than by smaller amounts. These results will be reported in a forthcoming paper.) This last experiment investigates what happen when spatial extent is varied from 7° to 70° (the full image) from a fixed temporal extent of 0-0.3. The 3rd motion is used and worst case relative image velocity error is varied from 0%-1.4%. The results show that increasing spatial extent also reduces output error. As well, table 4-6b shows what happen when both the spatial and temporal extents are varied (7° to 70° and 0-0.3 to 0-1 in 8 equal steps) simultaneously for fixed 1.4% worst case relative image velocity error. Again output error decreased with increased spatio-temporal extent. As we have seen from previous results, worst case image velocity error usually produces bad results. Obviously, we should use the largest feasible spatio-temporal in any motion and structure calculation.

5 Conclusions

Due to the limitations of space, we were not able to present more examples of our solution technique's performance. We have demonstrated that the addition of a temporal distribution of image velocity can increase the numerical stability of the solution technique. In addition, it allows us to analyze flow fields that are not be analyzable at one time. As well, increasing spatial extent also improves the algorithm's performance. Unfortunately, the greater the spatio-temporal extent the more likely the algorithm's underlying assumptions will be violated in realistic situations. It seems that the algorithm is quite sensitive to input image velocity error but relatively insensitive to initial guess error. We are able to report the existence of multiple solutions, a fact that is apparently overlooked by most other researchers.

The results of this paper show that the computation of motion and structure from image velocity fields is quite sensitive to noisy input. We expect our results would be duplicated for other reconstruction algorithms as well. We believe researchers in the Computer Vision community interested in computing motion and structure from real data must expend much more effort in analyzing their algorithms for noisy data and devise ways to overcome the poor performance inherent in such algorithms. In this paper we have shown that using a spatio-temporal distribution of image velocity (versus a purely spatial distribution of image velocity) is one way to improve performance.

Other research currently in progress involves comparing the use of local spatio-temporal image velocity data and the use of local spatio-temporal velocity derivative information (such as in [Waxman and Ullman 85]) [Barron et al 87a] and the improvement in algorithm performance resulting from using a least squares formulation [Barron et al 87b]. We have also formulated a binocular reconstruction algorithm that uses a spatio-temporal distribution of image velocities in left and right image sequences [Barron et al 87c]. Other results, including a complete analysis of the binocular algorithm and the use of spatio-temporal distribution of normal image velocity will be reported in future papers.

Acknowledgements

We gratefully acknowledge financial support from the National Science and Engineering Research Council of Canada and the Department of Computer Science at the University of Toronto.

Bibliography

(1) Adiv G., 1984, "Determining 3-D Motion and Structure from Optical Flow Generated by Several Moving Objects", COINS Technical Report 84-07, University of Massachusetts, April.

(2) Aloimonos J.Y. and I. Rigoutsos, 1986, "Determining the 3-D Motion of a Rigid Planar Patch Without Correspondence, Under Perspective Projection", Proc. Workshop on Motion: Representation and Analysis, May 7-9.

(3) Bandyopadhyay A. and J. Aloimonos, 1985, "Perception of Rigid Motion from Spatio-Temporal Derivatives of Optical flow", TR 157, Dept. of Computer Science, University of Rochester, NY, March.

(4) Barron, J.L., 1984, "A Survey of Approaches for Determining Optic Flow, Environmental Layout and Egomotion", RBCV-TR-84-5, Dept. of Computer Science, University of Toronto, November.

(5) Barron, J.L., 1987, "Determination of Egomotion and Environmental Layout From Noisy Time-Varying Image Velocity in Monocular and Binocular Image Sequences", 1987, forthcoming PhD thesis, Dept. of Computer Science, University of Toronto.

(6) Barron J.L., A.D. Jepson and J.K. Tsotsos, 1987a, "The Sensitivity of Motion and Structure Computations", AAAI-87, Seattle, Washington, July 13-17.

(7) Barron J.L., A.D. Jepson and J.K. Tsotsos, 1987b, "Using Least Squares to Compute Motion and Structure from Noisy Time-Varying Image Velocity Fields and from Noisy Spatio-Temporal Image Velocity Information: A Sensitivity Analysis", planned.

(8) Barron J.L., A.D. Jepson and J.K. Tsotsos, 1987c, "Determining Egomotion and Environmental Layout From Noisy Time-Varying Image Velocity in Binocular Image Sequences", IJCAI-87, Milan, Italy, August 23-28.

(9) Buxton B.F, H. Buxton, D.W. Murray and N.S. Williams, 1984, "3-D Solutions to the Aperture Problem", in **Advances in Artificial Intelligence**, T. O'Shea (editor), Elsevier Science Publishers B.V. (North Holland), pp631-640.

(10) Dreschler L.S. and Nagal H.-H., 1982, "Volumetric Model and 3-D Trajectory of a Moving Car Derived from Monocular TV-Frame Sequences of a Street Scene", CGIP 20, pp199-228.

(11) Fang J.-Q. and Huang T.S., 1984a, "Solving Three-Dimensional Small-Rotation Motion Equations: Uniqueness, Algorithms and Numerical Results", CVGIP 26, pp183-206.

(12) Fang J.-Q. and Huang T.S., 1984b,"Some Experiments on Estimating the 3-D Motion Parameters of a Rigid Body from Two Consecutive Image Frames", PAMI, Vol.6, No.5, pp545-554.

(13) Gibson J.J., 1957, "Optical Motions and Transformations as Stimuli for Visual Perception", Psychological Review, Vol. 64, No. 5, pp288-295.

(14) Gibson E.J., Gibson J.J., Smith O.W., and Flock H., 1959, "Motion Parallax as a Determinant of Perceived Depth", Journal of Experimental Psychology, Vol. 58, No. 1.

(15) Hay J. C., 1966, "Optical Motions and Space Perception: An Extension of Gibson's Analysis", Psychological Review, Vol. 73, No. 6, pp550-565.

(16) Horn B.K.P. and Schunck B.G., 1981, "Determining Optical Flow", AI 17, pp185-203.

(17) Kanatani K, 1985, "Structure from Motion without Correspondence: General Principle", Proceedings of IJCAI, pp886-888.

(18) Lawton D.T., 1983, "Processing Translational Motion Sequences", CGIP22, pp116-144.

(19) Longuet-Higgins H.C., 1981, "A Computer Algorithm for Reconstructing a Scene from Two Projections", Nature 293, Sept., pp133-135.

(20) Longuet-Higgens H.C. and K. Prazdny, 1980, "The Interpretation of a Moving Image", Proc. Royal Society of London, B208, pp385-397.

(21) Prazdny K., 1979, "Motion and Structure From Optical Flow", IJCAI-79, pp702-704.

(22) Roach J.W. and Aggarwal J.K., 1980, "Determining the Movement of Objects from a Sequence of Images", PAMI, Vol. 2, No. 6, Nov., pp554-562.

(23) Snyder M.A., 1986, "The Accuracy of 3-D Parameters in Correspondence-Based Techniques: Startup and Updating", Proc. Workshop on Motion: Representation and Analysis, May 7-9.

(24) Subbarao M. and Waxman A.M., 1985, "On the Uniqueness of Image Flow Solutions for Planar Surfaces in Motion", CAR-TR-114 (CS-TR-1485), Center for Automation Research, University of Maryland, 1985. (Also, 3rd Workshop on Computer Vision: Representation and Control, 1985.)

(25) Subbarao M., 1986, "Interpretation of Image Motion Fields: A Spatio-Temporal Approach", Proc. Workshop on Motion: Representation and Analysis, May 7-9.

(26) Thompson W.B. and Kearney J.K., 1986, "Inexact Vision", Proc. Workshop on Motion: Representation and Analysis, May 7-9, pp15-21.

(28) Tsai R.Y. and Huang T.S., 1984, "Uniqueness and Estimation of Three-Dimensional Motion Parameters of Rigid Objects with Curved Surfaces", IEEE PAMI, Vol. 6, No. 1, pp13-27.

(29) Ullman S., 1979, **The Interpretation of Visual Motion**, MIT Press, Cambridge, MA.

(30) Waxman A.M. and Ullman S., 1985, "Surface Structure and 3-D Motion from Image Flow Kinematics", Intl. Journal of Robotics Research, Vol.4, No.3, pp72-94.

(31) Waxman A.M. and Wohn K., 1985, "Contour Evolution, Neighbourhood Deformation and Global Image Flow: Planar Surfaces in Motion", Intl. Journal of Robotics Research, Vol.4, No.3, pp95-108.

(32) Webb J.A. and Aggarwal J.K., 1981, "Visually Interpreting the Motion of Objects in Space", IEEE Computer, Aug., pp40-46.

(33) Williams T.D., 1980, "Depth from Camera Motion in a Real World Scene", PAMI-2, No. 6, Nov., pp511-515.

HEURISTIC DESCRIPTION OF SPATIAL AND TEMPORAL BEHAVIOUR OF RAIN PATTERNS USING SIMPLE PHYSICAL MODEL

Alona Pawlina Bonati

CSTS - CNR Politecnico di Milano
P.zza L.da Vinci, 32,20133 Milano

Piero Mussio

Dipartimento di Fisica
Università Statale Milano

FORWORD

The goal of this paper is to present the case of image interpretation showing a design aspect of a tool developed for radar images.

Rain pattern signatures appearing on radar derived rain maps become for us an object of automatic interpretation since 1983. Initially the research was motivated by a strong need for an automatic extraction of individual rain patterns ("cells") required in the radio-propagation studies. This extraction was performed before quite coarsely [1] and succeedingly refined by adopting proper contouring and feature extraction techniques[2]. The long term goal extended then to become the interpretation of rain structures appearing on radar images. While the early attempt of automatic interpretation [3] is under further development we propose here another aspect of "understanding" of rain events. Rain event in our data refers to a sequence of rain maps collected at minimum time intervals.

Simple modelling of complex shapes present on the rain maps allows us:
(a) to give a very compact description of rain patterns for the further, application oriented, modeling[4];
(b) to study the meso-scale (some tens of kilometers) behaviour of patterns on the scene, e.g. predominant orientations, grouping along certain lines, random/regular spatial distribution etc.
(c) to track in time certain important (i.e. very severe and/or extended) rain formations or, more generally, to describe the development of rain event in time.

To this end we describe each rain pattern or cluster by means of inertia ellipse whose area is adjusted to the actual one. The model descriptors of rain pattern are used to identify its Predecessor/Successor in a sequence of scenes belonging to the same rain event.

The Predecessor/Successor pattern is identified combining the values of some "dynamic" parameters which quantise the change of selected attributes of compared patterns such as extension, ellipticity, inclination, barycenter and displacement.

Combining of these changes is performed through the Multi-Value Logical Trees (MVLT). The MVLT is used also for the characterisation of the pattern evolution. In both cases the MVLT formalises the heuristic reasoning of the expert. The resulting methodology is of the "plausible" kind[5].

RADAR RAIN MAPS AND MODELING OF RAIN PATTERNS

The meteorological radar system of Spino d'Adda next to Milan was used in full weather survelliance mode during the SIRIO experiment[6,7] in 1980 supplying a large collection of three-dimensional scans which were processed in order to obtain Ground Rain Maps - digital images of rain reflectivity on the horizontal plane.

Every GRM covers a region of 66 km x 120 km as in Fig.1; each pixel represents an average rain reflectivity for an area of 1 km x 1 km. A rain event is represented by a sequence of GRMs taken 6 minutes (full scanning cycle) apart. Fig. 1 shows a portion of rain event which occured June 9-th 1980 starting at 12:30 GTM ; the scenes are sampled there every 24 minutes.

Some examples of rain patterns are shown in detail in Fig.2 on the left. The rain reflectivity converted to the rain intensity, R, is quantized there in ten levels (classes).

Only three levels are considered in the interpretation process: Basic level characterized by R>5 mm/h, Core level characterized by R>20 mm/h, and Intense Core level characterized by R> 40 mm/h. These three limit levels are evidenced in Fig. 2 : the intensities exceeding them are marked by digits 0, 2 and 3, respectively.

Although the visual inspection of radar rain maps evidences a great variety and complexity of shapes the requirement for a compact description leads to the model simplification of shapes. The best model shape is represented by an ellipse. In fact the degree of "asymmetry" or "being elongated" in some direction may be measured once the original pattern is substituted by an equivalent "regular" shape.

In order to take into account also the spatial distribution of rain inside the pattern the "coloured" (i.e. weighted with the intensity values) inertia ellipse was adopted. For model purposes the dimension of the ellipse was then adjusted to match the area originally covered by the pattern. The report[6] gives formal definitions.

Fig. 2 shows a sample of patterns accompanied by their models. In case (a) and (b) we have actually a couple of clustered patterns modelled by one ellipse. The modeling is performed idependently at three levels: Basic structures (whole patterns) with the threshold "0" exceeded, Cores with the threshold "2" exceeded, and Intense Cores with the threshold "3" exceeded. This allows to trace separately the structures of interest, e.g. "whole" patterns or only the most severe formations - Intense Cores.

Describing Temporal Evolution of Patterns

Rain patterns change their shape and re-organize in time through basically two (opposite) processes : dissipation of old patterns and growing of new ones, both associated to the movement along some direction and/or the rotation. Thus it is improper to speak about "tracking" of rain patterns. Nevertheless at small time intervals the changes are not drammatic and some particular (usually strong and extended) rain formations persist, recognizable, in time. The information on their life time and trajectory are of utmost interest in the radio-meteorology.

Applying the elliptical model to the clusters in a sequence of scenes allows to follow easily in time the pattern development and also to recognise "the same" pattern in a pair of successive or close enough scenes.

Fig. 1 illustrates the above considerations showing an example of a sequence of scenes with a model representation of each scene. Observing the model sequence is possible to formulate some characteristics of the event not easily readible on the original scenes.

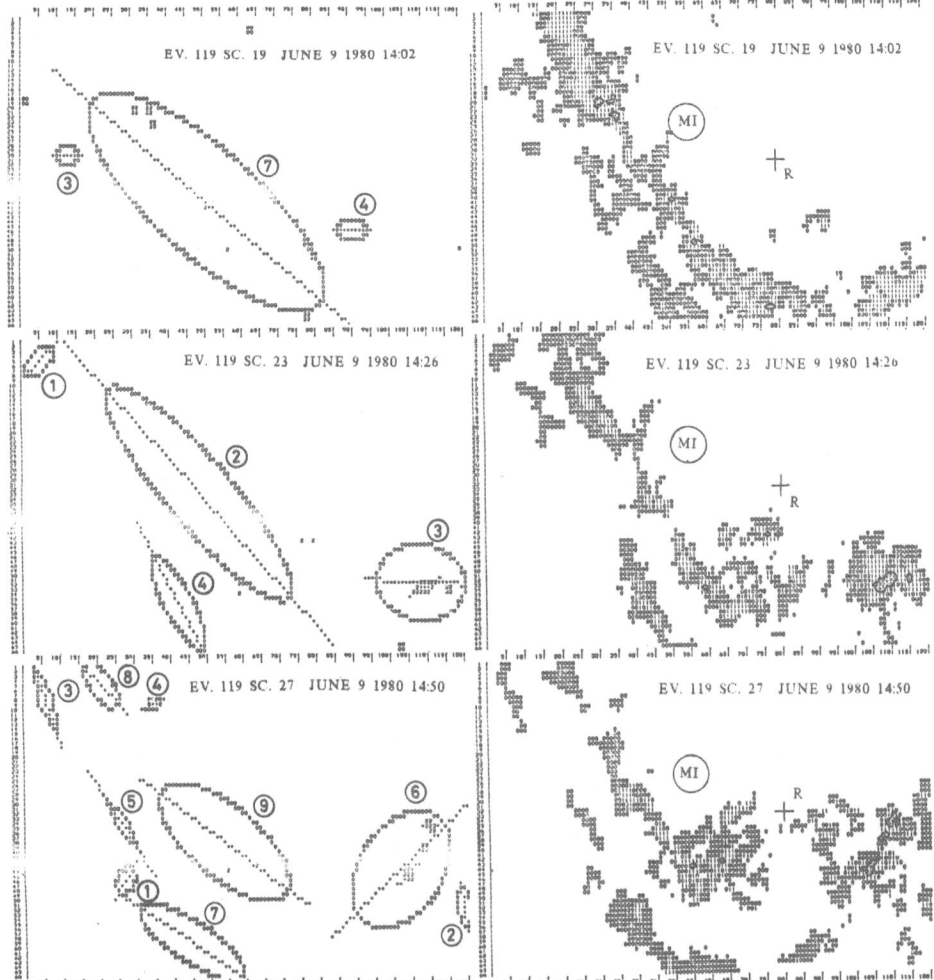

Fig. 1 - Time /space development of a rain event ; a sample of three
scenes in form of Grund Rain Map on the right and the
corresponding model representation on the left. A cross
indicates the radar site and a circle the town of Milan. Each
pixel corresponds to a square of 1 km X 1 km. The circled
numbers of model structures on the left point to their
description in Table 1.

The development of rain event may be in fact described very concisely
by giving the coordinates and parameters of model ellipses for every
scene in the sequence and then following the behaviour of the ellipses
in time and space. An example of the model-in-time description of the
event sequence of Fig.1 is shown in Table 1.
    The most important formation living through the event is clearly
recognisable in the sequence of elements : 4 of scene 15 (not shown), 7
of sc. 19, 2 of sc. 23, 9 of sc. 27, all shown in Fig. 1 on the left.
    It changes the extension quite quickly (1326-1412-806-512 km[2]), its
ellipticity factor oscillates around 0.3 (0.29-0.31-0.23-0.41) and it
keeps firm (at least in "fuzzy" sense [8]) the inclination (42-56-48-60°).

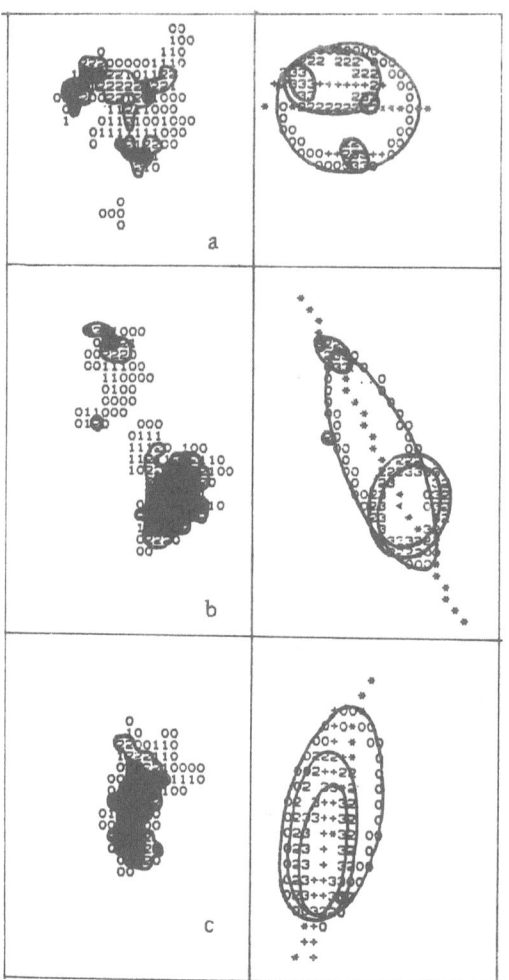

Fig. 2 –Modeling of rain patterns at three levels of rain intensity ;
on the left digital signature of a pattern with Cores
(contoured) and Intense Cores (black), on the right the
corresponding elliptical model. Pattern (c) is of "isolated"
type while patterns (a) and (b) are in fact clusters of
patterns.

Visual inspection of (modelled) objects gives in this case an immediate response in terms of " what is the successor of the pattern x ?".
It is much more hard to instruct the system how to answer this question. The concepts such as "the area, the inclination and the shape
combined together did not change too much" or ,more generally, "it
changed and moved within acceptable limits" can be "understood" and
evaluated only by tools able to consider the uncertainities in the
observed data and their descriptors which influence the interpretation
process. Such tools , allowing plausible reasoning, are Labeling
functions (l-functons) and MVLTs.

Table 1. Model-in-time description for a sequence of scenes.

event/scene	struct. number	struct. type	BX	BY	area	alfa	ellipt. factor
119 / 15	4	cluster	32	36	1326	42	0.29
119 / 19	3	isolated	30	13	14	83	0.58
119 / 19	4	chain	46	91	25	93	0.58
119 /19	7	cluster	40	51	1412	56	0.31
119 / 23	1	isolated	5	5	34	127	0.55
119 / 23	2	cluster	33	48	806	48	0.23
119 / 23	3	cluster	52	108	294	87	0.68
119 / 23	4	cluster	56	42	148	33	0.32
119 / 27	1	chain	46	27	18	140	0.82
119 / 27	2	isolated	52	119	22	10	0.52
119 / 27	3	chain	8	6	41	23	0.24
119 / 27	4	isolated	9	35	11	115	0.62
119 / 27	5	chain	34	26	37	39	0.20
119 / 27	6	cluster	44	102	472	126	0.68
119 / 27	7	cluster	59	45	211	63	0.26
119 / 27	8	cluster	5	21	55	50	0.51
119 / 27	9	cluster	39	54	517	60	0.41

DESCRIPTION of SPATIAL AND TEMPORAL BEHAVIOUR

A human expert judges (in our case heuristically) the importance of
each observed property in the process of interpreting a given structure.
For instance given a pair of modelled rain patterns ,Pt_1 and Pt_2, one
belonging to a scene s_i and other to a scene s_{i+k} of the same event the
expert estimates the difference in the patterns area, DA , giving it a
score[9] from a set SC, e.g. SC = {small,medium-small,medium-high,high}.

Assigning these scores depends on the interpretation context (small
for what purpose?); the expert may estimate e.g. the plausibility that
Pt_2 is a Successor of Pt_1.

Note that scores such as "medium-low" or "medium-high" may express the
expert's uncertainity about the interpretation, or represent transitory
verdicts.

In order to reproduce the scoring automatically it is necessary to
define an appropriate labeling function which maps numerical values of
the attribute onto the set of scores. In the example the DA values are
divided into four classes defined by the thresholds 10%, 20% and 40% and
given the labels L_{DA} = { 10, 10-20, 20-40, 40 }.

In different context the same attribute, DA may be mapped to a different set of labels : in fact when characterising the type of the evolu
tion effected by the pattern DA is mapped onto L_2 = { 30, 30 } which

is equivalent to the set of scores SC = { small, high }.

Once the labeling of all considered attributes is completed it is necessary to evaluate the combined meaning of observed properties.

This is performed using MVLTs. An MVLT is a combinatorial function which maps a combination of scores (labels) refering to different attributes onto a set of "global" labels which conclude the interpretation of the object.

MVLT is composed of multivalue operators. Multivalue operator maps a pair of labels, l_i, l_k of two attributes onto a set of new labels.

MVLT for evaluating Predecessor/Successor Relation

Given two modelled elements i and k (an element may represent an isolated pattern, a chain or a cluster of patterns) the following "dynamic" parameters are defined :

$$\text{Differential Area, DA} = \frac{|\ A_i - A_k\ |}{\max\ (A_i, A_k)}\ 100\ \%\quad [\ \%\ ]$$

where A_i, A_k are the element areas in square kilometers,

Differential Ellipticity, $DE = |\ ELL_i - ELL_k\ |$, ELL_i, ELL_k are ellipticity factors,

Differential Inclination, $DI = |\ ALFA_i - ALFA_k\ |\quad [^\circ\ ]$

Distance between Barycenters, $DB = d\ (\ B_i, B_k\ )\quad [\ km\ ]$

Average Area, $AA = (\ A_i + A_k\ )/2\quad [\ km^2\ ]$.

Each of these parameters is labelled by an appropriate l-function and then enters the MVLT.

The MVLT for evaluating the Predecessor/Successor relation, the P/S-MVLT in Fig. 3 a reflects the idea that reshaping (dissipation/growing process) is determinant for this relation while the linear displacement and/or rotation within certain limits are forseen. This is evaluated by combining the difference in area with differential ellipticity first and then the average area. The other attributes are used only to adjust the overall plausibility score.

The operators are given by the following set of tables:

Pump operator, P

DE \ DA	<10	10-20	20-40	>40
<0.3	S	S	QS	D
>0.3	QS	QD	D	D

Relative Size operator, RS

AA \ DA	<10	10-20	20-40	>40
<10	small	small	small	diff.
10-100	small	small	diff.	diff.
100-400	large	large	diff.	diff.
>400	large	large	large	diff.

Tr-DI operator

P/RS \ DI	<20	20-40	>40
QS	S	QS	QD
S	S	S	QS
QD	QS	D	D
D	D	D	D

P / RS operator

RS \ P	S	QS	QD	D
small	S	QD	QD	D
large	S	S	QS	D
diff.	QS	QD	D	D

Tr-DB operator

```
  \DB |  <4  |4-12 |12-20| >20
Tr-DI-------------------------
  QS |  S |  S  |  Q  |  D
   ------------------------------
   S |  S |  S  |  S  |  Q
   ------------------------------
  QD |  Q |  Q  |  D  |  D
   ------------------------------
   D |  D |  D  |  D  |  D
```

S - Similar
Q - Quasi-similar
D - Different

The final verdicts of the P/S-MVLT indicate the degree of plausibility that the two elements are linked by the Predecessor/Successor (in time) relation.

MVLT for the Evolution Characterisation, ECH - MVLT

In this tree shown in Fig. 3 b the same pattern attributes as before are used for a different purpose. Therefore the evaluation of their meaning changes both at the 1-function and the operator definition level.

The characterisation of the evolution is based on the following empirical assumptions :

(a) area and ellipticity concur in determining the pattern shape (SH_i)

(b) shape can be characterised only for extended structures; the ellipticity is not meaningful for very small patterns being strongly affected by a digital noise.

(c) shape differences are foundamental for the characterisation of the evolution of a pattern in time.

The Ech-MVLT operators are defined by the following tables :

SHape operator, SH_i

```
  \Ai    |  <10 |10-400| >400
E i------------------------
  <0.5  |  NE  |  E  |  VE
   ----------------------------
0.5-0.75|  NE  |  E  |  E
   ----------------------------
  >0.75 |  NE  |  NE  |  NE
```

SHape Difference operator, SHD

```
  \SH i |  NE  |  E  |  VE |
SH k  -------------------
   NE  |  NI  |  NI  |  NI |
   ----------------------
   E   |  NI  |  I  |  HI |
   ----------------------
   VE  |  NI  |  HI  |  I
```

Pump operator, P

```
  \DE   |  <0.3 |  >0.3
DA  ----------------
  < 30  |  CC  |  CD
   ----------------------------
  > 30  |  DC  |  DD
```

Tr-DI operator

```
  \DI    |  <20 |20-40| >40
SHD-----------------------
   I   |  F  |  SR  |  R
   ----------------------
  HI  |  Q  |  Q  |  R
   ----------------------
  NI  |  D  |  RD  |  RD
```

Out-Tr operator

Tr-DI \ P	CC	CD	DC	DD
F	S	S	V	DV
SR	S	S	S	S
R	R	RD	RV	RV
Q	S	D	V	M
RD	R	RD	RV	RD
D	D	M	M	D

S - similar, not changed
V - varied in area
DV - deformed & varied in area
R - rotated
RD - rotated & deformed
RV - rotated & varied in area
D - deformed
M - modified (very much)

CONCLUSIONS

A proposal for the formal definition of the relation Predecessor/Successor in a time sequence of rain patterns and for the characterisation of their evolution is presented, together with a relative evaluation tool. The use of such tool is based on the well defined description of the patterns.

The pattern description is done in terms of simple elliptical model, used both for the static and dynamic characterisation of the patterns.

This description consists in a set of crisp (precise) properties, but the high variability of patterns and the intrinsic fuzziness of the time evolution model implied : (a) imprecise measurments (labels) in ordinal scale to express the weight of a single property in the interpretation process, e.g. the l-function adopted for labeling of DA ; (b) adoption of combinatorial tools —such as MVLT— which do not require the definition of a distance on the set of input labels.

As regards the P/S-MVLT the comparison of the results of human and

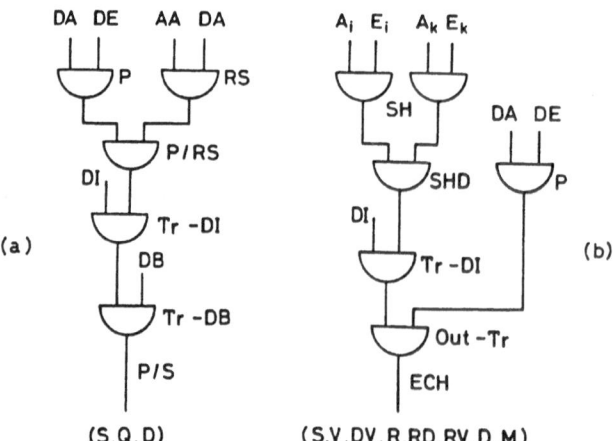

Fig. 3 - Multi-Value Logical Trees for two stages of classification process referred to the temporal evolution of rain patterns :
(a) for evaluating of the Predecessor/Successor relation and
(b) for the characterisation of pattern evolution ;
operators, input parameters and final verdicts are given in the text.

440

automatic interpretation seems quite satisfactory ; the set of l-functions and the MVLT are now considered for a successive step of confirmation on a new set of data [10].

The Ech-MVLT, on the contrary, does not behave satisfactorily. A further analysis is required before the confirmation.

In any case P/S-MVLT , Ech-MVLT and the associated l-functions represent the formal description of the heuristic process of interpretation. They are now offered to the radarmeteorology experts to be tested, criticised and improved. A deeper and more formal knowledge can be accumulated by this process toward better insight of the interpreted phenomena.

From the image interpretation point of view this case study represents another example of the system design which offers the possibility of better understanding of the attributes and properties selection and the definition of proper operators / decisional trees.

REFERENCES

[1] A. Pawlina "Some features of ground rain patterns measured by radar in north Italy", Radio Science, vol.19, n.3 May-June 1984
[2] D. Merelli, P. Mussio, M. Padula "An approach to the definition, description and extraction of structures in binary digital images", Computer Vision, Graphics, and Image Processing 31 , 19-49 (1985)
[3] A. Della Ventura, A. Maggioni, P. Mussio, A. Pawlina "Knowledge acquisition for automatic interpretation of radar images", Int. Conf. on Image Analysis and Processing, Rapallo, Sept. 1985
[4] C. Capsoni, A. Paraboni, A. Pawlina "A realistic model of the horizontal structure of rain cells", ESA Int. Symp. on Satellite Transmissions, Graz, Sept. 1985
[5] H. Prade " A computational approach to approximate and plausible reasoning with application to expert systems, IEEE Trans. Patt. Anal. Mach. Intell., vol. PAMI-7, May 1985 Transmissions, Graz, Sept. 1985
[6] Alta Frequenza , Special Issue on SIRIO Programme", Alta Frequenza, n. 4, vol. XLVII, April 1978
[7] C. Capsoni, M. Mauri, A.Paraboni, A.Pawlina "The experimental station at Spino, d'Adda: facilities,accuracy,types of data collection, amount and quality of data " Interim Report of ESTEC Contract n.4680/81NL/MS(SC) Nov. 1981
[8] L.A. Zadeh "Fuzzy sets and their application to pattern classification and cluster analysis" in "Classification and clustering", J. Van Ryzin ed., Academic Press, New York 1977
[9] S. Siegel "Non parametric statistics for the behavioural sciences", Mc Grow Hill, New York 1956
[10] U. Cugini, M. Dell'Oca, D. Merelli, P. Mussio "A computer aided system for interactive definition of digital images interpretation", Digital Image Analysis, S. Levialdi Editor,Pitman, 1984, pp. 270-279
[11] A. Pawlina Bonati, P. Mussio "Heuristic description of spatial and temporal behaviour of rain patterns using simple physical model", CNR-CSTS Politecnico di Milano, Rapporto Int. 87-1.

A FRAMEWORK FOR REGION CHARACTERIZATION IN REMOTE SENSING
IMAGES BY FRACTAL–BASED APPROACH

L. Giberti, L. Piccollo, S. Dellepiane, S. B. Serpico, and G. Vernazza

University of Genoa
Department of Biophysical and Electronic Engineering
Via All'Opera Pia 11/A, 16145 Genoa, Italy

ABSTRACT

The fractal–based technique is one of the most recent approach in texture analysis; the potentiality of this approach is under investigation in various fields, for the analysis of images, curves and natural 3–D structures.
In this work, we have verified the hypothesis of fractal behaviour of intensity surfaces of remote sensing images by appropriate tests. Then we have applied a specific method for estimation of fractal dimension (the so–called "blanket" method) for texture discrimination in remote sensing images (in particular, in satellite SAR images, Meteosat images and aerial photographs). By analyzing the fractal dimension histograms and splitting them appropriately, a segmentation of analyzed images was also implemented. Interesting results in texture discrimination and image analysis have been obtained for almost all the considered images.

INTRODUCTION

The improvements of thematic mapper (TM) sensors in spatial and radiometric resolution, the wider spatial coverage, and, at the same time, the introduction of artificial intelligence (AI) methods for a better image understanding, are creating a new interest in classification of earth morphological aspects.
Therefore, in the perspective of extensive application of synthetic aperture radar (SAR), which started at the end of this decade, and in consideration of the increasing attention on automatic territorial investigation (e.g., meteorological, geomorphological, pollution aspects, and so on), it seems useful to evaluate analysis and performances of new methods for region characterization in remote sensing images.
One of the most used parameters for region classification is image texture. Several works have been performed for texture analysis in remote sensing images, using classical statistical and structural methods, particularly on aerial photographs [1, 2, 3, 4, 5] and on SAR images [6].
In this paper we show that, for remote sensing images, the fractal model defined by Mandelbrot is applicable to gray level amplitude surface (i.e., the surface whose height in each point is represented by the gray level of the corresponding image pixel). This surface is also called "image–intensity surface". In order to assess the validity of the fractal model in this application, we carried out on each image the appropriate test, as specified by theory, to verify the fractal behaviour of intensity surfaces. Then we applied fractal methods for texture discrimination on aerial photographs, satellite SAR and METEOSAT images, obtaining interesting results.
Fractal techniques have already obtained appreciable results in texture segmentation and classification in application such as biomedical images. Several methods have been developed to compute fractal parameters of light–intensity surfaces [7], [9], [10], [12].

We chose the so−called "blanket" method for its good performances and easy implementation; this method was developed by S. Peleg et al. [7] and gave good results in texture classification on natural images.

The basic parameter for texture characterization is the "Fractal dimension" (D). The value of D for a certain image zone is a measure of the "roughness" or "complexity" of the surface represented by the image itself. Fractal surfaces have D in the range [2, 3); smooth surfaces are characterized by values of D close to 2, while very rough and complex surfaces have D close to 3. It must be pointed out that, for fractal surfaces, the roughness and complexity is present at all scales of observation, i.e., a perfect fractal surface presents the same topological characteristics when it is observed at every distance or magnification.

Fractals are very useful for texture analysis on natural images because most of natural surfaces are quasi−fractal. Mandelbrot [8] showed that a great number of physical processes (as water or wind erosion, orogenesis, etc.) produce surfaces which are fractal with a good approximation, i.e., they present the same topological characteristics (and then the same value of D) for a wide range of observation scales.

Furthermore, it has been demonstrated [9] that an image which represents a fractal or quasi−fractal surface is also fractal (with the same D value), if regarded as an image−intensity surface. As different textures are represented by light−intensity surfaces of different roughness, they can be discriminated by the values of D computed for the corresponding light−intensity surfaces.

When applicable, the fractal description of texture can be considered very powerful and concise, since different textures can be accurately discriminated on the basis of a single parameter. Texture analysis and discrimination can be obtained by computing the "fractal dimension" D for all the NxN pixel masks into which the image can be subdivided (generally N=4 or 8 to gain sufficient resolution in segmented images). The histogram of number of masks of a given fractal dimension versus such a fractal dimension is then calculated. Therefore, is possible to perform a segmentation by subdividing the analyzed image in some regions, each one characterized by a different kind of texture. Regions are singled out in the image by splitting the "D" histogram so that each region corresponds to a single histogram mode.

MATERIALS AND METHODS

The remote sensing images we used in this work can be classified in three main types:

− A "SEASAT SAR" (Synthetic Aperture Radar) image representing a 40x40 Km ground area of Algeria in a scale of approximately 1:250.000 (orbit 791 Aug. (1978, 25m resolution; frame centre N035−54−44 E 005−36−55);

− METEOSAT images representing Italy, executed in visible − close infrared band;

− Aerial photographs representing urban scenes of italian cities.

The system used for this work is composed of:

− A Hewlett−Packard 1000, 16 bit computer;

− A video memory TESAK VDC 501 for 512x512 pixel images.

The so−called "blanket" method [7] is based on the measurement of the light−intensity surface area. In this method we consider all the points in a 3D space at distance ε from the surface, thus covering the surface with a 2ε thick "blanket"; the surface area is then given by the blanket volume divided by 2ε .

The blanket is defined by its upper and lower surfaces u and b. At first, for $\varepsilon = 0$ and given the gray level function $g(i,j)$, we have: $u(0,i,j) = b(0,i,j) = g(i,j)$. For $\varepsilon = 1, 2, 3,.....$, blanket surfaces are defined as follows:

$$u(\varepsilon,i,j) = \max \left\{ u(\varepsilon-1,i,j)+1, \max_{|(m,n) - (i,j)| = 1} u(\varepsilon-1,m,n) \right\} .$$

$$b(\varepsilon,i,j) = \min \left\{ u(\varepsilon-1,i,j)-1, \quad \min_{|(m,n)-(i,j)|=1} \quad b(\varepsilon-1,m,n) \right\}$$

Pixels (m,n) having distance 1 from pixel (i,j) are the four nearest neighbours to (i,j); analogous expressions can be defined if we want to consider other pixel sets around pixel (i,j).
The above mentioned expressions ensure new surfaces u(ε) and b(ε) to be respectively at least 1 higher and lower from preceding surfaces u($\varepsilon-1$) and b($\varepsilon-1$).
Blanket volume v(ε) is calculated from u and b according to:

$$v(\varepsilon) = \sum_{ij} (u(\varepsilon,i,j) - b(\varepsilon,i,j))$$

As surface area measured at a distance ε is considered:

$$A(\varepsilon) = \frac{v(\varepsilon) - v(\varepsilon-1)}{2}$$

As we mentioned above, the surface area should be calculated by v(ε)/2ε ; in fact, subtracting v($\varepsilon-1$), we insulate those features changing from scale $\varepsilon-1$ to scale ε. When a real fractal surface is analyzed, both area definitions are identical because changes in surface features are independent from observation scale, and then measures at two different scales would provide the same fractal dimension; but, for nonfractal surfaces, we must disregard lower-scale features effects, and this is obtained by subtracting v($\varepsilon-1$). It has been shown [7] that the above definition provides reasonable results for both fractal and quasi-fractal surfaces.
Mandelbrot [8] showed that a fractal surface area behaves according to the following expression:

$$A(\varepsilon) = F\varepsilon^{2-D}$$

where D is the fractal dimension and F is the surface area, if this is computable by an analytic formula (nonfractal surface having D = 2).

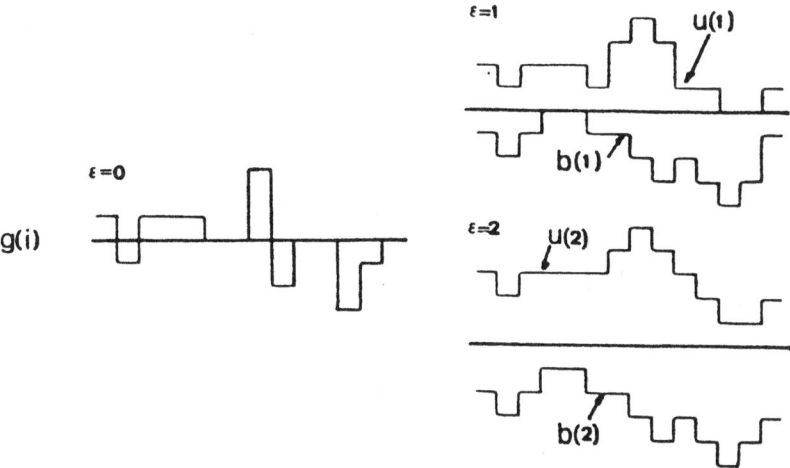

Fig.1. a monodimensional function g(i) and its covering blankets for $\varepsilon = 1$ and $\varepsilon = 2$.

If we represent A(ε) in a bilogarithmic graphic, we obtain a straight line having a slope equal to $2-D$ (this curve is not a straight line for nonfractal surfaces). For texture analysis, the image intensity surface area A(ε) is calculated for various values of ε ranging from 1 to M (e.g., M = 50). Then a linear regression is applied on the bilogarithmic graphic of A(ε) obtaining the slope value $2-D$. From this value, we can easily get the fractal dimension D of the image intensity surface corresponding to the analyzed texture.

The analysis of the bilogarithmic graphic of A(ε) is of great importance since it allows one to assess the validity of the fractal hypothesis. In particular the graphic correlation coefficient of log(ε) and log(A(ε)) gives a measure of the similarity between the image intensity surface and a fractal surface. When this correlation coefficient is close to 1, the fractal hypothesis is well verified, and the D value can be used as texture−discriminant feature.

For region classification purposes, we subdivide the image into a number of NxN pixel masks and calculate the fractal dimension D for each of them. An image segmentation in regions characterized by different textures is then performed as showed in the introduction. It has been observed, using synthetically generated fractal textures, that the values obtained for D become more exact when mask dimensions increase; yet this causes a loss of resolution in segmented images. Anyway, even by using masks of 4x4 pixels, good resolution in segmented images, without great errors in D calculation, can be obtained.

The correlation coefficient absolute value of the bilogarithmic graphic of A(ε) is extracted for each mask; in those regions where its value is close to 1, then the image intensity surface is fractal with a good approximation. This allows us to verify whether a texture originates from a fractal surface or not.

RESULTS

In this section, the results obtained by analyzing the remote sensing images described above are showed.

A METEOSAT digital image representing Italy and neighbouring regions is shown in fig.2. In this image several kinds of textures are present, related to sea, lands, mountains and clouds.

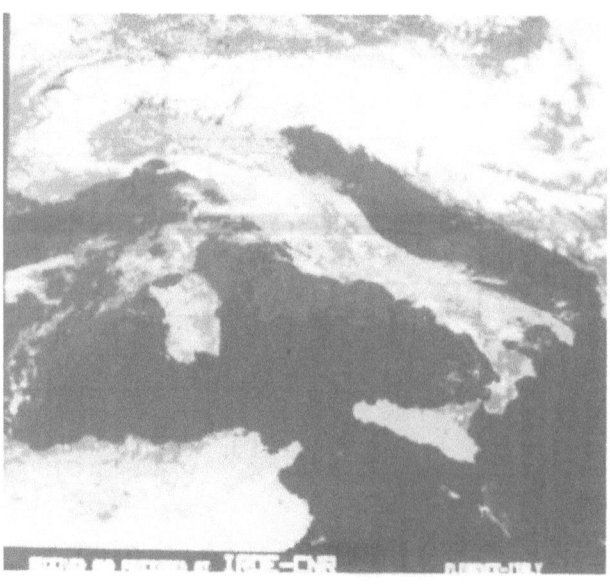

Fig.2. a METEOSAT image of Italy.

Fig.3 shows fractal dimensions D of the 4x4 pixel masks into which the image was subdivided. High fractal dimensions correspond to bright gray levels, and low fractal dimensions are represented by dark gray levels. We can see that coastlines and mountains present high values in fractal dimensions; this means those regions correspond to very rough and complex surfaces. Hilly lands and clouds have intermediate D values, while sea and planes show low values of fractal dimension. It may be pointed out that edge regions (e.g., coastlines) show very high fractal dimensions, often greater than 3; this implies that texture edges can be recognized by the use of fractal techniques.

The correlation coefficients for all the masks of the image in fig.3 are displayed in fig.4. Values close to 1 correspond to bright gray levels (this means that represented surfaces can be considered fractal with a good approximation), while dark gray levels correspond to low values for the correlation coefficient. As one can see, almost all areas show high values of the correlation coefficient, exept for some regions corresponding to the sea.

Since, for all the other considered images, the extracted correlation values show a quasi−uniform behaviour (generally, with values close to 1), in this paper we present only the above correlation coefficient image.

Finally, fig.5 shows a segmentation into four regions of the METEOSAT image; this segmentation was obtained by splitting the D histogram at the fractal dimension values 2.28, 2.57 and 2.98.

Fig.6 shows the SEASAT SAR image of a region in Algeria. Highly reflecting objects for radar signal are displayed in white (e.g., mountains ridges or perpendicularly−to−signal oriented slants), while horizontal plane surfaces (e.g., lakes) are black because they do not reflect any signal towards the radar antenna.

In fig.7, we can see fractal dimensions of 4x4 pixel masks which the preceding image has been subdivided into; one can observe that mountains and hills present highest fractal dimensions (bright gray tones), planes and agricultured lands have intermediate values of D, and the lake has the lowest fractal dimensions in the whole image (dark gray tones). The segmentation into three regions for this image is showed in fig.8; it was obtained by using the threshold values of 2.56 and 3.1. It must be pointed out that fractal dimensions greater than 3 result from regions rich in high spatial frequencies (e.g., edges) or from calculation errors due to the small dimensions of used masks.

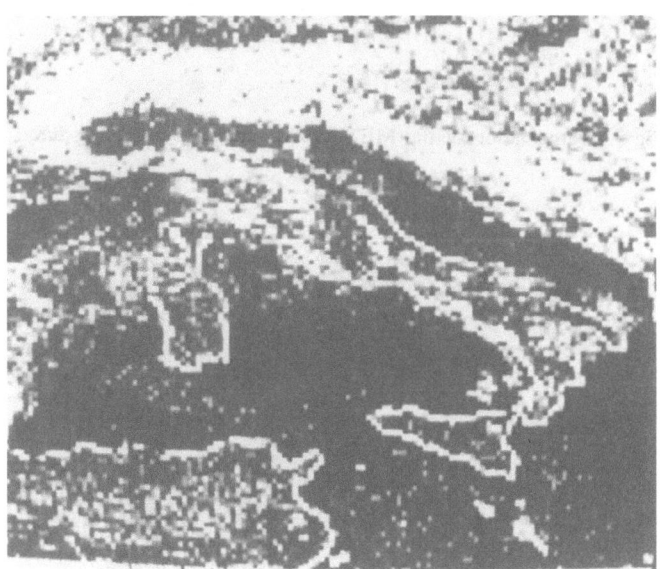

Fig.3. fractal dimensions of 4x4 pixel masks
obtained analyzing the METEOSAT image of Italy.

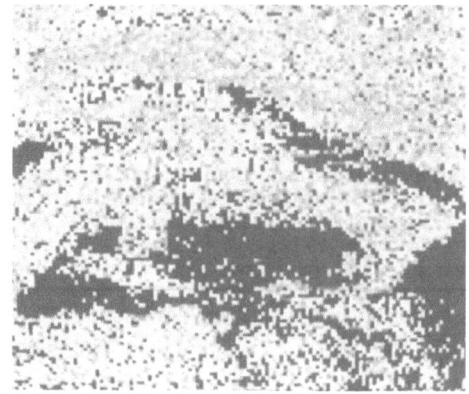

Fig.4. correlation coefficients of masks in fig.3.

Fig.5. segmentation of the METEOSAT image representing Italy.

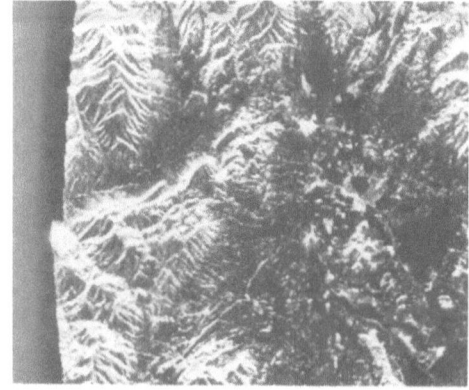

Fig.6. SEASAT SAR image of an Algeria region.

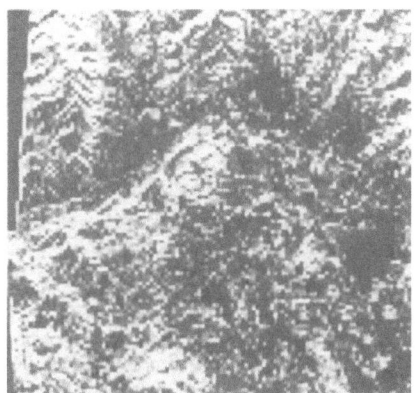

Fig.7. fractal dimensions of the 4x4 pixel masks
obtained from the analysis of image showed in fig.6.

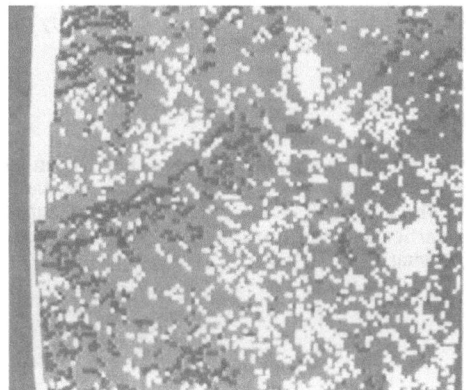

Fig.8. segmentation of SEASAT SAR image representing
an Algeria region

Fig.9. aerial photograph of an italian city.

As aerial photograph, we choose the image in fig.9, which represents an italian city. By examining fractal dimensions of the 4x4 pixel masks showed in fig.10, we can observe that built areas and road edges have the maximum values of fractal dimension; nonbuilt areas and large roads present low values for D.

The program for D calculation and image segmentation by the blanket method was written in FORTRAN 77; the processing time on a HP 1000 computer was of about 20 minutes employing 4x4 pixel masks, and 30 minutes in the case of 8x8 pixel masks.

CONCLUSIONS

In this work, we have proved that the fractal model well describes gray−level amplitude surfaces of remote sensing images. Consequently, the fractal dimension can be used as a powerful parameter to discriminate differently textured areas in such images.

In particular, the blanket method, for texture analysis by use of fractals, showed good performances. This method allows to obtain a sufficient precision in fractal dimension calculation, even if the considered surface area is small; then, it is possible to gain a good resolution in segmented images. Furthermore, fractal dimension can be used as textural feature for automatic segmentation and texture−based region classification.

Other features could be used for image segmentation, e.g., the "fractal signatures" (local slopes in bilogarithmic graphic of A(ε)), to obtain better results; yet, this would increase computation time. Automatic histogram splitting is also possible, provided that histograms present evident modes.

Further applications of blanket method have been developed at the Department of Biophysical and Electronic Engineering, University of Genoa, particularly in texture analysis and segmentation for biomedical and industrial images.

ACKNOWLEDGEMENTS

The METEOSAT and aerial photographs images have been supplied by the Electromagnetic Waves Research Institute (IROE) of National Research Council (CNR) of Florence (Dr. R. Carla'); the SEASAT SAR image has been supplied by University of Naples (Dr. S. Vetrella). Particular thanks are due to Prof. R. Viviani for useful help in SEASAT SAR image processing.

Fig.10. fractal dimensions of the 4x4 pixel masks
obtained from the analysis of image showed in fig.9.

REFERENCES

1. M. Nagao, T. Matsuyama: "A structural analysis of complex aerial photographs"; Plenum Press, New York (1980).
2. R. W. Conners, M. M. Trivedi, C. A. Harlow: "Segmentation of a high resolution urban scene using texture operators"; Computer Vision, Graphics and Image Processing, N. 25 (1984).
3. J. Weszka, C. Dyer, A. Rosenfeld: "A comparative study of texture measures for terrain classification"; IEEE trans. on SMC, vol. SMC−6 (1976).
4. H. Kaizer: "A quantification of textures on aerial photographs"; Boston University Research Laboratories, Tech. Note 121 (1955).
5. W. D. Stolemberg, T. G. Farr: "A Fourier based textural feature extraction procedure"; IEEE trans. on Geoscience and Remote Sensing, vol.GE 24, N.5 (1986).
6. F. T. Ulaby, F. Kouyate, B. Brisco, T. H. L. Williams: "Textural informations in SAR images"; IEEE trans. on Geoscience and Remote Sensing, vol.GE 24, N.2 (1986).
7. S. Peleg, J. Naor, R. Hartley, D. Avnir: "Multiple resolution texture analysis and classification"; IEEE trans on PAMI, vol. PAMI−6 (1984).
8. B. B. Mandelbrot: "The fractal geometry of nature"; Freeman, San Francisco (1982).
9. A. M. Pentland: "Fractal −based descriptions of natural scenes"; IEEE trans. on PAMI, vol. PAMI−6 (1984).
10. T. Lundhal, W. J. Ohley, W. S. Kuklinski: "Analysis and interpolation of angiographic images by use of fractals"; University of Rhode Island, Kingston, USA (1985).
11. B. B. Mandelbrot, B. J. Van Ness: "Fractional brownian motion, fractional noises and applications"; SIAM, vol.10, N.4 (1968).
12. T. Lundhal, W. J. Ohley, S. M. Kay, R. Siffert: "Fractional brownian Motion; a maximum likelihood estimator and its applications to image texture"; IEEE trans. on Medical Imaging, vol.MI 5, N.3 (1986).

ISSUES IN THE INTEGRATION OF SPATIALLY-DISTRIBUTED DATA

ANCILLARY TO REMOTELY SENSED IMAGES

Andrea G. Fabbri, Ko B. Fung, and Tonis Kasvand*

Canada Centre for Remote Sensing, Ottawa, Canada. K1A 0Y7
*Department of Computer Science, University of Ottawa
Ottawa, Canada. K1N 6N5

ABSTRACT

In remote sensing integration is the task of bringing together information for the systematic analysis of spatially-distributed data by digital image processing. This paper discusses some practical problems that are too often disregarded or underestimated when attempting to evaluate the cost and complexity of a data-integration task and its impact in a research project. Topics such as an application-independent approach to integration, the knowledge expected for a user of remotely-sensed data, data formats, media, metrics, and uncertainty, the cost of constructing an integrated data set, vector and raster data conversion, and some analytical requirements of integrated data, are discussed with an eye to the practical challenges of the present and of the future in remote sensing. This paper intentionally avoids emphasizing the power of special-purpose computer systems and comprehensive organizational schemes in order to focus on practical problems that still make data integration one of the most difficult and costly tasks in remote sensing.

INTRODUCTION

Our planet earth is being observed, measured, and recorded in many different disciplines. In geophysics, aeromagnetic, gravity and radio-metric properties are measured spatially and temporally. In remote sensing, spectral response and microwave backscatters are recorded. In survey and mapping, topography, topology, elevation, and other socio-economical information are collected. In climatology, wind direction, humidity, etc., are recorded. As the list goes on and on, we may notice that a lot of the data being recorded are interrelated. If these data can be brought together and analyzed in a systematic way, we can under-stand more precisely our environment and our resources.

We have termed "integration" the complex task of bringing together information for systematic analysis. Several questions can be asked regarding the integration of spatially-distributed data in the earth sciences: (i) What is currently being done by individual researchers, by governmental institutions and by private ones? (ii) How costly is it to capture, digitize and integrate ancillary data with remotely-sensed images? (iii) How likely are we to reuse the data? (iv) Are there any standards set for common tasks? (v) Does the user ·know the best procedure for integrating spatially-distributed data for the analysis to follow? (vi) What are the analytical requirements?

In addition to those questions that do not yet have satisfactory procedural answers, the impact of quantification in the geosciences can easily lead to "information overload" foreseeable by the increase in "black-box" sensor systems for gathering data (Griffiths, 1974, p. 88). For this reason it becomes important to distinguish between data and information: it can be done using subject-matter knowledge and the aid of a computer. In other words, the prolific gathering and the integration of spatially-distributed data requires a proper man-machine symbiosis and suitable representation of the problems into a computer domain.

The financial cost, man-time and intellectual effort in data integration can be very high. Attention must be given to procedural and data gathering bottlenecks, to processes that can be automated, to the construction of computational models, and to careful planning of application projects. This paper discusses some practical problems that are too often disregarded or underestimated when attempting to evaluate the cost of a data integration task and its impact on a research project.

THE GENERAL-PURPOSE APPROACH

Data ancillary to remotely-sensed imagery consists of manually produced terrain maps, manually and automatically contoured maps, manual and computer files with the location and description of point data. There is a definite advantage in focussing on application-independent approaches because one main target of integrating spatially-distributed data is teaching computers to perform the same kind of tasks that humans do so well in addition to the tasks that a computer performs quantitatively.

In order to say that we have the capability to "understand" we must have developed a computational strategy that is adaptable to whatever comes in front of our vision system. Then we can generalize our approach and isolate processes that could be most frequently used independently from the application and finally our application can be developed by constructing a strategy for those processes. What can be done with the data in general (irrespective of a specific application problem)? One answer to this question is: to provide the procedures and tools for integrating the data so that the information content can be fully exploited.

What can we do to integrate the image data? In essence, we have the following digital data:

(a) point data (0-D) for which spatial context is required;
(b) line segment data (2-D) for which spatial context is required;
(c) area data (2-D) for which we want to know shape, distribution and spatial context;
(d) digital elevation model data or DEM's (3-D), in which features such as gradients, trends, saddles, valleys, ridges, and autocorrelation functions have to be identified and analyzed; and
(e) multiple data sets (>3-D) in which individual images have to be put in registration and compared to establish spatial correspondence of identified features.

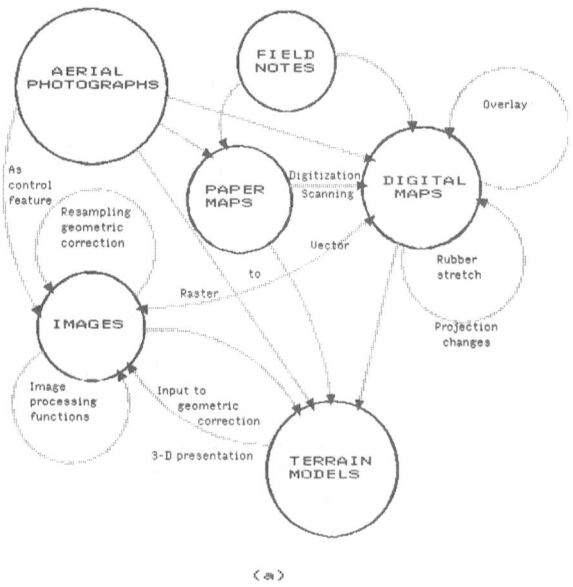

(a)

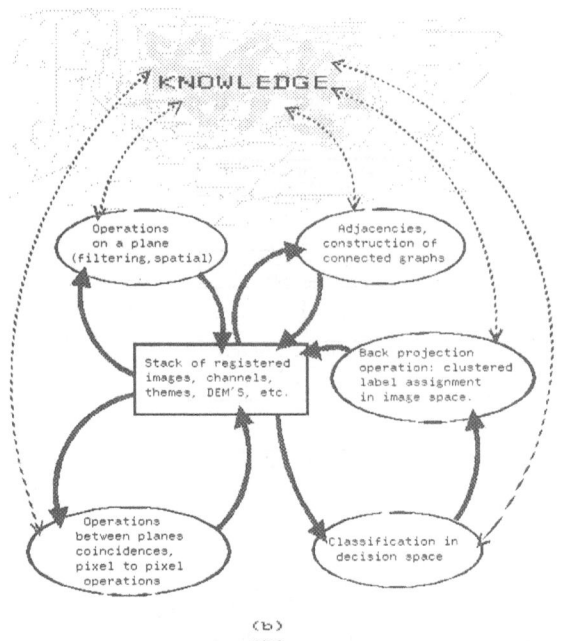

(b)
(b)

Figure 1. Inputs to data integration in remote sensing (a), and types of
image processing functions (b).

Figure 1 shows a synopsis of inputs to data integration and of types of processing function in remote sensing. The basic steps of a typical situation in the preprocessing of remote sensing images to obtain corrected standard products for analysis is illustrated in Figure 2. Well established procedures exist for those steps. Figure 3 describes the basic steps required to display, capture and preprocess ancillary data from maps and/or computer-processable files. In all cases, the end-product is a digital image that can be binary (black and white), a labeled image corresponding to a colored map (symbolic data), or a gray level image (corresponding to LANDSAT channel or a DEM) for continous data. Because data integration can be an expensive task, it may becme convenient to use a small microcomputer and inexpensive digitizing equipment for the integration task (see Kasvand et al., 1981; Fabbri, 1984).

It is difficult to decide on production-oriented integration in remote sensing, because the usability of many spatially-distributed data is not fully explored. The risk of high rebundancy in a data set and of poor correlation between different sets is still very high due to the complexity of analyzing and interpreting multiple, multisource and multiresolution data sets, and to the insufficient diffusion of image processing techniques among experts in conventional fields of application such as land planning, forestry, economic geology, resource assessment, geophysics, etc.

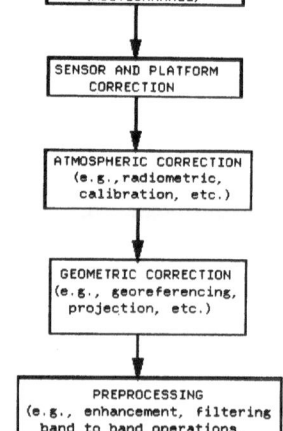

PREPROCESSING OF RASTER DATA
FOR REMOTE SENSING

REMOTE SENSING IMAGES
(MULTICHANNEL)

SENSOR AND PLATFORM
CORRECTION

ATMOSPHERIC CORRECTION
(e.g., radiometric,
calibration, etc.)

GEOMETRIC CORRECTION
(e.g., georeferencing,
projection, etc.)

PREPROCESSING
(e.g., enhancement, filtering
band to band operations,
combination of bands, grey
level stretching, etc.)

gray level images

Figure 2. Preprocessing of raster data for remote sensing.

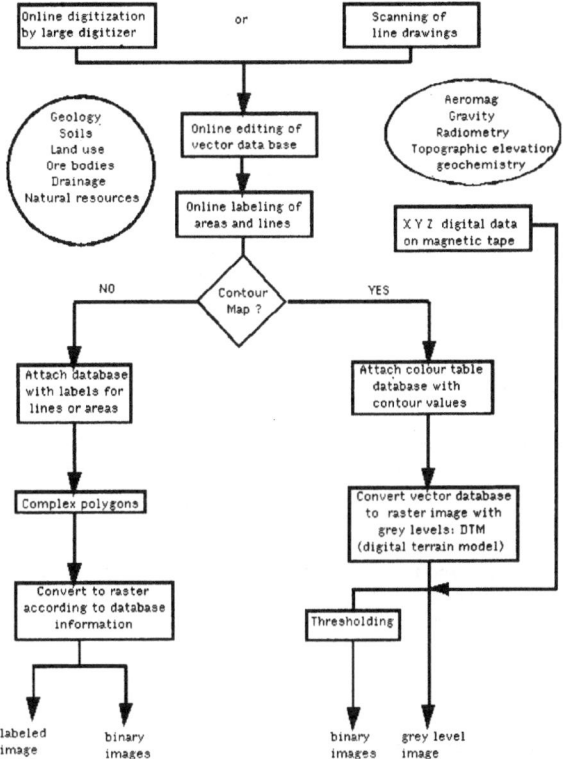

Figure 3: Capture and preprocessing of ancillary data from maps for remote-sensing applications.

THE USER OF DATA INTEGRATION

The users of multiple data sets can be generally divided into two categories: scientists and individuals interested in routine products. The former are interested in performing scientific research and to study the relationships between the sets. The latter assume to have all procedures in place: their interest lies in the routine production of visual and statistical outputs.

The first group consists mainly of earth scientists, environmentalists, land planners, and engineers. They have a physical concept in mind and some scientific confidence of the type of data to assemble. For them, the relationships among the data sets are the most important aspects to better interprete the analytical results in remote sensing applications. The parts of integration which concern physical representation, structure, and mechanisms of assembling the data, become less important than the operational procedures. In some rare occasions, he/she will find "truly friendly" hardware and software systems which have captured previous integration experience and which somehow guide in the integration. These are termed "knowledge-based systems" which, however, are still in an experimental stage of development. In general, the integrator must understand the system where each data set resides and the system in which the data integration will be done.

On the other hand, the second group of users will always deal with standard procedures. The error budget and the tolerance requirements, the data format and media of storage, and the data conversion schemes all should have been worked out beforehand. This group needs only an understanding of the hardware and software steps which can lead to the integrated data set. Then standard methods of analysis or production-oriented procedures are applied to obtain the desired results from the integrated set. Thus, the cost of integration is minimized.

The difficulty in remote-sensing studies is that frequently new original research is needed because of the interdisciplinary character of the applications and due to the complementary nature of remote sensing information to other kind of spatially-distributed data. Therefore, many problems in data integration cannot be fully anticipated.

FORMATS, MEDIA, METRIC AND UNCERTAINTY

The very first step to integrate data from different sources is to convert them together to common formats and common media. Historically, information is stored in maps and charts, but in the last few decades there have been changes to storing information in computers: even though, most information is still represented as points, lines, nodes, arcs, areas, and polygons, the storage media have extended to include tapes, disks, cartridges, etc. To confuse the storage issue further, there are numerous standards, formats, and structures to organize these data which may be machine dependent.

In the United States and in Canada the need to exchange cartographic data is well recognized. National committees for digital cartographic data standards have been set up in the United States and have involved government, industries, and academics to establish data exchange procedures (NCDCDS, 1985; 1986). In Canada, the Canadian Council on Surveys and Mapping has established a national standard for the exchange of topographic data in 1986, and the Canada Centre for Remote Sensing, CCRS, together with other government institutions has established a standard format for the exchange of geocoded vector information (EMR, 1984; Goodenough et al., 1983a, b). Remote sensing by satellite and airborne platforms is the latest technology introduced to acquire information about the earth surface. Multiple formats for the exchange of geocoded raster data have been proposed by Billingsley and Ureña (1984). The internationally accepted standard format is the LGSOWG (1979) proposed by the LANDSAT Ground Station Operation Working Groups.

The next obstacle is to bring the data into a common metric. On a high level the following must be addressed:

- accuracy requirement   - availability of data
- spatial scale and projection - time scale

The accuracy required shall first be established. The success of using an integrated data set depends on how much the end result is a function of the individual data source. This translates into the fact that the inherent accuracy of the source must be known and the requirement of accuracy on each data source must be determined. Once this is done, it may happen that some sources have to be eliminated due to insufficient accuracy. Other requirements may be relaxed by reducing data volume to minimize the integration task.

Directly related to the accuracy issue is data availability. It may happen that the model that we are dealing with requires a certain accuracy of data but the source cannot provide it. Relaxing the accuracy may invalidate the model. It is obvious that all sources of data need to be at the same scale and projection before the spatially-distributed information is overlayed and analyzed. However, it may also be important to check if the data are of the same time frame, since the comparison between data obtained at different times may be meaningless.

Another point, little considered in data integration, is the uncertainty often associated to spatially-distributed data on special-purpose maps. Duda (1980) and Duda et al. (1979) described different degrees of certainty assigned to map evidences and field evidences when integrating them in decisional metallogenic models for mineral exploration (see Fabbri, Kasvand and Yatabe, 1986). Another example of semi-probabilistic weights assigned to features in engineering geology maps is described by Varnes (1974). There, map units and transitions between pairs of map units can be assigned different degrees of importance, strength, or certainty using terms such as: rare, uncommon, absent in a particular context, may occur, inferred, logically improbable, essential, unique, inferential, etc. Obviously, various forms of identification of uncertainty for spatially-distributed data are necessary and means of analysis of such data are required.

THE COST OF INTEGRATING

The cost of integrating data sets is highly dependent on the objective for which the set is being constructed. Without knowing the objective and consequently the specification of a data set in terms of accuracy, tolerance, format, source of the original information, nature of the data, the hardware and software systems from which each set is derived, the target machine for the integration, etc., there is no way to estimate the cost, except for the approximate time required for some of the digitization of individual data sets.

For example, the digitization of a thematic map (geology, soil, etc.) containing symbolic data might take anywhere from 40 to 100 man-hours. Longer times might be required to digitize a contoured map such as topographic elevation or aeromagnetic anomaly where the contour lines may be many and very crowded. However, if a computer compatible tape, CCT, is available with the original quantitative data, then only 5 to 10 man-hours may be necessary to transform the data into a preferred processable format. Examples of integration cost estimates have been made by Lyon and Prelat (1978) and Kasvand et al. (1981). However, such estimates are only meaningful within a particular research environment. Perhaps, a more revealing estimate for the manual digitization of 1:50,000 aeromagnetic anomaly maps in Canada is of 5 man-hours per map: it simply covers the digitization by a graphic tablet of the intersections of flight lines and all the contour lines (Michael T. Holroyd, personal communication).

VECTOR AND RASTER INPUTS, DATA CONVERSION

Data in vector form can be stored in a computer by optical scanning or by manual digitization. By far, scanning is more convenient for large amount of data in a standard topographic map sheet where contour lines and other symbols are closely spaced. If a scanning device is not available, manual digitization is performed. When this is necessary, it is best to have an error budget before actual digitization starts. When the work asks for integration with remotely sensed data, it is important to estimate first the accuracy requirement of the following items:

(1) desired pixel size and resolution of remotely sensed data,
(2) position of ground control points and monuments,
(3) boundaries and areas,
(4) sensitivity of the dizitizing table,
(5) digital elevation or terrain model (if it is needed).

With these estimates, we are able to determine the spacing of digitization points along a line or along the boundary of a region. This estimate is important if the complexity of the features to digitize is high in relation to the accuracy needed, but if the lines are fairly straight and the curves follow more or less the circumference of a circle, the spacing can be sufficiently apart to greatly simplify the task and to make the estimate unnecessary.

If we expect to rasterize the manually digitized work in vector form and to overlay it on remotely sensed images, further rules must be observed.

(a) The remotely sensed data must be rectified to a common projection with the digitized work;
(b) The digitized work must be rasterized to the same pixel size of the remotely sensed data;
(c) The digitization process must observe the requirements imposed by the computer system (hardware and software) used to convert the data in raster form. For example, the Intergraph system at CCRS requires that the digitized work be "clean" before the individual polygons can be converted into complex polygons and then rasterized. By "clean", it is meant that:
(1) all those points where lines cross must be digitized points, "known as nodes";
(2) no overshoot or undershoot (i.e., free end-points) of boundary lines of any polygon should occur;
(3) no "double lines" or "double points" (i.e., no line or point on top of one another should be present; and
(d) any complete boundary of a polygon must be composed of at least two lines.

Data conversion is a key step to the success of integration between vector and raster data. A difficulty is that in each conversion we may suffer an information loss. We might need it because the storage of vector data is sometimes more compact, because of processing tradition, or simply because of the availability of a processing facility for a particular domain (raster techniques became available since TV screens became raster devices!). When raster data is converted into vector data, usually, the boundary between two homogeneous areas defines a line element. Because the basic unit for raster data is a pixel which defines a rectangular area, lines created by the conversion process may appear as "staircase" lines instead of as smooth straight lines. Several software packages are available to smooth out such "staircase" lines, however, due to the local implementation, they may not always agree with each other. Another problem that may occur in the conversion from raster to polygon form is that a line defined by a serie of pixels in the raster domain may be converted to an elongated polygon, known as "fat" line. This may have to be processed to form a single line to be merged with other line data (manipulated differently from the polygons).

Different algorithms are available for vector to raster conversion, however, the process is not without problems. For example, some systems provide three methods, as shown in Figure 4: (i) center dot encoding, (ii) predominant type encoding, and (iii) absence/presence encoding. There will be no problem for the resultant data if the conversion process if done for all the vectors at the same time with proper attribute linkage, so that the decision of assigning a theme value to a pixel is based on all polygons sharing the pixel. However, if the vector information is separated into multiple files which are converted to raster separately and later recombined, there will be disagreements among the boundaries even if all the processes use the same conversion scheme. This is usually not the case when the centre dot method is used, however, in some instances when the intersection of the polygon boundaries is at the centre, none of the converted layers will contain the pixel.

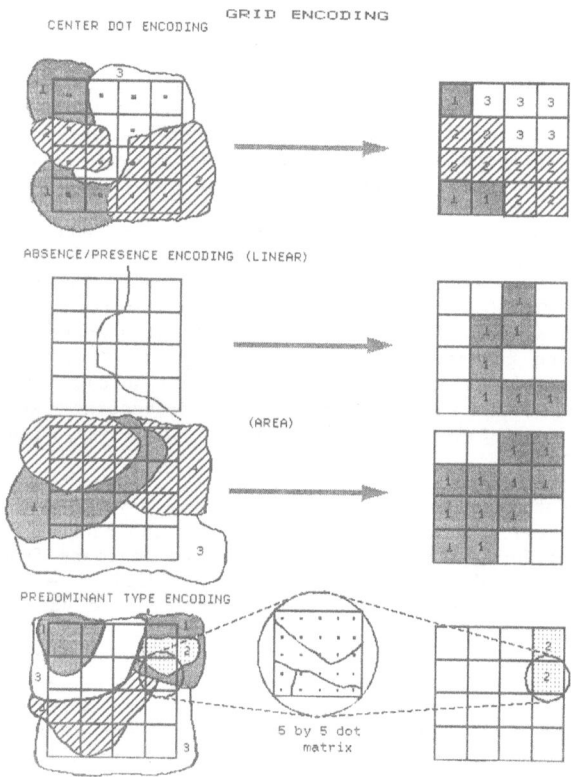

Figure 4: Three common methods of vector to raster conversion.

A similar situation exists in predominant type encoding. When the area to be converted into a pixel if equally divided among more than 2 classes of information, all separate raster conversion processes will not be able to assign a pixel to a predominant class. To avoid such problem, it is recommended that either the center dot encoding method or the absence/presence encoding method be used. All the methods may introduce an error of at most one pixel at boundaries, however, if different files of information from the same vector source-map are merged with the rule that where overlap occurs at the boundary, the last class always supercedes the previous one, then the boundary problem can be resolved.

Special requirements in data conversion may occur that have an impact on the processing to follow. For example, a thin 8-connected boundary line was needed as an individual theme for separating classes of areas, in a geological map-integration application by Fabbri (1984). Such lines simplified feature recognition, measurement and labeling. To satisfy such a requirement, images of area themes separated by thin 8-connected lines can be obtained by substracting·the independently rasterized binary images (center dot encoding) from the labeled image of all the area themes. Similar considerations will have to be made by the integrators when dealing with specific applications.

THE ISSUE OF DATA INTEGRATION SERVICES

Not only the spatially-distributed data to be integrated come from various disciplines, but also there are alternative ways of collecting the data in terms of both raster and vector formats, namely: (1) the data can be collected by the user, (2) by private industry, and (3) supplied by government agencies. Raster data can be generated by scanning photographs, converting a vector file, and directly storing digital outputs from spectral scanners. There are equipments available in the private industries which can generate raster data as described above. One can even contract some of these companies to do it. Recently, in Canada, the Canada Centre for Remote Sensing through a privatization scheme had aircrafts and spectral scanners made available through private companies. Anyone who needs to have an area scanned, can arrange projects with these private companies. There are also companies in Canada which provide raster processing services such as enhancement, geometric correction, image processing and analysis,.. etc. for customers.

Vector data can be handled in a similar way. One can either buy the equipment and generate vector data or can contract it out to private industry to perform tasks such as scanning of contours from topographic maps, creation of digital terrain models or digital elevation models, and generation of specific maps from surveyed field data. The Surveys and Mapping Branch of the Ministry of Energy, Mines and Resources is in the process converting their 1:50,000 and 1:250,000 scales topographic maps into digital form. There is research going on to revolutionize the storage of map information. Future maps may be able to provide the details at the scale requested by the user. Furthermore, they may be queried and may provide geographical information according to the topology specified. It seems that this kind of maps is an example of the KBGIS (Smith and Pazner, 1984) which the current research on GIS has been concentrated. However, at the present moment, we are still working with the GIS without KB (sophisticated knowledge base).

The user should be aware, though, that the standard products of these services have inherent limitations in terms of accuracy and representation. It remains important to consider carefully the requirements of a project, identify the suitability of the standard products, and the cost to convert the standard products to the specification of a project. This kind of cost may easily be underestimated during the planning phase of data integration. On the other hand, for non-standard products, the user may specify the accuracy, the format of representation, the format of storage and the media the data resides. Failure to observe the above may cause severe unexpeted costs.

CHALLENGES FOR THE FUTURE

We have seen some of the difficulties in exploring the usability of spatially-distributed data: not enough research, risk of high rebundancy, insufficient diffusion of image-processing techniques, etc. In remote sensing it is often difficult to separate a research task from a routine task: truly friendly hardware and software systems are rare, and knowledge-based systems are still too experimental. Too often original research is needed because applications are interdisciplinary, i.e., remote sensing data are complementary to other data. There are still unresolved problems with data formats, media and metric and map-feature uncertainties should be expressed in processable forms.

It is extremely difficult to estimate the cost of an integration task: no published reports are available on experimental work on cost evaluation. The implications of processing raster or vector data have to be considered in an integration task: conversion from vector to raster is a delicate process that requires an understanding of the types of processing to follow. Still much research is needed to provide comprehensive guidelines on the desirable data integration services to aid in remote sensing. The limitations of standard products at present may be too severe for many practical applications.

In the opinion of the authors of this paper, computer vision in remote sensing is the desirable follow-up to the integration of spatially-distributed data. The data collected form the various sources are of high dimensionality and their representations have to be consistent so that comparisons can be made. The representations must be in forms which allow problems to be solved with computers. In general, this "mass of data" is fundamentally beyond our innate abilities to "absorb and comprehend". To make use of the data we are forced to select and use only a few items and relations at a time. Seldom more that five to seven variables can be handled efficiently at one time. Computers do not suffer from this limitation in handling of multi-dimensional data. We have a whole "armoury" ("tool box") of techniques, ranging from statistical methods (discriminant analysis, principal component analysis, etc.) to all the aspects of image processing and pattern recognition (feature extraction, segmentation, clustering, learning with or without supervision, etc.). The problem is how to apply the techniques (what, when, where, and in what combinations) to obtain a solution to the problem at hand, while recognizing that the techniques and their combinations depend on the available data as well as on the problem to be solved.

Idealistically, there is no upper bound to the amount of knowledge required from the user that would help him to solve a particular problem. In practical terms, the lower bound to knowledge required from the user is reduced by the availability of good query languages and advisory expert systems that can help him to select suitable data, operations, computational sequences, that generate intelligible displays, and so forth. Such systems do not replace user intelligence, but should only help him to concentrate on the problem. We need far more research and testing of carefully constructed integrated data sets.

REFERENCES

Billingsley, F. C. and Ureña, J.L., eds. 1984, "Data Interchange Formats, Volume I: Workshop Proceesdings", February 7, 1984, Pasadena, California. NASA Jet Propulsion Laboratory, Publication JPL-D-1723.

Duda, R.O., 1980, The Prospector System for Mineral Exploration, SRI International, Final Report, April 1980, 120 p., 1980.

Duda, R.O., Hart, P.E., Barrett, B., Gashning, J.G., Konolige, K., Reboh, R. and Slocum, J., 1978, "Development of the Prospector Consultation System for Mineral Exploration", SRI International, Final Report Covering Oct. 1, 1976, Sept. 30, 1978, 193 p.

EMR, 1984, National Standards for the Exchange of Digital Topographic Data, Vol. I, 181 p., and Vol. II, 202 p., Canadian Council on Surveying and Mapping, Department of Energy, Mines and Resources, Surveys and Mapping Branch, Topographical Surveys Division, Ottawa, Canada, July 1984.

Fabbri, A.G., 1984 Image Processing of Geological Data, Van Nostrand-Reinhold, New York, 244 p.

Fabbri, A.G. and Kasvand, T., 1986, Automatic Integration of mineral resource data by Image processing and artificial intelligence, in Chung, C.F., Fabbri, A.G., and R. Sinding-Larsen, eds., Proc. NATO Adv. Study Inst. on "Statistical Treatments for Estimation of Mineral and Energy Resources", IL CIOCCO (Lucca), Italy, June 22-July 4, 1986 (Reidel, Dordrecht, in press).

Fabbri, A.G., Kasvand, T., and Yatabe, S.M., 1986, Map Logic for Data Integration by digital image processing. Proc. 15th Geochautauqua on "Computers in the Petroleum Industry: Integrated Approaches", Calgary, Alberta, October 2-4, 1986, Mathematical Geology (in press).

Goodenough, D.G., Plunkett, G.W., and Palimaka, J.J., 1983a, On the transfer of remote sensing classifications into polygon geocoded data bases in Canada. Proc. 6th Intl. Symp. on Automated Cartography, AUTO-CARTO SIX, October 16-21, 1983, Ottawa, Canada, p. 598-606.

Goodenough, D.G., Palimaka, J.J., Dickinson, K., and Murphy, J., 1983 b "Standard Format for the Transfer of Geocoded Information in Spatial Data Polygon Files", Ottawa, Canada. The Spatial Data Transfer Committee, 1979, Department of Energy, Mines and Resources, Statistics Canada, Environment Canada, and Agriculture Canada. Ottawa, Canada. Document Call FOR-SDP-0001, Revision B, Dec. 1983.

Kasvand, T., Fabbri, A.G., and Nel, L.D., 1981 Digitization and processing of large regional geological maps. Natl. Res. Coun. Canada, ERPB Report-938, 91 p.

Lyon, R.J.P., and Prelat, A.E., 1978 Cost analysis of emerging geophysical, geological, and LANDSAT data for mining exploration Proc. 2nd. Conf. on "Economics of Remote Sensing", Jan. 16-18, 1978, San Jose, Ca., p. 168-175.

NCDCDS, 1985-1986, "Issues in Digital Cartographic Data Standards", National Committee for Digital Cartographic Data Standards, Ohio State University, Numerical Cartography Laboratory, Columbus, Ohio, Reports 1-7, 1985, 1986.

Smith, T.R., and Pazner, M.I., 1984, A knowledge based system answering queries concerning geographic objects. Proc. IEEE 1984 IX Pecora Symp. on " Spatial Information Technologies for Remote Sensing Today and Tomorrow", p. 286-289.

Varnes, D.J., 1974, The logic of geological maps, with reference to their interpretation and use for engineering purposes. U.S. Geol. Survey Prof. Paper 837, 48 p.

RECONSTRUCTION AND REPRESENTATION

OF 3-D SURFACES FROM GEOPHYSICAL DATA

Maria F. Costabile

Dipartimento di Matematica
Università della Calabria
87036 Rende (CS), Italy

ABSTRACT

In this paper a method for reconstructing and representing 3-D surfaces from a set of parallel planar slices or cross sections is proposed. The method is quite general but we focus on an application in geophysics, that is 3-D reconstruction of geologycal layers, having as input a sequence of seismic sections. To generate a surface description the following three steps are involved. First, the surface contours at each slice are detected. A surface contour is the image of a curve representing the cross section of the surface with the plane of the slice. Second, a procedure for relating each surface contour with the correspondent one in the successive slice is performed and a triangulation process is accomplished over pairs of correspondent contours to generate the local bounding surface structure. Finally, indications about the surface structure are obtained by analyzing the orientation of adjacent triangular patches. In this paper we present the first step in details; the input seismic sections are converted into digital images and an algorithm for detecting the layer surface contours is described. Some initial results of 3-D surface reconstruction are also included.

INTRODUCTION

The real world is primarily composed of 3-D objects. One goal of computer vision research is to give computers humanlike visual capabilities so that they can sense the environment in their field of view, understand what is being sensed and take appropriate actions. The source data of most computer vision research performed during the past twenty years have been digitized gray-scale intensity images, namely matrices of numerical values obtained by digitizing black-and-white photographs representing a 3-D scene as seen from a particular point of view. People are able to identify, locate and quantitatively describe the objects in a black-and-white photograph. Computer vision systems try to give the computer a similar capability so that the machine can extract useful information about 3-D objects from the input 2-D images.

This paper focuses on 3-D surface reconstruction and representation. For determining the shape of a surface, techniques exploiting stereopsis and motion predominate[1,2]. When the illumination is directional, shading is also useful, while texture gradient is important if the surface is visibly textured[3,4]. Another source of shape information is surface contours, namely images of curves across a physical surface, as it is pointed out by Stevens[5]. In our work the surface contours are the curves obtained as intersection of a 3-D scene with a set of parallel planar slices. The only information about the 3-D objects consists of the intersections of their surfaces with the slices.

A technique for 3-D surface reconstruction has many applications. For instance, in biomedicine we can try to recover the 3-D structure of an organ, like a human heart, from a decomposition of that organ into a set of parallel planar slices. Each slice is characterized by a 2-D shape which defines the structure at that cross section.

The application considered in this paper is in geophysics. Interpreting seismic data is a very difficult task even for human beings. Sometimes, only the experts of a certain type of area are able to give a correct interpretation of data referring to that area. However, many data are now available and their information is so useful that it is worth trying to use the computer to help interpreting those data. Our goal is to give a method for layer surface reconstruction from seismic data.

The surface description is generated in three steps[6]. First, the surface contours at each slice are detected. Then, a procedure for relating each surface contour with the correspondent one in the successive slice is performed. In fact, on each slice we will have several contours of 2-D regions which are the intersections of the objects in the 3-D scene with that slice. It is very important to identify the contours of the same surface on two consecutive slices because the second step of our procedure performs a triangulation process over pairs of corresponding contours to generate the local bounding surface structure, as illustrated in Fig. 1. Finally, indications about the surface structure are obtained by analyzing the orientation of adjacent triangular patches. In this paper we present the first step in details and we show the results of some experiments. In the next section, the input data for the application in geophysics are described. In the third section we show how to detect the surface contours from the input images. Finally, in the last section, we give some initial results of 3-D surface reconstruction.

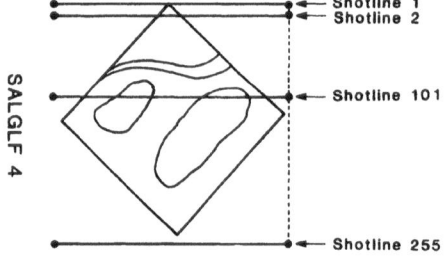

Fig. 1. Corresponding contours onto consecutive slices are triangulated to generate the local bounding surface structure.

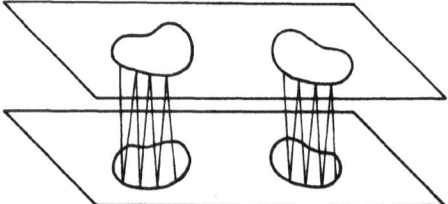

Fig. 2. Model position in the physical modeling tank. Data were collected along the indicated directions.

INPUT DATA DESCRIPTION

The goal of the 3-D surface reconstruction method presented in this paper is to describe the morphology of 3-D layer surfaces. For this purpose we used the physical model data provided by the Seismic Acoustic Laboratory (SAL) at the University of Houston, collected when a model is scanned in the physical modeling tank available at that laboratory. Each of these physical models is formed by a set of layers one on top of the other. The aim of our work is to describe the morphology of those layered surfaces.

To simplify the problem at the beginning, we decided to use a simple model formed by only one layer, with two hills in the center and a plateau in a corner. It is called SALGLF4. The 3-D data set across the model consists of 255 shots/line and 255 shotlines. They were collected by scanning the model in the tank along the directions indicated in Fig. 2. In Fig. 3 the section # 101 is shown (in Fig. 2 the position of this section on the model is also indicated). Each trace has 170 values obtained at a sampling rate of 4 ms. As we can see in Fig. 3, the upper dark band follows almost precisely the model profile giving a useful indication about the actual surface curvature at that section. We want the computer to detect a line from this upper band. This has to be done for each section of our model, so that we will have the surface contour at each section. Then, for this simple model we do not have to look for corresponding contours in successive sections because only one line is detected in each section; this line is the contour of the only surface forming the model. The surface contour detection is described in detail in this paper.

Each section of the type in Fig. 3 must be transformed into a digital image which will be the input of our procedure. Very shortly, a digital image is a matrix f whose values $f(x,y)$ indicate the brightness of each point (x,y) in the image. Image points are also called pixels. Each section is converted in a matrix in which the values of each trace form a column. Consequently, the matrix will have as many columns as the traces in the section and as many rows as the samples in a trace, namely 255 columns and 170 rows. We actually consider only the first 128 samples of each trace so that we obtain matrices of 128 rows. The samples in a trace are real numbers. We neglect the negative values (they are set to 0) and the positive values are compressed into the integer interval [0,63]. Fig. 4a is the digital image obtained after this transformation from the section in Fig. 3. It appears as the specular image of that section, since in Fig. 3 the traces proceed from left to right. Fig. 4b is another example. It is the image corresponding to the section # 115. From now on, each image will be denoted by the number of the corresponding section in the 3-D data set.

SURFACE CONTOUR DETECTION

The first step of the procedure for generating a surface description is the detection of the surface contour which is the intersection of the surface with the plane at each cross section. Looking at the images in Fig. 4, we see that the surface contours are actually

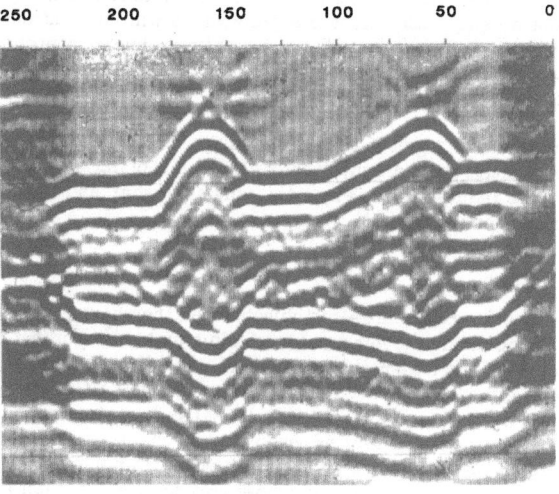

Fig. 3. Section # 101.

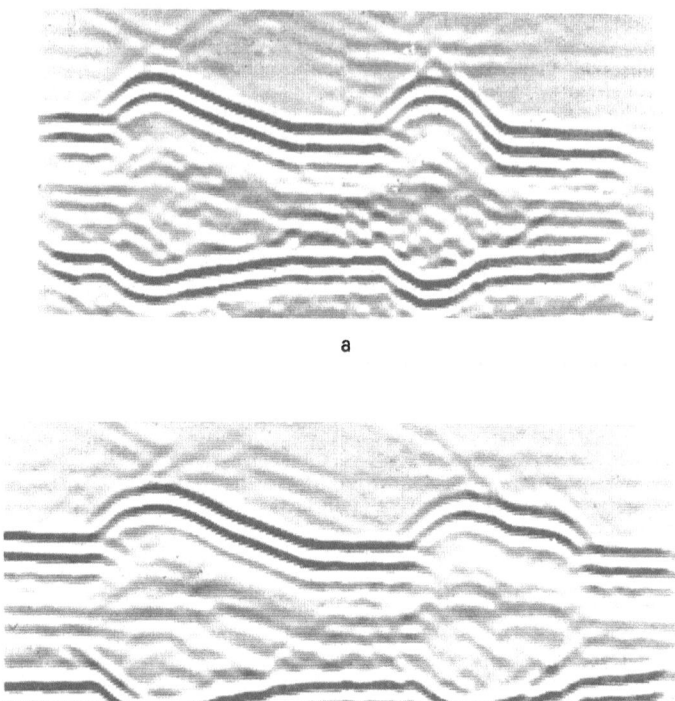

Fig. 4. (a) Digital image corresponding to section # 101; (b) digital image corresponding to section # 115.

displayed as bands whose pixel gray levels are darker than the background. We will perform a segmentation of this image in order to extract those bands and then, a thinning operation will reduce each band to a line. Of course, not all the lines we obtain will indicate surface contours; among those lines the machine will choose the lines representing surface contours. Generally this is not an easy task, but it is a problem we will not consider in this paper; in fact, we know that in our model the only line representing a surface contour is the upper line, that is, the line we will obtain by thinning the upper band in each image (see Fig. 4). All the other lines can be neglected.

We now discuss how to segment each input image, namely, how to extract the surface contour. Three consecutive steps are involved: 1) the bands of darker pixels are extracted, 2) the obtained bands are reduced to lines by a thinning algorithm, and 3) the line representing the actual surface contour is detected.

Band Extraction

A useful technique for extracting some objects when they have gray levels different than the background is thresholding the images at a certain gray level[7]. For instance, if the objects we want to extract are darker than the background, the histogram of the gray level distribution in the image should display two peaks corresponding to the two gray level ranges, one corresponding to the background (light gray levels) and one corresponding to the objects (dark gray levels). The picture can thus be segmented by choosing a threshold that separates these peaks, namely a gray level corresponding to the lowest point between two peaks, and eliminating from the images those elements whose gray level is less than the gray level chosen as threshold. Formally, this can be expressed as in the following. Let say $f(x, y)$ the input image and $f'(x,y)$ the thresholded image. If l_i is the gray level corresponding to the lowest point between two peaks in the histogram, the thresholded image will be for any pixel (x,y) in the image

$$f'(x,y) = \begin{cases} f(x,y) & \text{if } f(x,y) \geq l_i \\ 0 & \text{otherwise} \end{cases}$$

Our images do not have this kind of histogram. A typical histogram is shown in Fig. 5. In this histogram we can distinguish several peaks; we made some tentatives of thresholding by using gray levels at the bottom of any valley but we did not obtain very good results. The technique we finally adopted for segmenting our input images in order to obtain the darker bands uses two thresholds. It works in the following way:

1. Choose two gray levels $l_s > l_f$, so that only regions representing portions of darker bands will be detected when using l_s.

2. Threshold the image at level l_s.

3. For each level l_i from l_s to l_f, update the thresholded image by including only the pixels with gray level greater than l_i and being connected to the pixels already in the thresholded image.

Considering as input image the one in Fig. 4b and using $l_s = 30$ and $l_f = 15$, the result of this procedure is shown in Fig. 6a.

Band Thinning

At this point the images are not yet completely segmented. From each image we must extract a line which is the intersection of the model surface with the cross section defining that image. We now perform a "thinning" operation on the bands obtained in the images like the one in Fig. 6a. As a result, for each band B, we want to have a connected curve which is a reasonable approximation of B. The result of thinning B is called "skeleton" of B.

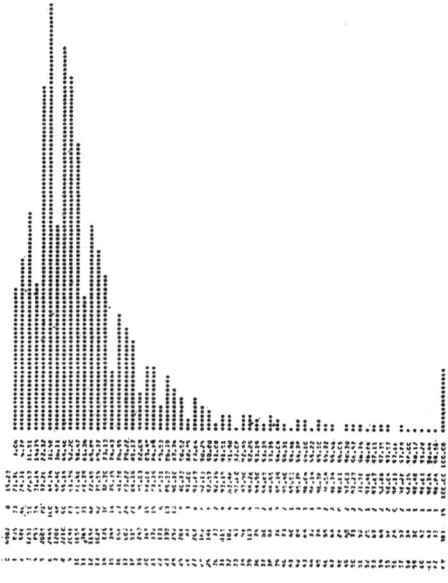

Fig. 5. Gray level histogram of image # 115.

469

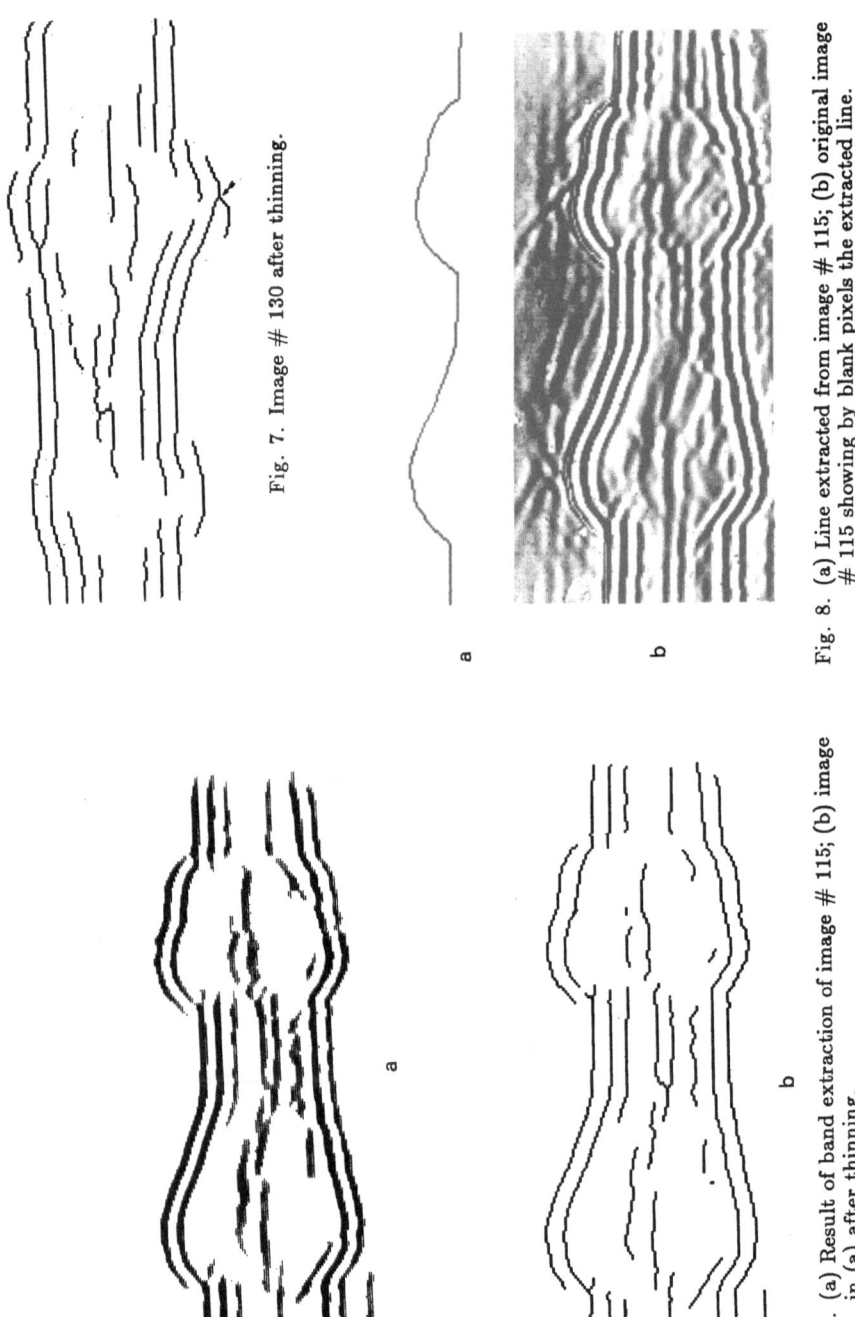

Fig. 7. Image # 130 after thinning.

a

b

Fig. 8. (a) Line extracted from image # 115; (b) original image # 115 showing by blank pixels the extracted line.

a

b

Fig. 6. (a) Result of band extraction of image # 115; (b) image in (a) after thinning.

The thinning algorithm we implemented is a specialized shrinking process which deletes from B, at each iteration, border points whose removal does not locally disconnect their neighboroods[7]. To prevent an already thin arc from shrinking at its ends, points having only one neighbor in B are not deleted. Furthermore, if we delete all such border points from B and B is only two points thick, e.g.,

$$1 \quad 1 \quad 1 \quad 1 \quad 1 \quad 1 \quad 1$$
$$1 \quad 1 \quad 1 \quad 1 \quad 1 \quad 1 \quad 1$$

then B will vanish completely. To avoid this, we delete only the border points that lie on a given side of B, i.e. that have a specific neighbor (north, east, south, or west) in $\overline{B}$, at a give iteration, being $\overline{B}$ the complement of B. To insure that the skeleton is as close to the "middle" of B as possible, we use opposite sides alternatively, e.g. north, south, east, west, and so on. Furthermore, a condition for removing a border point is that it is *simple*. A border point P of a subset B of a digital image f is called *simple* if the set of 8-neighbors of P that lie in B has exactly one connected component. Then, the thinning algorithm we used can be stated as follows: Delete all border points from a given side of a subset B if they are simple and not end points. Do this successively from the north, south, east, west sides of B and iterate until no further point is deleted. The border points must be deleted from a given side of B "in parallel", namely the conditions for deletion should be checked before any other points are deleted. In Fig. 6b the results of band thinning are shown for the image in Fig. 6a.

Surface Contour Identification

On images like that one in Fig. 6b we must now identify the lines which represent the actual surface contours. Because we are using a model formed by only one layer, we know that the only meaningful line is the upper line. This assumption narrows our problem a lot. But even in this simpler case the identification of a line is not an easy task for a computer. First of all what we are calling line is not always a connected line, as we can see in Fig. 6b. There are generally some gaps along the upper line, due to lack of signal in the original image. This happens because of the poor quality of the original data. Furthermore, there could be some ambiguous situations. For instance, there are points, like the one indicated by the arrow in Fig. 7, in which there are two possible directions where to go when traveling along the line.

The algorithm we used for identifying the upper line in the images obtained after the thinning works as follows:

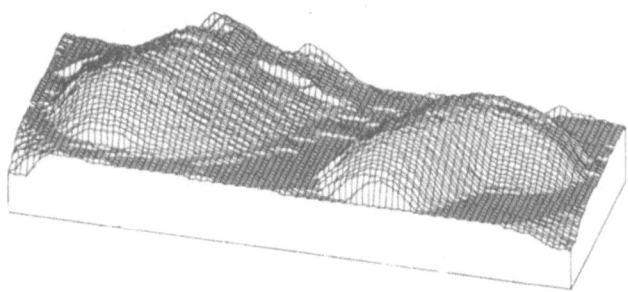

Fig. 9. Partial reconstruction of the SALGLF4 model. Only the sections from # 80 to # 150 have been processed.

1. Scanning the image from north, detect the first non zero pixel for each abscissa value. Several pieces of line will be obtained.

2. For each piece, extend it as far as possible. In other words, starting from each extreme, look for non zero neighbors not yet detetcted because there were some non zero pixels above them, and, if there are, extend that piece by including those pixels until no further pixel can be added. The reason of this operation is that some of the detected pieces will merge. This will happen, for example, in a situation like the one in Fig. 7, so that the noisy points will be eliminated.

3. If there are still several pieces because of the gaps, fill them by extrapolating with a parabola. This is done since the gaps are generally at the bottom of the two hills, as it is shown in Fig. 6 and Fig. 7.

Fig. 8 shows the line detected on the image # 115 by using this procedure.

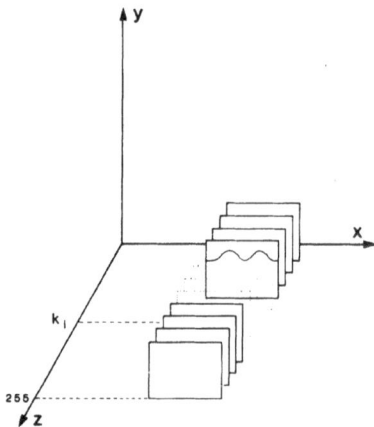

Fig. 10. Coordinate system for the 3-D model reconstruction.

RESULTS AND CONCLUSIONS

The procedure of surface contour detection previously described has been applied to all sections from # 80 to # 150 of the SALGLF4 model. Fig. 9 shows the results of some first experiments for reconstructing the 3-D model. In Fig. 9 we do not see the complete surface because we did not process all 255 sections of the SALGLF4 model, but only the sections from # 80 to # 150. The reconstructed surface is considered as a two-variable function and it is approximated by a matrix whose values are the function heights. That matrix is generated by using the surface contours detected on the seismic sections. In the sequel we explain how this is done.

Let us consider a matrix having as many rows as the seismic sections representing the model and as many columns as the traces on each section. In a 3-D coordinate system as shown in Fig. 10, the surface contours detected on the model sections are the intersections of the model surface with the planes $z = k_i$, $i = 1, 2, ..., 255$ for the SALGLF4 model. Let the rectangular grid of the matrix be the discrete approximation of the model base on the $z-x$

plane, with z-coordinates running along the matrix rows and x-coordinates running along the columns. Thus, the surface contour on the section $z = k_i$ will be stored on the matrix row i by setting the elements of that row to the contour pixel ordinates (on that section). The resulting matrix is the discrete approximation of the surface reconstructed from the detected contours.

A program for 3-D surface displaying[8] takes this matrix as input and plots the surface either on a graphics display or a plotter. In Fig. 9 the surface is plotted as the projection on the $x - y$ plane after a rotation counterclockwise of an angle of 30 degrees about the x-axis and a rotation clockwise of an angle of 15 degrees about the y-axis.

What is described in this paper is only the first step of our work of 3-D surface reconstruction. We are planning to analyze in the near future multiple layer models. They present more problems in the detection of the surface contours and they also present the additional problem of identifying cerresponding contours, namely contours of a same surface on two consecutive slices. Solving this problem is very important because, as we said in the introduction, a triangulation process will be performed over pairs of corresponding contours to generate the local bounding surface structure. In fact, our final aim is producing a surface description by using adjacent triangular patches.

ACKNOWLEDGEMENTS

The author wishes to acknowledge the Allied Geophysical Laboratories at the University of Houston for supplying the physical model data. She also thanks Dr. J. A. McDonald and Dr. K. K. Sakharan for useful information about seismic data and Mr. C. J. Wade for helping in manipulating the seismic data on the computer.

This work was done while the author was a visiting scholar at the Image Processing Laboratory at the University of Houston with a fellowship provided by the United States Agency for International Development. That support is greatly appreciated.

REFERENCES

1. D. D. Hoffman, Inferring shape from motion, M.I.T. A.I. Lab. Memo 592 (1980).
2. S. Ullman, "The Interpretation of Visual Motion", MIT Press, Cambridge, MA (1979).
3. B. K. P. Horn, Obtaining shape from shading information, in: "The Psycology of Computer Vision", P. H. Winston, ed., MacGraw-Hill, New York (1975).
4. A. P. Witkin, Recovering surface shape and orientation from texture, Artificial Intelligence, 17, 17-45 (1981).
5. K. A. Stevens, The visual interpretation of surface contours, Artificial Intelligence, 17, 47-73 (1981).
6. Y. F. Wang, and J.K. Aggarwal, Surface reconstruction and representation of 3-D scenes, Pattern Recognition, 19, 197-207 (1986).
7. A. Rosenfeld, and A. C. Kak, "Digital Image Processing", Academic Press, New York (1982).
8. M. F. Costabile, and S. B. McCraw, Visible 3D surface display, Image Processing Laboratory Annual Progress Review, 57-80 (1987).

HIPPARCOS PROJECT:

IMAGING APPROACH TO MULTIPLE STAR RECOGNITION

L. Borriello

CSATA
Valenzano - Bari - Italy

ABSTRACT

The data collected by HIPPARCOS satellite can be used in an imaging approach, like in VLBI field, to reconstruct the stars system image and to extract the astrometric parameters. The method is able to treat multiple star systems to determine the stars position and their intensities. It is not able to process variable stars, very faint stars and systems with high orbital motion. This paper is devoted to give a short description of the approach and to present a complete outline of the implemented algorithm. Results related to cpu time usage and to reduction results are given with particolar enphasis on detectable separation limits and parameters precision.

INTRODUCTION

The imaging approach to multiple star recognition is able to evaluate the stars position, respect to a sky reference point, the stars intensity, but is not able to determine the colour indices of the stars. The results are not significant for variable stars, for stars with large orbital motion and for faint stars.

THE IMAGING APPROACH

The HIPPARCOS measurements on a star system can be represented using a five parameters model frame by frame /1/

$$I(k) = b_0 + b_1 \cos(2\pi /s(\lambda t + \Phi_1)) + b_2 \cos(4\pi/s(\lambda t + \Phi_2)) \qquad (1)$$

where: b_0, b_1, b_2 are the modulation coefficients;
      Φ_1, Φ_1 are the phases;
      k is the frame index;
      λ is the satellite speed;
      s is the grid step.

These parameters obtained during the mission, calibrated to te referred to the same sky point, can be considered as an incomplete sampling of a bi-dimensional Fourier space /2,3/. The Inverse Fourier Transform (IFT) gives a degraded image of the observed star system:

$$S(n,m) = \frac{1}{N_K} \sum_{1}^{N_K} \{ b_{ok} + \sum_{1}^{N_H} b_{oh} \cos 2\pi h \left[\Phi_{hk} + \frac{1}{2N_{HF}} (n \cos\theta_k + m \sin\theta_k) \right])$$

$$-N/2 < n,m < N/2 \qquad (2)$$

where: N_K is the frames number;
N_H is the harmonics number;
N_{HF} is the extended Fourier space by null harmonics;
k is the frame index;
h is the harmonic index;
θ_k is the angle between the k-th frame and the reference axe;
NXN is the real space size: $DIMX = DIMY = N/2N_{HF}$;
n,m are the sample indices in the real space.

The degraded image (2) can be considered, as in the CLEAN approach to VLBI /4/, as the results of a Point Spread Function (PSF) convolved with the true Star Map. A plot of a Double Star Map is reported in Fig. 1. The PSF, in our case, can be defined on the base of simple considerations on the IFT of a single star located in the origin of the reference system in absence of noise:

$$I(n,m) = I_o + \frac{I_o}{N_K} \sum_{1k}^{N_K} \sum_{1h}^{N_H} M_h \cos(\pi h/N_{HF}) \left[n \cos\theta_k + m \sin\theta_k \right]$$

$$-N/2 \leq n,m \leq N/2 \qquad (3)$$

where: I_o is the star intensity;
M_h is the modulation coefficient.

Setting:

$$\bar{I}(n,m) = \frac{I_o}{N_K} \sum_{1k}^{N_K} \sum_{1h}^{N_H} M_h \cos(\pi h/N_{HF}) \left[n \cos\theta_k + m \cos\theta_k \right]$$

$$-N/2 \leq n,m \leq N/2 \qquad (4)$$

it is easy to demonstate, for two single stars, under the hypotesis that the modulation coefficients $M_{11}/M_{21} = M_{12}/M_{22}$:

$$\bar{I}_1(n,m) = \alpha \ \bar{I}_2 (n,m) \qquad (5)$$

Then we can define the PSF for the HIPPARCOS imaging system as:

$$PSF(n,m) = \bar{I} (n,m) / \bar{I}(0,0) \qquad (6)$$

In Fig. 2 is represented the Hipparcos Point Spreat Function. It is useful to describe the PSF and the Star Map properties. In particular, for the PSF, we have:

a - $PSF(0,0) = 1$ $\qquad\qquad\qquad\qquad\qquad\qquad (7)$

b - $\sum_{1}^{N} \sum_{1}^{N} PSF(n,m)/N^2 \Longrightarrow 0$ $\qquad\qquad$ if $N \Longrightarrow \infty$ $\qquad (8)$

and for the Star Map:

$$b - \sum_n^N \sum_m^N SM(n,m)/N^2 \implies \sum_1^{N_E} {}_q I_{oq} + \mu_N \qquad \text{if } N \implies \infty \qquad (9)$$

where: I_{oq} is the q-th star intensity in the star system;
μ_N is the expected noise mean value;
N_E is the star number.

The "b" properties are "true" if N is sufficiently large. These proper
ties allow us to use a deconvolution approach, like CLEAN /4/, applied
only on the variable part of the Star Map, to evaluate the multiple
stars parameters.

THE REDUCTION ALGORITHM

The algorithm is composed by three main steps:

- Generation of the PSF and of the Star Map;
- Deconvolution process;
- Parameters estimation.

Their characteristics are described in the following.

Images Generation

The images generation is performed using the IFT (2), to reduce
the aliasing effect and the secondary lobes. To reduce the computing
cost the algorithm take into account the PSF simmetry (PSF(n,m) =
= PSF(-n,-m)). In addition both for PSF and Star Map terms, such as
$\cos\theta_k$, $\sin\theta_k$, are calculated in advance and the two evaluation are
nested toghether.

Deconvolution Process

The deconvolution process is based on the hypotesis that Star Map
maxima are close to true stars position. The algorithm is composed by
the following four steps:

a - Determine the coordinates of the Star Map maximum;

b - Subtract the PSF multiplied by a fraction of the Star Map maximum from the Star Map image:

$$SM_i(n,m) = SM_i(n,m) - I_D(i)PSF(n-r_i, m-s_i)$$

$$\forall n,m = 1,\ldots,(N-1)/2$$

where: $I_D(i)$ is a fraction of the Star Map maximum;
r_i, s_i are the coordinates of the maximum;
i is the iteration number.

c - Analize the Star Map residual (SM_i):

1 - if SM_i contain a variable component to be atrtributed to a
star signal, return to "a";

2 - if SM_i is composed by a constant part plus noise, go to "d";

d - Estimate the multiple stars parameters.

<u>Star Map Parameters Evaluation</u>. The following values are eva-
luated:

- The maximum value and its coordinates $(M(i), r_i, s_i)$;
- The minimum value $(m(i))$;
- The mean value and the standard deviation $(\mu(i), \sigma(i))$.

<u>Star Map Subtraction</u>. The fraction of the Star Map maximum value,
to be used in the deconvolution process, is evaluated by the following
formula:

$$I_D(i) = (M(i) - \mu(i))(\mu(i) - \mu_N)/2\mu(i) \tag{10}$$

The properties of maximum intensity fraction are:

$$a - (\mu(i)) - \mu_N) \mu(i) \approx \text{constant:} \quad \text{for bright stars} \Rightarrow 1;$$
$$\text{for faint stars} \Rightarrow 0;$$

$$b - \text{1-st iteration } (p-1)I_B/2 < (M(i)-\mu(i))/2 < (p-1)N_E I_B/2$$

where: I_B is the intensity of the brightest star in the stars
system;

p is the sum of the modulation coefficients (M_1+M_2)
of the brightest star;

i-th iteration $(M(i) \mu(i))/2 \Rightarrow 0$.

The selection of the maximum intensity fraction is performed to allow
the subtraction of higher fractions for bright stars and lower fractions for faint stars in order to avoid that errors in the localization
of the stars position can affect secondary lobes in the Star Map.

<u>Star Map Residuals Analysis</u>. The parameters used to evaluate the
Star Map residuals (SM_i) to verify if it contains, or not, variable
part to be attribute to star intensity are:

- The Variability Ratio

$$V(i) = ((M(i)-m(i))/(M(1)-m(1))) \times 100\% \tag{11}$$

- The Average Residual

$$R(i) = ((\mu(i)-\mu_N - \sum_1^i I_D(1))/(\mu(i)-\mu_N)) \times 100\% \tag{12}$$

These two indices provide measurements of residuale after the i-th
iteration. The process is stopped if "V" or "R" are less than the fixed
thresholds.
The thresholds used are combination of the following terms:

- The threshold values must be higher if the compoent stars are faint:

$$TH1(i) \approx (\mu_N / \mu(i)) \times 100 \% \approx \text{constant} \tag{13}$$

- The threshold values must take into account that the error probability
increases with the iteration number:

$$TH2(i) \approx \sum_{11}^i \sigma(1)/\mu(1) \tag{14}$$

Multiple Stars Parameters Extraction

The parameters estimated are the positions and the intensities of the component stars.

Star Position. Starting from the hypothesis that the coordinates of the maximum values in the Star Map are close to the true stars position, each star position is determined by a cluster of these coordinates. The process uses a simple clustering algorithm to individuate the cluster related to each star. The evaluation of the stars position is then obtained as:

$$X_S = \sum_1^{N_S} I_{S1} X_{S1} / \sum_1^{N_S} I_{S1}, \quad Y_S = \sum_1^{N_S} I_{S1} Y_{S1} / \sum_1^{N_S} I_{S1} \qquad \forall\ S=1,\ldots,N_E \qquad (15)$$

where: N_E is the clusters number;

N_S is the point number in the S-th cluster;

I_{S1} is the intensity fraction $(I_D(\cdot))$ of the 1-th point in the S-th cluster;

X_{S1}, Y_{S1} are the coordinates associated to each cluster is given

by:

$$\hat{I}_S = \sum_1^{N_S} I_{S1} \qquad \forall\ S = 1,\ldots,N_E \qquad (16)$$

These values can be corrected taking into account that:

$$\mu = \mu_N + \sum_1^{N_E} I_S \qquad (17)$$

Setting:

$$\varepsilon = [\mu - (\mu_N + \sum_s^{N_S} I_S)] \ / \ N_E \qquad (18)$$

we have the estimation of the stars intensities:

$$I_S = \hat{I}_S + \varepsilon \qquad \forall\ S = 1,\ldots,N_E \qquad (19)$$

PROCEDURE EVALUATION

The testing activity referes both to cpu-time estimation and to results evaluation.

CPU TIME Estimation

An empirical formula was evaluated to estimate the cpu-time, for one scan and one frame, on an IBM 3081 computer:

$$T(sec) = 0.00001459 \times N^2 + 0.0033$$

where N^2 is the PSF size.

For a double star located at 45° of latitude (151 scans),if the 18 frames, for each can, can be averaged in two frames and the stars separation is not high (less than 0,5 grid steps),the cpu-time required is about 18.5 sec. If the same star system,under the same condition,has an higher separation (about 5 grid steps),the cpu-time required is about 3.74 min. This cpu-time can be reduced using a two step algorithm.In the first step a low resolution image can be processed. In the second step only two window, around the true stars position, of an high resolution Star Map and PSF must be processed to evaluate a better estimation of the parameters. In this way the cpu-time can be about 55.5 sec.

Experimental Results

The experimental analysis was devoted to determine both the limits of the algorithm and the quality of the results. To evaluate the algorithm an extensive test was performed. The testing critaria was:

- Analize all the double stars with magnitude difference of 0,1,1,3.
- Analize only the double stars with primary star up to magnitude 6 with color indices B-V=0.5.
- Analize all the separation ranging between 0.25 and 0.50 grid steps with increments of 0.05 grid steps.
- Employ a realistic sky observation. In particular was selected a sky point located at 45° of latitude.
- Employ a realistic observation strategy.
- Employ 9 frames (instead of 18) for each scan.
- Analise 50 complete simulation for each double star.

The simulation was too costly to be performed using the true simulation chain so a synthetic simulator was built up to reduce the cpu-time /5/. The performance of this simulator was worst respect to the true simulation chain. For this reason the obtained results must be considered as an upper limit of the true.
In Table 1 are reported the separation limits of the algorithm in function of the magnitude difference.

Table 1 - Separation limits and percentage of cases
correctly solved.

SEPARATION (grid step)

$M_p - M_s$	0.25	0.30	0.35	0.40	0.45	0.50
13-13	0	0	100	100	100	100
11-11	0	0	0	100	100	100
9- 9	0	0	0	100	100	100
7- 7	0	0	0	100	100	100
12-13	0	0	0	0	0	0
10-11	0	92	100	100	100	100
8- 9	0	0	100	100	100	100
6- 7	0	0	100	100	100	100
11-13	0	0	0	0	0	0
9-11	78	100	100	100	100	100
7- 9	0	100	100	100	100	100
10-13	0	0	0	0	0	0
8-11	100	100	100	100	100	100
6- 9	100	100	100	100	100	100

In Table 2 are reported the errors estimated for a limited set of the analyzed double stars.

Table 2 - Errors (RMS about true value) for double stars: B-V=0.5, 50 complete simulations at 45° of latitude; θ is the angle between the double star and the reference system.

M_p-M_s	separation (grid steps)	I_p (RMS %)	I_s (RMS %)	ρ (g.s.)	θ (degree)
9- 9	0.50	1.87	2.22	0.004	0.21
	0.45	1.88	3.11	0.007	0.36
	0.40	1.92	2.03	0.007	0.11
	0.35	/	/	/	/
	0.30	/	/	/	/
	0.25	/	/	/	/
7- 9	0.50	2.28	5.63	0.002	0.15
	0.45	2.27	4.30	0.004	0.19
	0.40	2.23	3.60	0.006	0.43
	0.35	2.65	7.40	0.003	1.52
	0.30	3.69	13.54	0.009	1.75
	0.25	/	/	/	/
9-11	0.50	3.65	8.17	0.007	0.66
	0.45	4.86	8.46	0.013	0.66
	0.40	3.94	8.91	0.015	0.98
	0.35	6.20	13.29	0.005	2.31
	0.30	5.48	16.55	0.026	3.50
	0.25	9.18	35.31	0.028	6.02

CONCLUSIONS

The imaging approach to multiple stars recognition is able to estimate the astrometric parameters without any a priori information about the stars.

Future work will be devoted to evaluate the influence of the parallax and proper motion effect as well as data calibration errors on results.

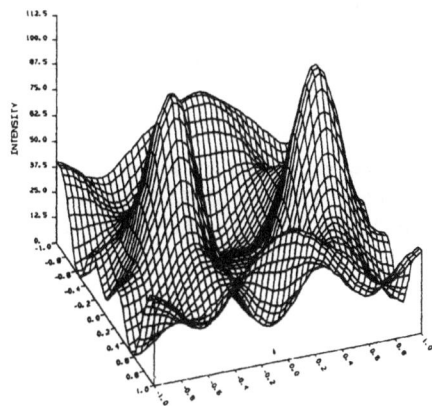

Fig. 1. Double Star Map.
Stars separation ρ = 0.5 grid step.
Grid step s = 1.208 arcsec.

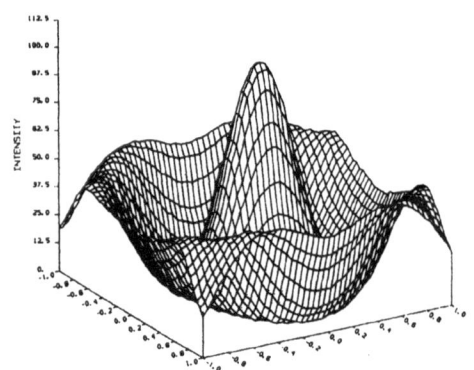

Fig. 2. Hipparcos Point Spread Function.

REFERENCES

1- E.CANUTO   Estimation Procedure. Nota Teonica CSS Torino - Doc.
   G.FASSINO   N. 3110/2-1984.

2- L.LINDEGREN   Imaging properties of HIPPARCOS and the observation
                   of multiple stars - Proc. of an International Collo-
                   quium on the Scientific Aspect of the HIPPARCOS Mis-
                   sion, Strasbourg, February 1982.

3- L.BORRIELLO   Imaging Approach to Multiple Star Recognition. FAST
                   Consortium - The Second FAST Thinkshop. Edited   by
                   J. Kovalevsky - CERCA Grasse, France. January 1985.

4- U.J.SCHWARZ   The Method 'CLEAN' - Use, measure and variation. C.
                   van Schooneveld (ed.) - Image formation from coher-
                   ence functions in astronomy. 1979, Reidel Pub. CO.

5- F.D'ALESSANDRO A Multiple Star Realistic Simulator.FAST Consortium
                   - The Third FAST Thinkshop. Edited by P.L.Bernacca-
                   CSATA Valenzano, Bari, Italy, November 1986.

ACKNOWLEDGEMENTS

This investigation has been supported by Piano Spaziale Nazionale   of
CNR, partially by contract n 2700 to CSATA-Valenzano-Bari-Italy.
I thank Dr. F. D'Alessandro, CSATA, for her contribution to this work.

SEGMENTING POSITRON EMISSION

TOMOGRAPHY (PET) IMAGERY

G. G. Pieroni[1], I. Rousseau[2] and N. Volkow[3]

[1]Dipartimento di Matematica, Università della Calabria
 87036 Rende (CS), Italy
[2]Image Processing Laboratory, University of Houston
 Houston, TX 77004, USA
[3]Department of Psychiatry and Behavioral Sciences
 University of Texas, Houston, TX, USA

ABSTRACT

PET provides images on the concentration of radioactivity throughout organs of patients. This information is relevant because it gives a measure of the functional activity of the various organ regions. The analysis of these images involves the delineation of anatomical regions of interest within the PET scan. An algorithm for automatically delineating regions of interest in PET images is presented in this paper.

INTRODUCTION

Increasingly, medical technology is providing non-invasive imaging procedures which map three-dimensional distributions. Most familiar, peraphs, is x-ray computed tomography (CT) which produces two-dimensional anatomical maps of the attenuation of x-rays by body tissues. Magnetic resonance excitation permits the mapping of proton density or information about the chemical environment in two-dimensional sections of the body. Similarly, measurement of ultrasonic pulse reflectance enables mapping of structural properties of tissues in two-dimensional planes of interest. Physiological and metabolic processes are imaged from measurements of the activity distribution of gamma-ray (single photon) or positron emitting radioactive tracer substances targeted at specific organs or sites. Radiation sensitive detectors surrounding the patient accumulate projections of the emitted radioactivity. The tomographic reconstruction of the projection data in single photon emission computed tomography (SPECT) and positron emission tomography (PET) allows quantitative mapping of the tracer distributions[1,2]. With all of the above modalities, the end product is a series of two-dimensional images mapping certain characteristics of neighboring and parallel body slices. By combining the information from all such slices, a representation of features in three-dimensional space is obtained with varying degrees of completeness. The full data set is thus four-dimensional and extents to five dimensions if time sequences of images are obtained.

Conventional display devices can only map in two spatial dimensions by using color or intensity to represent a third dimension. The problem of display is therefore how to provide sufficient information on the screen to allow the viewer to understand the three-dimensional spatial relationships between features. Various approaches are used in medical imaging to permit the physician to most effectively integrate his knowledge of three-dimensional anatomy and physiology with the displayed information in order to arrive at an interpretation or to select a volume of interest for numerical analysis. Commonly, the transverse slice maps are re-formed into orthogonal planes, termed sagittal and coronal according to anatomical convention. Special projection images, such as the bull's-eye display of cardiac nuclear medicine, or stereo pairs have also found application. Alternatively, surface points can be extracted from the data set and displayed using intensity of depth cueing or hidden surface removal to create the illusion of a three-dimensional object. In all of these cases, the process of display results in considerable loss of information to the viewer. Only the computer has access to all of the information and thus complete three-dimensional "vision". For this reason, computer analysis of the three-dimensional information has the potential of enhancing the utility of the images and augmenting the viewer's understanding and interpretation[3].

On the other hand, the physician viewer uses information to make an interpretation to which the computer does not normally have access. For example, knowledge of normal anatomy and physiology of the prevalence of desease states, of demographic information about the patient, of the results of other diagnostic tests, of the nature and limitations of the imaging modality, etc. are all used by the physician to aid in understanding the image and are eventually integrated to make a diagnostic decision. Any automated diagnostic system must therefore be able to access and correlate knowledge or data bases containing this crucial, non-image information.

The first step towards such a computer-aided diagnostic system is to be able to objectively and quantitatively analyze images.

Positron Emission Tomography allows to obtain information on functional activity of the brain of human subjects. PET provides images on the concentration of radioactivity throughout the brain. This information is relevant because the concentration of the compound is related to the degree of functional activity of the various brain regions. The PET data thus allow to start establishing normative data about the functional organization of the normal human brain. It also allows to detect defects in the brain of psychiatric and neurological patients. The analysis of these images involves the delineation of anatomical regions of interest within the PET scan. This is a lengthy procedure with poor reliability. An automatic method that will allow us to extrapolate the same type of procedure into various subjects is highly desirable. It could facilitate the accumulation of data that can be used as a normative sample to compare with patient populations. An automated delineation of the regions of interest could also allow itself to be used for three-dimensional reconstruction of the PET images. This is desirable since it will allow a more complete appraisal of regional function within the brain in a more integrated way. Until recently there has been a gap between the richness of the PET data and the methods for its analysis which have allowed to analyze only a small fraction of this information. This paper presents an automatic technique for segmenting brain images obtained by a PET scanner.

IMAGE SEGMENTATION

When a segmentation process of an image is performed, one expects to obtain a decomposition of the image into regions which should be uniform and homogeneous with respect to some characteristic such as gray tone or texture. Region interiors should be simple and without many small holes[4].

Adjacent regions of segmentation should have significantly different values with respect to the characteristic on which they are uniform. Boundaries of each segment should be simple, not ragged, and must be spatially accurate. Achieving all these desired properties is difficult because strictly uniform and homogeneous regions are tipically full of small holes and have ragged boundaries. Insisting that adjacent regions have large differences in values can cause regions to merge and boundaries to be lost.

If the image contains a bright object against a dark background and the measurement space is one-dimensional, measurement space clustering amounts to determining a threshold such that all points smaller than or equal to the threshold are assigned to one cluster and the remaining points are assigned to the second cluster. In the easiest cases a procedure to determine the threshold need only examine the histogram and place the threshold in the valley between the two modes. Unfortunately, it is not always the case that the two modes are nicely separated by a valley. To handle this kind of situation a variety of techniques can be used to combine the spatial information on the image with the gray tone intensity information to help in threshold determination.

Weszka et al.[5] suggest determining a histogram for only those pixels having high Laplacian magnitude. They reason that there will be a shoulder of the gray tone intensity function at each side of the boundary. The shoulder has high Laplacian magnitude. A histogram of all shoulder pixels will be a histogram of all interior pixels just next to the interior border of the region. It will not involve those pixels in between regions which help make the histogram valley shallow. It will also have a tendency to involve equal numbers of pixels from the object and from the background. This makes to two histogram modes about the same size. Thus the valley seeking method for threshold selection has a better chance of working on the new histogram.

Weszka and Rosenfeld[6] describe one method for segmenting white blobs against a dark background by threshold selection based on busyness. For any threshold, busyness is the percentage of pixels having a neighbor whose thresholded value is different from their own thresholded value. A good threshold is that point near the histogram valley between the two peaks which minimizes the busyness.

DATA DESCRIPTION AND ANALYSIS

The images of the brain analyzed in this paper are provided by a PET scanner. They are formed by matrices of 128×128 pixels, representing a numerical mapping of the tracer distribution. We performed a normalization of the values in the interval 0–63. By using a gray scale which maps 0 as white and 63 as black, we were able to obtain the representation given in Fig. 1a. We can easily see in that image two separated regions: a lighter region surrounding a darker one. In the darker region a more complicated structure can be observed. The central darker region represents the tracer distribution in a section of human brain at a given depth. The surrounding region is only background and will be discarded. For discarding the background a simple thresholding operation of the image is performed. Fig. 1b shows the result of that operation.

The goal of the procedure described in this paper is the extraction of a separation line between the central region R_1 and the pheriferal region of activity R_2, as shown in Fig. 2. This is only the first step of a more sophisticated procedure which consists of constructing a 3-D representation of the brain activity on the brain surface. In fact, the method takes advantage of a sequence of slices at several depth, each one similar to that shown in Fig. 1b, to extract a 3-D shape of the patient brain. After that, starting from the separation line shown in Fig. 2, a projection of the brain activity onto the brain surface is extracted. This is performed for every slice until a 3-D representation is constructed. For achieving this goal, a meaningful separation between the central region of each brain section and the pheriferal regions of activity must be extracted. We should emphasize that tracing this line is not a very difficult task for a physician having experience in this job. For instance the job can be executed via an interactive procedure with the help of a graphic display but:

a. The number of slices is normally very high. For such a reason a physician would spend his/her time sitting in front of a screen instead of performing a much more useful and interesting activity.

b. By performing the job via a human operator we introduce an important source of noise in the calculation of the lines and we cannot make sure that the response of the procedure is an objective and quantitative measurement of the phenomenon.

In order to get a standardized result, an automatic procedure has been implemented which is able to extract the lines above. Let us consider the brain section in Fig. 1b. Since

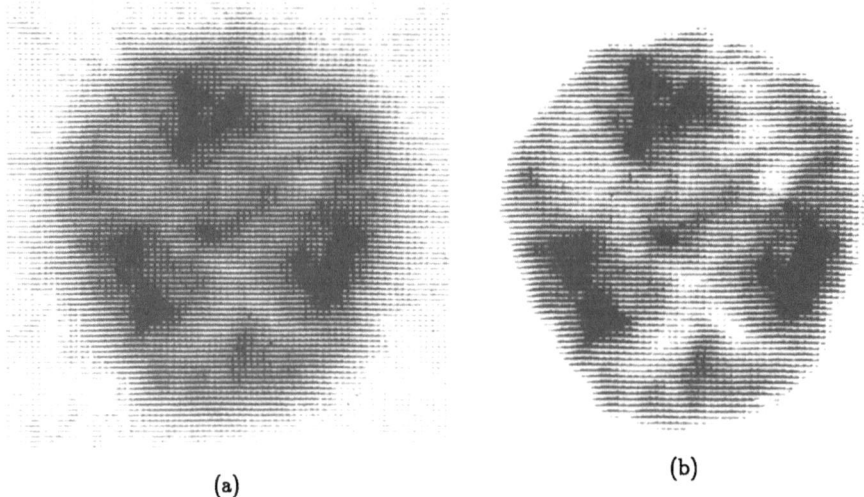

(a)

(b)

Fig. 1. (a) Original image showing a brain section; (b) image (a) after discarding
the background.

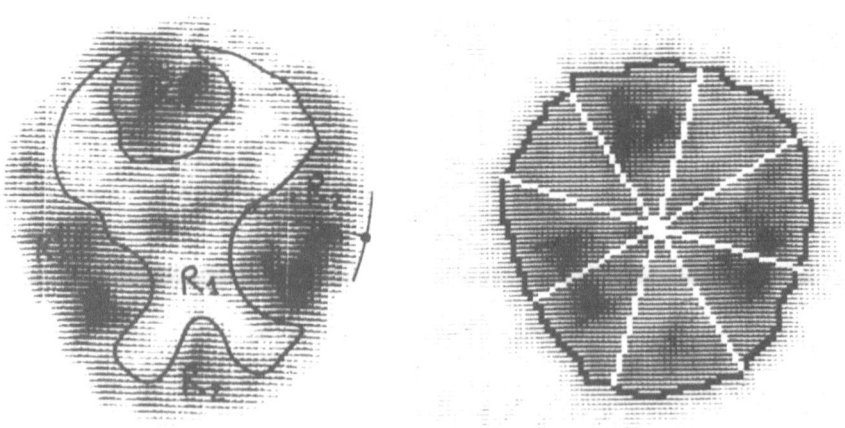

Fig. 2. Sketch of the region boundary in the   Fig. 3. Radius representation for $\Delta\theta = 45°$.
brain representation.

the histogram of this image does not separate any region in it, a simple single thresholding technique cannot provide a separation between the regions of the image.

IMAGE TRANSFORMATION AND REGION EXTRACTION

For achieving the goal of separating the region R_1 from the region R_2 (Fig. 2), we will use the following observations.

a. The region R_2 is formed by portions where the activity is clearly above a given threshold. These portions could be extracted by applying a convenient thresholding technique.

b. The separation line between R_1 and R_2 can be obtained by using as a cue the portion of the boundary line which is oriented toward the centroid of the image in each one of the portions of R_2; those portions will be called in the sequel "peripheral regions".

Considering the gray level of each pixel as elevation of a function $f(i,j)$ we transform each image by exploiting the following procedure. We first calculate the boundary of the brain region, then from the centroid of the configuration we trace a sequence of digital radii each one corresponding to one pixel of the boundary. The pixels which fall in each radius are represented horizontally in a new matrix named M_t.

In that way we construct a new image having number of rows equal to the number of radii (pixels of the boundary) and number of columns equal to the longest radius. Unfortunately the transformation does not guarantee a correspondence between the original image and the new one. In fact, successive radii can contain same pixels of the original image. This lack of correspondence is very misleading in the segmentation process. For overcoming this problem we decided to take the radii in a way that the number of overlapping pixels will be acceptable. We assume:

$$M = \max(Z_i) \qquad i = 1, N \qquad N = \text{number of pixels of}$$

$$\text{the boundary}$$

$$Z_i = \text{radius corresponding to}$$

$$\text{the pixel } i$$

We take the radii having the end points (X_E, Y_E), (X_C, Y_C) where

$$X_E = M\cos\theta + X_C \qquad\qquad \left. X_C \right\} \quad \text{Coordinate of}$$
$$Y_E = M\sin\theta + Y_C \qquad\qquad \left. Y_C \right\} \quad \text{the centroid} \qquad (1)$$

$$\theta = (\Delta\theta, 2\Delta\theta, \ldots, K\Delta\theta) \qquad \Delta\theta = \text{chosen step.} \qquad K = INT\left(\frac{360}{\Delta\theta}\right)$$

(see Fig. 3, $\Delta\theta = 45°$).

Then, we evaluated $\Delta\theta$ in order to have an acceptable overlapping of pixels. In effect the region where the maximum number of pixels overlap is located around the centroid of the image. The more we move away from the centroid, the more we can use a smaller $\Delta\theta$ and the more precise will be the representation of the image. We will see below how we will overcome the lack of correspondence in the region close to the centroid of the image.

The transformation explained above will be called "polar transformation" of the image, as well as the opposite transformation will be called "inverse polar transformation". In order to separate the central region of activity from the peripheral ones we should be able to detect in a precise way the configuration of the lighter region, valley, which is in between. The most reasonable idea seems to detect, by using some kind of edge detector, maxima and minima for each row of M_t. Then, once these points have been detected, an inverse polar transformation should give the separation lines between valleys and peaks in the original image.

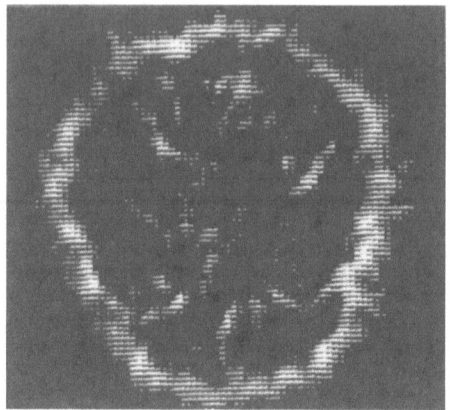

Fig. 4. Result of applying the gradient operator to the image shown in Fig. 1b.

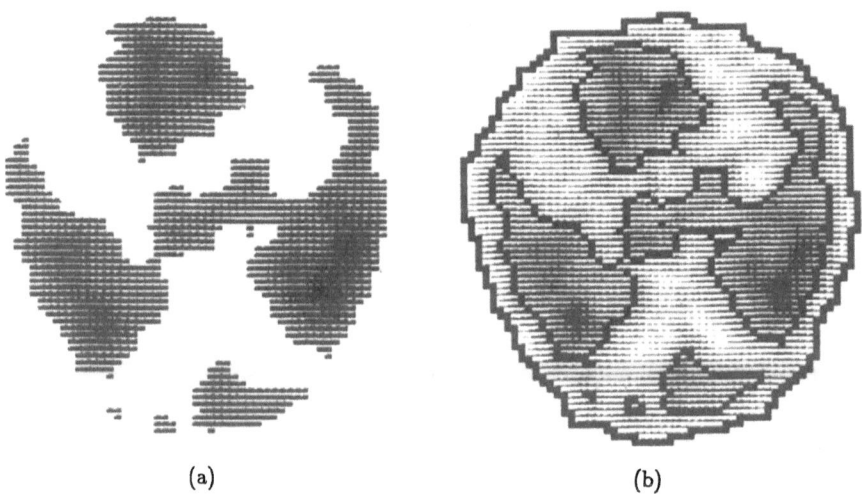

(a)   (b)

Fig. 5. (a) Image shown in Fig. 1b after the thresholding-averaging procedure; (b) boundary of the regions in (a) overlapped onto the original image.

We did not get that result, since the points, in such a way calculated, tourned out to be a cloud of pixels without spatial connection in the original image. On the other hand a procedure for detecting the edges based on local values of a difference operator gave the poor result shown in Fig. 4.

In order to get reasonable results a procedure of successive thresholding and averaging of the original image was performed, until a convenient delineation of the five regions, one central and four peripheral, was reached (Fig. 5a). Then we extracted the boundary lines of the configuration in Fig. 5a (Fig. 5b) and performed a polar transformation of those boundary lines (Fig. 6).

The structure of the image represented in Fig. 6 is essentially different from that one represented in Fig. 5b. In fact after the polar transformation each row in the image in Fig. 6 represents the interception points between a given radius belonging to the set indicated in equations 1 and the boundary lines represented in Fig. 5b. The following topological observations are made on the structure of that image.

   a. Normally the boundary of the central region is intercepted by the radius in one point only. The set of interception points forms a trace which is almost vertical in the polar representation; that trace is, almost everywhere, the first line we meet moving from left to right in the image (Fig. 6).

   b. When we move towards the right after the first trace we meet the traces of the interception points between the radii and the boundary of peripheral regions.

   c. Because of the proximity of the boundary line of the central region to the centroid, many points of that trace are just a duplicate of the previous one in the polar transformation. This effect is visible in the strokes appearing as segments in Fig. 6. This characteristic is very useful for our purposes since it provides a form of connectedness for the line. In fact, we can eliminate the trace of the central region from the image simply by scanning the image from the top to the bottom and deleting, starting from the left, the first connected, almost vertical, component among the lines present in the image.

   d. The residual lines should be closed lines, each one representing the boundary of one of the four peripheral regions.

A set of algorithms which take advantage of the previous topological observations were written and the result was the extraction of a set of four main lines frequently not closed and an archipelago of small isolated components. A succesive operations consisted of closing the four main lines by interpolating straight lines between the two extremes; then the four components were filled and a noise cleaning algorithm took care of eliminating the debris from the image. The result of the operation is shown in Fig. 7.

A major problem in extracting the four region boundaries was controlling connectivity lacks. For performing that, we defined the following algorithm:

For a given row i_0 of $M_t(i,j)$ (containing the image of Fig. 6) we do:

   a. Let us indicate $P_t(i_0, j_v)$ the pixel belonging to the almost vertical trace representing the boundary of the central region in Fig. 6.

   b. Let us assume as first point $P_f(i_0, j)$ the first pixel $\neq 0$ when we move right of the pixel $P_t(i_0, j_v)$, $j > j_v$. Then we look for the last point $P_e(i_0, j)$ that is the last pixel $\neq 0$ before the end of the row.

   c. If both first and last points are not present, the boolean *firstime* is true and we go ahead; otherwise we store the position of the previous first and last points.

   d. We eliminate the first point $P_f(i_0, j)$ if either the distance $d = \overline{P_f(i_0 - 1, j)P_f(i_0, j)} > T$ (T assigned threshold) or $j \leq j_v$.

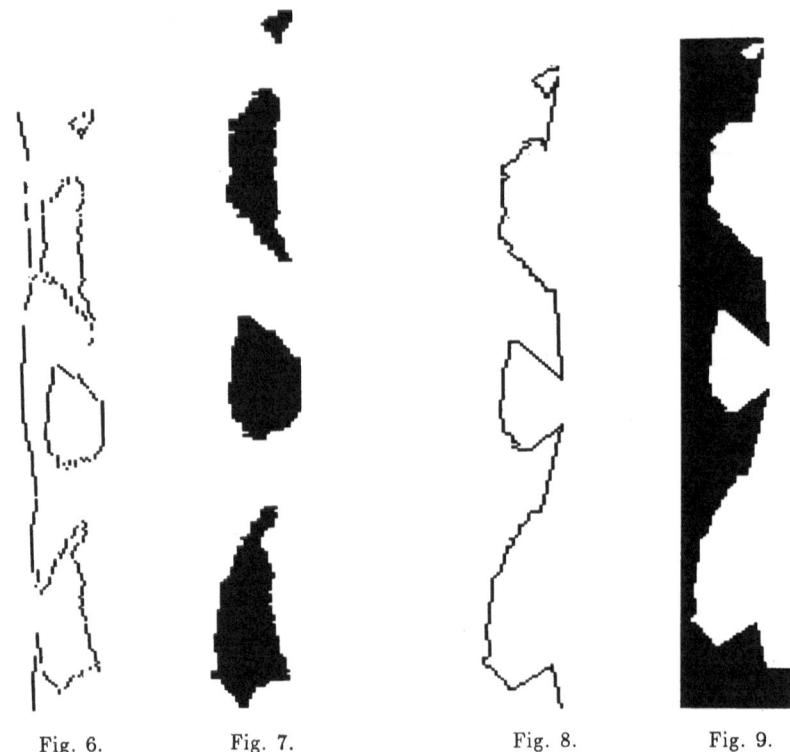

Fig. 6.   Fig. 7.   Fig. 8.   Fig. 9.

Fig. 6. Polar transformation of the boundary lines of the regions shown in Fig. 5.

Fig. 7. Representation of the four peripheral regions after performing the detection procedure.

Fig. 8. Boundary line between R_1 and R_2 regions represented in the transformed matrix M_t.

Fig. 9. Image shown in Fig. 8 after the filling procedure.

(a)

(b)

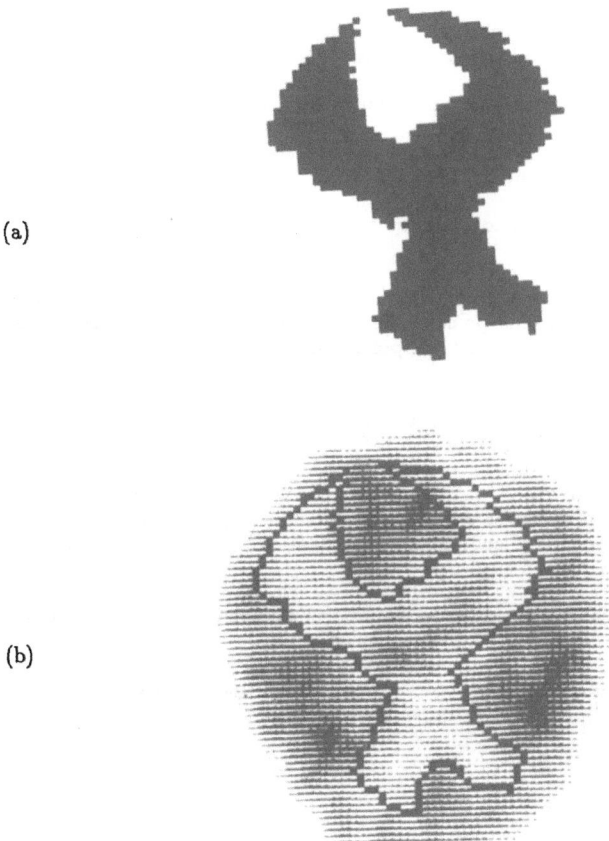

Fig. 10. (a) Image shown in Fig. 9 after the inverse polar transformation; (b) image
in (a) after the boundary smoothing procedure overlapped onto the original
image.

e. If *firstime* is true and there is no last point, we neglect this row and go ahead.

f. If there is a first point and no last point (or vice versa) we proceed to the successive row till we find both of them; then we extrapolate the missed points.

CONSTRUCTING THE LINE SEPARATING R_1 AND R_2; CONCLUSIONS

In order to construct the final separation line, we exploit the portion of the boundary of the four regions extracted above which is oriented toward the centroid. For doing that we scan the M_t matrix represented in Fig. 7 from left to right. In such a way we cross the boundary of each region in two points, P_l, P_r, which could sometimes be coinciding. We assume as belonging to the separation line all of the P_l points. Then we assume also a set of P_r points provided that the following conditions hold.

a. The set of P_r must be connected. They must form a portion of line called "$p_r - line$".

b. Every $p_r - line$ must form a "prolungation", without lacks, of a portion of line formed by P_l points and called "$p_l - line$".

c. The angle formed by a $p_l - line$ and a corresponding $p_r - line$ must be less than a given threshold.

By exploiting the observations above we extracted a number of portions of lines, then we joined each end point with the end point of the portion residing below in M_t obtaining the line showed in Fig. 8. After that, an inverse polar transformation of the image in Fig. 8 should provide the result.

But again, unfortunately, the inverse polar transformation breaks the connection and we were forced to perform another intermediate step before performing the final one. That step consists of filling the region which is between the left side of the image in Fig. 8 and the extracted line (corresponding to the R_1 region in the original image (Fig. 9)). That simple procedure allows us to perform the inverse transformation obtaining the image shown in Fig. 10a. Then, a procedure for boundary smoothing and a successive boundary extraction bring the result shown in Fig. 10b.

Several experiments are under way in order to perform a segmentation of different kinds of images. So far we have processed some images similar to the one shown in this paper; the results have always been acceptable. We do not know what could be the result when images having consistent structural differences are processed. We think that at least a calibration of the thresholds would probably be a necessary step in that case.

REFERENCES

1. N. A. Mullani, M. V. Ranganath, S. Adler, et al., 3D surface mapping of functional PET images for the heart and brain, Journal of Nuclear Medicine,27, 916 (1986).
2. N. Kehtarnavas, E. A. Philippe, and R. J. P. De Figueiredo, A novel surface reconstruction and display method for cardiac PET imaging, IEEE Trans. Medical Imaging, M1-3, 108-115 (1984).
3. J. K. Udupa, Display of 3D information in discrete 3D scenes produced by computerized tomography, Proc. IEEE, 71, 420-431 (1983).
4. R. M. Haralick, and L. G. Shapiro, Image segmentation techniques, Computer Vision, Graphics and Image Processing, 29, 100-132 (1985).
5. J. S. Weszka, J. S. Nagel, and A. Rosenfeld, A threshold selection technique, IEEE Trans. Computers, C-23, 1322-1326 (1974).
6. J. S. Weszka, and A. Rosenfeld, Threshold evaluation techniques, IEEE Trans. Systems Man and Cybernetics, SMC-8, 622-629 (1978).

DESIGN AND TESTING OF A CLASSIFICATION SYSTEM WHICH RECOGNIZES CORONARY
STENOSIS BY SITE AND RELATIVE SEVERITY, USING MYOCARDIAL Tl-201
SCINTIGRAMS

Krzysztof J. Cios* and Lucy S. Goodenday+

From the University of Toledo, Department of
Electrical Engineering* and Medical College of Ohio,
Cardiology Division, Department of Medicine+
Acknowledgements: MCO Foundation

The main theme of this paper is an implementation of fuzzy clustering in order to determine whether perfusion patterns exist which are
diagnostically specific for the location of stenosis in a particular
major coronary artery. During the past decade, Tl-201 myocardial
scintigraphy has proven its value for the detection of ischemic heart
disease, and, more recently, for establishing prognosis [1]. Other
clinically important data, however, such as the site of greatest
coronary artery obstruction and its relative severity compared to
stenoses in other arteries, has not been as readily derived from planar
Tl-201 scintigrams.

Methods

Study group. Preprocessed Tl-201 scintigraphic data from 41 patients who had undergone Tl-201 stress scintigraphy and coronary arteriography comprised the learning data set. Thirty patients had stenosis of
at least 70% diameter reduction in at least one of the major coronary
arteries; these patients form the group designated as abnormal. The
other 11 patients had normal coronary arteriograms and left ventricular
function, and their data comprise the normal group.

Collection and preprocessing of the data. Stress tests were performed according to the Bruce treadmill protocol. Tl-201, 2mCi, was
injected intravenously, one minute before stopping exercise. Scintigrams were obtained immediately thereafter, collecting for at least
300,000 counts at the 80Kev photopeak with a 20% window in each of three
views: anterior (ANT), left lateral (LAT) and left anterior oblique
(LAO).

The images were then preprocessed by a technique [1], which requires
an image analyst to manually define the cardiac region for each view.
The images from the normal patients provide count values for comparison
with the patients' images. Any normalized count values in the test
image which are > 2.5 S.D. below the normal mean are marked as being
poorly perfused. Percentages of poorly perfused points compared to
normal points are totaled for each view and for each of 10 regional
locations within each view. This results in a composite of 30 regional
percentage values for each patient. The 10 regions for each view are
shown in Fig. 1.

The location and severity (as a percentage of diameter obstruction) of coronary artery obstructions were determined by arteriography, which had been previously read and reported by two cardiologists. The set of abnormal patient studies were assigned for learning purposes to one of three groups, depending on the single coronary artery: left anterior descending (LAD), circumflex (CCX) and right coronary artery (RCA), having the greatest severity of stenosis.

Fuzzy clustering. Let X={x} be a space of objects. Then a fuzzy set A in X is a set of ordered pairs [2]

$$F = \{ (x,u(x)) \mid u(x)\varepsilon[0,1] , x\varepsilon X \}$$

where $u(x)$ is the grade of membership of x in A. Fuzzy partition space P_{fc}, where c is the number of clusters, is defined by:

$$P_{fc} = \{ U \varepsilon V_{cn} \mid u_{ik} \varepsilon [0,1] \forall i,k; \Sigma u_{ik}=1 \forall k; 0 < \Sigma u_{ik} <n \forall i \}$$

where V_{cn} is a set of real cxn matrices, n is a number of measurement vectors, $2<c<n$.

Fuzzy clustering algorithms have the following general form:

1. Choose the measure of similarity to calculate prototype vectors and membership functions:
   - choose initial fuzzy partition matrix $U=[u_{ik}]_{cn}$;
   - fix c, a number of clusters; fix m, a weight exponent;
2. Calculate prototype vectors;
3. Update membership functions;
4. If certain error criterion is met then stop, otherwise go to 2.

The choice of similarity measure is essential for all algorithms. In implemented supervised classification system Euclidean, diagonal and Mahalanobis metrics are used. A single universal method applicable to all different types of data does not exist [3]. Different methods generate different clusters and very often they cannot be compared against each other [4]. In all algorithms the central assumption is that at least some of the measured features are distinctive in the classification process.

Feature selection. In feature selection one searches for internal structure in measurement vectors $x\varepsilon X$ in order to find measurements which are redundant or irrelevant to class separability and at the same time reduce the dimension of measurement space. It is the lowest level of analysis [5]. There are two approaches to it called feature selection and feature extraction. In the former one achieves the goal by choosing subsets of original measurements and then performs classification. In the latter one finds a mapping into lower dimensional space, called feature space, taking into account all original measurements. Because the preprocessed Tl-201 data represent mathematical features [6], they were analyzed in both measurement and feature spaces [7]. The most general method for feature extraction, is Karhunen-Loeve transformation [6] which requires only that all measurement vectors have a zero or equal mean. There are some variations of this method which differ mainly in how a covariance matrix is estimated. These methods were implemented in this system.

Designing the classification system. The overall objective of the study was to develop a supervised mode classification system for analysis of preprocessed cardiac Tl-201 scintigrams to find which regions of the ventricular image where perfusion defects occurred are

responsible for specific coronary artery stenoses: LAD, RCA and CCX. A
sketch of the system is depicted in Fig. 2.

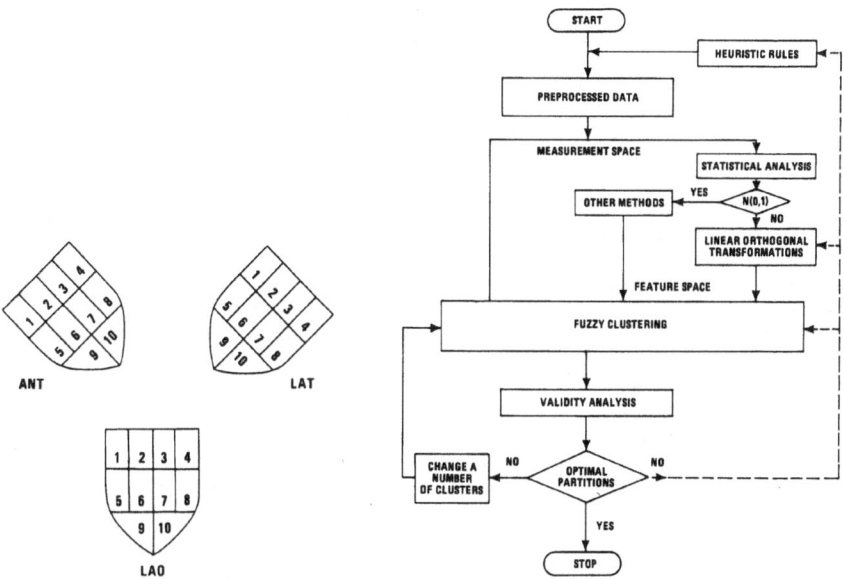

Fig. 1. The numbering scheme
for each region

Fig. 2. A sketch of the
classification system.

    The structure of this system utilizes all gathered and accessible
sources of information:
-   A priori knowledge about fuzzy pattern recognition techniques with
    its subjectivity and about the method of collecting data vectors,
-   A posteriori knowledge being derived from analysis of a learning
    data set,
-   Heuristic knowledge derived from cardiologists and analysis of
    normal patients' data.

    Previous research shows that all these types of knowledge are
essential in designing a successful system [1,2,3,6,]. A priori
knowledge was incorporated in the process of designing this classification system by using information obtained in recent research [8]. On
Fig. 4 continuous and dotted lines show feedback loops after checking
the condition for optimal partitions, as compared with the learning data
set. Without knowledge about the analyzed data it would be sufficient
to optimize the system by performing fuzzy clustering validity analysis.
Three such measures [2] are implemented: partition coefficient, partition entropy and partition exponent. In the described system however,
cluster validity analysis is combined with classification results
obtained on a learning data set utilizing both a posteriori and
heuristic knowledge, which is depicted as dotted lines on the right hand
side of Fig. 3. In the main part of the system fuzzy c-means algorithm
[2] is utilized.

In the search for the set of best features and best classification results, a slightly modified version of the method described by Bezdek [2] was used. The method helps in selecting the most meaningful features by first digitizing the original data and then performing fuzzy c-means clustering. The problem was to determine what threshold values should be used for digitizing the data. The plausible approach is to use as the threshold values the averages calculated from normal patients for each feature. Because of the limited sample size, however, instead of using average values for the features as the thresholds the maximum values of the features were used and the set of best features was found. Using maximal values can be treated as a heuristic rule as they were used [1] for excluding LAD stenosis using the same data set.

Results

The results giving the best outcome from the classification system, as compared with the learning data set, are discussed below. Emphasis was placed on specific recognition of patients with LAD stenosis as the ones with worst clinical prognosis as compared to patients with obstructions in only the CCX or the RCA. In all results shown below the weight exponent for the fuzzy c-means algorithm was set to two and the Euclidean measure was used as the measure of similarity. The two-digit numbers in the columns represent grades of membership for three clusters. They sum up to one in rows for each patient. Thus, every patient is considered to have ALL three vessels stenosed, but with different grades of membership. The decision to which group a patient belongs is made by comparing these three numbers for each patient. The patient is considered to belong to the group for which a certain stenosis prevails; that is, the site of stenosis is indicated by the largest grade of membership. The grades of membership do not indicate percentages of stenoses in those three arteries but only the severity of stenoses as compared one with another.

The problem of selecting subsets of best features was approached first in measurement space. First, the impact of all original features on final classification was studied. This is shown in Fig. 3, where all 30 original features were used for generating three clusters.

PATIENT #	GRADES OF MEMBERSHIP CORRESPONDING TO		
	LAD	CCX	RCA
1	0.00	0.96	0.03
2	0.02	0.45	0.53
CCX 3	0.00	0.00	0.99
4	0.00	0.01	0.99
5	0.01	0.97	0.02
6	0.01	0.02	0.96
7	0.00	0.00	1.00
8	0.00	0.00	1.00
9	0.06	0.11	0.84
10	0.29	0.15	0.56
11	0.00	0.01	0.99
12	0.59	0.09	0.32
13	0.00	0.00	1.00
LAD 14	0.00	0.00	1.00
15	0.92	0.04	0.04
16	0.47	0.13	0.40
17	0.77	0.09	0.14
18	0.83	0.09	0.07
19	0.01	0.02	0.97
20	0.91	0.04	0.05
21	0.00	0.01	0.99

	LAD	CCX	RCA
22	0.02	0.04	0.94
23	0.00	0.00	0.99
RCA 24	0.09	0.21	0.70
25	0.01	0.11	0.89
26	0.00	0.01	0.98
27	0.00	0.00	0.99
28	0.00	0.00	1.00
29	0.00	0.00	0.99
30	0.00	0.00	1.00

Fig. 3. All 30 original features considered.

The first 5 patients had arteriographically proven CCX stenosis, the next 16 had LAD stenosis and the remaining 9 had RCA stenosis. The above holds for all tables shown below.

Referring to the first two-digit column, 6 out of 16 patients with high-grade stenosis in the LAD were recognized correctly.In the second column, 3 out of 5 with CCX stenosis were correct recognitions and in the third column, 9 out of 9 with RCA stenosis were recognized properly. However, the specificity for RCA patients is very low, so in the third cluster (third column) all patients who were not recognized properly as having LAD or CCX stenosis were also included. Despite the fact that the sensitivity for LAD and CCX patients was low, at the same time no patients without these obstructions were classified improperly.

Next, classification was performed in the feature space. Karhunen-Loeve transformation was applied to original data and classification was performed using only few features in the transformed space. Different sets of features were subjected to transformation; 30 features from all three views and 10 features from each single view. This attempt did not improve results substantially so it is not shown.

One other approach was taken here. The whole digitized, as described above, data set was submitted to classification, as were data sets from separate views and combination of features. The best classification results were achieved from analysis of 10 features from the left lateral view, which is depicted in Fig.4.

PATIENT #	GRADES OF MEMBERSHIP CORRESPONDING TO		
	LAD	CCX	RCA
1	0.01	0.98	0.01
2	0.11	0.44	0.46
CCX 3	0.00	0.00	1.00
4	0.11	0.04	0.85
5	0.01	0.98	0.01
6	0.06	0.01	0.92
7	0.00	0.00	1.00
8	0.00	0.00	1.00
9	0.11	0.05	0.83
10	0.76	0.05	0.18
11	0.51	0.21	0.29
12	0.98	0.01	0.02
LAD 13	0.00	0.00	1.00
14	0.06	0.09	0.85
15	0.92	0.02	0.05
16	0.86	0.01	0.12
17	0.96	0.01	0.03
18	0.60	0.24	0.16
19	0.06	0.01	0.92
20	0.92	0.02	0.05

	21	0.86	0.01	0.12
	22	0.00	0.00	1.00
	23	0.18	0.03	0.79
	24	0.00	0.00	1.00
	25	0.09	0.09	0.82
RCA	26	0.00	0.00	1.00
	27	0.06	0.01	0.92
	28	0.00	0.00	1.00
	29	0.06	0.01	0.92
	30	0.06	0.01	0.92

Fig. 4. Ten features from left lateral view after digitizing.

Using these features, 9 out of 16 patients with LAD stenosis were recognized properly, 3 out of 5 with CCX stenosis and 9 out of 9 with RCA stenosis.

In the following tables the data were transformed by digitizing it using maximum values for the features. Using ten features from the left lateral view and asking for generating only two clusters gives the results shown in Fig. 6.

PATIENT #		GRADES OF MEMBERSHIP CORRESPONDING TO	
		LAD	CCX+RCA
	1	0.00	1.00
	2	0.00	1.00
CCX	3	0.00	1.00
	4	0.31	0.69
	5	0.00	1.00
	6	0.99	0.01
	7	0.00	1.00
	8	0.00	1.00
	9	0.00	1.00
	10	0.95	0.05
	11	0.99	0.01
	12	0.95	0.05
	13	0.00	1.00
LAD	14	0.00	1.00
	15	0.99	0.01
	16	0.95	0.05
	17	0.95	0.05
	18	0.99	0.01
	19	0.99	0.01
	20	0.99	0.01
	21	0.95	0.05
	22	0.00	1.00
	23	0.00	1.00
	24	0.00	1.00
RCA	25	0.00	1.00
	26	0.00	1.00
	27	0.99	0.01
	28	0.00	1.00
	29	0.99	0.01
	30	0.99	0.01

Fig. 6 Left lateral view with 2 clusters generated

Now 11 out of 16 patients with LAD are recognized correctly but at the same time 3 patients are incorrectly included in this group. Patients with obstructions in CCX or RCA are grouped in the second cluster

(except the three misclassified). From the above results two
corollaries can be drawn: first, that patients with LAD stenosis do form
a distinctive cluster as opposed to patients with CCX or RCA stenoses
who fall mainly into one cluster and second, that the sensitivity of
proper recognition of patients with LAD stenosis can be increased but at
the cost of specificity.

Discussion

Noninvasive cardiac imaging carries the advantage of little physical
risk or discomfort to the patient; the penalty paid may be one of less
diagnostic accuracy than more invasive techniques. The promise of
cardiac image analysis is that more information is often present in an
image than may be extracted without computer analysis techniques. The
goal of this study was to test the fuzzy clustering method in a supervised classification system to see whether it has potential use for
improving the diagnostic specificity of stress Tl-201 planar
scintigraphy, with regard to identifying the site and relative
severity of coronary artery stenosis. Planar scintigraphy was chosen
because it is more readily available to the majority of practicing
physicians than are the more sophisticated tomographic imaging
techniques. Previous attempts to identify the location of coronary
artery stenosis from Tl-201 planar scintigrams have met with variable
success [9-14].

The first step was to determine whether a natural clustering of data
into groups existed, and if so, into how many groups did the information
cluster. After defining the normal group as separate from patients
having coronary stenosis, results demonstrated that 3 natural clusters
existed, one for each of the stenosis groups; that is, one each for LAD,
CCX and RCA obstruction. The CCX and RCA obstruction groups did overlap
somewhat. This finding may be explained by coronary anatomy, in that
there is considerable variability among patients in the myocardial ter-
ritory supplied by the RCA and that supplied by the CCX. This finding
of intrinsic clustering structure confirms that the planar myocardial
scintigrams do contain data referable to the anatomic site of obstruction, and that this information should be extractable form the images.
The novelty of the proposed classification system is that it includes
some heuristic knowledge concerning analyzed data. Previously, classification systems relied heavily only on mathematical analysis of studied
data. Very often, however, in human studies, the data sets are not big
enough to allow such analysis. Therefore, current systems include
heuristic rules used by human beings for solving the same problems.

After using heuristics for digitizing the original data achieved
results enabled the detection of those regions which have the greatest
impact on finding a specific artery obstruction. In other words these
regions are sensitive to a specific kind of stenosis i. e. LAD, CCX or
RCA. Next, classification using these most sensitive features taken
together is performed. Obtained results are slightly above 50% sensitive
but very specific, especially for recognition of LAD patients. Patients
with CCX or RCA stenosis are more difficult to recognize with high
specificity. Another important finding which confirms correctness of
the methods used for the problem of classifying patients with specific
coronary artery stenosis is that although imperfect recognition is
achieved when compared against the learning data set,it does show that
only three clusters do exist in the data. Generating more than three
clusters results in much worse partitioning. Asking for just two clusters however enables grouping patients having LAD stenosis against those
having CCX or RCA which complies to previously achieved results [1].

The single view carrying the most information for detecting LAD stenosis is the left lateral view. The single regions from this view which gave the best results were regions 6 and 9. This result agrees with anatomical expectations as to what region should be influenced by which artery. Clinically, this result might also be anticipated, because in most patients this artery usually supplies more blood to the heart than the other two individually, so obstruction of the proximal LAD may cause a larger perfusion defect than stenosis of either of the other 2 vessels separately.

The most important aspect of this study is that the method shows promise for automating the process of classification of planar scintigrams into specific or multi-vessel disease according to the relative severity of each vessel's narrowing. Improvement in the performance of such a supervised classification system may be expected as more data are added to the learning data set, thus enabling work towards designing an unsupervised version of the system.

References

1. Goodenday LS, Nelson AD, Leighton RF, et al: Prediction of the site of coronary artery obstruction from thallium-201 scintigrams by a quantitative computer technique. IEEE Computer Society publication: Computers in Cardiology 277-279, 1981
2. Bezdek JC: Pattern Recognition with Fuzzy Objective Function Algorithms. New York, Plenum Press, 1981
3. Dubes R, and Jain AK: Clustering techniques: the user's dilemma. Pattern Recognition 8:247-260, 1976
4. Mucciardi AN, and Gose EE: A comparison of seven techniques for choosing subsets of pattern recognition properties. IEEE Trans Comp C-20:1023-1031, 1971
5. Watanabe S: Frontiers of Pattern Recognition. New York, Academic Press, 1972
6. Tou JT and Gonzales R: Pattern Recognition Principles. Reading MA, Addison-Wesley, 1974
7. Kittler J: Mathematical methods of feature selection in pattern recognition. Int J M-Machine Stud 7:609-637,1975
8. Cios KJ, Muswick G and Goodenday LS: Interpretation of Tl-201 images using fuzzy clustering. J Nucl Med 27:928, 1986(abstr)
9. Dunn RF, Freedman B, Bailey IK, et al: Localization of coronary artery disease with exercise electrocardiography: correlation with thallium-201 myocardial perfusion scanning. Am J Cardiol 48:837-843, 1981
10. Rigo P, Bailey IK, Griffith LSC, et al: Stress thallium-201 myocardial scintigraphy for the detection of individual coronary arterial lesions in patients with and without previous myocardial infarction. Am J Cardiol 48:209-216, 1981
11. Wainwright RJ, Maisey MN, Sowton E: Segmental quantitative analysis of digital thallium-201 myocardial scintigrams in diagnosis of coronary artery disease. Comparison with rest and exercise electrocardiography and coronary arteriography: Br Heart J 46:478-485, 1981
12. Wainwright RJ: Scintigraphic anatomy of coronary artery disease in digital Tl-201 myocardial images. Br Heart J 46:465-477, 1981
13. Hakki AH, Iskandrian AS, Segal BL, et al: Use of thallium scintigraphy to assess extent of ischemic myocardium in patients with left anterior descending artery disease. Br Heart J 45:703-709, 1981
14. Dunn RF, Freedman B, Bailey IK, et al: Exercise thallium imaging: location of perfusion abnormalities in single-vessel coronary disease. J Nucl Med 21:717-722, 1980

AUTOMATIC INTERPRETATION OF DIGITAL AUTORADIOGRAPH OF

DNA SEQUENCING GELS

D.Q. Xu   M.K-S. Tso* and W.J. Martin

Dept. of Instrumentation and Analytical Science
*Dept. of Mathematics
University of Manchester Institute of Science and Technology
P.O. Box 88, Manchester M60 1QD, U.K.

SUMMARY

An image processing system has been developed for sequencing DNA gels by digital autoradiography. A multi-wire proportional counter (MWPC) images DNA band patterns which form tracks on an electrophoresis gel. Algorithms have been developed to interpret the MWPC image to obtain a DNA sequence. The algorithms include: (1) a dynamic programming procedure for track detection; (2) a maximum entropy deconvolution algorithm for smoothing and sharpening the image, and (3) a procedure for assigning the correct band sequence. The sequence produced by this method can be confirmed by human operators working from conventional film autoradiographs. The algorithm is being evaluated on various gels and methods for incorporating the knowledge base are currently being investigated. With these improvements we expect the system will approach the performance of expert sequencers.

1 INTRODUCTION

The celebrated double helix, DNA, is the molecule of heredity which encodes genetic information by the order in which four bases adenine, guanine, thymine and cytosin (A,C,G,T) occur along its structure.

DNA sequencing aims to establish the serial order of the bases along the double helix. Sanger's method of sequencing[1] involves reactions which selectively cut the DNA molecule producing a set of fragments with different molecular weights. These fragments are then separated by gel electrophoresis into four tracks. The result is a two-dimensional pattern of bands, as represented in Fig.1. The image is formed by radio-labelling the bands and then using radiation sensitive film or a digital counter to map the distribution of activity. This procedure is known as autoradiography. The bands can be read cross-wise (see Fig.1) to reveal the base sequence. This process is called sequence interpretation. At present sequence interpretation is performed manually and is a very tedious and time-consuming task. It is however a classic problem of pattern analysis in which the knowledge base of the sequencer plays a part, for example, in resolving sequencing ambiguities.

We remark that automatic sequence interpretation is merely the final stage in an effort to automate the entire Sanger procedure[2]. Previous work in this area has been limited to the interpretation of film autoradiographs[3,4]. The ability to interpret blurred, digital MWPC images through software, thus avoiding the need for film, is a novel feature of the work presented here.

2. THE IMAGING SYSTEM

A multi-wire proportional counter (MWPC) is employed as the imaging system[5]. It detects and localises radiolabelled DNA bands on sequencing gels produced by electrophoresis. Fig.2 illustrates the structure of a MWPC. It has one anode wire plane and two cathode wire planes all enclosed in a chamber containing ionizing gas. The cathode planes are linked with electronic delay lines. The gel plate is placed in the top of the chamber. When a β–particle enters the upper cathode plane, it interacts with the gas and generates a cloud of free electrons. This cloud is drawn into the anode plane, resulting in an avalanche of electrons. The electronic disturbance is detected by the two cathode planes and by measuring the time delay the electronic system can assign position co-ordinates to this event, so the energy of the β–source can also be detected. With isotope S-35 as the emission element, and argon as the gas medium, the MWPC has a spatial resolution of 1.0mm fwhm.

Fig.3 (a) shows a typical MWPC image of a sample gel taken with an exposure time of 1½ hours. In Fig.3(b) we show the corresponding film image requiring typically 24 hours exposure. Compared with conventional film autoradiography, the use of a MWPC can reduce dramatically the time required for acceptable images. For subsequent automatic processing there is a clear advantage in sending digital data directly to a computer. However the MWPC image is clearly more blurred. This is because, unlike film, the detector planes of the MWPC cannot lie in contact with the gel and thus spatial resolution is lost. Also, because of the long range of particles within the gas of the detector, the system has a significantly greater point spread function (PSF) than does film. Thus processing a MWPC image calls for software for sharpening or deconvolution. Furthermore, the accummulation of radioactive counts is subject to Poisson noise so that some degree of smoothing is also required.

Data collection on the MWPC is at present controlled by an Apple II-e microcomputer. The gel forms two images, each of 512 x 512 pixels, with each pixel representing 0.226mm x 0.226mm. Pixel values can be as high as 4096 (12 bits), but a few hundreds are sufficient in most cases. The data are firstly collected on Apple formatted hard disk, then transferred to a mini-computer for subsequent processing. The processing software has been developed on a PRIME 9955, using FORTRAN77. In the near future, it is intended that the computer will be controlled by an IBMpc microcomputer, and the software adapted accordingly.

3. STEPS IN THE INTERPRETATION OF MWPC IMAGES

There are a number of difficulties in sequencing gels automatically. Firstly, the image requires segmentation into tracks which are generally not straight. Secondly the bands themselves exhibit curvature which must be corrected before sequencing. In addition to these geometric distortions, there is statistical variation in the position and intensity of bands. Finally there is the problem of incorporating a knowledge base that can for example discard "artefact" bands. In manual interpretation, these problems are solved by the pattern recognition ability and the accummulated experience or the "knowledge base" of the sequencer.

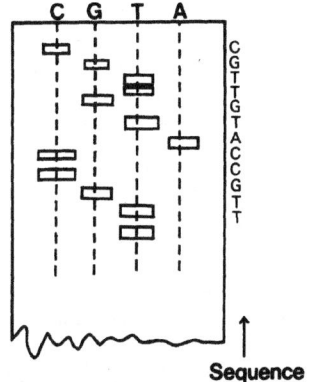

Fig. 1   Typical band pattern of DNA sequencing gels

MULTIWIRE PROPORTIONAL COUNTER

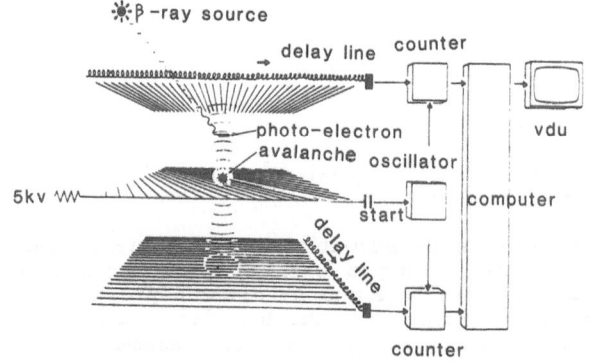

Fig. 2   MWPC Imaging system

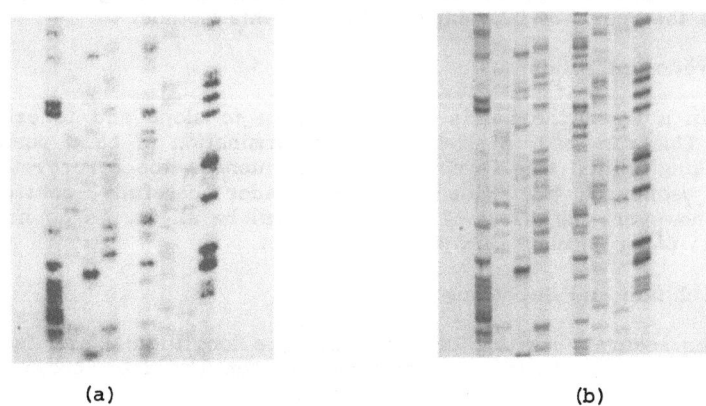

(a)   (b)

Fig. 3 (a)   Sample MWPC image

(b)   Corresponding film autoradiograph

3.1 Segmentation by dynamic programming

The first problem is to segment the image into tracks which can then be individually smoothed and sharpened. Although tracks are not straight, they run approximately parallel to the longest edge of a gel plate. The boundary between two tracks may therefore be approximated by a sequence of line segments, each parallel to a longest edge (see Fig.4). This boundary has the following characteristics.

(i)   The summed intensity along the boundary is low (boundary pixels appear white on an autoradiograph).

(ii)   Points sampled from regions on either side of a boundary will tend to have negatively correlated intensities (since these will have been sampled from different tracks).

(iii)   Adjacent line segments comprising a boundary will be displaced by at most one or two pixels (by continuity, assuming limited curvature).

Let x_j denote the position of the jth boundary segment as measured by its displacement from an edge. Criteria (i) and (ii) provide a measure of support $\gamma_j(x_j)$ for the jth segment to be at x_j. The entire boundary is found by the following optimization

$$\text{Minimize} \quad \sum_{j=1}^{n} \gamma_j(x_j)$$

$$\text{subject to} \quad |x_{j+1} - x_j| \leqslant 1 \quad j = 1, 2, \ldots, n$$

where $x_1, \ldots, x_n$ belong to some allowable range of integer values. This problem is equivalent to finding a shortest path through a n-stage network and the solution can be found by dynamic programming [3]. the constraint imposes the condition that the boundary should not shift by more than one pixel between neighbouring segments and forces continuity of the boundary. By optimising a criterion globally we average out the statistical errors of boundary detection in a single segment. Fig.5 shows the result of boundary finding by this method.

3.2 Registration of Bands

Within a track the bands are often seen to slope and to exhibit curvature. This can lead to an imprecise determination of band position along the longitudinal axis, which in turn influences the error rate of the final sequence. For gels prepared under carefully controlled conditions however, the effect of band slope will be slight. We do not in this present paper discuss this problem further.

3.3 Deconvolution and Smoothing

As the accurate location of bands along a longitudinal axis is the main requirement for obtaining a sequence, we sum the data across each track to obtain four one dimensional profiles. These four profiles form the basis of the final sequencing stage. (We assume that bands have had any slope removed by this stage so that no loss of precision results.)

Despite the improvement in signal to noise ratio achieved by summing the data across a track there is Poisson noise in the resulting profile.

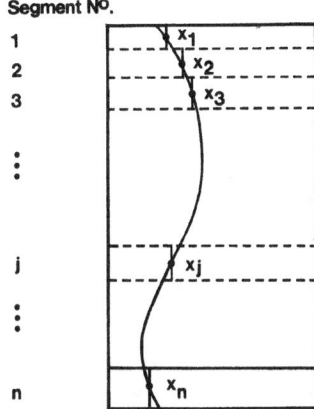

Segment Nº.

Fig. 4   Track boundary approximation Fig. 5   Results of track
          by straight line segments   segmentation by dynamic
                                                    programming

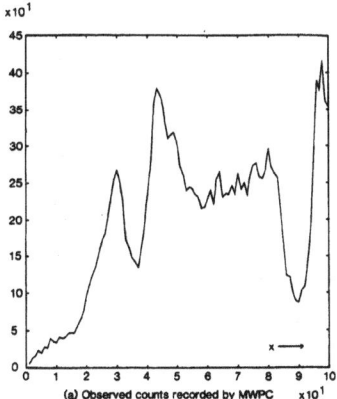

(a) Observed counts recorded by MWPC $\times 10^1$

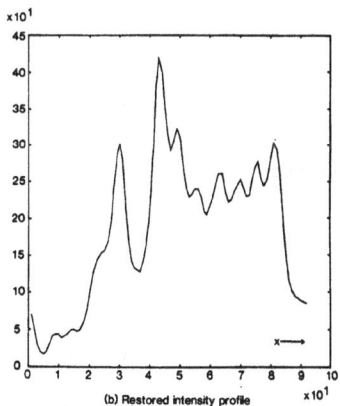

(b) Restored intensity profile $\times 10^1$

Fig. 6   MWPC data before and after deconvolution showing separation
          of seven peak cluster and appearance of shoulder on extreme
          left peak

This noise obscures the peaks which signify the presence of a band. The peaks are also broadened by a one-dimensional instrument point spread function. Image restoration is required which can be stated mathematically as the determination of a function $\underline{f}$ from an observed data vector $\underline{y}$ related to f by the following:

$$g = \underline{f} * \underline{h} \qquad \ldots\ldots (1)$$

$$y_i \sim \text{Poisson } (g_i) \quad i = 1,\ldots,N \qquad \ldots\ldots (2)$$

In (1) $\underline{f}$ is blurred by convolution with a known point spread function $\underline{h}$ to give a resultant function $\underline{g}$. In (2) we express the fact that y_i, the observed count at position i, is a random variable from a Poisson distribution with mean g_i.

Agard et al.[7] first applied deconvolution to electrophoresis gels. They used an algebraic reconstruction technique (ART) attributed to Jannson and van Cittert[8] which inverted the convolution relationship (1) whilst ignoring (2). Such a procedure may be applicable to film autoradiographs where the noise (resulting from optical densitometry) is low. However for MWPC images in which the Poisson noise is significant it is necessary to use a statistical procedure such as maximum entropy.

The use of maximum entropy for deconvolution of images was described by Gull and Daniell[9] and their approach was refined and developed into a more sophisticated algorithm by Skilling and Bryan[10]. We have developed an alternative solution specifically for one dimensional data contaminated by Poisson noise[11]. This solves the constrained optimisation problem

Maximise $\quad S = - \displaystyle\sum_{i=1}^{N} f_i \log f_i$

subject to $\quad x^2 = (\underline{y}-H\underline{f}) \ W^{-1}(\underline{y}-H\underline{f}) = N$

and $\quad \displaystyle\sum_{i=1}^{N} f_i = E$

where

N = the number of data in y

E = total counts $= \displaystyle\sum_{i=1}^{N} y_i$

W = diagonal matrix of weights

$\quad$ = diag($\underline{g}$) where $\underline{g} = H\underline{f}$

H = Toeplitz matrix representing convolution by $\underline{h}$

The solution is found by solving the set of non-linear equations obtained by differentiating the Lagrangean function for the above problem. Typical results are shown in Fig.6 and further details of the procedure are available in Tso[11].

3.4 Band detection

Bands show up as peaks in the profile which can be detected by finding the local maxima of f. We have calculated 5-point estimates of the first and second derivatives at any point by locally quadratic curve fitting. Applying the usual condition for a local maximum, we are able to detect isolated peaks quite successfully.

The location of "shoulders" or secondary peaks is more complicated. Following Elder[3], the sides of isolated peaks are examined for shoulders. A left side shoulder is indicated by a continuously positive first derivative and a sequence of changes in sign of the second derivative from positive to negative to positive again, and the shoulder position is taken to be where the second derivative is negative and minimum. A right side shoulder is detected similarly.

Every track contains some small amplitude peaks near a "background" level which are not signal bands. To avoid picking out these peaks, we introduce a threshold for every track, which is say 10% of the maximum intensity within each track. This step removes the most obvious noise peaks.

3.5 Reading the band sequence

A sequence is obtained by comparing the positions and intensities of the bands in every track. It may be assumed, however, that the complete list of bands detected at the previous stage contains some spurious peaks or "artefact" bands. Some discrimination may be achieved by examining the intensity of competing bands. Another piece of useful information is the knowledge that successive bands are separated by a characteristic unit of distance. To take these factors into account, we have adopted the following weighting function developed by Elder[3]:

$$w(h,x) = \begin{cases} (0.5 + 0.67x)h / \bar{h} & (0.30 \leqslant x < 0.75) \\ (0.75 - x) h / \bar{h} & (0.75 < x \leqslant 1.75) \\ 0 & \text{otherwise} \end{cases}$$

Here x is the band distance from the current peak position expressed as a proportion of the unit band separation, h is the band intensity (or the maximum peak height) and $\bar{h}$ is the mean intensity of the track. The band with maximum weight is chosen as the next base in the sequence.

Any algorithm based on the above requires the mean inter-band distance to be known. Knowing the total track length and also the approximate length of the sequence, this mean distance may be estimated. However, we are investigating methods for dynamically estimating the mean inter-band distance.

3.6 Results

Using the procedures described above, the following 59-base sequence (a) was obtained from the MWPC image shown in Fig.3. For comparison, the 64-base sequence (b) obtained by reading the corresponding film autoradiograph is also given below.

(a)   ATTATTTATTTTCCT AACATA CCAGA CAGCTAGG CGCTATC AGTAGA CTTAATTACATA

(b)   ATTATTT TTTTCCTCAACATAACGAGAACACACAGGGCGCTATCGCAGAGAATCAAATTCGATA

Comparing the two sequences, we can see that in the early part of the sequence where bands are well separated, the error rate is low. The later part of the sequence does not match well with the manually read sequence, the reason being that this part is in the very top of the gel where there are many artefacts and weak bands which are difficult even for an experienced sequencer to distinguish. Further improvements are being made to improve the sequencing of this part by using some of the knowledge base possessed by an experienced human sequencer.

4. DISCUSSION

We have analysed the key steps in procesing digital MWPC images of DNA sequencing gels. Our preliminary results indicate that the MWPC system is capable of producing an image of sufficient quality to enable sequencing.

By using algorithms for pathfinding based on dynamic programming and for image restoration by maximum entropy, a band sequence can be automatically obtained from the image data. The accuracy of sequencing is being compared to that achieved by manual methods based on film autoradiography. However further investigations are required to optimize the various stages. We believe that statistical methods and optimization techniques e.g. mathematical programming will play an important part in these future investigations. There is scope for improving the speed and reliability of the maximum entropy routine. Generalisation to two dimensions would also be useful. The use of dynamic programming for boundary detection raises the question of choice of criterion. Techniques for optimization with multiple criteria may be worthy of investigation in this context.

The final stage of sequencing involves using the knowledge base possessed by human sequencers who, for example, readily discard artefact bands. It is important to consider how this knowledge base should interact with measurements on, say, inter-band spacing to enable an automatic system to approach human performance levels. The framework of Bayesian decision theory may be suitable for consideration of such questions.

ACKNOWLEDGEMENTS

Thanks are due to Ed Bateman of Rutherford and Appleton Laboratory for development of the MWPC and to John Warmington of the department of Biochemistry and Applied Molecular Biology at UMIST for sequencing expertise.

REFERENCES

[1] F. Sanger, S. Niklen and A.R. Coulson, *DNA Sequencing with Chain-terminating Inhibitors.* Proc. Matl. Acad. Sci. 74, 5463-5467.

[2] W.J. Martin, J.R. Warmington, B.R. Galinski, M. Gallagher, M. Davies, M.S. Beck and S.G. Oliver. *Automation of DNA sequencing: A system to perform the Sanger Dideoxysequencing reactions.* Biotechnology 3: 911-915 (1985).

[3] J.K. Elder, D.K. Green, E.M. Southern, *Automatic reading of DNA sequencing gel autoradiographs using a large format digital scanner.* Nucl. Acids Res. 14: 417-424 (1986)

[4] D.Q. Xu, *Image processing of DNA autoradiographs.* UMIST Department of Instrumentation and Analytical Science Report, October 1985.

[5] J.E. Bateman, J.F. Connolly, and R. Stephenson, *High speed quantitative digital Beta autoradiography using a multi-step avalanche detector and an Apple II microcomputer.* Nucl. Instr. and Meth. A241: 275–289 (1985)

[6] E.V. Denardo, *Dynamic programming: models and applications.* Prentice-Hall Inc. (1982)

[7] D.A. Agard, R.A. Steinberg, and R.M. Teroud, *Quantitative analysis of Electrophoretograms: A Mathematical Approach to Super-resolution.* Anal. Chem. 53: 257–268 (1981).

[8] P.A. Jansson, R.H. Hunt, and E.K. Pyler, *Resolution Enhancement of Spectra.* J. Opt.Soc.Am. 60: 596–599 (1970).

[9] S.F.Gull, and G.J. Daniell, *Image Reconstruction from Incomplete and Noisy Data.* Nature 272: 686–690 (1978)

[10] J. Skilling, and R.K. Bryan, *Maximum Entropy Image Reconstruction: General Algorithm.* Mon. Not. R. Ast. Soc. 211: 111–124 (1984).

[11] M.K-S. Tso, *The restoration of a Poisson intensity function on the line.* UMIST Department of Mathematics Technical Report No. 181 (1986).

SUPPORTING DIAGNOSIS AND SURGICAL PLANNING BY

ANALYSIS AND 3D DISPLAY OF VOLUME IMAGES

J. Ylä-Jääski and O. Kübler

Institute for Communications Technology, ETH-Zentrum
CH-8092 Zürich, Switzerland

Introduction

Presently several 3D imaging methods such as computed tomography (CT) and magnetic resonance (MR) imaging have become standard clinical tools in medical diagnosis as well as in planning and monitoring of therapies. In normal clinical practice the data are evaluated by visually investigating suitable 2D slices of the volume. It requires, however, extensive training to visualize 3D phenomena based on pure slice representations so that the vast information amount in volume images is not fully exploited. The need for more powerful analysis methods has become even more pronounced after the development of fast data acquisition methods for MR imaging. An emerging standard of the order of 256^3 volume elements (voxels) can no more be evaluated by traditional visual methods in a reasonable amount of time.

The fundamental operations in extending the diagnostic possibilities of an analysis process are segmentation of objects from the raw volume data and visualization of the objects by projecting their surfaces on a plane such as a monitor screen. Surface shading can be carried out simply by coding the distance from the observer with varying brightness. Small surface details become more pronounced by combining the distance information with the surface orientation and direction of illumination [1,2].

In the so-called cuberille approach [3-5] the surfaces of the objects are first explicitly coded in terms of elementary voxel faces. Thereafter standard computer graphics algorithms can be used to render shaded views of the surfaces. The conversion from voxel to vector representation reduces the amount of data but introduces several disadvantages: the conversion is a time consuming preprocessing step which has to be repeated each time when object segmentation criteria are modified. Further, the surface representation cannot be combined with original gray values of the voxels as frequently desired.

Another class of display methods renders the surfaces from the voxel representation directly. For hidden surface removal two basic approaches can be used. In ray-tracing [6], projection rays are cast on the volume starting from the observer and followed until a voxel belonging to an object is met. In the back-to-front procedure [7,8] all the slices, rows, and columns of the volume are traversed in the order of decreasing distance to the observer. All voxels belonging to an object are projected onto a buffer overwriting the previous entry on that particular location which also results in hidden surface removal. Direct display methods are very time consuming, but this disadvantage has been overcome in the so-called prebuffer approach [9] in which the generation of an oblique projection is decomposed into two stages for time-efficiency. The algorithm operates directly on voxel data and can utilize both a ray-tracing and a back-to-front method for hidden surface removal. In this study the prebuffer approach has been used for reconstructing the visible surfaces from voxel data. The performance of the algorithm is demonstrated by an implementation in a parallel computer system suitable for clinical use.

For the segmentation of volume data widespread use is made of either simple thresholding or tedious interactive methods which require manual determination of the object contours in all slices. We have used a segmentation procedure in which edge contours are first determined from the the data using a 3D difference of Gaussians (DOG) operator. Regions to the interior of the contours are extracted to yield a binary image. Desired objects are then refined from this intermediate result using case dependent binary image procedures.

Applications of segmentation, 3D display and quantitative analysis of medical volume data to several clinical cases are presented. A qualitative analysis of a malformation in the spinal column has allowed to postpone a planned operation. By the segmentation and visualization of a complicated hip fracture the application of the developed methods to surgical planning is demonstrated. Finally, a fully quantitative computer analysis of a slipped femoral capital epiphysis has allowed the surgeon to determine the amount of the slip and the correct angle and shape for a wedge removal operation.

Segmentation Procedure

The segmentation procedure adopted in this study consists of three major steps: *image binarization, object refinement,* and *connected component labelling.* For image binarization either simple thresholding or a 3D implementation of the DOG operator with a ratio of 1.6 between the space constants of the two Gaussian functions has been used. The DOG is a good approximation to the Laplacian of Gaussian (LOG) operator proposed by Marr and Hildreth [10] for edge detection. The DOG operator projects intensity changes in images to zero crossings, the strength of an edge being described by the slope at a zero crossing. No thresholding with respect to the edge strength has been used which guarantees that the edge contours always are closed lines. Instead of the contours all the volume elements belonging to the objects have been extracted by thresholding the image at zero after applying the DOG operator.

We have applied the DOG operator to a variety of problems ranging from the extraction of bony structures in CT images to the distinction of different classes of soft tissue in MR data. The method is very robust requiring only one free parameter but produces excellent results even for complicated data. Due to the separability of the DOG also a time-efficient 3D implementation is possible.

A common problem encountered relates to an insufficient resolution; even though the transversal resolution in CT can be about 0.1 mm a slice represents an average over a thickness of $1-2$ mm and the interslice distance usually is $2-4$ mm. Isotropic resolution can be obtained in MR, but this is generally inferior to CT and so is also the signal to noise ratio. As a consequence, the desired objects often cannot be segmented simply by a straightforward application of an edge operator. We have complemented the edge detection with binary image operations tuned to a particular problem and incorporated *a priori* knowledge in these procedures.

A typical example of the segmentation procedure is illustrated in Figure 1. An original axial CT slice from the left half of a human hip is shown in **a** and a coronal (frontal) reconstruction in **b**; bones appear light and soft tissue dark in the images. For further analysis the hip and the femur indicated by 1 and 2 respectively need to be extracted as separate objects. Results after applying a DOG operator with space constants $\sigma_e = 1.05$ and $\sigma_i = 1.67$ pixels are shown in **c** and **d**.

The edge operator alone is not sufficient to distinguish between the two anatomically different parts in this case. An example of a failure of the operator is indicated by an arrow in **d**. In this case the main reason can be attributed to patient movement showing up as vertical irregularities in the coronal reconstruction. For the particular case of the hip area we have developed a procedure which utilizes the fact that the joint between the hip and the femur is close to spherical in shape. A matched filter is used by fitting a hemisphere in the joint region with the centre point coordinates and the radius as variable parameters. The result of the

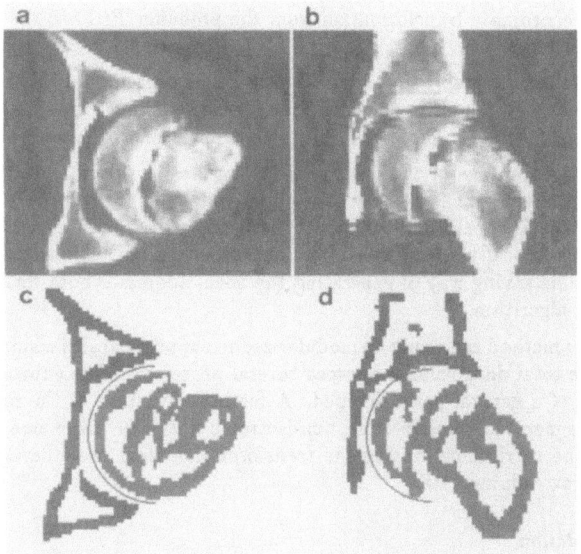

Figure 1. Example of the segmentation procedure to extract the hip (1) and the thigh bone (2) from a series of CT slices. Planar sections from original data are shown in **a** and **b** and the corresponding segmentation results in **c** and **d**

optimal fit is shown overlayed in the images in **c** and **d**. All object voxels coinciding with the hemisphere surface are subsequently removed to yield the final binary image.

A connected component labelling in 3D is finally carried out after object refinement to assign each object an individual identification label. We have allowed 6 nearest neighbours in 3D for the labelling procedure corresponding to 4-connectivity in 2D. The identification label is stored as an overlay to the original voxel gray value allowing a combination of the two during subsequent display algorithms.

Besides matched filters as described earlier we have incorporated user knowledge in procedures which use morphological operations in a predefined sequence. In many cases the objects can also be refined by an interactive procedure which simulates a surgeon's knife in 3D and utilizes a real-time roaming facility through the volume data in an oblique direction. After selecting a desired planar section the cursor is used to remove object labels up to a predetermined depth from the viewed section.

Shaded Surface Display

Algorithm

In projecting the surface of an object onto a viewing plane we limit ourselves to the particular case of parallel orthogonal projections. The prebuffer algorithm [10] is used for the projection due to its superior time-efficiency compared to other known algorithms. The basic idea of the prebuffer display algorithm is to decompose any parallel orthogonal projection into the following two steps:

1. A general oblique projection onto a plane $P(O)$ orthogonal to that of the main axes (x_1, x_2, x_3) on which the normal vector $\mathbf{w} = (w_1, w_2, w_3)$ has the maximum component. Without the loss of generality we can choose e.g. $|w_3| \geq |w_2|, |w_1|$. In other words $P(O)$ is parallel to planar sections orthogonal to x_3 denoted by $S(x_3)$. We shall call this intermediate result the prebuffer.

2. A linear 2D coordinate transformation from the prebuffer $P(O)$ to the final buffer $P(\Phi)$ perpendicular to **w**.

Evidently the second step in the algorithm is a mere resampling of the prebuffer $P(O)$ on $P(\Phi)$ with a generally non-orthogonal and less dense sampling raster. A hardware warper presents the appropriate tool to perform this type of transformation in real time.

The principal advantage of the decomposition resides in the size invariance property of the prebuffer due to the parallelism of the projection plane $P(O)$ and the planar sections $S(x_3)$. Thus in every section $S(x_3)$ the sampling raster resulting from the intersection of the straight rays with $S(x_3)$ is identical with the original raster except for a fractional offset $(\Delta x_1, \Delta x_2)^T$. This results in a time saving way of generating the voxel addresses both for a ray-tracing and for a back-to-front algorithm.

The prebuffer method can easily be modularized to support parallel computer architectures by subdividing the total data volume between several processors. The subvolumes are allowed to have the shape of a general parallelepiped. A further advantage of the method is that the prebuffer can be generated directly from non-isotropic data i.e. from non-cubic voxels and interpolation can be carried out during the transformation from prebuffer to the final buffer instead of on the raw volume data.

Parallel Implementation

To demonstrate the efficency of the prebuffer algorithm, the modularization and the interpolation concept we have implemented the algorithm into a system equipped with two fast general purpose 16 bit microprocessors operating in an MIMD architecture. Each processor has 8 Mbytes of external memory for storing the voxel data and a DMA interface for fast data retrieval. The system is also expandable in terms of the number of processors and the amount of memory. The DMA interface allows to read data by defining start address, address increment, and the number of words desired. This facility optimally supports the back to front method for the prebuffer algorithm. An equally efficient implementation of the ray tracing method would require the additional possibility to modify the increment according to a lookup table. Thus in this study the back to front method has been used throughout.

Table 1. Performance of the two processor system for the rendering of surfaces from voxel data.

Volume	Performance (s)		
(voxels)	prebuffer	2D transform	total
$256^2 \cdot 128$	4.0	1.6	5.6
$256^2 \cdot 40$	1.3	1.6	2.9
128^3	1.0	0.4	1.4
64^3	0.13	0.1	0.23

The performance of the system for rendering surfaces with distance coding directly from voxel representation is summarized in Table 1. The software implementation of the 2D transform includes an on-line interpolation facility but this step can be performed in real time using a hardware warper. Gradient shading is an additional operation on the distance buffer and requires less than 2 seconds for a full resolution image. The display algorithm can operate both on the original gray values performing on-line thresholding and on the overlay information on segmentation results together with clipping plane effects. For a given case optimal performance can be obtained by defining a region of interest in the volume since the response time depends linearly on the nuber of voxels to be investigated.

Analysis of Clinical Cases

Malformation in the spinal column

Investigation of anatomical malformations is one of the most promising medical application fields for surface display. We have investigated a a case of scoliosis in a 5-year old girl caused by a malformation in the spinal column in the form of a hemivertebra, an incomplete vertebral body, between the fourth and the fifth vertebrae. Investigation of slices from a CT investigation had given strong indication for a spinal fusion operation.

To assist the diagnosis shaded surface views have been reconstructed with results shown in Figure 2. The original CT data consists of 38 CT slices 256^2 elements each and for the reconstruction a volume of $256 \times 256 \times 224$ cubic volume elements with a spatial resolution of 0.3 mm has been used. The shading is based on the distance from the observer only. The hip bones are included in the coronal view shown in **a** and are removed in the sagittal view **b**.

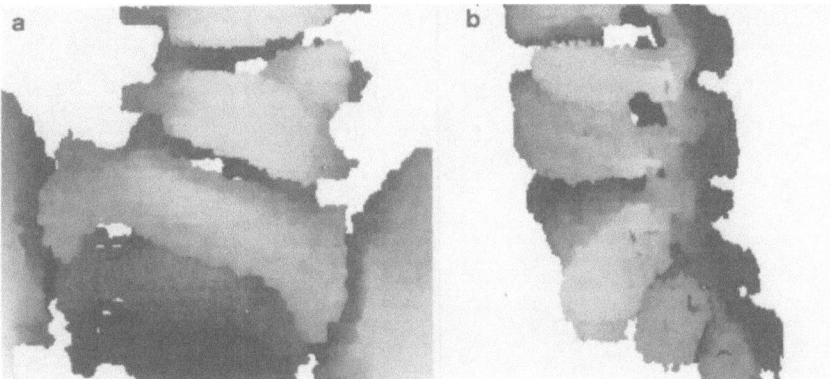

Figure 2. Shaded surface views for the visualization of a malformation in the spinal column.

The reconstruction clearly shows the incomplete vertebra, but it is also evident that, although possessing its own intervertebral foramen, the hemivertebra is fused together with the fifth vertebral body. Thus there are no unilateral growth plates indicating a good potential for an improvment of the scoliosis. In this particular case the shaded surface views have provided decisive additional information which have allowed to postpone an already planned operation of the patient.

Hip Fracture

Visualizing complicated fractures for surgical planning present another typical application field for 3D analysis. The visual investigation of slices which is widely used at present may require hours even from a specialist of the particular field to visualize the 3D relationship of the essential components. Time on the other hand is an important factor since an operation may have to be carried out immediately.

Results of the shaded surface display for the visualization of the hip fracture are shown in Figure 3. The original CT data set consists of 39 slices with a resolution of 256×256 pixels each. The data set used for segmentation and display with $256 \times 256 \times 139$ cubic voxels is generated by linear interpolation. For the display a region of interest of 128^3 elements has been selected from this volume.

Since the fracture extends down to the joint between the hip and the femur it is of importance to be able to extract and remove the femur from the display to facilitate a view of the fracture from inside the joint. For the segmentation a procedure consisting of edge detection by a DOG operator and subsequent fitting of a hemisphere in the joint has been applied as described earlier. A view with the femur included is shown in **a** and the same view after femur removal is shown in **b**. A combination of the distance buffer and the surface orientation is used for shading.

It is evident that the segmentation and display methods are of great advantage in the planning of a surgical operation. Shaded surface views with a preliminary segmentation by thresholding can be obtained within seconds after the data acquisition and the full segmentation procedure, practically automatic, requires approximately 10 minutes to complete.

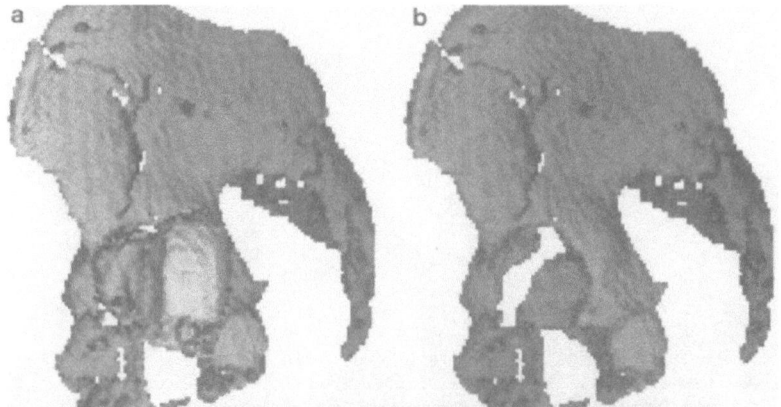

Figure 3. Shaded surface display for the visualization of a hip fracture. The femur is included in **a** and removed through segmentation in **b**.

Femoral Epiphysis

Besides mere qualitative analysis as required in the case of the malformation or semi-quantitative one for the hip fracture we have also developed tools for fully quantitative analysis. As an example of such applications we present an investigation of case of a slipped capital femoral epiphysis based on CT data. The segmentation procedure for this particular case has already been described earlier and is illustrated in Figure 1 together with original voxel data. Due to a slip in the femoral epiphysis there is a misorientation of the femoral bone with respect to the hip joint. In medical terms the the remedy is a wedge removal operation in which a part of the femur is removed to correct for the misorientation.

The analysis procedure to determine the misorientation and to simulate the surgical operation is summarized in Figure 4 with a direct comparison of the healthy side **a** and the injured one **b**. The hip and the femoral bone have been segmented and hollow regions within the bones filled. The illustrated cross sections are inclined about 35° from the axial direction to match the femoral bone axis which has been determined by a 3D medial axis transformation (MAT); the medial axis points are shown black in **a** and **b**. A straight line fit to the medial axis points within the femoral axis region is shown for the two sides in **c** and **d** respectively. The axis together with the hemisphere fitted to the hip joint finally yields the correct position of the femur and allows to determine the three components of the angular misorientation and the shape for the operation.

Discussion

The development of 3D imaging techniques such as CT and MR have opened up totally new possibilities for non-invasive investigation in several medical fields, but the vast amounts of data produced by these techniques also require an improvment in the analysis methods in order to fully exploit the available information. We have presented methods for an efficient visualization and analysis of volume data. The power of the developed methods has been demonstrated through practical applications ranging from a qualitative visualization of a malformation in the spinal column up to a fully quantitative analysis and a simulation of a reconstructive surgical operation in a case of a slipped capital femoral epiphysis.

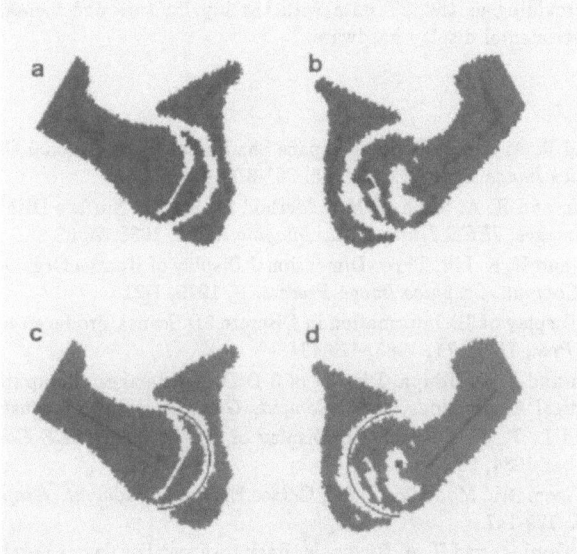

Figure 4. Illustration of a quantitative analysis procedure to determine and simulate the surgical correction of a slipped capital femoral epiphysis.

Fast and versatile display techniques to visualize the data are an essential part of any analysis procedure. The prebuffer algorithm offers an ideal tool for the generation of shaded surface views directly from voxel representation. Through a parallel implementation of the algorithm we have demonstrated both the modularization concept of the algorithm and the possibility to generate shaded surface views directly from non-cubic voxel data. By preserving original intensities throughout, the facility for their re-investigation as well as for the combination of voxel data with surface data is readily provided. Due to its extreme simplicity the prebuffer algorithm is easy to put in hardware to obtain a real time performance.

For an improved segmentation performance we have presented a method combining a preliminary segmentation based on edges determined by a 3D DOG operator with *a priori* knowledge about the desired objects in the form of case dependent binary image operations. The method is very robust and has proved out to provide good results even for complicated medical data. The DOG operator is also well suited for edge determination strategies that combine the information from different resolution levels while it offers an easy control over the desired details and a good scaling behaviour.

The methods presented in this study form a basis for the development of methods for quantitative analysis of medical volume data. First results of these methods have already been demonstrated by an analysis of the case of a slipped capital femoral epiphysis based on a 3D distance transform and subsequent skeleton extraction. The response of the pediatric surgeons supports our hope that these techniques will find a wide field of applications.

Acknowledgments

The authors wish to the thank Dr. U. Exner from the Department of Orthopedics, University Hospital Zürich for informing us of the interesting medical cases of the the spinal column and the femoral epiphysis. The Maurice Müller Institute of the University of Bern is gratefully acknowledged for providing us the CT data with the hip fracture and Siemens AG, Medical Systems for the experimental display hardware.

References

1. D. Gordon and R. A. Reynolds, Image Space Shading of 3-Dimensional Objects, *Comput. Vision Graphics Image Process.* **29**, 1985, 361-376
2. P. B. Heffernan and R. A. Robb, A New Method for Shaded Surface Display of Biological and Medical Images, *IEEE Trans. Med. Imaging* MI-4, 1985, 26-38
3. G. T. Herman and H. K. Liu, Three-Dimensional Display of Human Organs from computed Tomograms, *Comput. Graphics Image Process.* **9**, 1979, 1-21
4. J. K. Udupa, Display of 3D Information in Discrete 3D Scenes Produced by Computerized Tomography, *Proc. IEEE* **71**, 1983, 420-431
5. G. T. Herman and J. K. Udupa, Display of 3-D Digital Images: Computational Foundations and Medical Applications, *IEEE Comput. Graphics Appl.* **3** August 1983, 39-45
6. H. K. Tuy and L. T. Tuy, Direct 2-D Display of 3-D objects, *IEEE Comput. Graphics Appl.* **4**, October 1984, 29-33
7. D. Meagher, Geometric Modelling Using Octree Encoding, *Comput. Graphics Image Process.* **19**, 1982, 129-147
8. G. Frieder, D. Gordon, and R. A. Reynolds, Back-to-Front Display of Voxel-Based Objects, *IEEE Comput. Graphics Appl.* **5**, January 1985, 52-60
9. F. Klein and O. Kübler, Fast Direct Display of Discrete Volume Data, *Proc. 8th Int. Conf. Pattern Recognit.* Paris, France 1986, 633-635
10. D. Marr and E. Hildreth, Theory of Edge Detection, *Proc. R. Soc. Lond. B* **207**, 1980, 187-217

INDEX